Herber Drury

Hand-book of the Indian flora

Being a guide to all the flowering plants

Herber Drury

Hand-book of the Indian flora
Being a guide to all the flowering plants

ISBN/EAN: 9783741166464

Manufactured in Europe, USA, Canada, Australia, Japa

Cover: Foto ©berggeist007 / pixelio.de

Manufactured and distributed by brebook publishing software (www.brebook.com)

Herber Drury

Hand-book of the Indian flora

HAND-BOOK

OF

THE INDIAN FLORA;

BEING A GUIDE TO ALL THE
FLOWERING PLANTS
HITHERTO DESCRIBED AS INDIGENOUS TO THE
CONTINENT OF INDIA.

By

Lieut.-Colonel Heber Drury,
AUTHOR OF THE
USEFUL PLANTS OF INDIA

VOL. II.

Sold by Mr. J. HIGGINBOTHAM, Mount Road, Madras.

Travancore Sircar Press.
1866

PREFACE.

Without the guidance and assistance afforded by Wight and Arnott's Prodromus which rendered the compilation of the first volume a comparatively easy task, the present has required a far greater amount of labour. In it are included thirty-six Natural Orders all of which are now published in a collective form for the first time, and this circumstance the Author trusts will render it a more valuable possession to the Botanical Student than its predecessor.

ORDER LXXXVIII. COMPOSITÆ.

Herbaceous plants or shrubs: leaves alternate or opposite, usually simple, but often much divided, exstipulate: flowers (usually called florets) unisexual or hermaphrodite, collected in heads on a common receptacle and surrounded by an involucre: bracts, when present, situated at the base of the florets and known as the paleæ of the receptacle: calyx superior, closely adhering to the ovary, the limb either wanting or divided into bristles, paleæ, hairs or feathers, and called pappus: corolla monopetalous, superior, ligulate or funnel-shaped, if the latter 4 or 5-toothed: æstivation valvate: stamens equal in number to the teeth of the corolla and alternating with them, with the anthers cohering into a cylinder: ovary inferior, 1-celled, 1-ovuled: ovule erect: style simple: stigmas 3, distinct or united: fruit a dry pericarp, indehiscent, crowned with the limb of the calyx: seed solitary, erect: embryo with an inferior radicle; albumen none.

GENUS I. OIOSPERMUM.

Syn. Pol. Æqualis. *Sex: Syst:*

Deriv. From *Oios*, single, and *Sperma*, a seed.

GEN. CHAR. Capitula many-flowered: involucre imbricated; scales scariose, glabrous, acute, surrounded at the base with unequal leaves longer than the capitulum: receptacle broad, naked: limb of the corolla 5-cleft, scarcely distinct from the tube, lobes acuminated, shorter than the entire portion: anthers included: filaments smooth: achænium rounded at the apex, many-ribbed: pappus none.

O. WIGHTIANUM. *(Dec.)*

Ident. Dec. prod. V. p. 11.—Dcless. in Linn. p. 339.
Syn. Amphirephis Wightiana, *Wall. Cat. No.* 68.
Engrav. Deless. Ic. IV. t. 1.

SPEC. CHAR. Herbaceous; stem very villous; leaves alternate, sessile, ovate, acuminated, attenuated at the base, very villous on both sides, especially on the nerves; corolla glabrous; achænium very slender; style simple, glabrous: flowers very slender, pale-coloured.

Gathered by Dr. Wight, but the exact locality is not given.

GENUS II. ADENOON.

Syn. Pol. Æqualis. *Sur. Syst.*

GEN. CHAR. Capitula many-flowered: involucre imbricated, ovate, scales cartilaginous, hirsute, mucronate, 3-nerved, glandular; receptacle smooth, alveolate, fringed: corolla regular, tubulose, limb 5-cleft, segments linear-acute and somewhat 3 times shorter than the tube: anthers somewhat exserted: filaments short: style surrounded at the base with a cup-shaped gland, the divisions subulate, exserted; achænium cylindric, truncated at the apex, glabrous, 10-furrowed, glandular: pappus none.

(1) A. INDICUM. *(Dalz.)*

Ident. Dalz. in Hook. Journ. Bot. II. 344.—Bomb. flor. p. 121.

SPEC. CHAR. Erect branched plant, 1½ foot high; stem angular, scabrous, hispid: leaves alternate, sessile, broad-elliptic, acute at both ends, coarsely serrated, rough, glandular: panicles of flowers corymbose: flowers blue.

Phoonda Ghaut. Belgaum. Flowering in September.

GENUS III. ETHULIA.

Syn. Pol. Æqualis. *Sur. Syst.*

GEN. CHAR. Capitula many-flowered: achænium obpyramidal, 4-sided, 4-ribbed, glabrous, glandular, truncated at the apex: pappus minute, entire, fleshy, crown-like.—Erect ramous herbs; leaves alternate, pellucid-dotted; involucre terete, many series; scales subfoliaceous: receptacle naked; corolla rose-coloured or purple.

(1) E. CONYZOIDES. *(Linn.)*
Var. Indica.

Ident. Dec. prod. V. p. 12.—Linn. Sp. p. 1171.

Syn. E. ramosa, Roxb. fl. Ind. III. p. 413.

Engrav. Bot. Reg. 9. t. 695.

SPEC. CHAR. Stem and branches nearly erect, about 2 feet in height: leaves alternate, subsessile, elliptic, serrated; peduncles opposite the leaves, axillary or terminal, each supporting many small flowers: calyx imbricated: florets numerous, of a light purple colour: seeds 4-6-sided.

Interior of Bengal, flowering in the cold season.

GENUS IV. MONOSIS.

Syn. Pol. Segregata. *Sex: Syst:*

Deriv. From the Greek term meaning *Solitude*, alluding to its having only a single floret in the involucre.

GEN. CHAR. Capitula 1-flowered: involucres oblong, scales imbricated, obtuse, much shorter than the flower: corolla tubulose, 5-cleft: stigmas long-exserted: achænium glabrous, somewhat tapering: pappus 2-3-series, bristles stiffish, rough, equal.

M. WIGHTIANA. (Dec.)

Ident. Dec. prod. V. p. 77.—Wight's Contrib. p. 5.
Syn. Conyza Wightiana, *Wall. Cat. No.* 3028.
Engrav. Wight's Icon. t. 1085.—Spicil. II. 1. 105.

SPEC. CHAR. Large tree: branches velvety-tomentose: leaves alternate, obovate, cuneate at the base, somewhat entire, glabrous above and velvety-hirsute below: panicle much branched; capitules sub-corymbose, somewhat sessile at the tops of the branches: involucral scales tomentose at the back: flowers rose-coloured.

Neilgherries. Travancore mountains. Flowering in February and March.

GENUS V. VERNONIA.

Syn. Pol. Æqualis. *Sex: Syst:*

Deriv. Named in honor of Mr. Vernon, a Botanist and traveller.

GEN. CHAR. Capitula many-flowered: involucre imbricated, shorter than the flowers; interior scales longer: receptacle naked: corolla regular, 5-cleft: filaments smooth: achænium with a cartilaginous wart at the base: pappus often biserial, inner series bristly, much longer than the internal paleaceous one, the two series rarely equal.—Erect herbs, shrubs or trees, with alternate, rarely opposite leaves: inflorescence often scorpioid: flowers purple, rose or white, never yellow.

(1) V. ELLIPTICA. (Dec.)

Ident. Dec. prod. V. p. 22.—Wight's Contrib. p. 5.

SPEC. CHAR. Stem climbing, branched, somewhat pubescent: leaves short-petioled, exactly elliptic, quite entire, below with silky silvery adpressed hairs: panicles much branched, divaricate, bearing 3-5 few-flowered capitules on the ultimate branches: scales of the involucre ovate, somewhat acute, glabrous.

Neilgherries.

(2) V. ARBOREA. *(Ham.)*

Ident. Ham. in Linn. Soc. Trans. XIV. p. 218.—Dec. prod. V. p. 22.

Syn. Conyza arborea, *Wall. Cat. No.* 2.

SPEC. CHAR. Arboreous: leaves petioled, ovate-lanceolate, acuminated, quite entire, glabrous, somewhat scurfy below and on the petiole: cymes terminal, much-branched, panicle-shaped, with an angular rachis and scurfy branchlets: scales of the involucre ovate-oblong, afterwards opening: achænium turbinate, glabrous: bristles of the pappus one series, equal: florets purple, 6-7 in one capitulum.

Goalpara.

(3) V. PHYSALIFOLIA. *(Dec.)*

Ident. Dec. prod. V. p. 24.—Wight's Contrib. p. 6.

Syn. Conyza cinerea, *Wall. Cat. No.* 3008.

SPEC. CHAR. Stem herbaceous, erect, branched, striated: leaves petioled, broad-ovate, repandly toothed, obtuse, pubescent above, below and on the stem ashy-tomentose: corymb lax: heads of flowers pedicelled: scales of the involucre linear-lanceolate, villous at the back.

Madras.

(4) V. CINEREA. *(Less.)*

Ident. Less. in Linn. 1829. p. 291.—Wight's Contrib. p. 6.—Dec. prod. V. p. 24.

Syn. Conyza cinerea, *Linn.*—Serratula cinerea, *Roxb. fl. Ind.* III. p. 400.—Conyza mollis, *Willd.*—Cacalia rotundifolia, *Willd. phyt.* II. p. 6.

Engrav. Burm. Zeyl. t. 96. f. 1.—Rumph. Amb. VI. t. 14. f. 1.—Pluk. t. 177. f. 2.—Rheede Mal. X. t. 64.

SPEC. CHAR. Stem herbaceous, erect, branched, somewhat hoary with short hairs: leaves petioled, lower ones somewhat rounded, the rest obovate-oblong, somewhat toothed, ashy-white below: corymb lax, dichotomous: heads of flowers peduncled: scales of the involucre linear-lanceolate, acuminated, somewhat hairy outside: flowers small, light purple.

Common throughout the country, flowering from January to April.

(5) V. CONYZOIDES. *(Dec.)*

Ident. Dec. prod. V. p. 25.—Wight's Contrib. p. 6.
Syn. Conyza ovata, *Wall. Cat. No.* 3097.
Engrav. Wight's Icon. L. 829, 1076.

SPEC. CHAR. Stem herbaceous, erect, striated, with very short pubescence: leaves ovate or oblong-lanceolate, acuminated, attenuated into a short petiole, serrated, glabrous above, softly villous below: corymb composite, many-headed, naked: scales of the involucre lanceolate-linear, softly villous, acuminated.

Neilgherries.

(6) V. ALBICANS. *(Dec.)*

Ident. Dec. prod. V. p. 26.—Wight's Contrib. p. 6.
Syn. Conyza albicans, *Wall. Cat. No.* 3072.

SPEC. CHAR. Whole plant hoary-villous with adpressed silky tomentum: leaves ovate-rhomboid, cuneate at the base, quite entire, coarsely toothed above, acuminated: corymb somewhat panicled, naked: involucral scales linear, villous, acuminated.

Coromandel Coast.

(7) V. ANTHELMINTHICA. *(Willd.)*

Ident. Willd. Sp. III. p. 1634.—Dec. prod. V. p. 61.—Wight's Contrib. p. 7.
Syn. Conyza anthelminthica, *Linn.*—Serratula anthelminthica, *Roxb. fl. Ind.* III. p. 405.—Baccharoides anthelminthica, *Moench.*—Ascarivia Indica, *Cass.*
Engrav. Rheede Mal. II. t. 24.—Burm. Zeyl. t. 95.

SPEC. CHAR. Stem hispidly pubescent, branched above: leaves ovate or oval-oblong, acuminated at both ends, coarsely serrated, puberulous: capitula corymbose, many-flowered: involucral scales lanceolate-linear, acute, outer ones somewhat open, leafy, obovate-linear: achænium oblong: outer series of the pappus very short, paleaceous; flowers purplish.

Common everywhere in the Peninsula, flowering in the cold season. This species varies in the stem and leaves being more or less hirsute or glabrous.

(8) V. PECTINATA. *(Dec.)*

Ident. Dec. prod. V. p. 31.
Syn. Eupatorium pectinatum, *Wall. Cat. No.* 3112.—V. pectiniformis, *Dec. l. c.*—*Wight's Contrib.* p. 6.
Engrav. Wight's Icon. t. 1077.—Spicil. II. t. 103.—Rheede Mal. X. t. 62.

SPEC. CHAR. Fruticose: leaves short-petioled, ovate, pectinately acute and long-serrated, somewhat hairy on both sides: cyme terminal, corymbose, naked: capitula long-peduncled, many-flowered, ovate-cylindric; involucral scales dry, ovate-oblong, very obtuse, 3-nerved: achænium glabrous; outer row of the pappus very short.

Neilgherries. Pulney Hills.

(9) V. ASPERA. *(Ham.)*

Ident. Ham. in Trans. Linn. Soc. XIV. p. 219.—Dec. prod. V. p. 31.

Syn. Eupatorium asperum, *Roxb. β. Ind.* III. p. 415.—V. Roxburghii, *Less.*—Conyza platylephis, *Spreng.*

Engrav. Plnk. Alm. t. 393. f. 7.

SPEC. CHAR. Stem herbaceous (?) branched, somewhat villous: leaves elliptic-oblong, shortly petioled, attenuated at the base, acuminated at the apex, serrated, scabrous above, reticulated below, covered with reddish hairs, and glandular between the nerves: panicles somewhat corymbose: capitula many-flowered: inner scales of the involucre oblong, mucronate, outer ones smaller, very acuminate: achænium oblong-cylindric, angled, hairy; flowers small, purplish blue.

Bengal, flowering in the cold season.

(10) V. MULTIFLORA. *(Less.)*
Var. latifolia.

Ident. Dec. prod. V. p. 31.—Less. in Linn. 1831. p. 642.

Syn. Conyza multiflora, *Wall.*—C. divergens, *Wall.*—Eupatorium versicolor, *Wall. Cat. No.* 3167.

SPEC. CHAR. Stem herbaceous, erect, branched: leaves oval-lanceolate, acuminated at both ends, serrated from the middle, rough above, thinly pubescent below: panicle somewhat corymbose, many-headed: involucral scales lanceolate, acute, pale: achænium glabrous: pappus somewhat universal.

Rajmahal Hills.

(11) V. SALVIFOLIA. *(R. W.)*

Ident. Wight's Icon. vol. III.

Engrav. Wight's Icon. t. 1079.

SPEC. CHAR. Shrubby, tomentose: leaves long, narrow-lanceolate, rugose, glabrous above, densely white-tomentose beneath: corymbs axillary and terminal, naked or with a few scattered leaves: capitula numerous, densely aggregated, subsessile, many-flowered: involucre sub-campanulate, tomentose: scales lanceolate, submucate, callous at the apex: achænium glabrous, somewhat 4-angled, the sides sprinkled with glutinous points: exterior pappus palcaceous. Courtallum.

(12) V. Neilgherrensis. *(Dec.)*

Ident. Dec. prod. V. p. 32.
Syn. Eupatorium polyanthum, *Wall.*
Engrav. Wight's Icon. t. 1078.

Spec. Char. Annual, herbaceous, roundish, sub-puberulous: leaves short-petioled, ovate, acuminate, prickly-serrated, rough above, tawny-coloured beneath: cymes terminal, panicle-shaped, branches very ramous, polycephalous: capitula ovate, crowded, 2-5-flowered: scales of the involucrum dry, oval-oblong, acute, glabrous, pilose at the apex: achænium glabrous: external series of the pappus very short, deciduous: flowers pale pink.

Common in hedges on the Neilgherries, flowering in March and April.

(13) V. acuminata. *(Dec.)*

Ident. Dec. prod. V. p. 32.
Syn. Conyza acuminata, *Wall.*

Spec. Char. Stem fruticose, branched: younger branches covered with velvety hairs: leaves short-petioled, elliptic-obovate, long-cuneate, cuspidately acuminate at the apex, somewhat quite entire, glabrous, principal nerves below velvety, panicles elongated, naked, with many-headed branches: involucral scales oval, obtuse, velvety at the back, imbricated: pappus white, outer bristles irregular, short, easily deciduous: achænium glabrous, attenuated at the base.

Silhet mountains.

(14) V. Punduana. *(Dec.)*

Ident. Dec. prod. V. p. 32.
Syn. Conyza Punduana, *Wall.*

Spec. Char. Stem herbaceous? erect: leaves short-petioled, lanceolate, very long, attenuated at the base, quite entire, acuminated at the apex, coarsely and rarely serrated, glabrous: middle nerve beneath very slightly puberulous: panicles elongated, naked, with many-headed branchlets: involucral scales oblong, obtuse, inner ones longer, at length stellately spreading: achænium glabrous: pappus white, outer bristles short, irregular.

Pundua, Silhet.

(15) V. vagans. *(Dec.)*

Ident. Dec. prod. V. p. 32.
Syn. Conyza vagans, *Wall.*

[Spec. Char. Stem barbarous, branched, climbing, very shortly puberulous: leaves short-petioled, ovate-lanceolate, acuminated, quite entire, glabrous: flowering branches axillary, longer than the leaf, many-headed and arranged in a large panicle: involucral scales lanceolate, acute, clliolate, imbricated: pappus rufous, outer bristles few, linear, paleaceous, ciliated, the rest in many rows, scabrous: flowers deep purple, fragrant.

Silhet, flowering in the cold season.

(16) V. SCANDENS. *(Dec.)*

Ident. Dec. prod. V. p. 32.

Syn. Conyza scandens, *Wall. Cat. No.* 3060.

Spec. Char. Stem scandent; younger branches velvety, older ones glabrous; leaves short-petioled, oval, acuminate, quite entire, glabrous on both sides: flowering branches axillary, leafy, somewhat panicled: capitula pedicelled, ovate: involucral scales ciliated, glabrous at the back, outer ones ovate, acute, inner ones oblong; pappus somewhat rufescent, many rows of bristles, equal, outer row scarcely any, deciduous: flowers dark purple.

Silhet.

(17) V. LONGICAULIS. *(Dec.)*

Ident. Dec. prod. V. p. 33.

Syn. Eupatorium longicaule, *Wall. Cat. No.* 2169.

Spec. Char. Herbaceous: stem simple, striated, somewhat hairy-pubescent: leaves very shortly pedicelled, oblong or oval, cuneate at the base, acuminated at the apex, serrated, glabrous, with elevated glands in the middle of the areolar spaces above: cymes terminal, panicle-shaped, branches many-headed, somewhat leaf-bearing: involucral scales dry, ovate, pointed-mucronate: achænium glabrous: outer row of the pappus very short, silvery.

Pundua, Silhet.

(18) V. SALIGNA. *(Dec.)*

Ident. Dec. prod. V. p. 33.

Syn. Conyza saligna, *Wall. Cat. No.* 3061.

Spec. Char. Stem herbaceous, erect, minutely puberulous, paniculately branched: leaves oblong-lanceolate, acuminated at both ends, here and there serrated, roughish above, somewhat pubescent beneath: panicles many-headed, the branches sparingly leafy, elongated, 7-9-headed: involucral scales somewhat membranaceous, oblong, acute: pappus very white: flowers white.

Silhet.

(19) V. BRACTEOLATA. (*Dec.*)

Ident. Dec. prod. V. p. 62.

Syn. Conyza bracteolata, *Wall. Cat. No.* 3036.

SPEC. CHAR. Stem herbaceous? erect, hirsutely villous at the apex, corymbose: leaves subsessile, obovate-cuneate, shortly acuminated, distantly serrated, glabrous on both sides: middle nerve beneath velvety: peduncles of the fastigiate corymb sparingly branched, the capitulum laden with long-awned imbricated bracteoles: involucral scales in many rows awned, somewhat villous, inner ones broader: achænium villous: pappus rufescent in many rows, outer row short, irregular: flowers whitish.

Bengal.

(20) V. SUBSESSILIS. (*Dec.*)

Ident. Dec. prod. V. p. 62.

Syn. Conyza divergens, *Wall. Cat. No.* 3000.

SPEC. CHAR. Stem fruticose, erect, branched, branches pubescently-villous at the apex: leaves subsessile, oblong, quite entire, subattenuated at the base, long cuspidately-acuminated at the apex, glabrous, pale beneath with the middle nerve pubescent: corymb laxy capituls many-flowered, long-peduncled: peduncles bearing somewhat winged bracteoles under the capitulum: involucral scales in many rows, outer ones shorter, subulate, squamose: inner ones longer, smooth, lanceolate-linear, acute: pappus rufous, long, outer one irregular, short.

Silhet.

(21) V. DINDIGULENSIS. (*Dec.*)

Ident. Dec. prod. VII. p. 263.

SPEC. CHAR. Fruticose: branches villously-pubescent: leaves oblong-lanceolate, subacuminate at the base, attenuated into a short villous petiole slightly bispid above, more or less pubescent below, serrated, the serratures mucronate: corymb composite, naked: scales of the cup-shaped involucre oblong, somewhat pointedly-mucronate, hirsutely-pubescent towards the apex: achænium somewhat glabrous: outer pappus short, palaceous.

Dindigul hills.

GENUS VI. DECANEURUM.

Syn. Pol. Æqualis. *Sus. Syst.*

Deriv. From *Deca*, ten, and *Neuros*, a nerve or rib, alluding to the ribbed achænium.

GEN. CHAR. Capitula many-flowered; achænium usually glabrous, marked with 10 prominent ribs: pappus 1-series, bristles thick, rigid, densely barbellate.—Herbaceous, or suffruticose: leaves alternate: involucre imbricated, often surrounded with foliaceous bracts: receptacle flat, alveolate: flowers regular, 5-cleft, purplish.

(1) D. RETICULATUM. (*Dec.*)

Ident. Dec. prod. V. p. 66.—Wight's Contrib. p. 7.

Syn. Amphirephis Indica, *Wall. Cat. No.* 3956.—*Less.* in *Linn.* 6. p. 686.

Engrav. Wight's Icon. t. 1080.—Spicil. II. t. 104.

SPEC. CHAR. Stem suffruticose, erect, ramous, everywhere rough with bristly hairs: leaves sessile, ovate, mucronate, and mucronately sub-dentate; rough above, densely white-tomentose beneath: nerves and veins scabrous, reticulated: peduncles few, axillary and terminal: capitula closely embraced by numerous foliaceous bracts: interior scales of the involucre scariose, glabrous, longer than the bractess: flowers purple.

Banks of streams on the Neilgherries, flowering nearly all the year.

(2) D. MOLLE. (*Dec.*)

Ident. Dec. prod. V. p. 67.

Syn. D. scabridum, *var.* and D. epilejum, *var. Wight's Contrib.* p. 7.—Amphirephis mollis, *Wall. Cat. No.* 3957.

Engrav. Rheede Mal. X. t. 63.—Wight's Icon. t. 1082.

SPEC. CHAR. Stem herbaceous, erect, here and there bristly-scabrous, tomentose at the apex: leaves attenuated into the petiole, ovate-lanceolate, acuminated, coarsely and irregularly serruted, bristly-scabrous above or glabrous, white-tomentose below: peduncles axillary, 1-headed: bracts foliaceous approximating the capitulum, oblong, tomentose below: involucral scales chartaceo-scariose, glabrous, subacute.

Courtallum. Bombay. Southern Peninsula.

(3) D. SILHETENSE. (*Dec.*)

Ident. Dec. prod. V. p. 67.

Engrav. Wight's Icon. t. 1083.

SPEC. CHAR. Stem herbaceous, erect, ramous: leaves short-petioled, oblong-lanceolate, acuminated at both ends, remotely bristly-serrated, above glandularly-scabrous, below along the nerves scabrous: capitula terminal, usually solitary, subcorymbose: interior scales of the involucrum oblong, mucronate: exterior ones filiform, subulate, subpatulous, with a few scattered on the peduncle.

Silhet. Courtallum.

(4) D. DINDIGULENSE. (*Dec.*)

Ident. Dec. prod. V. p. 67.—Wight's Contrib. p. 7.
Syn. Conyza Indica, *Wall. Cat. No.* 3042.

SPEC. CHAR. Stem somewhat suffruticose, tuberous at the base, erect, sparingly ramous, striated at the base and pubescent, tomentose at the apex: leaves subsessile, broad-ovate, or subrotund, coriaceous, toothed, glabrous above, longish, densely white-tomentose below, uppermost ones oval, acute, small: corymb fastigiate, composite: involucrum ovate, scales oval, mucronate, imbricated, tomentose, woolly outside, quite glabrous within.

Dindigul Mountains.

(5) D. COURTALLENSE. (*R. W.*)

Ident. Wight's Icon. Vol. III.
Engrav. Wight's Icon. t. 1081.

SPEC. CHAR. Stems scabrous, suffruticose, erect, ramous: leaves attenuated into the petiole, ovate-lanceolate, obtuse, slightly mucronate-dentate, rough above, softly whitish-tomentose below, at first uniformly white, afterwards reticulately veined: peduncles axillary, 1-headed: capitula closely bound by several ovate-obtuse mucronate 3-nerved bracts: interior scales of the involucrum scarioser, glabrous, longer than the bracts.

Courtallum, flowering in February.

(6) D. MICROCEPHALUM. (*Dalz.*)

Ident. Dalz. Bomb. Flor. p. 122.—Hook. Jour. Bot. III. p. 281.

SPEC. CHAR. Stem branched, scabrous, pubescent: leaves petioled, elliptic-acuminata, gradually attenuated into the petiole, pubescent above, hoary-tomentose beneath: capitula solitary at the apices of the branch: involucral scales scariose, hoary and tomentose beneath, exterior ones lanceolate-acuminate, bristle-pointed, ciliated: achænium without ribs: pappus very caducous.

Parwar Ghaut, flowering in November. This species forms a remarkable exception in having an unribbed achænium, so contrary to the usual generic character. The plant fresh gathered smells like Chamomile. (*Dalz.*)

(7) D. PHYLLORCHNUM. (*Dec.*)

Ident. Dec. prod. VII. p. 264.

SPEC. CHAR. Stem herbaceous, erect: leaves ovate-acuminate, coarsely serrated, sparingly hairy above, hoary-tomentose below, upper ones very shortly attenuated into the petiole: capitula ses-tile at the apices of the branches and much bracteated: bracts and outer involucral scales ciliated at the base, inner ones acuminated.

Near Bombay. This approaches very near in its characters to *D. mollis.*

GENUS VII. ELEPHANTOPUS.

Syn. Fol. Segregata. *See : Syst :*

Deriv. From *Elephas*, an elephant, and *Pous*, a foot, alluding to some of the leaves which resemble an elephant's foot.

Gen. Char. Capitula of several (3-5, usually 4) equal flowers, densely congested into a glomerulus, enclosed by leaves: achænium slightly compressed, many ribbed, pilose : pappus 1-series, some of the bristles dilated at the base.—Perennial pilose herbs: with alternate sessile leaves and terminal : glomerulus involucre compressed, scales in a double series, alternately flat and conduplicate : rachis naked; corolla palmate, limb 5-cleft, one of the fissures deeper.

(1) E. scaber. (*Linn.*)

Ident. Dec. prod. V. p. 86.—Wight's Contrib. p. 8.—Roxb. Fl. Ind. III. p. 445.

Engrav. Rheede Mal. X. t. 7.—Wight's Icon. t. 1086.

Spec. Char. Stem dichotomously branched, strigose; radical leaves rough, crenated, wedge-shaped, long-attenuated towards the base; stem leaves lanceolate : floral ones broadly cordate-ovate, acuminate, hoary; flowers purple.

Common under the shade of trees in the Peninsula. Silhet, Concans. Flowering in the cold season.

GENUS VIII. AGERATUM.

Syn. Fol. Æqualis. *Ser. Syst.*

Deriv. From *A*, not, and *Geras*, age, alluding to the constancy of the colours.

Gen. Char. Capitula many-flowered : achænium somewhat 5-angled, attenuated at the base: pappus 5-10, free paleaceous scales aristato-acuminate, or pectinate-obtuse.—Annual erect herbs, with opposite leaves : involucre imbricated : receptacle naked : flowers blue or white.

(1) A. conyzoides. (*Linn.*)

Var. cordifolium.

Ident. Dec. prod. V. p. 108.—Wight's Contrib. p. 6.

Syn. A. cordifolium, Roxb. flor. Ind. III. p. 415.

Engrav. Hook. Exot. flor. t. 15.

Spec. Char. Stem branched, hispid: leaves ovate, acuminated, rhomboid or cordate, with longish very hispid petioles: pales of the pappus dilated at the base, serrulate, long-awned at the apex, somewhat equalling the corolla: flowers small, pale-purple.

Belgaum, Bombay, and other localities in the Peninsula. The whole plant has a strong unpleasant smell.

GENUS IX. ADENOSTEMMA.

Syn. Pal. Æqualis. *See: Syst:*

Deriv. From *Aden*, a gland, and *Stemma*, a crown, alluding to the glandular apex of the achænium.

Gen. Char. Capitula many-flowered: achænium obovate-oblong, surmounted by 3–5 rigid bristles, glandular, globose, or clavate at the apex.—Herbs, with opposite leaves and corymbosely-panicled capitula: involucre campanulate, somewhat shorter than the flowers: scales 1-series, foliaceous, oblong: receptacle flat, naked, foviolate: corolla white; stigmas long exserted, thickened at the point and coloured.

(1) A. microcephalum. (*Dec.*)

Ident. Dec. prod. V. p. 111.

Syn. Lavenia viscida, *Ham. ex. Wall. Cat. No.* 332.

Spec. Char. Stem rooting at the base, ascending, very minutely downy: leaves short-petioled, ovate, denticulate, obtuse, glabrous: panicles lax, with small capitula: involucral scales oblong, sub-acute, somewhat glabrous: achænia scarcely muricated.

Goalpara. Sibet.

(2) A. rivale. (*Dals.*)

Ident. Dals. in Hook. Journ. Bot. III. p. 231.—Wight's Contrib. p. 9.—Bomb. flor. p. 122.

Syn. Ageratum aquaticum, *Roxb. flor. Ind.* III. p. 416.—Adenostemma leiocarpum, *Dec. prod.* V. p. 113.

Spec. Char. Stem erect, round, glabrous: leaves linear-lanceolate, long-attenuated at the base, serrately toothed, glabrous: panicle corymbose, lax, few-headed: scales of the involucre linear or linear-spathulate, obtuse: flowers small, white.

Borders of the Circars. Margins of rivulets in the Peninsula, Bengal. Flowering in February and March.

(3) A. madurense. (*Dec.*)

Ident. Dec. prod. V. p. 113.—Wight's Contrib. p. 9.

Syn. Lavenia macrophylla, *Wight's Herb.*

Engrav. Burm. Zeyl. t. 42.

Spec. Char. Stem erect, younger ones very shortly puberulous: leaves petioled, broadly ovate-rhomboid, acuminated at both ends, coarsely and acutely toothed; corymb shortly exserted above the ultimate leaves, branches puberulous; involucral scales glabrous, subacute; achænia smoothish.

Madura.

(4) A. RUGOSUM. (*Wight.*)

Ident. Dec. prod. V. p. 112.—Wight's Contrib. p. 8.

Syn. A. aquaticum, *Don. prod. fl. Nep. p.* 81.

Spec. Char. Stem erect, somewhat 4-sided at the base, puberulous or rough: leaves broad-ovate, subcordate, irregularly 3-lobed, coarsely toothed: nerves below reticulately wrinkled, glabrous above: panicle lax, many-headed: involucral scales oblong, scabrous at the back: achænia muricately tubercled.

Peninsula.

(5) A. RETICULATUM. (*Dec.*)

Ident. Dec. prod. V. p. 113.—Wight's Contrib. p. 8.

Engrav. Wight's Icon. t. 1088.

Spec. Char. Stem erect, somewhat 4-cornered, glabrous, sparingly glandulosely puberulous: leaves ovate, coarsely toothed, rigid, principal nerves beneath reticulated, puberulous: panicle much divaricated: involucral scales linear-oblong, obtuse, subpubescent at the back: achænia smooth.

Dindigul.

(6) A. LATIFOLIUM. (*Don.*)

Ident. Don. prod. p. 181.—Dec. prod. V. p. 113.

Syn. Lavenia latifolia, *Spreng.*

Engrav. Wight's Icon. t. 1087.

Spec. Char. Stem erect, branched, pubescent towards the apex: leaves broadly ovate-rhomboid or subcordate, coarsely serrated: panicle corymbose, hirsute, many-headed: involucral scales rough on the back: achænia muricately tuberoled: flowers white.

Common in the Concan jungles. Neilgherries.

(7) A. ELATUM. (*Don.*)

Var. laxum.

Ident. Don. prod. flor. Nep. p. 181.—Dec. prod. V. p. 112.

Syn. Lavenia carnosa, *Wall. Cat. No.* 329.—L. elata, *Spreng. Syst.* III. p. 449.—L. alba, *Wall. Cat. No.* 330.

Spec. Char. Erect, obtusely 4-sided, glabrous, branched: leaves ovate-oblong, subsessile, attenuated at the base, crenato-dentate, glabrous: panicle many-headed, divaricated: involucral scales linear-oblong, obtuse, glabrous: achænia sparingly muricated.

Silhet.

GENUS X. EUPATORIUM.

Syn. Pol. Æqualis. *Ses : Syst:*

Deriv. Named after *Eupator*, king of Pontus, who is said to have first used the plant in medicine as a counter-poison.

GEN. CHAR. Capitula many-flowered: receptacle smooth, naked: involucral scales 1–2 or many serial, equal or unequal, loosely or firmly imbricated: orifice of the corolla scarcely dilated: anthers included: divisions of the style exserted, cylindric, obtuse.—Herbs or under-shrubs: leaves usually opposite, rarely alternate or verticillate: heads of flowers mostly corymbose or panicled.

(1) E. PUNDUANUM. *(Wall.)*

Ident. Wall. Cat. No. 3170.—Dec. prod. V. p. 179.

SPEC. CHAR. Stem herbaceous, erect, branched, pubescently-velvety: leaves scarcely petioled, lanceolate, acuminate, sharply serrated from the middle, downy on both sides, somewhat 3-nerved: branches somewhat arranged in a panicle, almost naked and closely corymbose at the apex: capitula 4-flowered: involucral scales glabrous, 5–7, oblong-linear, obtuse.

Pundas in Silhet.

(2) E. NODIFLORUM. *(Wall.)*

Ident. Wall. Cat. No. 3166.—Dec. prod. V. p. 179.

SPEC. CHAR. Stem herbaceous, erect, hirsute: leaves sessile, oblong-lanceolate, mucronately acute, coarsely serrated, sparingly downy, hispid on the nerves below: corymb composite, closely packed: capitula 4-flowered, congested towards the tops of the branches: involucral scales 5–6, unequal, pubescent at the back.

Silhet mountains.

(3) E. DIVERGENS. *(Roxb.)*

Ident. Roxb. flor. Ind. III. p. 414.—Dalz. Bomb. flor. p. 123.

Syn. Decaneurum divergens, *Wight's Contrib.* p. 8.—*Dec. prod.* V. p. 68.

Engrav. Wight's Icon. t. 1084.

SPEC. CHAR. Shrub, 6–8 feet: branches diverging: leaves short-petioled, elliptic, recurved, serrato-dentate, rugose, downy when young, scabrous when old, 1–6 inches long: corymbs ternical, compound: heads of flowers 5-8, flowers very numerous, purple.

Common on the Western Ghauts, Mysore. Flowering in February.

GENUS XI. CALLISTEPHUS.

Syn. Pol. Superflua. *Sex. Syst:*

Deriv. From *Kallistos*, prettiest, and *Stephos*, a crown.

Gen. Char. Capitula many-flowered: of the ray ligulate, female; of the disk tubular, hermaphrodite: achaenium obovate, cuneiform, compressed, roughish: pappus double, each of 1-series; exterior short, palaeaceously setaceous, cohering into a crown: interior long, filiform, rough, deciduous.—Erect ramous herbaceous annuals: branches with a single capitulum on the apex: involucre 3-4 series: scales ciliate, obtuse, embraced by foliaceous bracts, shorter than the involucre: receptacle slightly alveolate.

(1) C. Wightianus. (*Dec.*)

Ident. Dec. prod. V. p. 275.
Syn. Polos Wightiana, *Wight's Contrib.* p. 10.
Engrav. Wight's Icon. t. 1089.

Spec. Char. Leaves sessile, oblong-linear, entire or somewhat serrate, shortly mucronate; branches leafy, compressed at the apex, minutely puberulous: outer scales of the involucre leafy, linear-oblong, not ciliated, scarcely larger than the inner ones: flowers large, yellow.

Dindigul. Common in the Deccan.

(2) C. concolor. (*Dals.*)

Ident. Dals. in Hook. Jour. Bot. 11. p. 344.—Bomb. flor. p. 128.

Spec. Char. Small glaucous plant: branches ascending, radical: leaves oblong-obovate, toothed towards the apex: leaves of the stem linear-oblong, toothed or entire, glandular-dotted, glabrous: outer scales of the involucre linear-obtuse, inner ones erect, foliaceous, glandular-dotted; peduncles terminal, solitary, 1-headed: flowers yellow.

In rocky ground near Malwan, flowering in September and October.

GENUS XII. ERIGERON.

Syn. Pol. Superflua. See: Syst:

Deriv. From *Er*, spring, and *Geros*, an old man, alluding to the plants becoming old at the beginning of the season.

GEN. CHAR. Capitula many-flowered, radiate: ligulæ linear, female, several series; disk tubular, either all hermaphrodite, or with the exterior florets female: achænium compressed, beakless: pappus 1-series.—Herbaceous or suffruticose; leaves alternate: capitula hemispherical: involucre 2–3-series: receptacle naked, foviolately punctate: flowers of the ray white, blue, or purple: disk yellow.

(1) E. LESCHENAULTII. *(Dec.)*

Ident. Dec. prod. V. p. 292.

SPEC. CHAR. Whole plant bispid with hairs: stems herbaceous, erect, striated: leaves sessile, oblong-linear or ligulate, mucronate, quite entire: peduncles axillary, longer than the leaf, 1-headed, higher ones shorter, few-headed, subcorymbose: involucral scales linear, shorter than the disk: achænium villous.

Neilgherries.

(2) E. HISPIDUM. *(Dec.)*

Ident. Dec. prod. V. p. 292.—Wight's Contrib. p. 9.

Syn. Aster Bengalensis, *Heyne Herb.*—E. hirsutum, *Wall. Cat.*

SPEC. CHAR. Stem erect, branched, hispid: leaves oblong, here and there coarsely toothed, subpuberulous, lower ones obtuse, attenuated at the base, upper ones acute, dilated at the base, cordately stem-clasping: capitula 1-2 at the apex of the branches: Involucral scales linear-acuminate: ligulæ numerous, very narrow-linear, elongated, in 1-2 rows: female flowers in many rows, scarcely ligulate, central ones hermaphrodite, 5-toothed.

Bengal. Sandy places on the Coromandel Coast.

(3) E. SUBLYRATUM. *(Roxb.)*

Ident. Dec. prod. V. p. 292.—Wight's Contrib. p. 9.

Syn. Conyza hirsuta, *Wall.*

SPEC. CHAR. Stem ascending, erect, softly hispid: leaves villous on both sides, obovate, coarsely cut and toothed, obtuse, mucronate, lower ones attenuated at the base, cut, sublyrate, upper ones cordately stem-clasping: corymb terminal, 7–9-headed: capitula pedicelled: involucral scales linear-acuminate, somewhat glabrous.

Negapatam and Tanjore. Rampore in Rohilcund.

(4) E. MULTICAULE. *(Dec.)*
Var. Madurense.

Ident. Dec. prod. V. p. 292.—Wight's Contrib. p. 10.

SPEC. CHAR. Stems herbaceous, numerous, erect, scabrous: leaves scabrous with short scattered pubescence; lanceolate, attenuated at the base, sessile, acute, upper ones quite entire: panicle terminal, lax: flowering branches elongated: capitula solitary: involucral scales linear-acuminate; ligulæ linear, in one row, a little longer than the involucre: female flowers in many-series; males about 20: flowers white.

Madùra.

(5) E. WIGHTII. *(Dec.)*

Ident. Dec. prod. V. p. 286.

Engrav. Wight's Spicil. t. 106.—Icon. t. 1090.

SPEC. CHAR. Stem erect, shortly branched: leaves oblong, lower ones attenuated at the base, subserrated, somewhat obtuse, upper ones entire, acute, all puberulous on both sides: capitula pedicelled, subracemose: involucral scales rough on the back, linear-subulate, equalling the disk: ligulæ very slender, longer than the disk: achænium glabrous: flowers purple.

Neilgherries, flowering in the rainy season.

GENUS XIII. MYRIACTIS.

Syn. Pol. Superfluum, *Star Syst.*

Deriv. From *Murios*, a myriad, and *Aktis*, a ray.

GEN. CHAR. Capitula heterogamous: flowers of the ray 2, or many series, female: ligulæ very narrow; of the disk hermaphrodite: achænium compressed, beakless, often glanduliferous at the apex: pappus none.—Erect dichotomously-ramous herbs with alternate leaves: peduncles long, 1-cephalous, paniculate: capitula globose: involucre 1-2-series: receptacle naked: florets white or yellow.

(1) M. WIGHTII. *(Dec.)*

Ident. Dec. prod. V. p. 308.—Wight's Contrib. p. 10.

Engrav. Wight's Spicil. II. t. 107.—Icon. t. 1091.

SPEC. CHAR. Stem sparingly pilose: radical leaves ovate, attenuated into the petiole, the inferior cauline ones cuneate at the base, coarsely cut and serrate, the upper ones oblong, entire, sessile: capitula terminal, solitary: involucre somewhat hairy, reflexed after flowering: ligulæ white in about two rows.

Neilgherries in dry pastures.

GENUS XIV. AMPHIRHAPIS.

Sym. Pol. Æqualis. *Ste. Syst.*

Deriv. From *Amphi*, around, and *Raphis*, a needle, alluding to the pappus-like hairs at the base of the fruit.

GEN. CHAR. Capitula radiate: ligulæ 10–20, the same colour as the disk, narrow: disk 5-toothed, hermaphrodite: schænium linear-oblong, slightly compressed, villous or pubescent, the villi sometimes simulating an exterior pappus: pappus 1-series, setaceous, rigid.—Perennial erect herbs with alternate leaves and corymbose capitula; involucre imbricated; receptacle alveolate; flowers yellow.

(1) A. HETEROTRICHA. *(Dec.)*

Ident. Dec. prod. V. p. 343.

Syn. Solidago heterotricha, *Wall. Cat. No. 2275.*

SPEC. CHAR. Stem fruticose: branches densely bristly-villous: leaves very shortly petioled, acuminated at both ends, rarely callosely-serrulate, sparingly hairy above, below with adpressed bristly villi: corymb composite, many-headed: involucral scales oblong-linear, sub-obtuse, villous: ligulæ about 10: schænium villous.

Pundua, Silhet.

GENUS XV. BLEPHARISPERMUM.

Sym. Pol. Segregatum. *Ses. Syst.*

Deriv. From *Blepharis*, an eyelash, and *Sperma*, a seed.

GEN. CHAR. Capitula 4-flowered, numerous, combined into a globose glomerulus: flowers tubular, 2 exterior female, 3-toothed, 2 interior male, 5-toothed: style not exserted, scarcely bifid: anthers apiculate, subcordate at the base: schænium oval-oblong, compressed, subtetragonous: two of the angles nerve-like, glabrous, 2 marginal, ciliate: pappus of the female flowers 3–5, palæa; of the male none.—Glabrous shrubs, leaves alternate; glomerules globose, one or more on the ends of the branches: involucre double: receptacle of the capitula narrow, with a paleæ between the flowers.

(1) B. PETIOLARE. *(Dec.)*

Ident. Dec. prod. V. p. 368.—Wight's Contrib. p. 12.

Engrav. Wight's Icon. t. 1092.

SPEC. CHAR. Leaves petioled, ovate-lanceolate, acuminated: glomerules three, long-peduncled: involucral bracts small, membranaceous, shorter than the capitula.

Courtallum.

(2) B. SUBSESSILE. *(Dec.)*

Ident. Dec. prod. V. p. 368.—Wight's Contrib. p. 12.
Syn. Leucoblepharis subsessilis, *Arn. in Mag. Bot.* 11. p. 422.
—*Dals. Bomb. flor.* p. 123.
Engrav. Wight's Icon. t. 1093.
SPEC. CHAR. Herbaceous, glabrous, simple: leaves alternate, subsessile, elliptic-obtuse, or attenuated at both ends, entire, 3-nerved: clusters of flowers globose, terminal, subsessile: bracts leafy, longer than the flowers: seeds black, shining, plano-convex.

Western Ghauts. Bellary in dry arid soil.

GENUS XVI. DICHROCEPHALA.

Syn. Pol. Superflua. *Ser: Syst:*

Derio. From *Dicras*, double, and *Kephalos*, a head, alluding to the flowers.

GEN. CHAR. Capitula heterogamous: florets all tubular; marginal ones, female, many series, 3–4-toothed; central, hermaphrodite, or male, few, 4-toothed: achænium compressed, beakless; of the females bald; of the hermaphrodites crowned with 1–2 bristles.—Annuals with alternate leaves and small globose capitula, spreading involucre, and conical naked receptacles.

(1) D. CHRYSANTHEMIFOLIA. *(Dec.)*

Ident. Dec. prod. V. p. 372.—Wight's Contrib. p. 11.
Syn. Cotula chrysanthemifolia, *Blume.*
Engrav. Wight's Icon. t. 1095.—Spicil. II. t. 108.
SPEC. CHAR. Erect, branched, the whole plant rough from short hairs: lower leaves lyrately pinnatifid, upper ones oblong, cordately stem-clasping, coarsely serrated: peduncles much longer than the capitula: flowers pink.

Common on the Neilgherries, flowering nearly all the year.

(2) D. LATIFOLIA. *(Dec.)*

Ident. Dec. prod. V. p. 372.—Wight's Contrib. p. 11.
Syn. Cotula bicolor, *Roth.*—C. latifolia, *Pers.*—Grangea latifolia, *Lam.*—C. sinapifolia, *Roxb. fl. Ind.* III. p. 347.
Engrav. Wight's Icon. t. 1096.—Lam. Ill. t. 699. f. 1.
SPEC. CHAR. Stem erect, sparingly pilose: leaves obovate, attenuated into the petiole, coarsely toothed, often inciso-pinnatifid at the base: flowering branches ramous, nearly naked: pedicels rigid, divaricated, longer than the globose capitula: flowers yellowish-white.

Mysore. Neilgherries. Flowering nearly all the year.

(3) D. Senxion. *(R. W.)*

Ident. Wight's Icon. vol. IV.
Engrav. Wight's Icon. t. 1610.
Spec. Char. Procumbent, diffusely ramous, glabrous: leaves obovate, cuneate, slightly toothed at the apex: capitula globose, sessile, axillary and terminal.

Neilgherries, on the banks of dry ditches near the Ootacamund lake.

(4) D. gracilis. *(Dec.)*

Ident. Dec. prod. V. p. 371.
Spec. Char. Quite glabrous or sub-pubescent at the very apex: stem ascending, slender, sparingly branched: leaves exauriculate, pinnately divided, lobes 2-3 on both sides somewhat distant, oblong, few-toothed at the apex: capitula long-pedicelled, few, racemose.

North-West Provinces.

(5) D. grangeæfolia. *(Dec.)*

Ident. Dec. prod. V. p. 371.
Spec. Char. Stem erect, hairy: cauline leaves glabrous, shortly auricled at the base, half-stem-clasping, narrow-oblong, pinnatifid; lobes few, scarcely sub-dentate, ovate-oblong: upper ones linear, somewhat entire: peduncles 1-headed, slightly leafy: capitulum obovate.

North-West Provinces.

GENUS XVII. SPILÆBANTHUS.

Nym. Pol. Segregata. *Sea: Syst.*

Deriv. From *Sphaira*, a globe, and *Anthos*, a flower, alluding to the globular heads of flowers.

Gen. Char. Capitula heterogamous, densely aggregated into a globose glumerule: flowers tubular; females thickened at the base, 3-toothed, several series in the circumference; males 5-toothed; style in the centre; florets undivided: achænium beakless, hairy: pappus none.—Herbs with decurrent leaves, winged or wingless, 1-headed peduncles: scales of the partial involucre numerous, 2-3-series: general receptacle bracteolate: flowers purple.

(1) S. hirtus. *(Willd.)*

Ident. Dec. prod. V. p. 369.—Wight's Contrib. p. 11.
Syn. S. hirtus, *Blume.*—S. Indicus, *Roxb.*
Engrav. Burm. Zeyl. t. 94. f. 3.—Wight's Icon. t. 1094.

Spec. Char. Annual, winged: leaves sessile, decurrent, long-obovate, bristly-serrated, downy and glutinous: heads of flowers solitary, leaf-opposed or terminal: peduncles winged; flowers rose-coloured or purple.

Common in the Peninsula in rice-fields, flowering nearly all the year. The whole plant is remarkably fragrant.

(2) S. MOLLIS. *(Roxb.)*

Ident. Roxb. flor. Ind. III. p. 448.—Dec. prod. V. p. 369.
Syn. S. hirtus, *Burm.*
Engrav. Rheede Mal. X. t. 43.

Spec. Char. Annual, winged: leaves lanceolate, decurrent, serrate, villous: peduncles shortly winged: involucral scales cuneate, smooth: female florets about fifteen, gibbous: flowers purplish.

Peninsula. Hindostan. Bengal. Flowering in the cold season.

GENUS XVIII. OLIGOLEPIS.

Syn. Pol. Necessaria. *Nec: Syst:*

Deriv. From *Oligos*, few, and *Lepis*, a scale.

Gen. Char. Capitula numerous, heterogamous, about 5-flowered: flowers all tubular: females few, about 4 in the circumference, pedicelled, 3-toothed; hermaphrodites solitary, sessile, 5-toothed: style 2-cleft in the female: achænium beakless, of the females terete, hairy, of the disk obsoletely 4-sided, glabrous: pappus none.

(1) O. AMARANTHOIDES. *(R. W.)*

Ident. Wight's Icon. vol. III.
Syn. Sphæranthus amaranthoides, *Burm. flor. Ind.* p. 186.—Wight's *Contrib.* p. 11.—*Dec. prod.* V. p. 370.
Engrav. Wight's Icon. t. 1149.

Spec. Char. Annual, glabrous: leaves lanceolate, quite entire or serrated, glabrous, shortly decurrent: capitula ovate-globose, subsessile: peduncles very short, wingless: bracts very much acuminated: involucre usually one scale to each flower, that of the hermaphrodite much larger and forming a common involucre to the caphulum, those of the females linear-obtuse, mucronate or truncated, folded round the flower and adhering to the pedicel.

Rice fields near Negapatam.

GENUS XIX. GRANGEA.

Syn. Pol. Superflua. *Ses: Syst:*

Deriv. Called after Grange, probably some friend of Adanson's.

GEN. CHAR. Capitula heterogamous: florets all tubular: two
or more outer series female, very slender, 3-toothed: centre ones
hermaphrodite, 5-toothed: styles of the disk undivided: achænium
somewhat compressed with a cup-shaped pappus, somewhat fimbriate on the margin.—Procumbent herbs, with sinuately pinnatifid
alternate leaves: capitules solitary, globose, terminal: involucre 2-series: receptacle hemispherical.

(1) G. MADERASPATANA. *(Pair.)*

Ident. Wight's Contrib. p. 12.—Dec. prod. V. p. 373.

Syn. G. Adansonii, *Cass.*—Artemisia Maderaspatana, *Roxb. flor. Ind.* III. p. 422.—Cotula Maderaspatana, *Willd.*

Engrav. Rheede Mal. X. t. 49.—Wight's Icon. t. 1097.

SPEC. CHAR. Stems procumbent or diffuse, villous: leaves sinuately pinnatifid, lobes obtuse: peduncles terminal or leaf-opposed: heads of flowers subglobose, solitary, yellow.

Rice fields in the Peninsula. Flowering in the cold season.

GENUS XX. CYATHOCLINE.

Syn. **Pol. Necessaria.** *See: Syst.*

Deriv. From *Cyathos*, a cup, and *Cline*, a bed, alluding to the shape of the disk.

GEN. CHAR. Capitula heterogamous, marginal florets many-series, female, 3-toothed; central ones male, 5-toothed, achænium attenuated at both ends, with a short beak: pappus none.—Erect herbaceous viscid heavy-smelling plants with alternate bipinnatifid leaves: capitula racemose, at the ends of long peduncles: involucre of many series: receptacle ebracteolate.

(1) C. STRICTA. *(Dec.)*

Ident. Dec. prod. V. p. 374.—Daiz. Bomb. flor. p. 124.

Syn. Tanacetum viscosum, *Wall. Cat. No.* 3232.

SPEC. CHAR. Erect, branched: leaves pinnately-divided, lobes somewhat linear, coarsely serrated: flowers corymbose, terminal, purple.

Banks of streams in the Concan jungles.

(2) C. LYRATA. *(Cass.)*

Ident. Dec. prod. V. p. 374.—Wight's Contrib. p. 12.

Syn. Tanacetum gratum, *Wall. Herb.*—Artemisia hirsuta, Spr. *Syst.* III. p. 490.

Engrav. Wight's Icon. t. 1098.

Spec. Char. Herbaceous: lower leaves lyrate, upper lobes of the leaves larger, obovate.

Orange valley on the Neilgherries, flowering in July.

(3) C. Lawii. *(Dalz.)*

Ident. Dalz. Bomb. flor. p. 124.

Syn. C. lutea, *Wight's Icon.* vol. III.

Engrav. Wight's Icon. t. 1150.

Spec. Char. Leaves nearly all radical, minute, sub-pinnatifid, pubescent; stems slender, erect, dichotomously branched, often with a capitulum in the fork; flowers yellow.

The Ghauts, Tannah district near Bombay.

GENUS XXI. THESPIS.

Syn. Pal. Necessaria. *Ses. Syst.*

Deriv. The name alludes to Thespis, a hero in mythology, the father of fifty daughters. In these plants there are fifty female florets.

Gen. Char. Capitula heterogamous: marginal flowers many series, females very slender; style scarcely exserted, central ones few; male with a short tube and campanulate limb; anthers included; achænium compressed, glabrous: pappus of 7-8 barbellate bristles.—Erect ramous annuals: leaves alternate; peduncles axillary and terminal: involucre of several series, shorter than the disk: receptacle naked, flat, punctulate; pappus at first white, afterwards reddish.

(1) T. divaricata. *(Dec.)*

Ident. Dec. prod. V. p. 375.

Syn. Cotula divaricata, *Wall. Cat. No.* 3238.

Spec. Char. Stem much branched from the base; branches divaricate; peduncles shorter than the leaf; capitula small, subcorymbose: flowers yellow.

Silhet.

(2) T. erecta. *(Dec.)*

Ident. Dec. prod. V. p. 375.

Syn. Cotula sinapifolia, *Wall. Cat. No.* 3227.

Spec. Char. Stem with a simple base, straight, erect, corymbosely branched at the apex: branches obliquely erect; peduncles longer than the leaf: flowers yellow.

Silhet.

GENUS XXII. BERTHELOTIA.

Syn. Pol. Necessaria. *See : Syst :*

Deriv. Named after M. Berthelot, who illustrated the Flora of the Canary Islands.

GEN. CHAR. Capitula heterogamous: flowers of the circumference many-series, female 2–3-toothed; of the disk few, 5-toothed: achænium tapering, without a beak: pappus 1-series, paleaceous, cohering at the base, scabrous at the apex.—A shrub with alternate sessile leaves : capitula forming thick corymbs on the buds of the branches : receptacle naked.

(1) B. LANCEOLATA. *(Dec.)*
Var. Indica.

Ident. Dec. prod. V. p. 376.
Syn. Conyza lanceolata, *Wall. Cat. No.* 2991.

SPEC. CHAR. Small shrub, erect, branched, covered with soft hoary pubescence: leaves alternate, sessile, lanceolate, mucronate, quite entire, or here and there sharply serrated at the apex: capitula densely corymbose at the tops of the branches : flowers purplish.

Banks of the Ganges near Cawnpore.

GENUS XXIII. CONYZA.

Syn. Pol. Superflua. *See : Syst :*

Deriv. From *Konis*, dust, because it is supposed to have the power when powdered and sprinkled of driving away flies, whence the English name of flea-bane or fly-bane.

GEN. CHAR. Capitula heterogamous: florets of the circumference many-series, female, 3–5-toothed, of the disk few, 5-toothed: achænium compressed: pappus 1-series: bristles filiform. —Erect ramous herbs with alternate leaves and corymbose pedicelled capitula : involucre many-series : receptacle ebracteolate.

(1) C. STRIATA. *(Wall.)*

Ident. Dec. prod. V. p. 383.

SPEC. CHAR. Glabrous or minutely pubescent above: stem herbaceous, erect, striately nerved, branched : leaves lanceolate or oblong-linear, acuminated at both ends, quite entire or scarcely subserrated : branches leafy, arranged in a panicle and terminating in a many-headed close corymb : involucral scales puberous at the back, lanceolate-linear, acuminated, shorter than the disk.

Silhet.

(2) C. Travancorensis. (*Wall.*)

Ident. Dec. prod. V. p. 384.—Wall. Cat. No. 3067.

Spec. Char. Stem herbaceous, erect, very finely striated and puberulous: leaves sub-petioled, lanceolate, acuminated at both ends, subserrulate, glabrous, with the middle nerve below puberulous: branches corymbose, few-headed: capitula pedicelled: involucral scales 1-2-series, lanceolate, acuminate, somewhat equalling the flowers.

Travancore.

(3) C. Asteroides. (*Wall.*)

Ident. Dec. prod. V. p. 382.—Wall. Cat. No. 3052. (not Linn.)

Spec. Char. Stem ascendent, erect, hairy, ramous: leaves oblong, hairy; lower ones attenuate at the base, obtuse, thickly fewtoothed; upper ones half-stem-clasping, acute, subentire, branches lax, sub-panicled; heads of flowers solitary or few, aggregate: involucral scales linear, cuspidate-acuminate, equalling the flowers; receptacle ring-shaped in the circumference, convex in the centre: pappus pale red: female flowers filiform.

Monghyr. Hurdwar.

(4) C. Absinthifolia. (*Dec.*)

Ident. Dec. prod. V. p. 383.—Wight's Contrib. p. 16.

Syn. C. cuneata, *Ham. Herb.*—C. montana et stricta, *Heyne. Herb.*—C. pinnatifida, var. *Roxb. C. C. p.* 61.—C. trifida, *Ham. ex. Don.*—Erigeron pinnatifidum, var. *Don. prod.* 172.—C. pinnatifida, *Roxb. flor. Ind.* III. p. 430.

Spec. Char. Pubescent with small hairs, erect, much branched: leaves obovate-oblong, mucronate, cuneate-attenuate at the base; lower ones thickly few-toothed; upper ones quite entire: panicle corymbose, much branched, many-headed: involucral scales linear, acuminate, shorter than the flowers: female flowers most numerous, scarcely ligulate: achænium glabrous, lenticular: pappus rufescent.

Dindigul Hills. Neilgherries. Nellore.

(5) C. Adenocarpa. (*Dalz.*)

Ident. Dalz. Bomb. flor. p. 125.

" Spec. Char. Stem much branched from the base, pubescent and scabrous: leaves linear or spathulate, stem-clasping, auricled at the base, entire or distantly toothed, mucronulate: corymbs terminal, dichotomous, 8-flowered: hermaphrodite florets 10-12: achænium glabrous, covered with yellow resinous glands: involucral scales lanceolate-acute, scabrous with membranous margins, rough and hairy: flowers small, yellow.

Hurser fort, near Jooneer. Hurrychunder.

(6) C. LANCEOLARIA. (*Roxb.*)

Ident. Roxb. flor. Ind. III. p. 432.

SPEC. CHAR. Perennial, erect: leaves lanceolar, serrated, smooth; petioles sub-pinnatifid, winged, not decurrent: flowers panicled, sub-ovate.

Silhet, flowering in February and March.

(7) C. DIFFUSA. (*Roxb.*)

Ident. Roxb. flor. Ind. III. p. 429.

SPEC. CHAR. Annual, erect: lower branches long, spreading or drooping, downy: inferior leaves sub-lyrate: superior ones sessile, oblong, serrately-toothed: flowers terminal.

Bengal, flowering in the rainy season.

(9) C. ANDRYALOIDES. (*Dec.*)

Ident. Dec. prod. V. p. 377.

SPEC. CHAR. Somewhat stemless: whole plant hoary with downy tomentum: leaves subradical, petioled, obovate, obtuse, subtruncate at the base, coarsely and sparingly toothed: scapes 1-headed, three times longer than the leaves and bearing a few acute entire linear leaflets: involucral scales in two rows, linear, acute, puberous at the back, scarcely longer than the achænium.

North-West Provinces.

(10) C. ROYLEI. (*Dec.*)

Ident. Dec. prod. V. p. 381.

SPEC. CHAR. Stems herbaceous, ascending, simple, velvety-hirsute: radical leaves petioled, oval, cauline ones attenuated at the base, obovate, all coarsely cut and toothed, villosely sub-hispid: capitula racemose, pedicelled, subround: involucral scales, oblong-lanceolate, acute: achænium compressed, glabrous on the disk, subciliated at the margin; male florets yellowish.

North-West Provinces.

(11) C. PUBESCENS. (*Dec.*)

Ident. Dec. prod. V. p. 381.

SPEC. CHAR. Stems herbaceous, simple, ascending and with the leaves and involucres, velvety-pubescent: lower leaves attenuated into a long petiole, obovate-oblong, upper ones shortly attenuated at the base, oval-oblong, acute, all sparingly and minutely toothed: capitula racemose, shortly pedicelled: involucral scales oblong, mucronate: achænium compressed, slightly hispid.

North-West Provinces.

(12) C. BUYA. (*Wall.*)

Ident. Dec. prod. V. p. 384.—Wall. Cat. No. 3079.

SPEC. CHAR. Stem herbaceous, erect, simple, manifold from the base, hairy: leaves villous, oblong, obtusely and unequally toothed, obtuse, lower ones attenuated at the base into the petiole, upper ones cuneate at the base, sessile: panicle corymbose: capitula pedicelled: involucral scales ovate, broadly white-margined, in two rows, cuspidate.

Assufghur.

(13) C. WALLICHII. (*Dec.*)
Var. serrulata.

Ident. Dec. prod. V. p. 384.

Syn. Conyza disticha, *Wall. Cat. No.* 3929.

SPEC. CHAR. Stem herbaceous, erect, ramosely-corymbose above, glabrous: ultimate branchlets covered with powdery pubescence: leaves elongately linear-lanceolate, acuminate, very slightly serrulated: panicles corymbose, many-headed, with the branches often bifid: capitula sessile, closely collected at the apex of the branchlets: involucral scales linear-acuminate, somewhat longer than the disk.

Oude.

GENUS XXIV. BLUMEA.

Syn. Pol. Superflua. *Ser: Syst :*

Deriv. Called after Dr. Blume, a Botanist, formerly resident in Java.

GEN. CHAR. Capitula heterogeneous; flowers of the circumference many-series, truncated or 2-3-toothed: of the disk few, cylindrical, 5-toothed: achænium terete: pappus 1-series; bristles capillary.—Herbaceous or suffruticose plants with panicled or loosely corymbose inflorescence: involucre imbricated, many-series, scales linear, acuminated: receptacle flat, quite naked or sometimes hairy: flowers yellow or purple.

(1) B. AMPLECTENS. (*Dec.*)

Ident. Dec. prod. V. p. 433.—Wight's Contrib. p. 13.

Syn. Erigeron obliquum, *Linn. Mani.*—Conyza amplexicaulis, *Lam. Dict.* II. p. 83. (?)

SPEC. CHAR. Stem herbaceous, hairy, dichotomously branched; younger leaves villous, adult ones subglabrate: leaves of the stem ovate, half-stem-clasping, acutely toothed; lowest ones here and there cut out at the base: peduncles few, subterminal, a little longer than the leaf, 1-headed: scales of the involucre linear, bearded, longer than the disk: female flowers many-series: stigmas exserted, red: male flowers 15-20; anthers exserted.

Common on the road sides and in sandy places at Madras and Negapatam.

(2) B. ARENARIA, (Dec.)

Ident. Dec. prod. V. p. 433.—Wight's Contrib. p. 13.
Syn. Conyza villosa, *Wall. Cat. No.* 3105.

SPEC. CHAR. Glabrous at the base, everywhere silky, villous at the apex: stem spreading from a branching base: leaves oblong, sessile, with a callous point, acute, entire or sharply denticulate: peduncles solitary, 1-headed, hirsute, axillary and terminal: scales of the involucre linear, acuminate, hirsute at the back, ciliate at the apex: male corollas bearded at the apex; female flowers slender: style bifid, long-exserted.

In sandy places near Negapatam.

(3) B. TENELLA. (Dec.)

Ident. Dec. prod. V. p. 433.—Wight's Contrib. p. 13.

SPEC. CHAR. Stem herbaceous, many-headed, puberulous, sparingly branched; leaves oblong, half-stem-clasping at the base, acute; distantly sharply and callously-toothed, scabrous with small pubescence: branches subentire: peduncles terminal and opposite-leaved, puberous, 3 times as long as the leaf: scales of the involucre linear, acuminate: flowers yellow, females very numerous, slender, males 8-9: anthers included.

In sandy places at Negapatam.

(4) B. ANAGALLIDIFOLIA, (Dec.)

Ident. Dec. prod. V. p. 433.—Wight's. Contrib. p. 13.
Syn. Conyza anagallidifolia, *Wall.*
Engrav. Rheede Mal. t. 18. (?)—Pluk. t. 19. f. 4. (?)

SPEC. CHAR. Stems manifold, diffuse, branched, puberulous at the apex; stem-leaves oval-oblong, sessile, here and there sharply toothed: leaves of the branches lanceolate, entire, puberulous: heads of flowers long-pedicelled, solitary, axillary and terminal, disposed into a leafy lax subcorymbose panicle: scales of the involucre linear, acute, longer than the disk: female flowers numerous, males 10-12, somewhat bearded at the apex of the corolla.

Common in moist places at Negapatam.

(5) B. CLINOCEPHALA. *(Dec.)*

Ident. Dec. prod. V. p. 434.—Wight's Contrib. p. 13.

Syn. Conyza amplexicaulis, *Wight in Wall. Cat.*

SPEC. CHAR. Sparingly puberulous: many-stemmed: lower leaves obovate-oblong, cuneate at the base; upper ones oblong-oval, sessile, half-stem-clasping: all remotely and sharply-toothed at the apex: peduncles 1–3 from the apex of the stems, villous, 1-headed, longer than the leaf: scales of the involucre linear, acute, equalling the disk: male flowers 15–20: anthers exserted.

Madras.

(6) B. PUBIFLORA. *(Dec.)*

Ident. Dec. prod. V. p. 434.

Syn. Erigeron asteroides, *Roxb. flor. Ind.* III. p. 432.

SPEC. CHAR. Stem herbaceous, erect, branched: hairs soft, more or less hispid: leaves sagittate, half-stem-clasping, oval-oblong, acute, subentire or acutely toothed: branchlets flower-bearing, 1-headed: scales of the involucre linear, acuminate, equalling the flowers: styles of the female flowers long-exserted: flowers large, yellow.

Madras. Concans. Flowering in the cold season.

(7) B. BIFOLIATA. *(Dec.)*

Ident. Dec. prod. V. p. 434.—Wight's Contrib. p. 14.

Syn. Conyza foliolosa, *Wight.*—C. bracteolata, *Kottl.*—C. bifoliata, *Roxb. flor. Ind.* III. p. 430.

Engrav. Pluk. Alm. t. 177. f. 1.

SPEC. CHAR. Stem herbaceous, ascending, ramous, particularly puberulous at the apex: leaves acutely toothed; lower ones ovate, long-attenuated at the base, glabrous; upper ones sessile, oblong, cuneate, downy: peduncles axillary, often carrying two sub-opposite leaves above the base: pedicels elongated, 1-headed: scales of the involucre linear, acute, nearly equalling the disk: male corollas 8-16, yellowish; anthers enclosed.

In shady moist places in Tanjore and Travancore, and many other places in the Peninsula, flowering in the cold season.

(8) B. BARBATA. *(Dec.)*

Ident. Dec. prod. V. p. 434.—Wight's Contrib. p. 14.

Syn. Conyza barbata, *Wall. Cat. No.* 3090.

SPEC. CHAR. Stem herbaceous, erect, simple, somewhat hairy: leaves lanceolate, attenuated at the base, acutely and distantly serrate, acute, above sparingly downy, below thickly silky-villous: branches axillary shorter than the leaf; small leaves, racemose at the apex; upper ones corymbose: scales of the involucre linear, acute, glabrous, as long as the disk: flowers yellowish, females numerous, style exserted, bifid: male flowers 5: corollas glabrous at the apex.

Peninsula.

(9) B. WIGHTIANA. (Dec.)

Ident. Dec. prod. V. p. 435.—Wight's Contrib. p. 14.

Syn. Conyza Wightiana, *Wall. Cat. No.* 3093.

Spec. Char. Stem herbaceous, erect, striated, puberulous; lower leaves petioled, oval, cuneate at the base, obtusely and unequally toothed; upper ones cuneate, oval-oblong: heads of flowers collected into a terminal interrupted racemose panicle; scales of the involucre linear, acute; lower ones glabrous, longer than the disk; male corollas outwardly tubercled with glands.

Madras. Pondicherry.

(10) B. LYSCHENAULTIANA. (Dec.)

Ident. Dec. prod. V. p. 435.

Spec. Char. Stem erect, herbaceous, hairy: leaves oval, serrate, puberulous; lower ones obtuse, attenuated at the base; upper ones sessile, acute: flowering branchlets very hirsute, few-headed: lower ones axillary, shorter than the leaf; upper ones arranged into a short panicle; involucral scales linear, hirsute, equalling the disk: female flowers numerous: style exserted, bifid, purple: males 10–11, glabrous: anthers half-exserted.

Peninsula.

(11) B. LACERA. (Dec.)

Ident. Dec. prod. V. p. 35.—Wight's Contrib. p. 14.

Syn. Conyza lacera, *Roxb. fl. Ind.* III. *fp.* 428.

Engrav. Burm. Ind. t. 59. f. 1.

Spec. Char. Whole plant tomentose or covered with white hirsute pubescence: stem herbaceous, erect: leaves obovate-oblong, thickly toothed, often divided by incisions at the base or pinnatifid: panicle oblong or corymbose, compressed or lax: involucral scales linear, acuminate, equalling the flowers: flowers yellow, females numerous: style bifid: males about 20.

In rice fields in the Peninsula. Silhet. Flowering in the cold season.

(12) B. GLANDULOSA. (Dec.)

Ident. Dec. prod. V. p. 438.—Wight's Contrib. p. 14.

Spec. Char. Whole plant covered with short glandulose hairs: stem herbaceous, erect, striated: cauline leaves shortly petioled, ovate, subacuminate, irregularly toothed, divided with lyrate incisions at the base: leaves of the branch cuneate at the base, obovate-oblong: heads of flowers constituting a leafy panicle at the apex of the branches: pedicels ramous, glandulose, shorter than the leaf: involucral scales linear, acute, equalling the disk: flowers yellowish, males about 10.

Negapatam.

(13) B. PHYLLOSTACHYA. . (*Dec.*)

Ident. Dec. prod. V. p. 438.

Syn. Conyza paniculata, *Herb. Madr. ex. Wall. Cat. No.* 3090.

Spec. Char. Stem herbaceous, erect, glabrous at the base, hispid with soft spreading hairs at the apex together with the branches and petioles: leaves petiolate, lyrate, pubescent; inferior lobes few, small, oblong-linear, acute, or 3-toothed; terminal one greatest, ovate-obtuse, sub-lobate at the base, acutely toothed: panicle compressed: branches short, bearing obovate, or oval acutely-toothed leaves intermixed with the heads of flowers: involucral scales linear, acute, glabrous, longer than the disk: female flowers numerous; males 5.

Peninsula.

(14) B. RUNCINATA. (*Dec.*)

Ident. , Dec. prod. V. p. 438.

Syn. Conyza runcinata, *Wall. Cat. No.* 3087.

Spec. Char. Stem herbaceous, erect, ramous, covered with velvety pubescence: lower leaves produced into a dilated auricled petiole at the base, oblong, runcinate, irregularly toothed, puberulous above, below villous: panicle elongated: branches bearing at the apex heads of flowers heaped together and interruptedly spiked: involucral scales linear, acute, villous at the back, longer than the disk: female flowers slender: style scarcely exserted: males 12–15.

Hurdwar.

(15) B. LONCHIFOLIA. (*Dec.*)

Ident. Dec. prod. V. p. 438.

Syn. Conyza lonchifolia, *Ham. in Wall. Cat.*

Spec. Char. Whole plant covered with viscid villous pubescence: stem erect, simple, striated: lower leaves petioled, lyrate; lower lobes few, oblong, acute, terminal one greatest, triangular, all thickly and unequally toothed; upper ones divided or toothed;

panicle terminal, elongated: lateral peduncles twice as long as the leaf: involucral scales linear, acute, longer than the disk; female flowers with the style long-exserted, deeply bifid; males 3-4: corolla shortly ciliated.

Patna.

(16) B. SENECIOIDES. (*Dec.*)

Ident. Dec. prod. V. p. 439.

Syn. Conyza senecioides, *Ham. in Wall. Cat. No.* 3102.

SPEC. CHAR. Stem herbaceous, erect, a little striated, very shortly sub-puberulous: leaves petioled, oval, sub-acute, long-cuneate at the base, distantly callous-toothed, membranaceous, subhirsute on the petiole and nerve; panicle lax, terminal: heads of flowers long-pedicelled: involucral scales linear, acuminate, ciliate, somewhat equalling the disk: flowers yellow; styles bifid; males about 15.

Nawaubgunge.

(17) B. VIRENS. (*Dec.*)

Ident. Dec. prod. V. p. 439.—Wight's Contrib. p. 15.

Syn. Conyza virens, *Wall. Cat. No.* 3037.

SPEC. CHAR. Stem herbaceous, erect, ramous, glabrous: leaves rigid, sessile, sparingly puberulous on both sides: cauline ones elongated, cuneate at the base, sessile, trifid: lobes lanceolate, mucronate, regularly-serrated; terminal one longer: leaves of the branchlets obovate, cuneate at the base, sparingly serrate: panicle lax: pedicels axillary, longer than the leaf, 1-3-headed; heads of flowers pedicellate: inner scales of the involucre linear, longer than the disk, mucronate: flowers purplish, females numerous, males 5.

In sandy places on the sea-shore at Negapatam.

(18) B. LAPSANOIDES. (*Dec.*)

Ident. Dec. prod. V. p. 440.—Wight's Contrib. p. 15.

SPEC. CHAR. Glabrous: stem herbaceous, erect, ramous: leaves petioled, membranaceous; stem ones sinuate-repand, here and there mucronately serrulate: lobes mucronate, lateral, sub-triangular, terminal one ovate: leaves of the branches obovate, scarcely subserrate: panicle lax, few-headed: branches lateral, leafy: heads of flowers pedicellate: involucral scales linear, mucronate, longer than the disk: flowers yellowish; females numerous, males 10-12.

Courtallum.

(19) B. HEYNEANA. (*Dec.*)

Ident. Dec. prod. V. p. 441.—Wight's Contrib. p. 15.

Syn. Conyza Heyneana, *Wall. Cat. No.* 3089.—C. aurita, *Heyne, Herb.*

SPEC. CHAR. Stem herbaceous, erect, much branched, finely
striated and puberulous: leaves oblong, attenuated at the base,
obtuse at the apex, here and there bearing sharp exserted teeth,
glabrous on both sides; lower ones lyrate-auricled at the base:
panicle leafy, lax: branches the length of the leaf, subcorymbose
at the apex: scales of the involucre linear, acuminate, sub-puberulous at the back, a little longer than the disk: flowers yellowish,
females numerous: style exserted: males about 12: pappus white.

Exact locality not specified.

(20) B. PAUCIFOLIA. *(Dec.)*

Ident. Dec. prod. V. t. 440.

SPEC. CHAR. Stem herbaceous, erect, everywhere very minutely puberulous; sub-viscid at the base: leaves petioled, obovate-
oblong, cuneate at the base, here and there serrated; above in the
adult ones glabrous; below and in the young ones on both sides
softly clothed with villous pubescence: panicle elongated: involucral scales linear, acuminate, equalling the flowers: flowers yellowish; males 12-18.

Silhet.

(21) B. SUBSIMPLEX. *(Dec.)*

Ident. Dec. prod. V. p. 441.

Syn. Conyza subsimplex, *Wall. Cat. No.* 3003.—C. paniculata, *Willd.*

SPEC. CHAR. Stem shrubby at the base, slender, glabrous;
branches herbaceous, elongated, somewhat naked, covered with
villous pubescence; lower leaves obovate, petioled, cuneate at the
base, rarely and sharply callous-toothed; above glabrous, below
pale and pubescent: leaves of the branches less, petioled, often
divided at the petiole: heads of flowers shortly 2-3-pedicelled from
the upper axils, almost constituting an interrupted spike: invo-
cral scales linear, acuminate, puberulous at the back, glabrous at
the apex, archly reflexed: flowers pale-yellowish brown.

Silhet.

(22) B. HIERACIFOLIA. *(Dec.)*

Ident. Dec. prod. V. p. 442.—Wight's Contrib. p. 15.

Syn. Conyza hieracifolia, *Spr. Syst.* III. p. 814.—Erigeron
hieracifolium, *Don. prod. fl. Nep.* p. 172.—C. communis, *Wall.
Cat. No.* 3018.

Engrav. Wight's Icon. t. 1099.

SPEC. CHAR. Whole plant hirsute: stem herbaceous, erect,
simple leaves callous-toothed: lower ones obovate, obtuse, atten-

nated into the petiole, upper ones oval or oblong, acute, sessile or half-stem-clasping: heads of flowers sessile, collected together and disposed into an ovate-oblong thyrsus: involucral scales linear, acuminate, longer than the disk: female flowers numerous; style exserted, undivided; males 3: ovaries pubescent.

Silhet, Carnatic.

(23) B. PURPUREA. *(Dec.)*

Ident. Dec. prod. V. t. 443.

Syn. Conyza purpurea, *Wall. Cat. No.* 3076.

SPEC. CHAR. Whole plant sparingly villous: stem erect, herbaceous, simple: leaves ovate, acute, serrate, entire at the base, attenuated: branches many, somewhat naked, paniculately disposed and bearing sessile-spiked heads of flowers: involucral scales linear, acute: exterior ones short, ashy-green: inner ones coloured, a little longer than the flowers: involucres purplish: male flowers about 20: pappus white.

Rajmahal hills.

(24) B. GLOMERATA. *(Dec.)*

Ident. Dec. prod. V. p. 443.—Wight's Contrib. p. 15.

Syn. Conyza glomerata, *Wall. Cat. No.* 3053.—C. fistulosa, *Roxb. flor. Ind.* III. *p.* 420.

SPEC. CHAR. Whole plant villous and somewhat hirsute, ash-coloured: stem herbaceous, erect, branched: branches leafy, arranged in a panicle and bearing at the apex sessile glomerated interruptedly-spiked heads of flowers: leaves lanceolate, serrate, acute at both ends: involucral scales many-series: outer ones smallest; inner ones linear, exceeding the flowers: female flowers scarcely more numerous than the males, truncated: stigmas bifid, purple: male flowers 30, bifid: anthers yellow, exserted.

Sea-beach at Negapatam.

(25) B. BELANGERIANA. *(Dec.)*

Ident. Dec. prod. V. p. 444.

SPEC. CHAR. Stem herbaceous, erect, ramous from the base: the base sparingly hispid: the apex softly villous; lower leaves obovate, somewhat obtuse, attenuated into a short petiole; upper ones subsessile, all sharply serrate, above sparingly hairy, below thickly silky-villous: heads of flowers heaped in the upper axils, almost interruptedly-spiked: involucral scales linear, glabrous, ciliate at the apex: female flowers few, males 8-10; receptacle narrow, naked: anthers exserted.

Malabar Coast.

(26) B. ERIANTHA. *(Dec.)*

Ident. Dec. prod. V. p. 444.—Wight's Contrib. p. 15.

Spec. Char. Stem herbaceous, erect, glabrous at the base, and very softly hispid at the apex: leaves membranaceous, sessile, oval, sharply and abruptly dentate, softly pubescent; upper ones cordate stem-clasping: peduncles axillary and terminal, 1-headed, solitary or few, aggregate, at last much elongated; involucral scales linear, ciliate, acuminate: male and female corollas bearded at the apex: male flowers 25-30.

Southern Peninsula.

(27) B. OXYODONTA. *(Dec.)*

Ident. Dec. prod. V. p. 444.—Wight's Contrib. p. 15.

Syn. Conyza oxyodonta, *Wall. Cat. No.* 3015.

Spec. Char. Herbaceous, covered with long silky hairs: stems many, slender, simple, diffuse; leaves sharply and thickly serrate, below more silky-villous, lower ones obovate, petiolate, upper ones elliptic, subsessile; heads of flowers few, racemosely corymbose at the apex of the branches, subsessile: involucral scales linear, acuminate, a little longer than the disk: female flowers many-series: males about 15-20: pappus whitish.

Pondicherry. Common on the sea-coast near Negapatam.

(28) B. PROCERA. *(Dec.)*

Ident. Dec. prod. V. p. 445.

Syn. Conyza procera, *Wall. Cat. No.* 3050.

Spec. Char. Stem herbaceous, erect, paniculately ramous, glabrous below, at the apex and on the branches covered with villous down: cauline leaves, glabrous, sessile, oval-lanceolate, acuminate, subserrate: panicle broad, divaricate: heads of flowers aggregate at the tops of the branchlets: involucral scales covered with hirsute pubescence at the back, linear, exceeding the flowers: receptacle villous: flowers purple: style bifid: pappus white.

Silhet.

(29) B. MYRIOCEPHALA. *(Dec.)*

Ident. Dec. prod. V. p. 445.

Syn. Conyza squarrosa, *Wall. Cat. No.* 3025.

Spec. Char. Stem herbaceous, erect, below with the leaves glabrous, above puberulous with the branches of the panicle: leaves oblong, acuminate, long-attenuated at the base into the petiole, subserrulate: panicle much branched; heads of flowers lateral, sessile and terminal; involucral scales linear, acuminate, recurved, puberulous at the back, a little longer than the disk; female flowers numerous, truncated; style bifid; males 5-9; pappus pale reddish.

Silhet.

(30) B. SPECTABILIS. (*Dec.*)

Ident. Dec. prod. V. p. 445.

SPEC. CHAR. Nearly the whole plant glabrous, with sub-hirsute branchlets; stem herbaceous, erect, panicled, ramous; leaves smooth on both sides, oval-lanceolate, long-attenuate at the base, acuminate at the apex, here and there somewhat divided, remotely and callously serrate, shining above; panicles lax, branches velvety, lateral, longer than the leaf, many-headed, involucral scales linear, acuminate, glabrous at the back, ciliate at the apex, almost exceeding the flowers: receptacle alveolate, glabrous.

Peninsula.

(31) B. PTERODONTA. (*Dec.*)

Ident. Dec. prod. V. p. 448.—Wight's Contrib. p. 15.

Syn. Conyza alata, *Wall. Cat. No.* 3039.

Engrav. Wight's Icon. t. 1100.

SPEC. CHAR. Stem herbaceous, ramous, scarcely sub-puberulous, clammy at the apex; leaves elliptic-oblong, glabrous, sub-serrate, produced into a wing, deeply and acutely toothed, and almost divided; branches leafy, sub-panicled, 3-or few-headed at the apex; pedicels naked; exterior involucral scales oblong, leafy, short, inner ones scarious, linear, acute, a little exceeding the flowers.

Common on the shores of the Coromandel Coast.

(32) B. AURITA. (*Dec.*)

Ident. Dec. prod. V. p. 449.—Wight's Contrib. p. 16.

Syn. Conyza aurita, *Linn. Suppl.* 367.—*Wall. Cat. No.* 3069 and 3086.

SPEC. CHAR. Hirsuto-villous; stem erect, ramous; leaves lyrato-pinnatipartite, lobes oblong, acute, toothed, terminal one greater, lower ones decurrent at the base, 1-2-eared together with the stem; panicle terminal, lax; heads of flowers pedicellate, erect; involucral scales lanceolate, acuminate, outer ones hirsute at the back, inner ones somewhat scarious, a little longer than the disk; male flowers 5-8; styles exserted.

On the road sides about Madras. Coromandel Coast.

(33) B. ALATA. (*Dec.*)

Ident. Dec. prod. V. p. 448.

Syn. Conyza alata, *Roxb. Cat.* p. 61.—C. alata, var. *Wall. Cat. No.* 3039.—Erigeron alatum, *Don. prod.* 171.—Vernonia alata, *Heyne. Herb.*

Engrav. Wight's Icon. t. 1101.—Spicil. II. t. 109.

SPEC. CHAR. Stem herbaceous, erect, ramous, and with the leaves covered with a short reddish velvety pubescence; leaves elliptic-oblong, toothed, produced together with the stem into quite entire wings; peduncles axillary, 1-or few-headed, racemosely panicled; heads of flowers suberect; exterior scales lanceolate, leafy, squarrose, somewhat downy; inner ones linear, scariose, equalling the flowers; flowers purple, males 10–12.

Concans.

(34) B. LACINIATA. *(Dec.)*

Ident. Dec. prod. V. p. 436.
Syn. Conyza laciniata, Roxb. flor. Ind. III p. 427.

SPEC. CHAR. Stem herbaceous, erect, branched, puberulous, softly and sparingly hairy at the apex; lower leaves pinnately-partite, somewhat scabrous on both sides, lobes four, oblong, oval, the higher ones confluent with the unequal lobe, all somewhat expand, acutely and unequally toothed, upper ones oblong, sessile, here and there serrated; panicle many-headed, branches elongated, thyrsoid at the apex; capitula pedicelled: involucral scales linear-acuminate, longer than the disk.

Bengal, flowering towards the end of the year.

(35) B. MUSRA. *(Dec.)*

Ident. Dec. prod. V. p. 435.
Syn. Conyza musra, *Ham. in Wall. Cat. No.* 3005.

SPEC. CHAR. Whole plant ash-coloured and softly villous: stem herbaceous, erect, branched; leaves oval-oblong, awned and serrated, attenuated at the base, lower ones petioled, lyrately appendiculate below, lobed, narrow, acute, sometimes distinct from the edge, at other times confluent with it: branches of the panicle erect, closely corymbose at the apex; involucral scales scarcely acute, villous, the length of the disk: male florets bearded at the apex; flowers yellowish.

Holasgunge.

(36) B. CERNUA. *(Dec.)*

Ident. Dec. prod. V. p. 436.
Syn. Conyza nutans, *Ham. in Wall. Cat. No.* 3080.

SPEC. CHAR. Stem herbaceous, erect, somewhat simple, puberulous at the base, velvety-villous at the apex: leaves obovate, attenuated at the base, coarsely and acutely toothed, puberulous above, villous below, lower ones long-extended at the base, upper ones oblong, villous on both sides: branches of the elongated panicle longer than the leaf, bearing at the apex subsessile glomerated nodding capitula: involucral scales linear, acute, equalling the flowers.

Bhimtghur, flowering in February.

(37) B. TRICONA. (*Dec.*)

Ident. Dec. prod. V. p. 437.

SPEC. CHAR. Whole plant glandulosely puberous: stem herbaceous, erect, 3-sided below, above terete, sparingly branched: leaves petioled, oval, acute, irregularly toothed, inciso-lyrate at the base, upper ones somewhat entire, acuminate: panicle subracemose, leafy, straight: involucral scales linear, acute, glabrous: receptacle glabrous: achænium slightly terete, glabrous.

North-West Provinces.

(38) B. FLAVA. (*Dec.*)

Ident. Dec. prod. V. p. 439.

Syn. Conyza fasciculata, *Wall. Cat. No.* 3017.

SPEC. CHAR. Whole plant glabrous, herbaceous, erect: stem smooth, branched: lowest leaves obovate, attenuated at the base, subterrate, obtuse, cauline ones cordately half-stem-clasping, acuminate, coarsely calloso-serrated: flowering branches axillary, bearing a close-packed corymbose raceme and loosely arranged as a panicle: involucral scales linear, cuspidately-acuminate: flowers yellow.

Silhet.

(39) B. BALSAMIFERA. (*Dec.*)

Ident. Dec. prod. V. p. 417.

Syn. Conyza balsamifera, *Linn. Wall. Cat. No.* 2998.—Pluchea balsamifera, *Less.*—C. odorata, *Rumph.*—Baccharis Salvia, *Lour. Coch.* II. p. 603.

Engrav. Rumph. Amb. 6. t. 24. fig. 1.

SPEC. CHAR. Stem suffruticose at the base: branches woolly-villous: leaves oblong or elliptic-lanceolate, duplicato-dentate, villous above, silky-woolly beneath, the veins wrinkled, lobes linear-lanceolate, appendiculate: corymb subpanicled, divaricate: involucral scales linear, acute, hirsute: flowers small, yellow.

The Concans. Assam. Flowering in February and March.

(40) B. DENSIFLORA. (*Dec.*)

Ident. Dec. prod. V. p. 446.

Syn. Conyza aromatica, *Wall. Cat. No.* 2997.

SPEC. CHAR. Stem herbaceous, erect and with the branches softly woolly: leaves elliptic-oblong, acuminate, attenuated at the base, here and there sharply cut and toothed, scabrous above, woolly below, a few lyrate at the base: panicle much branched, compact: involucral scales linear, acuminated, somewhat equalling the corollas: female florets numerous, males about five.

Pundua, Silhet.

(41) B. MURALIS. (*Dec.*)

Ident. Dec. prod. V. p. 440.—Dalz. Bomb. flor. p. 125.
Syn. Conyza muralis, *Wall. Cat. No.* 3077.

SPEC. CHAR. Glabrous below, glandular and hairy at the apex: stem herbaceous, erect, simple: leaves lyrate, pinnately lobed, narrowed into the petiole, lobes coarsely toothed: panicle elongated, lax: flowers yellow, shorter than the involucre: involucral scales linear-acuminate, slightly coloured.

Bassein fort, Bombay. North-West Provinces.

(42) B. HOLOSERICEA. (*Dec.*)

Ident. Dec. prod. V. p. 442.—Dalz. Bomb. flor. p. 125.
Syn. Conyza holosericea, *Wall. Cat. No.* 3103.

SPEC. CHAR. Whole plant clothed with long silky hairs: stem round, erect, scarcely branched: leaves oblong, attenuated at the base, acute, deeply and acutely serrated, covered with silky close-pressed hairs: panicle elongated, interrupted, the lower branches longer than the leaves, racemose at the apex, upper ones shorter, closely many-headed: involucral scales linear, puberous, ciliated, longer than the disk: flowers purplish.

Concans.

GENUS XXV. PLUCHEA.

Syn. Pol. Necessaria. *See : Syst :*

GEN. CHAR. Capitula heterogamous: florets of the circumference many-series, female, truncated or 2–3-toothed: central ones few, male, 5-toothed: anthers bicaudate at the base: achænium beakless, cylindrical or furrowed: pappus 1-series, bristles filiform, scarcely rough.—Herbaceous or suffruticose plants with alternate leaves: capitules corymbose: involucre many-series, imbricated: receptacle naked or hairy.

(1) P. TOMENTOSA. (*Dec.*)

Ident. Dec. prod. V. p. 450.—Wight's Contrib. p. 16.
Syn. Conyza tomentosa, *Wall. Cat. No.* 3031.

SPEC. CHAR. Shrubby; branchlets and leaves covered on both sides with short grey-velvety, scarcely tomentose, pubescence: leaves obovate or oval, cordately half-stem-clasping, toothed, subdecurrent: peduncles rising from the upper axils, longer than the leaf, bearing pedicelled heads of flowers and arranged in a short collected panicle: involucral scales oblong-linear, acute, a little shorter than the disk: style scarcely exserted: male flowers 5.

Mysore.

(2) P. INDICA. *(Less.)*

Ident. Dec. prod. V. p. 451.—Less. in Linnæa. 1831. p. 150.
Syn. Baccharis Indica, *Linn. Sp.*—Conyza Indica, *Blume.*—C. corymbosa, *Roxb. fl. Ind.* III. p. 428.
Engrav. Wight's Ill. II. t. 131.

SPEC. CHAR. Stem shrubby, tapering, corymbose, above ramous, glabrous below: branchlets, pedicels and young leaves minutely covered with dusty pubescence: leaves obovate, short-petioled, distantly and sharply serrated: corymbs terminating the branches; outer scales of the involucre ovate, downy at the back, inner ones linear, glabrous: anthers shortly caudate: flowers pale rose.

On the banks of the Ganges. Chittagong. Flowering in the cold season.

(3) P. WALLICHIANA. *(Dec.)*

Ident. Dec. prod. V. p. 451.
Syn. Conyza sessilifolia, *Wall. Cat. No.* 8029.

SPEC. CHAR. Stem somewhat shrubby, ramous: leaves and peduncles scabrous with very short powdery pubescence: leaves oval or obovate, acute or obtuse-serrulate, or quite entire, small; corymbs terminal, trichotomous: heads of flowers shortly pedicelled, oval-oblong: involucral scales lanceolate, acute, the length of the flowers: female flowers truncated: style bifid: anthers very shortly caudate at the base.

Sreenaghur.

GENUS XXVI. EPALTES.

Syn. Pol. Superflua. *Ass: Syn:*

Deriv. From *Epalthes*, healing, alluding to its medicinal qualities.

GEN. CHAR. Capitula heterogamous: florets of the circumference many-serias, female, of the disk sterile: anthers included: styles of the disk scarcely exserted, undivided: achænium of the ray obovate, subangled, glabrous: pappus none.—Ramous shrubs: leaves alternate, decurrent: peduncles opposite the leaves: involucre two, or several series, imbricated: receptacle naked, convex.

(1) E. DIVARICATA. *(Cass.)*

Ident. Dec. prod. V. p. 461.—Wight's Contrib. p. 16.
Syn. Ethulia divaricata, *Linn. Mant.* 110.
Engrav. Burm. Ind. t. 58. fig. 1.—Pluk. t. 160. fig. 5 and t. 78. fig. 2.

SPEC. CHAR. Stem herbaceous, erect, divaricating, much branched; leaves linear-oblong, attenuate at the base, decurrent, remotely denticulate; peduncles much shorter than the leaf: scales of the involucre much acuminated, much longer than the disk: flowers pink.

Common in rice fields in the Peninsula and other parts of the country.

(2) E. PYGMÆA. *(Dec.)*

Ident. Dec. prod. V. p. 461.

SPEC. CHAR. Stem herbaceous, dwarf, branched; leaves linear-oblong, somewhat denticulate, very narrowly decurrent; peduncles ebracteate, scarcely exceeding the leaves: involucral scales adpressed, scarcely acute, not exceeding the flowers.

Coromandel Coast.

GENUS XXVII. INULA.

Syn. Pol. Superflua. *Sm. Syst.*

Deriv. A corruption of *Helenium.*

GEN. CHAR. Capitula heterogamous: flowers of the ray one-series, female, usually ligulate, of the disk hermaphrodite, tubular, 5-toothed; anthers with two bristles at the base; schænium beakless, roundish: pappus uniform, bristles capillary.—Herbaceous perennials: capitula at the tops of the branches solitary, often corymbose; involucre many-series, imbricated; receptacle naked; flowers yellow.

(1) I. ROYLEANA. *(Dec.)*

Ident. Dec. prod. V. p. 464.

SPEC. CHAR. Whole plant softly villous and slightly tomentose: stem erect, simple, 1-headed: leaves ovate, somewhat denticulate, lower cauline ones produced into a winged petiole which is dilated and half-stem-clasping at the base, higher ones few, sessile, cordately stem-clasping.

North-West Provinces.

(2) I. EUPATORIOIDES. *(Dec.)*

Ident. Dec. prod. V. p. 469.

Syn. Conyza eupatorioides, *Wall. Cat. No.* 2993.

Spec. Char. Stem fruticose: branches striated, rufous, velvety: leaves elliptic, acuminated, attenuated at the base, short-petioled, rarely and minutely serrated, above sparingly umbrous, below with rufous ashy pubescence: panicle thyrsoid, many-headed, composite: lower branches axillary: involucral scales oblong, somewhat hirsute: ligulæ few, somewhat longer than the involucre: achænium villous.

Silhet mountains.

(3) I. ACUMINATA. *(Dec.)*

Ident. Dec. prod. V. p. 471.

Spec. Char. Stem erect, branched, puberulous: cauline leaves quite entire, dilated at the base, somewhat auricled, cordately half-stem-clasping, lanceolato-acuminate at the apex: capitula solitary at the tops of the branches: involucral scales linear, leafy, rather lax: achænium glabrous.

North-West Provinces.

GENUS XXVIII. VICOA.

Syn. Pol. Superflua. *Ser: Syst.*

Gen. Char. Capitula heterogamous: florets of the ray 1-series, narrow, ligulate, of the disk tubular, 5-toothed; achænium beakless: pappus of the ray none, of the disk capillary, 1-series.— Erect ramous annuals with alternate auriculately-sagittate sessile leaves: involucre imbricated; receptacle ebracteolate.

(1) V. INDICA. *(Dec.)*

Ident. Dec. prod. V. p. 474.—Wight's Contrib. p. 10.

Engrav. Wight's Icon. t. 1148.

Spec. Char. Leaves auricled at the base, lanceolate, acuminate, serrate or subentire, more or less puberulous on both sides: ligulæ twice as long as the disk; flowers yellow.

Common throughout the Peninsula, flowering nearly all the year.

(2) V. CERNUA. *(Dalz.)*

Ident. Dalz. Bomb. flor. p. 126.

Spec. Char. Leaves oblong-lanceolate: heads of flowers drooping, with 1-2 bristles as pappus to the ray flowers.

Elevated parts of the Concans, flowering in October.

GENUS XXIX. FRANCŒURIA.

Syn. Pol. Superflua. *Ser; Syst;*

GEN. CHAR. Capitula many-flowered, heterogamous: ray florets ligulate, 1-series, female, disk ones tubular, 4-toothed, hermaphrodite; achænium beakless; pappus caducous, 1-series; bristles rough below, plumose above; aginilinated at the base into a short ring, and crowned with minute palese or bristles.—An erect undershrub, with alternate leaves: involucre campanulate, imbricated; receptacle naked, flat.

(1) F. CRISPA. *(Cass.)*
Var. Indica.

Ident. Dec. prod. V. p. 475.
Syn. Inula quadrifida, *Wall. Cat. No.* 73.
SPEC. CHAR. Suffruticose, erect: branches 1-headed, diverging: leaves smooth at the margin, half-stem-clasping, oblong, toothed; flowers yellow.
Province of Behar, near the banks of the Ganges.

GENUS XXX. PULICARIA.

Syn. Pol. Superflua. *Ser; Syst;*

Deriv. From *Pulex*, a flea, a kind of flea-bane.
GEN. CHAR. Capitula many-flowered, heterogamous: ray florets 1-series, female, ligulate, disk ones tubular, 5-toothed, hermaphrodite; achænium beakless, not compressed: pappus 2-series, exterior one crown-like, dentate, short; interior one of rough bristles.— Erect ramous villous odorous plants, with cordato-sagittate leaves: peduncles 1-headed: involucre loosely imbricated: receptacle naked, areolate, flat.

(1) P. FOLIOLOSA. *(Dec.)*

Ident. Dec. prod. V. p. 480.
Syn. Conyza foliolosa, *Wall. Cat. No.* 3281.
SPEC. CHAR. Whole plant ashy-villous: stems herbaceous, much branched, softly hispid at the apex: leaves oblong, acute, puberulous on both sides: lower ones attenuate at the base, subdentate; upper ones entire, auricled at the base, half-stem-clasping: involucral scales linear, subulate, villous: flowers of the ray tubular, very slender, the rest shorter: exterior pappus crown-shaped, toothed, interior, 5–10-bristled: flowers yellow.
Patna. North-West Provinces.

(2) P. Hydrabadensis. *(Lindl.)*

Ident. Dec. prod. V. p. 480.

Syn. Aster Hydrabadensis, *Wall. Cat. No.* 2973.

Spec. Char. Stem herbaceous, erect, branched, somewhat downy: leaves lanceolate, puberulous, lower ones obtuse, toothed, attenuated at the base into the petiole, upper ones sessile, narrow, acute, sub-entire: branches sparingly leafed, 1-headed at the apex: involucral scales linear, acuminate, exterior ones foliaceous, puberulous at the back, interior ones dry, glabrous.

Hyderabad. Seringapatam.

(3) P. angustifolia. *(Dec.)*

Ident. Dec. prod. V. p. 479.

Spec. Char. Whole plant somewhat ashy-pubescent: stem suffruticose at the base, branched: leaves sessile, oblong-linear, quite entire or here and there toothed: peduncles 1-headed, three times longer than the leaf: involucral scales leafy: ligulæ longer than the involucre: exterior pappus denticulate, inner 10–12-bristles.

North-West Provinces.

GENUS XXXI. CÆSULIA.

Syn. Pal. Segregata. *See : Syst :*

Deriv. From *Cæsus*, beaten, appearing if trampled upon.

Gen. Char. Capitula numerous, aggregated in a general head: proper involucre 2-valved, at length cohering with the ovary: corolla tubular, spreading: anthers caudate: tops of the styles exserted: achænium compressed, apparently 2-winged; pappus none. —Herbaceous diffuse plants, with alternate leaves, amplexicaul above: flowers sessile, surrounded by 2–3 orbicular leaves, and some small bracts.

(1) C. axillaris. *(Roxb.)*

Ident. Dec. prod. V. p. 482.—Roxb. flor. Ind. III. p. 447.— Wight's Contrib. p. 11.

Syn. Moyera orientalis, *Don. prod. fl. Nep.* 180.

Engrav. Roxb. Cor. I. p. 64.—And. Bot. Rep. t. 431.— Wight's Icon. t. 1102.

Spec. Char. Glabrous, ascending: leaves lanceolate-linear, acuminate at both ends, distantly serrate, dilated at the base at the origin of the flowers, half-stem-clasping: floral branchlets axillary, very short: flowers pale violet, or white.

In moist places on the Coromandel Coast. In rice fields near Tanjore, flowering nearly all the year.

GENUS XXXII. ECLIPTA.

Syn. Pol. Superflua. *See : Syst :*

Deriv. From *Ekleipo*, to be deficient; the seed, crown and wings are wanting.

GEN. CHAR. Capitula many-flowered, heterogamous: florets of the ray 1-series, female, shortly ligulate, of the disk tubular, 4-toothed: achænia of the ray triangular, of the disk compressed, tuberculated on the sides.—Herbaceous plants with opposite leaves and solitary axillary peduncles: involucre 2-series, scales 10-12; receptacle flat, covered with lanceolate paleæ.

(1) E. ERECTA. *(Linn.)*

Ident. Dec. prod. V. p. 490.—Wight's Contrib. p. 17.

Syn. Verbesina alba, *Linn.*—Cotula alba, *Linn.*—Micrelium asteroidæa, *Forsk.*—Eclipta appressa, *Moench.*

Engrav. Rumph. Amb. VI. p. 18. fig. 1.—Pluk. Alm. t. 109. f. 1.

SPEC. CHAR. Stem erect, with adpressed hairs; leaves oblong-lanceolate, acuminate at both ends: capitule five times longer than the solitary or twin pedicels; flowers white.

In grassy, damp, or clayey soils, flowering nearly all the year. Dr. Roxburgh says, "*Verbesina prostrata* and *alba*, and I suspect *Eclipta erecta*, *punctata* and *prostrata*, are only one and the same species, or at most but accidental varieties from age, soil, and situation.

(2) E. PARVIFLORA. *(Wall.)*

Ident. Dec. prod. V. t. 490.

SPEC. CHAR. Stem prostrate, with adpressed hairs; leaves oblong-lanceolate, acuminated at both ends, sub-entire, somewhat scabrous; peduncles scarcely twice as long as the capitule.

Very common in moist places. This has the habit of *E. prostrata* and the leaves of *E. erecta*, but is different from both in its rooting stems. Almost all the species of the genus throw out roots in a wet soil, from near the bottom of the stems. The shape of the leaves is very inconstant.

GENUS XXXIII. BLAINVILLEA.

Syn. Pol. Superflua. *Sex: Syst:*

GEN. CHAR. Capitula heterogamous: florets of the ray few, 1-series, expanding into a broad short irregularly 3-cleft ligulæ, of the disk tubular, 5-cleft; style of the disk included: achænium of the ray triquetrous, bearing 3 ciliate bristles, of the disk compressed, biaristate.—Herbaceous plants with opposite triplinerved leaves: peduncles axillary, monocephalous: involucre 1-2-series, exterior one foliaceous: receptacle narrow: paleæ embracing the flowers.

(1) B. LATIFOLIA. (Dec.)

Ident. Dec. prod. V. p. 492.—Wight's Contrib. p. 17.

Syn. Eclipta latifolia, *Linn.*—Verbesina dichotoma, *Wall. Cat. No.* 314.—V. Lavenia, *Roxb. flor. Ind.* III p. 442.

Engrav. Pluk. t. 382. fig. 6.

SPEC. CHAR. Leaves rhomboid, acuminate, thickly toothed, or both sides sparingly pubescent: branches and petioles roughish, opposite, upper ones dichotomous: peduncles generally much shorter than the petiole: flowers white.

Rajahmundry. Dindigul hills. Madura. Monghyr. Flowering in the cold season.

GENUS XXXIV. SIEGESBECKIA.

Syn. Pol. Superflua. *Sex: Syst:*

Deriv. Named after Siegesbeck, a German Physician.

GEN. CHAR. Capitula heterogamous: florets of the ray 1-series, ligulate or deformed, of the disk tubular, 3-5-toothed: achænium somewhat 4-sided, curved within: pappus none.—Herbaceous dichotomous plants, upper parts viscid: Involucre 2-series, the 5 exterior ones linear, spathulate, spreading, interior ones viscidly pilose on the back: receptacle flat.

(1) S. ORIENTALIS. (Linn.)

Ident. Dec. prod. V. p. 495.—Wight's Contrib. p. 17.—Roxb. flor. Ind. III. p. 439.

Engrav. Pluk. Amalth. p. 58. t. 380. f. 2.—Linn. Hort. Cliff. t. 23.—Wight's Icon. t. 1103.

SPEC. CHAR. Leaves ovate, cuneate, acuminate at the apex, thickly toothed, upper ones oblong-lanceolate: scales of the external involucre twice as long as the interior one: flowers yellowish.

Dindigul hills, at 2500 feet.

GENUS XXXV. XANTHIUM.

Monœcia Pentandria. *Sex: Syst:*

Deriv. From *Xanthos*, yellow, an infusion of the plant is said to stain that colour.

GEN. CHAR. Capitula monoicous: male involucre subglobose, many-flowered, scales free, 1-series: receptacle paleaceous: corolla elavate, 5-lobed; filaments scarcely adnate to the corolla: anthers free; female, 2-flowers, enclosed within a 2-beaked prickly involucre; corolla filiform: stigmas 2, diverging: achaenium, one in each cell of the afterwards hardened involucre.—Herbaceous plants with variously divided leaves; heads of flowers irregularly glomerate, spicate: males above, females below.

(1) X. INDICUM. (*Roxb.*)

Ident. Dec. prod. V. p. 523.—Wight's Contrib. p. 17.
Syn. X. orientale, *Linn.*—X. Indicum, *Wall. Cat. No.* 291.
Engrav. Wight's Icon. t. 1104.

SPEC. CHAR. Involucre fruit-bearing; oval, pubescent between the prickles and at the base of the beaks; beaks hooked at the apex: flowers small, white.

Common on the road-sides in most parts of the country, flowering in February and March. Decandolle remarks, "*Genus inter compositas valde abnorme.*" It has been allotted by other Botanists both to *Verbenaceæ* and *Cucurbitaceæ*.

GENUS XXXVI. MOONIA.

Syn. Pol. Necessaria. *Ser: Syst.*

Deriv. Named after Mr. Moon, a Ceylon Botanist, author of a useful Catalogue of the Plants of that Island.

GEN. CHAR. Capitula monoicous; florets of the ray 1-series, female, ligulate, 3-cleft, of the disk, male; branches of the style of the female linear, revolute; style of the male included, simple, or slightly cleft at the apex: achaenium obovate, somewhat compressed, entire, or shortly bicornate at the apex.

(1) M. ARNOTTIANA.

Ident. Wight's Icon. vol. III.
Engrav. Wight's Icon. t. 1105.—Spicil. II. t. 110.

SPEC. CHAR. Erect ramous shrub: leaves opposite, unequally pinnatifid, the terminal lobe longer, deeply 3-cleft; pinnæ lanceolate, acuminate, coarsely cut and serrated, glabrous; flowers of the ray numerous; achaenium entire at the apex.

Neilgherry and Pulney hills.

GENUS XXXVII. WEDELIA.

Syn. Pol. Superflua. *Ser: Syst.*

Deriv. Named in honor of G. W. Wedel, a German Botanist.

GEN. CHAR. Capitula heterogamous: florets of the ray 1-series, ligulate, female, of the disk hermaphrodite, 5-toothed: branches of the style of the hermaphrodite flowers ending in a cone: achænium obovate or compressed, beakless, with a crown or calyx-like pappus: receptacle paleaceous.

(1) W. CALENDULACEA. (*Less.*)

Ident. Wight's Contrib. p. 17.—Dec. prod. V. p. 539.
Syn. Verbesina calendulacea, *Linn.*—Jægreia calendulacea, *Spr. Syst.* III. p. 500.
Engrav. Wight's Icon. t. 1107.—Burm. Zeyl. t. 22. fig. 1.— Rheede Mal. X. t. 42.
SPEC. CHAR. Herbaceous: leaves oblong-lanceolate, attenuated at the base, strigosely hairy on both sides, rarely serrated at the apex: peduncles 1-headed, solitary, axillary, three times longer than the leaf: outer involucral scales oblong, sub-acute, longer than the disk: calyculus, of the achænium somewhat stalked, denticulate: flowers yellow.

Concans. Coromandel. Bengal. In moist situations, flowering nearly all the year.

(2) W. URTICÆFOLIA. (*Dec.*)

Ident. Dec. prod. V. p. 539.—Wight's Contrib. p. 18.
Engrav. Wight's Icon. t. 1106.
SPEC. CHAR. Herbaceous, somewhat climbing: leaves petioled, ovato-lanceolate, coarsely and unequally serrated, hispid on both sides, acuminate, often incurved at the apex: peduncles solitary, 1-headed: involucral scales 10, in 2-series, acuminate, rough on the back: paleæ of the receptacle much acuminated: achænium surrounded by a short denticulate calyculus.

Neilgherry, Pulney and Shevagherry mountains.

GENUS XXXVIII. WOLLASTONIA.

Syn. Pol. Superflua. *Sex. Syst.*

Deriv. In honor of Dr. Wollaston, a celebrated Natural Philosopher.

GEN. CHAR. Capitula heterogamous: florets of the ray ligulate, 1-series, female, of the disk hermaphrodite: corolla articulated above the ovary: achænium thick, turbinate or compressed, umbilicate at the apex or bearing 5 slender aristæ.

(1) W. BIFLORA. (Dec.)

Ident. Dec. prod. V. p. 546.

Syn. Verbesina biflora, *Linn.* Roxb. *flor. Ind.* III. p. 440.— Acmella biflora, *Spreng.*—Wedelia biflora, *Wight's Contrib.* p. 18.

Engrav. Wight's Icon. t. 1108.—Rheede Mal. X. t. 40.

SPEC. CHAR. Annual: leaves petioled, ovate, shortly at the base and long-acuminated at the apex, sharply serrated, above sparingly rough and hairy, smoother below: peduncles 1-3, monocephalous, one-terminal, 1-2-springing from the upper axils: involucral scales in two rows, long-lanceolate: achænium bald or one-awned: flowers largish, yellow.

Coromandel. Common on the Ghauts. Negapatam. Flowering nearly all the year.

(2) W. SCABRIUSCULA. (Dec.)

Ident. Dec. prod. V. p. 517.

Syn. Eclipta scabriuscula, *Wall. Cat. No.* 3212.—Magern capitata, *Spreng.*—Verbesina dichotoma, *Heyne.*—V. biflora, *Blume.*

SPEC. CHAR. Perennial; stem herbaceous, furrowed and angular, smoothed: leaves petioled, ovate-lanceolate, acuminate, cuneate at the base, triplinerved above the base, coarsely serrated, above sparingly strigillose, puberulous below: corymbs terminal, composite, trichotomous: involucral scales in two rows, oval-oblong, acute: achænium thick, umbilicate at the apex, not awned: flowers yellow. Silhet.

GENUS XXXIX. GUIZOTIA.

Syn. Pol. Superflua. *Syst: Syst:*

Deriv. In honor of M. Guizot, an eminent French Statesman.

GEN. CHAR. Capitula heterogamous: ray ligulate, female: disk hermaphrodite, tubular, 5-toothed: base of the tube of the corolla furnished with a ring of thick-jointed hairs: achænium smooth.

G. OLEIFERA. (Dec.)

Ident. Dec. prod. V. p. 551.

Syn. Polymnia Abyssinica, *Linn. f. Suppl.*—P. frondosa, *Bruce.* —Verbesina sativa, *Roxb. flor. Ind.* III. p. 441.—Parthenium luteum, *Spreng.*—Heliopsis platyglossa, *Cass.*—Tetragonotheca Abyssinica, *Ledeb.*—Jægera Abyssinica, *Spreng.*—Ramtilla oleifera, *Dec.*—Buphthalmum Ramtilla, *Buch.*—R. oleifera, *Wight's Contrib.* p. 18.

Engrav. Wight's Ill. II. t. 132.—Bot. Mag. XXVI. t. 1017.

SPEC. CHAR. Stem pubescent at the apex: leaves half-stem-clasping, subcordate or ovate-lanceolate, remotely serrated, roughish; exterior scales of the involucre broadly-ovate, leafy; flowers large, bright yellow.

Lower Bengal. Deccan. Mysore. Commonly cultivated for the sake of its oil, flowering in the cold season.

GENUS XL. BIDENS.

Syn. Pol. Frustranea. *Sext Syst*

Deriv. From *Bis*, twice, and *Dens*, a tooth, in allusion to the seeds.

GEN. CHAR. Capitula either homogamous, discoid or radiate, with the florets of the ray ligulate, neuter; achænium surmounted by an aculeate beak, ending in 2–5 rigid bristles.

(1) B. NODIFLORA. (*Linn.*)

Ident. Dec. prod. V. p. 595.
Engrav. Dill. Elth. LII. t. 44. f. 52.

SPEC. CHAR. Stem dichotomous, hairy; leaves ovate or oblong, 1–2-toothed at both ends; capitula sessile at the wings or tops of the branches between the last pair of leaves, solitary, discoid; involucre cylindric; achænium bearing 4 awns, (2 long and 2 short.)

Bengal.

(2) B. WALLICHII. (*Dec.*)

Ident. Dec. prod. V. p. 598.
Syn. B. Chinensis, *Wall. Cat. No.* 3189.—*Wight's Contrib.* p. 19.—B. bipinnata, *Roxb. flor. Ind.* III. p. 411.—Agrimonia Moluccana, *Rumph.*
Engrav. Rumph. Amb. VI. t. 15. f. 2.

SPEC. CHAR. Smoothish or somewhat puberous; lower leaves pinnately divided, upper ones ternately divided; segments ovate-acuminate, cut and serrated, or toothed, lateral ones obliquely attenuated at the base; heads of flowers long-peduncled, loosely corymbose; achænium linear, angular, striated, glabrous, 5-awned at the apex.

Deccan. Bengal. Dindigul hills. Flowering nearly all the year.

(3) B. DECOMPOSITA. (*Wall.*)

Ident. Wall. Cat. No. 298.—Dec. prod. V. p. 603.
Syn. Coreopsis corymbifolia, *Ham. ex Wall.*

52

Spec. Char. Stem glabrous, 4-sided, branched: leaves bipinnately-partite: segments lanceolate, cuneate at the base, mucronate, scabrous on both sides: capitula long-peduncled, loosely subpanicled, discoid: involucre in two rows, afterwards reflexed: achaenia elongated, 2-awned, some short, strigossly-scabrous, others glabrous at the base, somewhat strigose at the apex in the same head of flowers.

Monghyr.

GENUS XLI. SCLEROCARPUS.

Syn. Pol. Frustranea. *Sax: Syst.*

Deriv. From *Scleros*, rough, and *Karpos*, fruit.

Gen. Char. Capitula many-flowered, heterogamous: florets of the ray 3, neuter, small, irregularly ligulate, situated before the inner scales of the involucre, of the disk 8–12, tubular, hermaphrodite, central ones often abortive: involucre double, outer scales leafy, sub-petiolate, larger, inner ones three, less, distant: receptacle small, convex, paleae of the disk encircling the flowers and shorter than them; corolla of the disk pubescent, 5-toothed, persistent: anthers scarcely concrete, produced at the apex into a subulate process: divisions of the style subulate, hispid: achaenia of the ray abortive, very slender, of the disk obovate, smooth, bald, closed in between the hardened somewhat curvedly folded palese, gaping within and contracted at the apex into a neck.

(1) S. Africanus. (*Jacq.*)

Ident. Dec. prod. V. p. 566.—Dals. Bomb. flor. p. 129.

Engrav. Jacq. Ic. Rar. I. t. 176.

Spec. Char. Annual, erect: stem rough: leaves broadly-ovate, acutely serrated, petioled, alternate, 3-nerved: heads of flowers yellow, solitary, subsessile at the apex of the branchlets.

North-West Provinces. High hills round Jooneer, (Bombay.) Flowering in July and August.

GENUS XLII. SPILANTHES.

Syn. Pol. Æqualis. *Sys: Syst.*

Deriv. From *Spilos*, a spot, and *Anthos*, a flower, alluding to the original species having yellow flowers and a brown disk.

Gen. Char. Capitula sometimes heterogamous: florets of the ray ligulate, female, sometimes homogamous, all the florets hermaphrodite, tubular, 4–5-toothed: branches of the style of the hermaphrodite florets truncated, penicillate: achaenia of the disk compressed, often ciliate on the edges.

(1) S. ACMELLA. (*Linn.*)

Ident. Dec. prod. V. p. 925.—Linn. Syst. Veg. p. 610.

Syn. Verbesina acmella, *Linn. Mant.* p. 475.—Acmella Linnæi, *Cass. Dict.* 24. p. 330.—S. Pseudo-acmella, *Linn.*—S. calva, *Dec. prod.* V. p. 625.—*Wight's Contrib.* p. 19.

Engrav. Wight's Icon. t. 1109.—Rumph. Amb. 6. t. 65.—Pluk. Alm. t. 159. fig. 4.

SPEC. CHAR. Stem slightly rooting at the base, erect or ascending; leaves petioled, ovate-lanceolate, 3-nerved, entire or toothed, rather glabrous: peduncles three times longer than the leaf: heads of flowers ovate; ligulæ 5–6, very small: achænia ciliated, 1-2-awned: flowers small, yellow.

Peninsula. Mysore. Travancore. Flowering nearly all the year.

GENUS XLIII. XIMENESIA.

Syn. Pol. Superflua. *Ser. Syst.*

Deriv. In honor of Joseph Ximenes, a Spanish Apothecary.

GEN. CHAR. Capitula many-flowered, heterogamous: florets of the ray female, ligulate, 1-series, of the disk hermaphrodite, tubular, 5-toothed; tube of the corolla hispid: achænia compressed, winged, deeply emarginate, somewhat hairy, with 2 aristæ cohering with the sides of the wings.

(1) X. ENCELIOIDES. (*Cav.*)

Ident. Dec. prod. V. p. 627.—Wight's Contrib. p. 19.

Syn. Pallasia serratifolia, *Sm. in Rees' Cycl.*

Engrav. Cav. Icon. II. p. 60. t. 178.

SPEC. CHAR. Annual, herbaceous: leaves opposite or alternate, often attenuated into a somewhat winged petiole auricled at the base, ovate and ovate-oblong, toothed: achænia of the disk somewhat villous, surrounded on all sides by a wing, emarginate at the apex, of the ray wingless, wrinkled; ligulæ deeply 3-toothed: flowers large, bright yellow.

Palamcottah. It is still doubtful whether this is a native of India or not.

GENUS XLIV. CHRYSANTHELLUM.

Syn. Pol. Superflua. *Ser. Syst.*

Deriv. A diminutive of Chrysanthemum.

GEN. CHAR. Capitula radiate, ligulate, 1-series, female, linear, bidentate; disk hermaphrodite, 5-toothed: branches of the style in the hermaphrodite florets appendiculate: achænia bald, exterior ones oblong, emarginate, or slightly winged on each side.

(1) C. INDICUM. *(Dec.)*

Ident. Dec. prod. V. p. 631.

SPEC. CHAR. Annual, herbaceous, glabrous, very small: branchlets somewhat naked, 1-headed at the apex: leaves of different shapes, radical ones oval, cuneate at the base, upper ones oblong-linear, 3-toothed at the apex: achænia somewhat compressed, very shortly emarginate at the apex, callous at the margin, at one place smooth and level, at another convex and striated: flowers bright yellow.

Sakanaghur.

GENUS XLV. GLOSSOCARDIA.

Syn. Pol. Superflua. *Ses: Syst:*

Deriv. From *Glossa*, a tongue, and *Kardia*, a heart. The exact application is not apparent.

GEN. CHAR. Capitula few-flowered: florets of the ray 1-2 or none, female, of the disk, hermaphrodite, tubular, 4-toothed: achænia of the disk obcompressed, very villous on the 4 angles, somewhat 2-winged and crowned with 2 stiff bristles from the lateral angles.

(1) G. BOSVALLEA. *(Dec.)*

Ident. Dec. prod. V. p. 631.—Wight's Contrib. p. 19.

Syn. Verbesina Bosvallea, *Linn. Suppl.* p. 340.—*Roxb. fl. Ind.* III. p. 443.—G. linearifolia, *Cass. Dict.* 19. p. 62.

Engrav. Wight's Icon. t. 1110.

SPEC. CHAR. Small annual, diffuse: leaves alternate, much divided, linear at the base: heads of flowers solitary, yellow, on short naked peduncles: involucre oblong, imbricated: receptacles covered with narrow membranaceous paleæ.

Coimbatore in dry pastures. Kulludgee, and common elsewhere on the Western Coast.

GENUS XLVI. GLOSSOGYNE.

Syn. Pol. Superflua. *Ses: Syst:*

Deriv. From *Glossa*, a tongue, and *Gyne*, female; alluding to the elongated styles.

Gen. Char. Capitula heterogamous: florets of the ray ligulate, female, of the disk hermaphrodite: styles of the hermaphrodite short, elongated into 2 long rough stigmas: achænia linear, angled, bearing two retrorsely-hispid bristles.—Erect glabrous herbs, with alternate pinnatifid leaves, congested near the base: capitula without bracts: involucre short, 2-series: receptacle paleolate: flowers yellow.

(1) G. PINNATIFIDA. (Dec.)

Ident. Dec. prod. V. p. 632.—Wight's Contrib. p. 19.

Syn. Bidens pinnatifida, *Wall.*—Zinnia Bidens, *Retz.*—*Roxb. flor. Ind.* III p. 435.

Spec. Char. Erect: stems dichotomous: leaves alternate, crowded, short, pinnately divided: lobes linear, acute, entire; uppermost ones undivided: heads of flowers erect: awns on the seeds widely spreading.

Goruckpore. South Mahratta country. Samulcottah, Chinglepput. Flowering in the rainy season.

GENUS XLVII. ENHYDRA.

Syn. Pal. Superflua. *Ser: Syst:*

Deriv. From *En*, in, and *Hydor*, water, alluding to the habit of the plant.

Gen. Char. Capitula heterogamous: all the florets wrapped in a folded palea: exterior ones many-series, female, Interior hermaphrodite, sterile: tube of the ray florets filiform, of the disk subconical: limb of the ray subligulate, 3–4-cleft, of the disk 5-cleft: stigmas exserted, revolute, rough towards the apex: achænia bald, beakless, enclosed in the palea.—Aquatic rooting oppositeleaved herbs, with axillary capitula: receptacle small, flattish: paleæ as if 2-valved, connivent: involucre 4-leaved, 2 opposite, larger: flowers small, white.

(1) E. HINGCHA. (Dec.)

Ident. Dec. prod. V. p. 637.

Syn. R. heloncha, *Dec. l. c.*—Meyera beloneba, *Wall.*—Hingtcha repens, *Roxb. flor. Ind.* III. p. 442.

Spec. Char. Stem glabrous at the base, slightly hairy at the apex: leaves somewhat hastate at the base, scarcely petioled, serrated, glabrous: capitula sessile.

In moist places and pools in Bengal. Dumdum, Goalpara, Silhet and Assam. Flowering nearly all the year.

GENUS XLVIII. PYRETHRUM.

Syn. Pol. Superflua. *Sex: Spl:*

Derio. From *Pyr*, fire, alluding to the hot taste of the roots.

GEN. CHAR. Capitula heterogamous: ray ligulate, 1-series, female; disk hermaphrodite, 4-5-toothed: achænia wingless, furnished with a crown-like pappus, equal to its diameter.—Annuals or perennials, with alternate, dentate or lobed leaves: involucre campanulate, scales scariose on the margin: receptacle flat, naked or sometimes bracteolate: disk yellow: ray white or yellow.

(1) P. INDICUM. *(Cass.)*

Ident. Dec. prod. VI. p. 62.—Cass. Dict. XL. IV. p. 149.— (not Roxb.)

Syn. Chrysanthemum Indicum, *Linn.*—C. Japonicum, *Thunb.*—C. tripartitum, *Sweet.*

Engrav. Sweet. Brit. fl. Gard. I. 193.—Rheede Mal. X. t. 44.—Rumph. Amb. V. t. 91. f. 1.—Bot. Reg. XV. t. 1287.

SPEC. CHAR. Stem fruticose, branched: branches pubescent at the apex: leaves petioled, ovate, cut or pinnatifid, frequently toothed, flaccid, uppermost ones quite entire: involucral scales very obtuse, broadly scariose at the margin: ligulæ a little longer than the involucre: flowers smallish, yellow.

Dindigul mountains, flowering nearly all the year. Decandolle suggests that it may probably have been introduced by cultivation.

GENUS XLIX. ARTEMISIA.

Syn. Pol. Superflua. *Sus: Spl:*

Derio. *Artemi*, the Greek appellation for Diana; so named from the plant having been used for bringing on precocious puberty.

GEN. CHAR. Capitula discoid, homogamous or heterogamous: florets of the circumference 1-series, usually female, 3-toothed, with a long exserted bifid style, of the centre 5-toothed, hermaphrodite or male by abortion: achænia bald, obovate, with a small epigynous disk.—Herbaceous or suffruticose plants: leaves alternate, variously lobed: capitula spicate or racemose: involucre imbricated: scales dry on the margin: receptacle naked or hairy: flowers yellow or purple.

(1) A. GLABRATA. *(Wall.)*

Ident. Wall. Cat. No. 413.—Dec. prod. VI. p. 100.
Engrav. Wight's Icon. t. 1111.

Spec. Char. Suffruticose, erect: branchlets and younger leaves somewhat villous below: stem-leaves stipellately cut at the base, lanceolate-cuneate, acutely 3-cleft at the apex: racemes very straight, subsecund, arranged in a panicle: capitula somewhat drooping, pedicelled, small, globose: involucral scales ovate, murginate; inner ones membranaceous at the margin.

Pondua, Silhet.

(2) A. PARVIFLORA. (Roxb.)

Ident. Roxb. flor. Ind. III. p. 420.—Dec. prod. VI. p. 100.

Spec. Char. Stem erect, simple, panicled: branches of the panicle divaricate: leaves villosely tomentose below, lower ones cuneate, 3-7-toothed at the apex, cauline ones stipellately cut at the base, lanceolate, quite entire: capitula racemose, panicled, globose, minute: involucral scales ovate, glabrous, scariose at the sides: flowers small, green.

Khasia mountains, flowering in May.

(3) A. OBATA. (Wall.)

Ident. Wall. Cat. No. 404.—Dec. prod. VI. p. 114.

Syn. A. vulgaris, Burm. fl. Ind. p. 177.

Engrav. Rheede Mal. X. t. 45.

Spec. Char. Suffruticose, erect: leaves white-tomentose below, pinnatifid, upper ones trifid, uppermost and branched ones undivided, and with the lobes oblong, obtuse, mucronate: lobes of the lower ones somewhat crenated: capitula spicately panicled, oblong: panicle leafy, patent: younger racemes nodding: outer scales of the younger pubescent involucre leafy, acute, of the inner ones scariose, obtuse: flowers small, greenish-white.

Peninsula, flowering in February.

(4) A. ROYLEANA. (Dec.)

Ident. Dec. prod. VI. p. 115.

Spec. Char. Herbaceous, very straight, entirely glabrous: cauline leaves pinnati-partite from the base, 3-4 pair with an odd one: lobes broad-linear, acute, quite entire: branches of the panicle erect: capitula racemose, secund, somewhat spreading and drooping, sub-globose: involucral scales ovately subrotund, scariose at the margin.

North-West Provinces.

(5) A. CARNIFOLIA. (Wall.)

Ident. Wall. Cat. No. 409.—Dec. prod. VI. p. 119.—Roxb. for. Ind. III. p. 422.

SPEC. CHAR. Erect, quite glabrous: leaves tri-pinnatisect: lobes
cut, linear-filiform: branches of the panicle patent, simple: capitula globose, nodding, racemose as far as the branches: involucral
scales obovate, scariose: flowers small, greenish-yellow.

Khasia mountains, flowering in March and April.

(6) A. PALLENS. (*Dec.*)

Ident. Wall. Cat. No. 412.—Dec. prod. VI. p. 120.

Syn. A. orientalis, *Herb. Madr.*

SPEC. CHAR. Herbaceous, canescent, erect, branched: lower
leaves sub-pinnatisect, upper ones pinnatisect: lobes and floral
ones undivided, linear-cuneate, obtuse: capitula globose, sub-erect,
few, racemose, sub-panicled, long-pedicelled: bracteoles hooded,
leafy, exceeding the scariose elliptic scales.

Peninsula.

(7) A. CUNEIFOLIA. (*Dec.*)

Ident. Dec. prod. VI. p. 126.

SPEC. CHAR. Stem herbaceous, erect, simple, ashy, and covered
loosely with villous pubescence: lower leaves long, upper ones
short, cuneate, sub-petiolate, with a linear stipule at the base on
both sides, dilated and three-cleft or three-parted at the apex: lateral lobes acutely 3-toothed at the apex, middle one trifid with
smaller lobes acutely 3-toothed.

North-West Provinces.

GENUS L. MYRIOGYNE.

Syn. Fed. Superflua. *Ser.* i *Syst.*

Deriv. From *Murios*, a myriad, and *Gyne*, female; alluding to
the numerous female florets.

GEN. CHAR. Capitula heterogamous: marginal florets many-
series, female, tubular; central ones few, hermaphrodite, 4-toothed,
with a short tube and campanulate limb: achenia angled, wingless,
without pappus.—Very ramous diffuse herbs: leaves alternate,
obovate, dentate: capitula small, first terminal, afterwards lateral;
involucrum 2-series: receptacle naked, convex: flowers yellow.

(1) M. MINUTA. (*Less.*)

Ident. Dec. prod. VI. p. 139.—Wight's Contrib. p. 20.

Syn. Cotula minuta, *Forsk.*—C. minuta et cuneifolia, *Willd.*—
Grangea minuta et cuneifolia, *Poir.*—G. decumbens, *Desf.*—Artemisia minima, *Thunb.* (not *Linn.*)

Var. Lanuginosa, *Dec. l. c.*—Artemisia sternutatoria, *Roxb.* fl.
Ind. III. p. 423.—Cotula minima, *Blume.*—C. sternutatoria, *Wall.*

SPEC. CHAR. Decumbent or ascending: slightly glabrous or somewhat velvety at the apex: leaves oblong, cuneate at the base, serrated at the apex, somewhat obtuse or sub-acute: flowers small, yellow.

In moist places in Bengal and the Peninsula, flowering in February and March. In the variety the stems are more or less covered with hoary wool, and the leaves are shorter.

GENUS LI. SPHÆROMORPHÆA.

Syn. Pol. Superflua. *Sar: Syst:*

Deriv. From *Sphaira*, a globe, and *Morphe*, form; alluding to the shape of the flowers.

GEN. CHAR. Capitula heterogamous: marginal florets many-series, female, scarcely dentate; central few, 4-toothed: style bulbous at the base: achænia cylindrical, striated, without pappus.—Decumbent herbs with alternate leaves: capitula globose, depressed, axillary: involucrum campanulate, 2-3-series, longer than the flowers; receptacle naked.

(1) S. RUSSELIANA. *(Dec.)*
Var. glabrata.

Idem. Dec. prod. VI. p. 140.
Engrav. Delem. Ic. Sel. 4. t. 49.

SPEC. CHAR. Stems somewhat hairy: leaves glabrous, obovate, toothed, shortly attenuated at the base: capitula somewhat twin: involucre glabrous: flowers numerous.

North-West Provinces.

GENUS LII. MACHLIS.

Syn. Pol. Superflua. *Sar: Syst:*

Deriv. A Greek term, alluding to the female florets.

GEN. CHAR. Capitula heterogamous: marginal florets many-series, not furnished with a corolla? or very minute; central ones numerous: corolla obconical, 4-toothed: stigmas bearded at the apex: achænia terete, subangled, subglandulose, of the marginal florets bidentate, of the centre truncated at the apex.—Herbs with alternate pinnatifid leaves: capitula pedicelled, axillary: involucrum 2-series: receptacle naked, punctulate.

(1) M. HEMISPHÆRICA. *(Dec.)*

Ident. Dec. prod. VI. p. 140.

Syn. Artemisia hemisphærica, *Roxb. flor. Ind.* III. p. 422.—Cotula hemisphærica, *Wall. Cat. No.* 3236.

Engrav. Deless. Ic. Sel. 4. t. 50.

SPEC. CHAR. Annual, many-stemmed, branched, erect, villous at the tops of the branches: lobes of the leaves linear, acutely mucronate: capitula solitary, somewhat convex: flowers small, yellow.

Silhet. Bengal, in dry rice-fields, flowering in February and March.

GENUS LIII. HELICHRYSUM.

Syn. Pol. Superflua. *Ser. Syst.*

Deriv. From *Helios*, the sun, and *Chrysos*, gold; in allusion to the brilliant flowers.

GEN. CHAR. Capitula sometimes homogamous: florets all hermaphrodite, 5-toothed, sometimes heterogamous, marginal florets often very few, 1-series, female: achænia beakless, sessile, with a terminal areola; pappus 1-series of roughish bristles.—Herbs or shrubs with alternate leaves: involucrum imbricated: scales scariose, interior ones connivent or radiant: receptacle flat, naked, areolate or fimbrilliferous: involucres white, purple or yellow: corolla yellow or purple.

(1) H. BUDDLEOIDES. *(Dec.)*

Ident. Dec. prod. VI. p. 201.—Wight's Contrib. p. 20.

Engrav. Wight's Icon. t. 1103.—Spicil. II. t. 111.

SPEC. CHAR. Stem suffruticose, erect, branched, woolly, especially at the apex: leaves sessile, oval-lanceolate, acuminated, 7-9-nerved, glabrous above, hoary-tomentose below, quite entire: corymb many-headed, composite at the tops of the stem and branches: capitula ovate, clustered: involucral scales oval, obtuse, somewhat equal, a little longer than the disk: flowers yellow.

Neilgherries.

GENUS LIV. GNAPHALIUM.

Syn. Pol. Superflua. *Ser. Syst.*

Deriv. From *Gnaphalon*, soft down; alluding to the woolly covering of the plants.

GEN. CHAR. Capitula heterogamous: florets all tubular, marginal ones many-series, female, disk ones hermaphrodite: achænia somewhat tapering, subpappilose: pappus 1-series, scarcely rough. —Herbaceous or suffruticose, generally woolly or tomentose plants: capitula often disposed in glomerules, terminal or axillary, fascicled, corymbose or spicate: involucres white, red, purple or yellow.

(1) G. RAMIOSUM. *(Dec.)*

Ident. Dec. prod. VI. p. 222.

SPEC. CHAR. Stem suffruticose at the base, branched, sub-glabrous: branches white-tomentose: leaves linear, sessile, quite entire, white-tomentose on both sides: capitula 8–10, collected into a sub-globose glomerule: involucral scales yellowish, scarious, oblong, sub-acute.

North-West Provinces.

(2) G. MULTICEPS. *(Wall.)*

Ident. Dec. prod. VI. p. 222.—Wall. Cat. No. 8949.

SPEC. CHAR. Woolly: stems many, herbaceous, ascending, erect, simple: lower leaves lanceolate, acute, flat: cauline ones sessile, scarcely sub-decurrent, uppermost ones linear: capitula terminal, densely glomerate: involucral scales oblong, elliptic, obtuse.

Silhet.

(3) G. HYPOLEUCUM. *(Dec.)*

Ident. Dec. prod. VI. p. 222.—Wight's Contrib. p. 21.

Engrav. Wight's Icon. t. 1114.

SPEC. CHAR. Stem erect, terete, scabrous below, above branched, tomentose: leaves linear, acuminate, subrevolute at the margin, slightly scabrous above, below white-tomentose with the middle nerve a little scabrous, half-stem-clasping at the adnate base, sub-decurrent: capitula congested at the tops of the branches, subsessile: glomerules corymbosely panicled: involucral scales yellow, scarious, oval-oblong, somewhat obtuse, a little longer than the disk.

Neilgherries.

(4) G. HURDWARICUM. *(Wall.)*

Ident. Dec. prod. VI. p. 231.—Wall. Cat. No. 2931.

SPEC. CHAR. Woolly, very small: stem erect, bifid: leaves oblong, attenuated at the base, obtuse, mucronate, scattered on the stem, aggregated round the glomerules: capitula densely clustered in globose glomerules at the bifurcation and tops of the branches: involucral scales woolly outside at the base, otherwise scarious, quite glabrous, brownish, acuminated.

Hardwar.

(5) G. PALLIDUM. *(Lam.)*

Ident. Dec. prod. VI. p. 230.—Wight's Contrib. p. 22.—Wall. Cat. No. 2953.—Lam. Dict. II. p. 750.

Syn. G. orixense, *Roxb. flor. Ind.* III. p. 425.

SPEC. CHAR. Woolly: stems many, herbaceous, ascendent, erect, simple: lower leaves lanceolate, sub-acute, flat, cauline ones broadly sessile, uppermost ones linear: flowering branchlets sub-umbellate at the apex: capitula sub-aggregated in glomerules at the axils and tops of the branches: involucral scales oblong, elliptic, obtuse, woolly quite at the base, otherwise quite glabrous, reddish straw-coloured.

Circar mountains.

(6) G. INDICUM. *(Linn.)*

Ident. Dec. prod. VI. p. 231.—Wight's Contrib. p. 22.

Syn. G. multicaule, *Willd. Roxb. flor. Ind.* III. p. 425.—G. polycaulon, *Pers.*—G. pluricaule, *Poir.*—G. strictum, *Roxb. fl. Ind.* III. p. 424.

SPEC. CHAR. Stems herbaceous, many, diffuse, tomentose: leaves obovate-oblong or oblong-linear, mucronate, more or less tomentose, lower ones attenuated at the base, upper ones sessile: capitula aggregated into an interrupted simple or branched spike: involucral scales linear, obtuse, scariose, straw-coloured or reddish: flowers very small, yellow.

Peninsula. Patna. Silhet. Flowering in February and March.

(7) G. MARCESCENS. *(R. W.)*

Ident. Wight's Icon. vol. III.

Engrav. Wight's Icon. t. 1115.

SPEC. CHAR. Shrubby, somewhat diffuse at the base: branches ascending, the lower parts clothed with persistent withered leaves, which are revolute on the edges and linear-subulate, green ones narrow-lanceolate, acute, glabrous above, tomentose beneath: flori-ferous branches umbellate at the apex: heads of flowers aggregated at the tops of the branchlets: involucral scales ovate-lanceolate, woolly at the base, white-scariose towards the apex: marginal florets 2-series: styles not exserted: achænia obovoid, puberulous: pappus, uniform, scabrous.

Neilgherries.

GENUS LV. FILAGO.

Syn. Pol. Superflua. *Ser : Syst :*

Deriv. From *Filum*, a thread; the plant appears as if covered with cotton or down, whence the name Cotton Rose.

GEN. CHAR. Capitula heterogamous; marginal florets numerous on an elongated filiform receptacle, the apex only dilated and bearing a few male or hermaphrodite flowers; achænia papillose, terete; pappus of the central ones bristly, filiform, of the margin wanting or dissimilar.—Tomentose herbs: capitula axillary or aggregated on the ends of the branches, small.

(1) F. PROSTRATA. (*Dec.*)

Ident. Dec. prod. VI. p. 248.—Wight's Contrib. p. 22.

Syn. Gnaphalium prostratum, *Bomb. Herb.*—G. depressum, *Roxb. fl.* III. p. 425.

SPEC. CHAR. Stems many, diffuse, prostrate: whole plant woolly, much branched: leaves elliptic, mucronate, narrowed into the petiole; capitula densely corymbosely-aggregated, floral leaves stalked; involucral scales mucronate: flowers small, yellow.

Negapatam. Patna. Circars. Flowering in February and March.

GENUS LVI. ANTENNARIA.

Diœcia Pentandria. *&c. Syst.*

Derив. So called in reference to the down of the pappus, which is like the antennæ of some insects.

GEN. CHAR. Capitula diœcious or subdiœcious, with an alveolate convex receptacle; female flowers filiform, 5-toothed; male anthers half-exserted; achænia terete: pappus 1-series; bristles of the female flowers filiform, of the male clavate.—Herbaceous or suffruticose plants: capitula corymbose: involucrum imbricated, variously coloured at the apex or scariose, never yellow: flowers yellow.

(1) A. CINNAMONEA. (*Dec.*)

Ident. Dec. prod. VI. p. 270.

Syn. Gnaphalium cinnamoneum, *Wall. Cat. No.* 2944.

SPEC. CHAR. Stem erect, herbaceous, covered with darkish cobweb-like wool; leaves linear-lanceolate or linear-acuminate, half-stem-clasping at the base, 3-nerved, glabrous above, rufous tomentose beneath: corymb composite, involucral scales pure white, obtuse.

Pundua, Silhet.

(2) A. SEMIDECURRENS. (*Dec.*)

Ident. Dec. prod. VI. p. 271.

Syn. Gnaphalium semidecurrens, *Wall. Cat. No.* 2947.—G. subdecurrens, *Wight's Contrib.* p. 21.

SPEC. CHAR. Stem herbaceous, erect, whitish-tomentose: leaves linear, acuminated, mucronulate, 1-nerved, cobweb-like above, tomentose below, shortly decumbent at the base; corymb composite: involucral scales pure white, sub-obtuse.

Dindigul hills.

(3) A. CONTORTA. (*Don.*)

Ident. Dec. prod. VI. p. 271.—Don. prod. fl. Nep. p. 175.

Syn. Gnaphalium simplicicaule, *Wall. Cat. No.* 3046.—G. contortum, *Spreng. Syst.* III. p. 490.

Engrav. Don. in Bot. Reg. t. 605.

SPEC. CHAR. Stem suffruticose, branched, erect: leaves linear, mucronate, sessile, younger ones tomentose, incurred, revolute at the margin, adult ones twisted: capitula aggregated in somewhat capitate simple or composite corymbs: involucral scales oblong, slightly obtuse.

Pundua, Silhet.

GENUS LVII. ANAPHALIS.

Syn. Pol. Superflua. *Ser. Syst:*

Deriv. The name of a plant classed by the ancient Greeks next to *Gnaphalium.*

GEN. CHAR. Capitula heterogamous, discoid; marginal florets few or many-series, female, disk ones hermaphrodite, sterile, 5-toothed: styles of the marginal florets long-exserted, bifid, of the disk undivided: achenia glabrous, sessile, beakless: pappus 1-series, uniform, bristles rough.—Herbaceous or suffruticose, woolly or tomentose plants; stems sometimes 1-headed, oftener many-headed, corymbose: involucral scales niveo-scariose, lanceolate, several-series, the middle ones with a brownish claw: receptacle convex, alveolate.

(1) A. NEILGHERRYANA. (*Dec.*)

Ident. Dec. prod. VI. p. 272.—Wight's Contrib. p. 21.

Syn. Gnaphalium Neilgherryanum, *R. W.*

Engrav. Wight's Icon. t. 478.

SPEC. CHAR. Stem suffruticose, many-headed, low: flowering branches erect, tomentose; lower leaves very densely clustered, imbricated backwards, linear, subobtuse, lower ones glabrous, uppermost ones as far as the flowering branches erect, tomentose, acute, somewhat distant; capitula closely clustered in a terminal corymb: involucral scales oblong-linear, subacute, longer than the disk, white-coloured.

Neilgherries.

(2) A. ROYLEANA. (*Dec.*)

Ident. Dec. prod. VI. p. 272.

SPEC. CHAR. Herbaceous, many-stemmed at the base: branches erect, simple, hoary-tomentose; leaves sessile, very shortly subadnate, oblong-linear, mucronate, quite entire, above powdery, scabrous, below cobweb-like, whitish: corymb somewhat simple, 7-9-headed; peduncles covered with hoary wool: involucral scales glabrous, pure white, outer ones ovate, shorter, middle ones ovate-oblong, exceeding the flowers, innermost ones linear-oblong, equaling the disk; capitulum pure white, brown-spotted at the base outside; florets yellow.

North-West Provinces.

(3) A. POLYLEPIS. (*Dec.*)

Ident. Dec. prod. VI. p. 272.

SPEC. CHAR. Stem suffruticose at the base, erect, branched; flowering branches and peduncles hoary with tomentose wool: lower leaves subspathulate, the rest linear, quite entire, the younger ones terminated by a short thick mucro, glabrous above, white-woolly below: corymb somewhat simple, 7-9-headed: involucral scales about 70-80, scarious, glabrous, outer ones somewhat shorter, purple, the rest pure white, oval-oblong, longer than the disk.

North-West Provinces.

(4) A. LINEARIS. (*Dec.*)

Ident. Dec. prod. VI. p. 174.

SPEC. CHAR. Stem simple (?) erect, hoary, covered with cobweb-like wool; leaves sessile, linear-acuminate, quite entire, cobweb-like above, covered with dense hoary down below: margin subrevolute; corymb composite, the branches 8-10, bearing at the top 5-6 densely clustered capitula: involucral scales pure white, glabrous, oval-oblong, the length of the disk: flowers yellow.

North-West Provinces.

(5) A. ARANEOSA. (*Dec.*)

Ident. Dec. prod. VI. p. 275.

SPEC. CHAR. Stem herbaceous (?) erect, branched, somewhat angled, here and there cobweb-like, branches covered with hoary villi at the apex: leaves decurrent, slightly glabrous, lanceolate-linear, acuminate, quite entire: capitula densely collected at the tops of the branches; corymb subcomposite; involucral scales pure white, glabrous, oblong, scarcely exceeding the flowers: flowers yellow.

North-West Provinces.

(6) A. NOTONIANA. *(Dec.)*

Ident. Dec. prod. VI. p. 273.
Syn. Gnaphalium Notonianum, *Wall. Cat. No.* 2952.—Helichrysum Notonianum, *Dec. in Wight's Contrib. p.* 20.
Engrav. Wight's Icon. t. 1116.

SPEC. CHAR. Whole plant woolly; stem suffruticose: branches leafy at the top: leaves broad-linear or oblong, obtuse, with revolute margins, very thick with wool, sessile, sub-decurrent: corymb terminal, composite, dense: scales of the ovate involucre many-series, imbricated, white-scariose, acute, somewhat crisp at the top, afterwards stellately patulous.

Neilgherries.

(7) A. LEPTOPHYLLA. *(Dec.)*

Ident. Dec. prod. VI. p. 273.
Syn. Helichrysum leptophyllum, *Dec. in Wight's Contrib. p.* 20.

SPEC. CHAR. Whole plant silky and hoary: stem branched, erect: leaves distant, sessile, linear, acutely-mucronate, quite entire, spreading: corymb terminal, composite, sub-umbellate: capitula pedicelled: involucral scales oblong-linear, sub-acute, white-coloured, squarrosely-reflexed.

Neilgherries.

(8) A. WIGHTIANA. *(Wall.)*

Ident. Wall. Cat. No. 2940.—Dec. prod. VI. p. 273.
Syn. Gnaphalium Wightianum, *Dec. in Wight's Contrib. p.* 21.
Engrav. Wight's Icon. t. 1117.

SPEC. CHAR. Stem suffruticose at the base, erect, covered with scabrous hairs as far as the leafy base, woolly at the apex: leaves sessile or subadnate, oblong-linear, obtuse, with scabrous hairs above, hoary with wool below, uppermost ones callosely subuncinate at the apex: corymb terminal, truly composite but very thick and many-headed: involucral scales oblong, acutish, white-coloured, a little longer than the disk: flowers yellow.

Neilgherries.

(9) A. ELLIPTICA. *(Dec.)*

Ident. Dec. prod. VI. p. 274.
Syn. Gnaphalium ellipticum, *Wight's Contrib. p.* 21.—A. oblonga, *Dec. l. c.*
Engrav. Wight's Icon. t. 1118.

Spec. Char. Whole plant covered with pure white down: stem suffruticose at the base, low, branched: leaves elliptic, mucronulate, quite entire, downy, 5-7-nerved, adnate or shortly decurrent: capitula collected into an ovate terminal corymb thickly surrounded by leaves: involucral scales oval, acute, scariose at the apex, reddish-white.

Neilgherries.

(10) A. ARISTATA. *(Dec.)*

Ident. Dec. prod. VI. p. 274.
Syn. Gnaphalium aristatum, *Wight's Contrib.* p. 21.
Engrav. Wight's Icon. t. 1119.

Spec. Char. Stem suffruticulose at the base, branched, erect, scabrous from the apex to the leafy base, tomentose at the apex: leaves linear, elongated, aristately mucronate: margins subrevolute, dilated and cordate at the base, shortly decurrent, scabrous above, with the middle nerve somewhat tomentose, below hoary-tomentose, with the middle nerve scabrous: capitula condensed into a very thick composite terminal corymb: involucral scales oval, obtuse, white and rose-coloured, glabrous.

Neilgherries.

(11) A. ADNATA. *(Wall.)*

Ident. Wall. Cat. No. 2948.—Dec. prod. VI. p. 274.

Spec. Char. Whole plant white with woolly down: stem suffruticose at the base, ascendent, erect, somewhat simple: lower leaves oval-oblong, attenuated at the base, upper ones linear or lanceolate, adnate, acute; lower flowering branchlets axillary, shorter than the leaf, bearing a small corymb at the apex, upper ones longer, bearing a full composite corymb: involucral scales oblong or oval, obtuse, white-coloured, inner ones longer.

Pundua, Silhet.

GENUS LVIII. CARPESIUM.

Syn. Pol. Æqualis. *Bess: Spel:*

Deriv. From *Carpesion*, a bit of straw; alluding to the appearance of the involucral leaves.

Gen. Char. Capitula heterogamous, discoid, marginal florets female, several series, disk ones hermaphrodite: anthers caudate: achænia oblong, compressed, rostrate: beak beset with viscid glands: pappus none.—Herbaceous erect plants with alternate leaves: capitula solitary on the ends of the branches: involucrum many-series, imbricate, outer ones somewhat leafy: flowers yellow.

(1) C. NEPAULENSIS. (Less.)

Ident. Dec. prod. VI. p. 222.—Wight's Contrib. p. 22.
Engrav. Wight's Icon. t. 1120.—Spicil. II. t. 112.

SPEC. CHAR. Stem hirsutely villous; leaves elliptic-lanceolate, acuminate, dentate, attenuated into the petiole: capitula somewhat drooping, campanulate, interior scales of the involucre somewhat acute.

Ootacamund.

(2) C. CILIATUM. (Wall.)

Ident. Wall. Cat. No. 3214.—Dec. prod. VI. p. 281.

SPEC. CHAR. Slightly scabrous: lower leaves oval, long-attenuated at the base, dentate, upper ones distant, oblong, subsessile, almost quite entire: capitula erect, campanulate, outer scales of the involucre ovate, erect, distinctly ciliated, inner ones membranaceous, subacute.

Travancore.

GENUS LIX. GYNURA.

Syn. Pol. Æqualis. *Ber: Syst:*

Deriv. From *Gyne*, female, and *Oura*, tail; alluding to the appendage of the style.

GEN. CHAR. Capitula discoid, homogamous: base of the tube of the corolla horny, branches of the style produced into a long hispid appendage, usually exserted: achænia striated, beakless: pappus many-series, filiform, scarcely barbellate.—Herbaceous or shrubby plants: capitula corymbose: involucrum cylindrical, 1-series, calyculate at the base with subulate bracts: receptacle flat, alveolate: flowers white.

(1) G. ANGULOSA. (Dec.)

Ident. Dec. prod. VI. p. 298.
Syn. Cacalia angulosa, *Wall. Cat. No.* 3152.

SPEC. CHAR. Glabrous: stem herbaceous, erect, furrowed and angled, branched: leaves oval-oblong, serrated, attenuated at the base, cordately somewhat half-stem-clasping: branches of the lax panicle elongated and bearing 3-7-headed corymbs: involucre subcylindric, equal to the flowers, 2-3 times longer than the subulate bracteolæ: flowers purplish.

Pundua, Silhet.

(2) G. NITIDA. *(Dec.)*

Ident. Dec. prod. VI. p. 299.—Wight's Contrib. p. 24.
Syn. Cacalia incana, *Heyne. Wall. Cat. No.* 3158.
Engrav. Wight's Icon. t. 1121.

SPEC. CHAR. Glabrous: stem thickish at the base: branches elongated, somewhat angled, rather naked at the apex: leaves lanceolate, attenuated at both ends, coarsely serrated: corymb terminal, 5—7-headed: involucre cylindric, a little shorter than the flowers, four times longer than the subulate bracteoles.

Dindigul mountains.

(3) G. PSEUDO-CHINA. *(Dec.)*

Ident. Dec. prod. VI. p. 299.
Syn. Senecio Pseudo-China, *Linn.*
Engrav. Dill. Elth. 345. t. 258. f. 335.

SPEC. CHAR. Rhizome thick, fleshy: leaves somewhat radical, oval-oblong, attenuated into the petiole, coarsely toothed or lobed at the base, puberulous on the nerves below: peduncles scapiform, very long, bearing 1—3 pedicellate heads of flowers: involucre surrounded at the base by linear-subulate bracteoles: receptacle shortly fimbrilliferous: flowers golden-yellow.

Coromandel Coast.

(4) G. LYCOPERSICIFOLIA. *(Dec.)*

Ident. Dec. prod. VI. p. 300.—Wight's Contrib. p. 24.
Syn. Cacalia laciniata, *Wall. Cat. No.* 3135.

SPEC. CHAR. Stem herbaceous, erect, branched, striated, slightly glabrous: leaves pinnately lobed, widely and roundly auricled at the base, velvety on both sides with very short pubescence: lobes lanceolate, here and there toothed: capitula 3—6 corymbose at the tops of the branches, pedicellate: involucre cylindric, glabrous, somewhat equalling the florets, three times longer than the subulate bracteoles.

Courtallum. Dindigul hills at 2000 feet. It sometimes varies with the auricles being small and acute.

(5) G. NUDICAULIS. *(Dec.)*

Ident. Dec. prod. VI. p. 301.—W. Arnot. pug. pl. Ind. Or. No. 109.

SPEC. CHAR. Herbaceous: root tubariferous: stems somewhat simple, striated: leaves thickish, nearly all radical, lyrately pinnatifid, younger ones hoary-pubescent, adult ones subglabrous; cauline ones few, linear-oblong, pinnatifid: corymb few-headed, contracted: peduncles hoary-pubescent; bracteoles few, subulate: involucre sub-campanulate, nearly equalling the florets, twice as long as the bracteoles.

Peninsula.

(6) G. WALKERI. (R. W.)

Icon. Wight's Icon. vol. III.
Engrav. Wight's Icon. t. 1122.

SPEC. CHAR. Shrubby, erect: stems naked at the base, terete, marked with numerous scars of fallen leaves, leafy at the apex: leaves long-petioled, ovate-lanceolate, acuminated, entire or slightly crenulate: corymbs terminal, large, lax, many-headed: involucre cylindrical, shorter than the flowers, much longer than the slender subulate bracteoles: flowers white.

Neilgherries, usually in moist soil near streams. Flowering nearly all the year.

(7) G. SIMPLEX. (*Dals.*)

Icon. Dals. Bomb. Flor. p. 130.

SPEC. CHAR. Tall, erect, unbranched, glabrous: stem thick at the base, angular: leaves oblong-obovate, sessile, coarsely sinuate-toothed: stem distantly covered with leaves to the apex: corymb terminal with 5–7 capitula: bracteoles linear-acute: flowers deep orange-colour.

Highest hills round Joonees, Bombay.

GENUS LX. EMILIA.

Syn. Pol. Æqualis. *Ser: Syn:*

GEN. CHAR. Capitula homogamous, discoid: florets 5-lobed: branches of the style ending in a cone: achænia oblong, pentagonal; angle ciliate, hispid: pappus several-series, filiform, scarcely barbellate.—Herbs with a few subcorymbose pedicelled capitula: involucrum cylindrical, 1-series, ecalyculate: flowers reddish, purple or orange-coloured.

(1) E. SONCHIFOLIA. (*Dec.*)

Icon. Dec. prod. VI. p. 302.—Wight's Contrib. p. 24.

Syn. Cacalia sonchifolia et glabra, *Wall. Cat. No.* 3144–5.— *Roxb. for. Ind.* III. p. 413.—E. purpurea, *Cass.*—Senecio sonchifolia, *Moench.*—Crassocephalum sonchifolium, *Less.*

SPEC. CHAR. Sparingly puberous or glabrous, subglaucous, erect or diffuse: lowest leaves petioled, sublyrate, cauline ones sagittate or cordate-stem-clasping, subdentate; corymbs few-headed: capitula long-pedicelled; involucre cylindrical, nearly equalling the florets; flowers 30–50, outer ones erect, small, bright-purple.

Peninsula. Bengal. Flowering in the cold season. This is a very variable plant in respect to both stem and leaves.

(2) E. SCABRA. (*Dec.*)

Ident. Dec. prod. VI. p. 303.—Wight's Contrib. p. 24.
Engrav. Wight's Icon. t. 1123.

SPEC. CHAR. Stem ascending, leafy and densely hairy at the base, naked, smooth at the apex: lowest leaves lyrate, cauline ones cordately half-stem-clasping, ovate-lanceolate, rather obtuse, dentate, rough on both sides from scattered hairs: corymbs terminal, 5–7-headed: involucre scarcely shorter than the flowers, enclosing about 100 flowers.

Peninsula—not uncommon.

(3) E. PRENANTHOIDEA. (*Dec.*)

Ident. Dec. prod. VI. p. 303.
Syn. Cacalia teres, *Wall. Cat. No.* 3164.—Prenanthes sarmentosa, *Wall. Cat. No.* 3262. *ex Herb. Madr.*

SPEC. CHAR. Stem erect, glabrous, somewhat simple, naked at the apex: leaves sagittately stem-clasping, long-lanceolate, acuminated, nearly quite entire, somewhat glabrous above, scabrous with thickish pubescence below: capitula few, long-peduncled: involucre ovate-cylindric: florets about 50.

Silhet.

(4) E. ANGUSTIFOLIA. (*Dec.*)

Ident. Dec. prod. VI. p. 303.
Syn. Cacalia angustifolia, *Wall. Cat. No.* 3163.

SPEC. CHAR. Glabrous: stem erect, much branched, twiggy: leaves distant, cordately sagittate, long-lanceolate, acuminated, nearly quite entire: panicle lax, few-headed: pedicels long, naked: involucre cylindrical, a little shorter than the flowers.

Pundua, Silhet.

GENUS LXI. DORONICUM.

Syn. Pol. Superflua. *Ser: Syst:*

Deriv. Called from its Arabic name *Doroniji.*

GEN. CHAR. Capitula radiate, heterogamous: ray florets 1-series, ligulate, female or sterile by abortion; disk hermaphrodite: achænia beakless, oblong, turbinate, furrowed, of the ray bald, of the disk pappose: pappus setaceous, several series.—Herbaceous plants, with solitary or several capitula; involucrum few series, scales linear: receptacle convex, ehracteolate: flowers yellow.

Almost the only distinction between this genus and Senecio consists in the marginal florets, in this having no pappus, in that being furnished with pappus similar to those of the disk. (R. W. in Calc. Journ.)

(1) D. ROYLEI. (*Dec.*)

Ident. Dec. prod. VI. p. 321.

SPEC. CHAR. Stem herbaceous, erect, somewhat hispid with spreading scattered bristles: upper leaves sessile, half-stem-clasping, ovate-lanceolate, denticulate, sparingly hispidulous: branches few, elongated, somewhat naked, 1-headed: involucre emlyculate, hispid, scales lanceolate-linear, acuminated, a little longer than the disk: achænia of the ray glabrous, without pappus, of the disk somewhat hispid, pappose.

North-West Provinces.

(2) D. LINIFOLIUM. (*Dec.*)

Ident. Dec. prod. VI. p. 322.

Syn. Aster odontophyllus, *Wall. Cat. No.* 3285.

SPEC. CHAR. Quite glabrous: stem erect, branched: leaves scattered, linear, acute, 1-nerved, slightly rigid, here and there coarsely toothed: panicle corymbose, lax: pedicels bearing scattered subulate bracteoles; involucral scales 15–20, somewhat in one series shorter than the disk: achænia quite glabrous: pappus cinnamon-red.

Pundua, Silhet.

(3) D. WIGHTII. (*Dec.*)

Ident. Dec. prod. VI. p. 322.

Syn. Madaractis glabra, *Dec. prod.* VI. p. 440.

Engrav. Wight's Icon. t. 1124.

SPEC. CHAR. Glabrous: stem erect, subsimple, angularly striated at the base: leaves lanceolate, acute, coarsely-toothed, sub-revolute at the margin, lowest ones attenuated at the base, upper ones half-stem-clasping: corymb few-headed: pedicels bracteolate at the apex: involucral scales linear, subacute: ligulæ 8–10, flat: achænia glabrous.

Neilgherries, in pastures and near the banks of watercourses, flowering towards the end of the rainy season.

(4) D. ARNOTTII. *(Dec.)*

Ident. Dec. prod. VI. p. 322.
Syn. Madaractis polycephala, *Dec. prod.* VI. p. 440.
Engrav. Wight's Icon. t. 1125.

SPEC. CHAR. Stem simple, erect, striated, rough below, glabrous: leaves close set towards the base, cordately half-stem-clasping, oblong, 3-6-nerved, obtuse, dentate, rough on both sides: corymb terminal, few-flowered: bracteoles linear-subulate: involucre sub-calyculate: scales about 15, linear; ligulæ 10, oval, 6-nerved: achænia glabrous: florets very numerous.

Northern slopes of the Neilgherries, near Nedawuttem, flowering in October and November.

(5) D. LESSINGIANUM. *(W. & A.)*

Ident. Dec. prod. VI. p. 322.—W. & A. pug. pl. Ind. Or. No. 106.
Syn. Madaractis glabra, *Dec. l. c.* p. 439.
Engrav. Wight's Spicil. II. t. 113.

SPEC. CHAR. Stem long, striated, hairy: leaves cordately stem-clasping, oblong-lanceolate, few-nerved, cut and serrated: corymb few-headed, terminal: involucre hemispherical: scales linear-subulate, inner ones oblong-lanceolate, muricately-hispid; ligulæ 8-10, narrow-oval, about 9-nerved.

Neilgherries, flowering after the rainy season.

(6) D. CANDOLLEANUM. *(Arn.)*

Ident. Dec. prod. VI. p. 322.—W. & A. pug. l. c. No. 105.
Syn. Madaractis pinnatifida, *Dec. prod.* VI. p. 439.
Engrav. Wight's Icon. t. 1127.—Spicil. II. t. 114.

SPEC. CHAR. Suffruticose, ramous: branches striated, nearly glabrous, few-headed; leaves whitish, hispidly pubescent, pinnatifid: lobes short, oblong, acute, occasionally shortly dentate: peduncles minutely bracteolate at the apex: involucre 1-series: scales lanceolate, whitish, hispid on the back: ligulæ 8-10, narrow-oval, 3-5-nerved.

Neilgherries, frequent in pastures.

(7) D. RUPESTRE. *(R. W.)*

Ident. Wight's Icon. vol. III.
Engrav. Wight's Icon. t. 1128.

SPEC. CHAR. Suffruticose: branches near the base naked, above leafy: leaves long-petioled; limb lobed or somewhat pinnatifid, attenuated into a long slender petiole: nerves beneath bristly: pedicels short, leafy at the base, closely beset towards the apex with minute subulate bracteoles: involucre 1-series, calyculate: leaflets linear, acuminate: ligulæ 8, linear-lanceolate, obtuse, 4-nerved.

Shevagberry mountains, flowering in August and September.

(8) D. TENUIFOLIUM. *(R. W.)*

Ident. Wight's Icon. vol. III.

Syn. Senecio tenuifolius, *Burm. Fl. Ind.*—S. multifidus, *Willd.*—S. laciniosus, *Arn.*

Engrav. Wight's Icon. t. 1129.

SPEC. CHAR. Herbaceous, erect or ascending, branched, glabrous: leaves pinnatifid or bipinnatifid; divisions linear-acute, variously toothed or lobed, glabrous: corymb few-headed: capitula peduncled: leaflets of the involucre linear-lanceolate, acute: ligulæ about 8, broad-oval, obtuse, 4-nerved.

Neilgherries, and widely distributed both on the subalpine plains and mountains of Southern India.

(9) D. TOMENTOSUM. *(R. W.)*

Ident. Wight's Icon. vol. III.

Engrav. Wight's Icon. t. 1151.

SPEC. CHAR. Stem herbaceous, erect, sub-tomentose, at first simple, leafy, afterwards corymbosely branched: branchlets nearly naked: leaves rough, lower ones elliptic, tapering to the base: upper ones sub-ovate-lanceolate, auricled and nearly stem-clasping, coarsely and unequally dentate, rough and slightly araneose above, densely white-tomentose beneath: corymb lax: peduncles bracteolate: ligulæ about 14, sterile: disk florets numerous, 5-cleft: pappus setaceous, hispid: achænium ribbed, hairy: flowers yellow.

North-Western slopes of the Neilgherries, flowering in September and October.

(10) D. RETICULATUM. *(R. W.)*

Ident. Wight's Icon. vol. III.

Engrav. Wight's Icon. t. 1151. (B.)

SPEC. CHAR. Herbaceous, erect, branched, leaves somewhat rhomboidal, coarsely and unequally dentate, teeth mucronate, rough with cobweb-like pubescence above, tomentose between the veins beneath: capitula lax, corymbose on longish pedicels: bracts subulate: ligulæ 10–12, sterile, throat hairy, pappus none: disk florets numerous, tube contracted, throat dilated, campanulate: pappus paleaceous, hispid: achænium ribbed, conical, hairy.

Tannah district, Bombay.

(11) D. HEWRENIE. *(Dalz.)*

Ident. Dalz. Bomb. flor. p. 130.

SPEC. CHAR. Stem erect, striated, pilose; leaves oblong or lanceolate, attenuated at the base, auricled, coarsely toothed, pubescent above, ciliated on the margin, hispid on the prominent nerves beneath: peduncles axillary or terminal, long, slender, 1–3-headed; ligules 3-flowered, very small, oval: disk florets about twelve, small, yellow.

Rocky places about Jooneer (Bombay), flowering in July and August.

GENUS LXII. MADACARPUS.

Sym. Pol. Superflua. *Serv Syst:*

Deriv. From *Madas*, to be bald, and *Karpos*, fruit.

GEN. CHAR. Capitula radiate, heterogamous: ray flowers 1-series, ligulate, sterile; disk ones numerous, hermaphrodite: achænia beakless, oblong, furrowed, hairy, without pappus.—Herbaceous plants: capitula corymbose: involucrum campanulate, 1-series: scales linear, lanceolate, mucronate: receptacle convex, foveolate; corolla subinfundibuliform, ribs of the achænia bispid.

(1) M. BELGAUMENSIS. *(B. W.)*

Ident. Dalz. Bomb. flor. p. 130.
Engrav. Wight's Icon. t. 1152.

SPEC. CHAR. Annual, erect, hirsute: leaves ovate, crenato-dentate, auricled at the base, pubescent above, tomentose beneath; ligules about 8-nerved: style and stigma wanting: achænia 10-nerved, nerves hispid: pappus none.

Belgaum. Neilgherries.

GENUS LXIII. SENECIO.

Sym. Pol. Superflua. *Ser: Syst:*

Deriv. From *Senex*, an old man; the receptacle resembles a bald head.

GEN. CHAR. Capitula homogamous, discoid or heterogamous, radiate: flowers of the ray ligulate, female: branches of the style of the hermaphrodite flowers truncated, the point only penicillate: achænia beakless, terete or angularly furrowed: pappus pilose, several series, caducous.—Herbaceous or shrubby, sometimes climbing plants, with solitary or corymbose inflorescence: involucrum 1-series, sometimes naked, sometimes calyculate, with accessory squamellæ, often with the points of the scales sphacellate; receptacle naked, alveolate.

(1) S. ROYLEANUS. *(Dec.)*

Ident. Dec. prod. VI. p. 366.

SPEC. CHAR. Glabrous or very minutely pubescent: stem herbaceous, erect; upper leaves subpetioled, pinnately lobed: lobes lanceolate, serrulate: terminal one elongated, more serrated, all acuminate: corymb composite, many-headed: involucres oblong, 6-leaved, glabrous: bracteoles linear-setaceous, calyculate: flowers in the capitulum 16–17, of which 4–5 are ligulate, oblong, and 3–4-nerved: achænia glabrous.

North-West Provinces.

(2) S. SISYMBRIIFORMIS. *(Dec.)*

Ident. Dec. prod. VI. p. 366.

SPEC. CHAR. Whole plant puberulously-hirsute: stem herbaceous, erect; leaves lyrate: all the lobes unequally and acutely toothed; middle ones oblong, lower ones dilated, auricle-shaped, cut, terminal one wide, ovate, lobed: corymb composite: peduncles sub-ebracteolate: scales of the campanulate involucre 10–12, lanceolate, hirsute on the back, dark at the margin: flowers about 60–70: ligulæ about 12, oblong: achænia smooth, somewhat contracted at the apex, glabrous.

North-West Provinces.

(3) S. ANALOGUS. *(Dec.)*

Ident. Dec. prod. VI. p. 366.

SPEC. CHAR. Stem herbaceous, erect, slightly glabrous: leaves lyrate, muriculate above, araneosely-tomentose below: lower lobes wide, auricle-shaped, middle ones few, oblong; terminal one wide, ovate, all coarsely and acutely calloso-dentate: corymb terminal, 14–17-headed: pedicels sparingly bracteolate: scales of the campanulate involucre subpuberulous at the back, scarcely subsphacellate at the apex: flowers about 30–40: ligulæ 10–12: achænia glabrous.

North-West Provinces.

(4) S. CORYMBOSUS. *(Wall.)*

Ident. Dec. prod. VI. p. 364.—Wall. Cat. No. 3121.

Engrav. Wight's Icon. t. 1130.—Spicil. II. t. 115.

SPEC. CHAR. Stem climbing, appearing as if covered with cobwebs: leaves petioled, without stipules, cordately suborbicular, shortly acuminated, subserrate, glabrous above, densely tomentose beneath, 5–7-nerved at the base: corymbs axillary and terminal, compactly many-headed: involucre 6-leaved, bracteolate at the base: ligulæ none: achænia glabrous.

Neilgherries, climbing over lofty trees.

(5) S. Walkeri. *(Arn.)*

Ident. Arn. pug. pl. Ind. Or. No. 103.—Dec. prod. VI. p. 364.

Engrav. Wight's Icon. t. 1131.

Spec. Char. Climbing, araneose towards the extremities: leaves exstipulate, petioled, heart-shaped, acute, calloso-dentate, glabrous, above flosculosely araneose: peduncles axillary, longer than the leaves, corymbosely many-headed: capitula discoid, 6–7-flowered; involucral scales 8, with a few smaller ones at the base.

Neilgherries.

(6) S. Neilgherrianus. *(Dec.)*

Ident. Dec. prod. VI. p. 368.—Wight's Contrib. p. 23.

Engrav. Wight's Icon. t. 1132.

Spec. Char. Stem erect, suffruticose, roughly striated at the base: leaves linear-lanceolate, acute, hirsutely tomentose beneath, rough above, the lower ones attenuated at the base and half pinnatifid, the middle ones sessile, dentate, the upper auriculately stem-clasping, nearly entire: corymbs few-headed: pedicels bracteolate at the apex: involucral scales linear, scarcely acute: ligulæ 12–14, flat; achænia glabrous.

Neilgherries, in moist pastures near springs and watercourses.

(7) S. lavandulæfolius. *(Wall.)*

Ident. Dec. prod. VI. p. 368.—Wall. Cat. No. 3130.—Wight's Contrib. p. 23.

Engrav. Wight's Icon. t. 1133.

Spec. Char. Stem erect, hirsutely striated: leaves crowded, oblong-linear, entire, revolute on the margin, hairy or hispid above, tomentose beneath, the upper ones linear, distant: racemes corymbose, simple: peduncles bracteoled; involucre 15-leaved, calyculate: flowers about 40: ligulæ 15, long, spreading, 4-nerved: achænia glabrous.

Common in pastures on the Neilgherries, flowering in the cold season.

(8) S. araneosus. *(Dec.)*

Ident. Dec. prod. VI. p. 364.

Syn. S. arachnoideus, Wall. Cat. No. 3136.

Spec. Char. Stem climbing, araneose: leaves petioled, exstipulate, heart-shaped, acuminated, distantly calloso-dentate, glabrous above, araneose below: panicles axillary, many-headed, and with the bracts araneose; involucre 8-leaved, almost ecalyculate: ligulæ none? achænia glabrous.

Silhet mountains.

(9) S. ramosus. *(Wall.)*

Idem. Dec. prod. VI. p. 365.—Wall. Cat. No. 3129.

Spec. Char. Glabrous: stem erect, much branched, striated: leaves pinnati-partite; lobes oblong-linear, slightly obtuse, subdentate; lower ones slightly attenuated at the base, cauline ones eared and stem-clasping; panicle lax; pedicels scarcely bracteolate: involucral scales linear-acuminate: ligules few, minute, scarcely distinguishable from the disk: achænia a little scabrous.

Silhet.

(10) S. spectabilis. *(Wall.)*

Idem. Dec. prod. VI. p. 366.—Wall. Cat. No. 3127.

Spec. Char. Slightly glabrous: stem erect, corymbose at the apex; cauline leaves eared and stem-clasping, sometimes shortly decurrent, lyrate, deeply and sharply cut at the base, ovate at the apex, coarsely toothed; corymb many-headed: pedicels bearing subulate bracteoles: involucral scales lanceolate-acuminate, white-membranaceous at the margin and with the ligulæ 10-12, narrow: achænia glabrous.

Silhet.

(11) S. obtusatus. *(Wall.)*

Idem. Dec. prod. VI. p. 367.—Wall. Cat. No. 3133.

Spec. Char. Glabrous: stem herbaceous, somewhat simple: radical leaves obovate, obtuse, subernate, cuneate at the base, attenuated into the petiole; cauline ones few, oblong, acute, cut and toothed, very upright; corymb terminal, 7-8-headed: bracteoles subulate: scales of the calyculate involucre elliptic, acute, 3-nerved: ligulæ 10-12, spreading, 4-nerved: achænia roughish.

Silhet.

(12) S. Arnottianus. *(Dec.)*

Idem. Dec. prod. VI. p. 367.

Spec. Char. Stem herbaceous, erect, striated, glabrous, branched: leaves linear-lanceolate, acute, glabrous above, araneose below, subrevolute at the margin and bearing remote callous teeth: panicle lax: pedicels elongated: involucre subcalyculate, glabrous, about 15-leaved; scales acuminated: ligulæ 9-12: achænia terete, glabrous.

Peninsula.

(13) S. intermedius. *(R. W.)*

Idem. Wight's Icon. vol. III.

Engrav. Wight's Icon. t. 1135.

SPEC. CHAR. Climbing: leaves petioled, glabrous, triangular, acuminated, unequally crenate or dentate: petioles auricled at the base, with a large kidney-shaped stipule: panicles corymbose: bracts linear-subulate: pedicels divaricate: capitula many-flowered: involucre calyculate: ligulæ 12–14, oblong-lanceolate, obtuse: achænia papillose.

Neilgherries, flowering in February and March.

(14) S. ANOCLOSUS. *(Wall.)*

Ident. Dec. prod. VI. p. 369.—Wall. Cat. No. 3117.
Syn. S. cappa, *Don.* prod. *flor. Nep.* p. 179.

SPEC. CHAR. Stem fruticose at the base ! branches angled, and with the peduncles involucres and leaves underneath hoary-tomentose: leaves very shortly petioled, oval-lanceolate, acuminated at both ends, serrated, slightly glabrous above: racemes axillary, branched, shorter than the leaf, ultimate ones collected into a panicle: involucre oval, torulose, accompanied at the base with subulate bracteoles: ligulæ 10, 4–5-nerved: achænia glabrous.

Pundua, Silhet.

(15) S. CANDICANS. *(Wall.)*

Ident. Dec. prod. VI. p. 309.—Wight's Contrib. p. 22.
Engrav. Wight's Icon. t. 1134.

SPEC. CHAR. Climbing; everywhere clothed with white tomentum: branches striated: leaves petioled, auricled, with kidney-shaped stipules, cordate, acute, serrated, araneose above, pure white below: panicle corymbose: bracts linear-subulate: pedicels diverging: involucre white, campanulate, sparingly bracteoled at the base: ligulæ 6, oblong, flat: achænia glabrous.

Neilgherries.

(16) S. STIPULATUS. *(Wall.)*

Ident. Dec. prod. VI. p. 370.—Wall. Cat. No. 3122.

SPEC. CHAR. Stem climbing, angular, glabrous: petioles araneose, appendiculate with a kidney-shaped stipule on both sides: leaves cordate-acuminate, sharply toothed, slightly glabrous above, araneosely whitish below: corymbs few-headed: pedicles bracteolate: involucre oblong, sparingly calyculate at the base: ligulæ 8–9, oblong-linear: achænia glabrous.

Coromandel Coast.

(17) S. WIGHTIANUS. *(Dec.)*

Ident. Dec. prod. VI. p. 370.—Wight's Contrib. p. 23.
Syn. Cacalia Wightiana, *Wall. Cat.*
Engrav. Wight's Icon. t. 1135.

SPEC. CHAR. Glabrous: branches scandent, angularly striated: leaves petioled, ovate or elliptic-lanceolate, acuminate, serrated: limb obtuse at the base or shortly cuneate: petioles with a small auricle at the base: panicle divaricating: pedicels bracteolate at the apex.: capitula small, 8-10-flowered: ligulæ 3-4, small: achænia puberulous.

Neilgherries. Forests of Malabar.

GENUS. LXIV. NOTONIA.

Sym. Pol. Æqualis. *Sex: Syst:*

Deriv. Named after Mr. Benjamin Noton of Bombay, an indefatigable collector of Indian plants.

GEN. CHAR. Capitula discoid, homogamous: florets 5-toothed: branches of the style ending in a short hispid cone: achænia terete, many, striated, glabrous: pappus many series, bristles filiform, barbellate.—Shrubby succulent plants with corymbose few-headed inflorescence: capitula large: involucre cylindrical, 1-series, ecalyculate, the peduncle furnished with a few bracteaceous scales: receptacle alveolate, naked or slightly fimbrilliferous: flowers longer than the involucre, pale-yellow or cream-coloured.

(1) N. GRANDIFLORA. (*Dec.*)

Ident. Dec. prod. VI. p. 442.—Wight's Contrib. p. 24.

Syn. Cacalia grandiflora, *Wall. Cat. No.* 3147.—N. corymbosa, *Dec. l. c.*

Engrav. Wight's Icon. t. 484.—Deless. Ic. Sel. IV. t. 61.

SPEC. CHAR. Shrubby, stem thick, round, marked with scars of fallen leaves: leaves oblong or obovate, quite entire: corymb few-headed: pedicels much, longer than the capitulum: flowers terminal, pale-yellow.

South Travancore. Neilgherries. High rocky places in the Deccan. Flowering in the cold season. I am inclined to agree with Dr. Wight that there is no real specific difference between this, and N. corymbosa.

(2) N. BALSAMICA.* (*Dalz.*)

Ident. Dalz. Bomb. Flor. p. 133.

SPEC. CHAR. Suffruticose, glaucous and perfectly smooth: leaves somewhat fleshy, petioled, lanceolate, long-attenuated at the base: flowers terminal, corymbose at the apex of the tall nearly naked stem: branches of the corymb few, simple, 2-3-flowered: pedicels short, club-shaped, angular or slightly winged, with two lanceolate bracts at the apex: involucre of 5–7 unequal linear-acute leaflets, tubular: florets about 15, a half longer than the involucre; achænia cylindric, mutic, striated, smooth.

Inland Ghauts of the Deccan. The achænium has a strong balsamic odour.

GENUS LXV. ECHINOPS.

Syn. Pol. Segregata. *Ses: Syst:*

Deriv. From *Echinops*, a hedgehog; alluding to the heads of flowers.

Gen. Char. Capitula numerous, aggregated on a naked globose receptacle, the centre one opening first: corolla tubular, 5-cleft: anthers ecaudate: achænia cylindrical, silky-villous: pappus short, crown-like, the hairs somewhat fimbriated.—Erect prickly plants: glomerules spherical: capitula inserted on a circular horny areola: partial involucre 3-series, the interior scales the longest, linear, acuminated, carinate: flowers blue or white.

(1) E. ECHINATUS. (*Roxb.*)

Ident. Roxb. fl. Ind. III. p. 447.—Dec. prod. VI. p. 526.— Wight's Contrib. p. 24.

Spec. Char. Erect, much branched: leaves pinnatifid, pubescent and viscid above, hoary-tomentose beneath: divisions ovate-lanceolate, waved, smooth: heads of flowers terminal, solitary, globose, spinous; florets pale lilac; tube slender; divisions linear-acute, revolute: pappus short, brush-like; ovary very hairy.

Guzerat. Deccan. Mysore. Hyderabad. Mahableshwur. Flowering in February and March.

(2) E. CORNIGERUS. (*Dec.*)

Ident. Dec. prod. VI. p. 525.

Spec. Char. Leaves araneose above, below and with the stem hoary-tomentose, pinnati-partite: lobes broad-lanceolate, cut and furnished with strong spinous teeth: bristles of the partial involucre almost exceeding the middle scales, inner scales two-shaped, some short and shortly ciliated, others long, horn-shaped, glabrous, subconvolute, lowest ones concrete into a tube.

North-West Provinces.

GENUS LXVI. APLOTAXIS.

Syn. Pol. Æqualis. *Ses: Syst:*

Deriv. From *Aploos*, simple, and *Taxis*, row or series; alluding to the pappus.

GEN. CHAR. Capitula homogamous: corolla slender; throat ventricose: anthers ending in long appendices, with two ciliate. bristles at the base: stigmas long, diverging, continuous with the style: achænia glabrous: pappus 1-series, plumose.—Herbs with entire leaves: capitula usually corymbose: involucre many-series: receptacle fimbrillate or paleaceous: corolla purple or white, never yellow.

This genus ought not to have been separated from Saussurea, the only difference being that the pappus is 1-series.

(1) A. CANDICANS. *(Dec.)*

Ident. Dec. prod. VI. p. 540.

Syn. Cnicus candicans, *Wall.*—Carduus heteromallus, *Don. prod. fl. Nep.* p. 166.—Cirsium heteromallum, *Spreng. Syst.* III. p. 372.

SPEC. CHAR. Stem erect, branched, somewhat downy; leaves covered below with pure white down, slightly glabrous above, lowest ones petioled, lyrate; terminal lobe the largest, middle ones attenuated into the petiole, sinuately pinnatifid, uppermost ones oblong, undivided, all more or less denticulate, unarmed: capitula long-peduncled, terminal: scales of the downy subglobose involucre linear-acuminate, all nearly equal.

Hurdwar.

(2) A. CARTHAMOIDES. *(Ham.)*

Ident. Dec. prod. VI. p. 540.

Syn. Cnicus carthamoides, *Wall. Cat. No.* 2890.—Serratula carthamoides, *Roxb. H. B. flor. Ind.* III. p. 407.

SPEC. CHAR. Stem erect, branched, somewhat downy: leaves a little scabrous above, white-woolly below; lowest ones petioled, pinnati-partite; lobes sinuately toothed, middle ones sessile, pinnatifid, uppermost ones few, narrow, sinuate or quite entire: capitula subcorymbose: scales of the almost glabrous involucre linear, imbricated, acuminated, somewhat pointed; pappus very tender.

Silhet.

(3) A. ROYLEI. *(Dec.)*

Ident. Dec. prod. VI. p. 538.

SPEC. CHAR. Stem erect, simple, 1-headed, leafy, hoary-villous: leaves araneose, hoary-villous below, pinnatifid: lobes somewhat triangular, mucronate; lower ones directed downwards and deeper, radical ones petioled: cauline ones sessile, uppermost ones long-acuminate, entire, those next to the capitulum bract-shaped: capitula erect: involucral scales lanceolate-linear, acuminated, somewhat villous on the back: florets dark-purple.

North-West Provinces.

(4) A. DISCOLOR. (*Dec.*)

Ident. Dec. prod. VI. p. 511.

SPEC. CHAR. Stem erect, slightly glabrous, simple; leaves glabrous above, white tomentose beneath, elliptic, acuminate, attenuated at the base, half-stem-clasping, coarsely toothed: teeth apiculated: capitula many, corymbose: involucral scales villous at the back, acuminated: tails of the anthers short, villous, lacerated.

North-West Provinces.

(5) A. CLASSIOIDES. (*Dec.*)

Ident. Dec. prod. VI. p. 540.

Syn. Carduus lanatus, *Roxb. fl. Ind.* III. p. 408.

SPEC. CHAR. Stem erect, branched at the apex, slightly downy; leaves covered with white wool below, araneosely woolly above, everywhere armed with strong spinous teeth; lower ones subpetioled, sinuately pinnatifid, upper ones sessile, somewhat sinuately toothed: capitula globose, terminating the branches: involucral scales mucronately pointed: outer ones shorter, adpressed, inner ones at length slightly recurved at the apex: flowers large, pale purple.

Bengal, flowering in February and March.

GENUS LXVII. AMBERBOA.

Syn. Pol. Necessaria. *Bess. Syst.*

GEN. CHAR. Capitula several-flowered, heterogamous; marginal florets larger, sterile: achænia compressed or turbinate, tetragonal, with a lateral or basilar areola; pappus paleaceous: paleæ obovate, spathulate.—Herbaceous plants: involucre many-series, scales various, rarely spinescent: flowers blue or reddish-purple.

(1) A. GONIOCAULON. (*Dec.*)

Ident. Dec. prod. VI. p. 558.

Syn. Goniocaulon glabrum, *Cass.*

SPEC. CHAR. Stem erect, branched, quite glabrous, angled with ribs; stem-leaves half-stem-clasping, linear, acute, subdenticulate; capitula fascicled.

Tranquebar.

(2) A. INDICA. (*Dec.*)

Ident. Dec. prod. VI. p. 558.

Syn. Serratula Indica, *Klein. Willd.*—Athanasia Indica, *Roxb. fl. Ind.* III. p. 417.—Centaurea Indica, *Less.*

Engrav. Wight's Icon. t. 479.

SPEC. CHAR. Stem erect, branched, furrowed and angled, naked at the apex, and with the leaves glabrous or slightly scabrous: leaves lanceolate, coarsely toothed, upper ones few, distant, linear, entire: flowers largish, purplish rose-coloured.

Peninsula. Guzerat. Flowering in February and March.

GENUS LXVIII. TRICHOLEPIS.

Syn. Pol. Æqualis. *Ses : Syst :*

Deriv. From *Thrix*, hair, and *Lepis*, a scale.

GEN. CHAR. Capitula homogamous: corolla sub-regular, 5-fid: achænia with a lateral areola crowned at the apex with a circular margin: pappus setaceous, concrete at the base, or palaeaceous, not concrete, or wanting.—Herbaceous plants: capitula ebracteolate: involucre many-series, scales linear, setaceous, recurved at the points: receptacle fimbrilliferous.

(1) T. ELONGATA. (*Dec.*)

Ident. Dec. prod. VI. p. 563.

Syn. Carduus elongatus, *Wall.*

SPEC. CHAR. Stem somewhat simple, leafless at the apex, 1-headed: leaves scabrous on both sides, serrated, cut, lanceolate, attenuated at the base: involucral scales all slightly glabrous.

Bengal.

(2) T. RADICANS. (*Dec.*)

Ident. Dec. prod. VI. p. 564.—Wight's Contrib. p. 25.

Syn. Carduus radicans, *Roxb. fl. Ind.* III. p. 408.

SPEC. CHAR. Stem branched, rooting at the base, angular, leaves oblong-linear, sharply and coarsely awned and serrated, glabrous, pointed: scales of the ovate involucre araneose: flowers middle-size, lilac.

Mysore. Malabar. Deccan. Flowering in the rainy season.

(3) T. GLABERRIMA. (*Dec.*)

Ident. Dec. prod. VI. p. 564.

SPEC. CHAR. Whole plant quite glabrous: stem erect, branched, angular: leaves linear-lanceolate, acuminate, stem-clasping, distantly spotted with black specks: involucral scales produced into a subulate spinescent appendage: flowers terminal, purple.

Deccan. Concan.

(4) T. ANGUSTIFOLIA. (Dec.)

Ideal. Dec. prod. VI. p. 564.—Wight's Contrib. p. 25.

Syn. Serratula Indica, *Willd.*—Carduus nitidus, *Wall. Cat. No.* 2908.

SPEC. CHAR. Glabrous: stem branched, angled: leaves long-linear, acuminated, rarely awnedly serrated: capitula ovate, glabrous: pappus none.

Travancore mountains.

(5) T. PROCUMBENS. (*R. W.*)

Ideal. Dalz. Bomb. flor. p. 151.

Syn. T. Candolleana, *Wight.*—Carduus ramosus, *Roxb. fl. Ind.* III. p. 407.

Engrav. Comp. Bot. Mag. I. t. 4.—Wight's Icon. t. 1139.

SPEC. CHAR. Stem flexuose, short, branched: branches diffuse, procumbent, angularly striated, somewhat glabrous: leaves shortly pubescent; cauline ones lyrate, of the branches sinuately pinnatifid: lobes spinosely mucronate: involucre ovate: scales araneous, terminating in a prickly appendage: flowers purple.

Guzerat, Mysore, Coimbatore. Flowering in October and November.

(6) T. MONTANA. (*Dalz.*)

Ideal. Dalz. Bomb. flor. p. 131.

SPEC. CHAR. Leaves obovate-oblong, very coarsely toothed, sometimes pinnatifid: stigmas long, slender.

Western Ghauts.

GENUS LXIX. CIRSIUM.

Sym. Pol. Æqualis. *Sex Syst:*

Deriv. From *Kirsos*, a swelled vein, from its supposed healing properties.

GEN. CHAR. Capitula homogamous, hermaphrodite or dioicous: tube of the corolla short, throat oblong, 5-cleft: anthers ecaudate: stigmas connected: achænia oblong, compressed, glabrous, membranaceous, ecostate, with a fleshy terminal areola.—Herbaceous thistle-like plants: involucre imbricated, scales more or less prickly pointed: receptacle fimbrilliferous: flowers purple or yellow.

(1) C. ARGYRACANTHUM. *(Dec.)*

Ident. Dec. prod. VI. p. 640.—Wight's Contrib. p. 23.
Syn. Carduus argyracanthus, *Wall. Cat. No.* 2903.
Engrav. Wight's Icon. t. 1137-8.—Spicil. II. t. 116.

SPEC. CHAR. Leaves half-stem-clasping, sinuately pinnatifid, ciliated with spines, lobes produced into strong spines, below and with the stem araneosely villous: capitula paniculately heaped together: bracts many-cleft, very spiny: involucral scales produced into long spines: flowers purplish.

Neilgherries. Pulney hills. Flowering in August and September.

GENUS LXX. SERRATULA.

Sym. Pol. Æqualis. *Sex: Syst:*

Deriv. From *Serra*, a saw, the leaves being edged with cutting teeth.

GEN. CHAR. Capitula usually homogamous, sometimes by abortion 1-sexual, or ray female: corolla 5-cleft, sub-irregular: filaments pilose: stigmas diverging: achænia oblong, compressed: pappus unequal, hairs many-series, rough not annulate.—Unarmed or prickly herbs: involucre ovate, scales imbricated, the exterior ones shorter, spinulose, interior long, scariose at the apex: receptacle fimbrillate: corolla purple or white, never yellow.

(1) S. PALLIDA. *(Dec.)*

Ident. Dec. prod. VI. p. 670.
Syn. Centaurea pallida, *Wall. Cat. No.* 2983.

SPEC. CHAR. Stem terete, striated, puberulous, 1-headed, simple: leaves a little glabrous on both sides or minutely puberulous, lyrate or pinnatifid, attenuated into the petiole: lobes acuminated, mucronate, somewhat entire: involucral scales coriaceous, glabrous, outer ones ovate, mucronate; innermost ones long, lanceolate-linear: lobes of the corolla linear, sub-callous at the apex: florets purplish.

North-West Provinces.

GENUS LXXI. AINSLIÆA.

Sym. Pol. Æqualis. *Sex: Syst:*

Deriv. Named after Dr. Whitelaw Ainslie of Madras, Author of the Materia Medica of Hindostan.

GEN. CHAR. Capitula 3-flowered, homogamous: corolla tubular, bilabiate, exterior lip 3, interior 2-cleft: anthers appendiculate, with long spurs at the base: stigmas exserted, obtuse, glabrous, often by abortion unequal, acute or one altogether abortive: achænia terete, villous, not beaked: pappus 1-series: bristles elegantly plumose.—Herbaceous perennials with simple and erect stems: involucre cylindrical: scales lanceolate, acuminated, imbricate; receptacle naked: flowers purple.

(1) A. PTEROPODA. (*Dec.*)
Var. Silhetensis.

Ident. Dec. prod. VII. p. 14.

Syn. Liatris latifolia, *Don. prod. p.* 169.—Vernonia lobelioides, *Wall. Cat. No.* 2927.

SPEC. CHAR. Leaves ovate, subcordate, acuminate, scarcely denticulate below, and with the stem covered with villous down: capitula spreading: involucres sparingly downy.

Silhet.

GENUS LXXII. BERNIERA.

Syn. Pol. Æqualis. *Sex: Syst:*

Deriv. Named after Francis Bernier, a French traveller in India.

GEN. CHAR. Capitula many-flowered, homogamous: receptacle naked, alveolate: involucre subcampanulate, 2–3-series, scales lanceolate-linear, acuminate: florets tubulose at the base, bilabiate: outer lip 3-toothed, erect, inner one bipartite, revolute: anthers with lanceolate appendages, bearded at the apex: style somewhat included: lobes obovate, slightly puberulous at the apex behind: achænium scabrous, angled, shortly beaked: pappus many-series: bristles stiffish, subscabrous.

(1) B. NEPAULENSIS. (*Dec.*)

Ident. Dec. prod. VII. p. 18.

Syn. Chaptalia maxima, *Don. prod. flor. Nep. p.* 166.—Tussilago macrophylla, *Wall. Cat. No.* 2989.

Engrav. Deless. Icon. IV. t. 77.

SPEC. CHAR. Perennial, stemless: radical leaves long-petioled, cordate-sagittate, membranaceous, quite glabrous above, araneose below: peduncles or scapes longer than the leaves, naked, 1-headed, and with the base of the petioles covered with cobweb down: flowers purplish, very fragrant.

Silhet.

GENUS LXXIII. DICOMA.

Syn. Pol. Æqualis. Ser. Syn.

Deriv. From *Dis*, double, and *Kome*, a tuft of hair, probably in allusion to the biserial pappus.

GEN. CHAR. Capitula homogamous or heterogamous: corolla of the disk regular, 5-parted, of the ray, in heterogamous capitula, neuter, ligulate, bilabiate or roundish tubulate: anthers long, caudate, caudæ bearded: branches of the style short, erect, obtuse, hispidulous at the apex: achænia turbinate, often 10-ribbed: pappus two or more series.—Suffruticose or herbaceous: leaves alternate: capitula solitary: involucre campanulate: scales absolutely many-nerved, sometimes pungent: receptacle alveolate: flowers white or purple.

(1) D. LANUGINOSA. *(Dec.)*

Ident. Dec. prod. VII. p. 36.—Wight's Contrib. p. 26.

Syn. Xeropappus lanuginosus, *Wall. Cat. No.* 2980.—Onopordon lanatum, *Herb. Madr.*

Engrav. Wight's Icon. t. 1146.

SPEC. CHAR. Erect, much branched, downy: scales of the ovate involucre a little glabrous outside: palea of the pappus serrated,: fruit very villous: flowers whitish.

Madras. Trichinopoly.

GENUS LXXIV. PICRIS.

Syn. Pol. Æqualis. Ser. Syn.

Deriv. From *Picros*, bitter.

GEN. CHAR. Capitula many-flowered, achænia terete, attenuated at both ends, rugulose with a terminal areola, beak none or short: pappus of the disk plumose, 3-series, exterior shortest. Herbaceous plants: capitula pedunculed: involucre 2-series: receptacle naked.

(1) P. HIERACIOIDES. *(Linn.)*
Var. Indica.

Ident. Wight's Contrib. p. 26.—Dec. prod. VII. p. 129.

Engrav. Wight's Icon. t. 1147.

SPEC. CHAR. Stem erect, scabrous with glochidiate hairs, corymbosely branched at the apex: leaves half-stem-clasping, lanceolate, coarsely toothed, scabrous: corymb divaricate: outer involucral scales lax, oblong.

Courtallum.

GENUS LXXV. LACTUCA.

Sys. Pol. Æqualis. *Sex: Syst:*

Deriv. From *Lac,* milk, on account of the milky juice which exudes from the plants when broken.

GEN. CHAR. Capitula few or many-flowered: achænia compressed, wingless, abruptly terminating in a filiform beak.—Herbaceous; heads of flowers panicled: involucre cylindrical, calyculately imbricated, 2–4-series: receptacle naked.

(1) L. GRACILIS. *(Dec.)*

Ident. Dec. prod. VII. p. 140.

Syn. Chondrilla gracilis, *Wall. Cat. No.* 377.

SPEC. CHAR. Glabrous: stem erect, furrowed, leafy in the lower part, naked at the apex, dichotomously panicled: leaves linear-lanceolate, acuminated at both ends, quite entire, 1-nerved: panicle lax, subcorymbose at the apex: capitula 8–10-flowered: involucre cylindric, minutely calyculate.

Pundua, Silhet.

GENUS LXXVI. TARAXACUM.

Sys. Pol. Æqualis. *Sex: Syst:*

GEN. CHAR. Capitula many-flowered: involucre double; outer scales small, adpressed, spreading or reflexed, inner ones 1-series, erect, all often callosely horned at the apex: receptacle naked: achænia oblong, striated, muricated with the ribs or spinulose at the apex, produced into a long beak: pappus pilose, many-series, quite white.—Stemless perennials: leaves all radical, oblong, entire, sinuate or runcinately pinnatifid, usually glabrous: scapes very often 1-headed, piped: capitula yellow; outer ligulæ often reddish.

(1) T. WALLICHII. *(Dec.)*

Ident. Dec. prod. VII. p. 147.

Syn. Leontodon glaucescens, *Wall. Cat. No.* 356.

SPEC. CHAR. Glabrous, glaucescent: leaves radical, obtusely pinnati-partite: lobes subdentate: scapes ascending about the length of the leaf: involucral scales callosely horned at the apex, outer ones small, squarrosely reflexed: achænia pale, spinulosely muricated at the apex, three times shorter than the leaf.

Oude.

GENUS LXXVII. IXERIS.

Syn. Pol. Æqualis. *Ser. Syst.*

GEN. CHAR. Capitula many-flowered; achænia oblong, acutely 10-ribbed, beaked; pappus pilose, 1-series.—Herbaceous; stems naked at the apex, corymbose; involucre ovate, 1-series, with 3–5 calyculate scales; receptacle naked.

(1) I. POLYCEPHALA. (*Cass.*)

Ident. Cass. Dict. XXIV. p. 50.—Dec. prod. VII. p. 151.
Syn. Chondrilla tenuis, *Ham. in Wall. Cat. No. 3274.*
SPEC. CHAR. Stem simple or branched at the apex; leaves linear-subulate, lower ones close together, half-stem-clasping, sagittate; acutely and somewhat retrorsely toothed, a little hairy beneath, upper ones very acutely sagittate at the base, remote.

Goruckpore.

GENUS LXXVIII. BRACHYRAMPUS.

Syn. Pol. Æqualis. *Ser. Syst.*

Deriv. From *Brachys*, short, and *Ramphos*, a beak; alluding to the achænia.

GEN. CHAR. Capitula 10–15-flowered; achænia oblong, muricate, suddenly attenuated into a short beak, neither angled nor costate; pappus many-series.—Herbaceous; capitula racemosely spicate; involucre oblong, imbricated, the scales scariose on the margin; receptacle naked.

(1) B. SONCHIFOLIUS. (*Dec.*)

Ident. Dec. prod. VII. p. 177.
Syn. Cacalia sonchifolia, *Wall. Cat. No. 3144.*—Lactuca remotiflora, *Wight's Contrib. p. 26.*
SPEC. CHAR. Glabrous; stem erect, leafy at the base, naked at the apex, sparingly branched; leaves membranaceous, stem-clasping, obovate, sinuately subruncinate, bristle-ciliated at the margin; capitula, together with the branches, remotely spicato-racemose; pedicels very short, somewhat scaly.

Common in the Peninsula.

(2) B. HEYNEANUS. (*R. W.*)

Ident. Wight's Icon. vol. III.
Syn. Lactuca Heyneana, *Dec. prod. VII. p. 140.*
Engrav. Wight's Icon. t. 1146.

SPEC. CHAR. Stem erect, glabrous, naked below: leaves rigid, subradical, runcinate, coarsely bristle-ciliate, stem-clasping: capitula short-pedicelled, remotely fascicled along the branches: achænia compressed, striated, slightly muricated, shortly beaked.

Coimbatore and elsewhere, by wall sides and hedges, flowering in the rainy season.

GENUS LXXIX. MICRORHYNCHUS.

Syn. Pol. Æqualis. *Ser: Spri*

Deriv. From *Micros*, small, and *Rhynchos*, a beak; alluding to the achænia.

GEN. CHAR. Capitula several-flowered: achænia 4, rarely 5-angled, subrostrate at maturity; beak wanting in the ovary: culm thick, sub-rugose: pappus pilose.—Herbaceous perennials: involucre cylindrical, calyculate, imbricate at the base: receptacle naked; flowers yellow.

(1) M. PATENS. *(Dec.)*

Ident. Dec. prod. VII. p. 181.
Syn. Prenanthes patens, *Wall. Cat. No.* 308.
SPEC. CHAR. Subglaucescent: stems somewhat erect, patently ramoso-paniculate, sparingly leafy at the base: leaves runcinate, furnished with white scariose teeth, radical ones narrowed into the petiole, stem ones auritely half-stem-clasping: capitula together with the branches shortly pedicelled: pedicels bracteate.

Patna. Onde.

(2) M. ASPLENIFOLIUS. *(Dec.)*

Ident. Dec. prod. VII. p. 181.
Syn. Prenanthes asplenifolia, *Willd.*—Hieracium dichotomum, *Roxb. Fl. Ind.* III. p. 401.—P. dichotoma, *Wall.*
SPEC. CHAR. Glabrous: stems many, somewhat naked, ascending, dichotomous: leaves radical, subsessile, pinnatifid or runcinate; lobes obtuse, denticulate: panicle lax, somewhat naked.

Bengal. Coromandel Coast in sandy places.

(3) M. SARMENTOSUS. *(Dec.)*

Ident. Dec. prod. VII. p. 181.
Syn. Prenanthes sarmentosa, *Willd.* — Lactuca sarmentosa, *Wight's Contrib.* p. 27.
Engrav. Wight's Ill. II. t. 133.

Spec. Char. Stems twiggy, filiform, procumbent, here and there bearing roots and leaves: leaves close together, sinuately pinnatifid: lobes obtuse or subacute: peduncles 1-headed, somewhat shorter than the leaf, bearing at the apex subimbricated scaly bracts, scarious at the margin: flowers yellow.

In sandy places near Negapatam.

(4) M. glabra. (R. W.)

Ident. Wight's Icon. vol. III.
Syn. Lactuca glabra, *Dec. Wight's Contrib.* p. 26.
Engrav. Wight's Icon. t. 1145.—Splcil. II. t. 116.

Spec. Char. Glabrous: stem naked, dichotomously branched: leaves long-linear, acute, entire or toothed: capitula corymbose, long-pedicelled, cylindrical, 7–8-flowered: scales of the involucre 5–6, linear-lanceolate, somewhat scariose on the margin: achænia 5-angled, obscurely beaked: flowers yellow.

Neilgherries, flowering all the year.

GENUS LXXX. SONCHUS.

Syn. Pol. Æqualis. *Ser: Syd:*

Deriv. From the Greek *Sonchos*, the English sowthistle.

Gen. Char. Capitula many-flowered: achænia wingless, compressed, erostrate, longitudinally costulate: costulæ often transversely tuberculato-muricate: pappus soft, most slenderly filiform.—Herbaceous polymorphous plants: involucre imbricated: receptacle naked: flowers yellow.

(1) S. ciliatus. (*Lam.*)

Ident. Dec. prod. VII. p. 185.
Syn. S. oleraceus, *Linn. Wight's Contrib.* p. 27.—*Roxb. flor. Ind.* III. p. 402.
Engrav. Wight's Icon. t. 1141.—Gaetrn. fr. II. t. 158.

Spec. Char. Stem erect, glabrous or rarely hairy glandulose at the apex: cauline leaves stem-clasping, sharply dentato-ciliate, runcinate or undivided: auricles acuminate: involucres and pedicels slightly glabrous: achænia transversely muricately wrinkled: flowers large, yellow.

Negapatam and other places in the Peninsula, flowering in February and March.

(2) S. orixensis. (*Roxb.*)

Ident. Roxb. fl. Ind. III. p. 402.—Dec. prod. VII. p. 190.

SPEC. CHAR. Stem erect, glabrous, glandulosely hairy at the apex: cauline leaves stem-clasping, sagittate, lanceolate, smooth, slightly sublobed, toothed: peduncles subumbellate, hairy: involucres tomentose: flowers large, yellow.

Samulcottah, flowering in February and March.

(3) S. WALLICHIANUS. *(Dec.)*

Ident. Dec. prod. VII. p. 185.
Syn. S. longifolius, *Wall. Cat. No.* 3251.
Engrav. Delesa. Ic. IV. 1. 97.

SPEC. CHAR. Glabrous: root perpendicular: stem erect: leaves elongated, obtusely auricled, half-stem-clasping, pinnatifid, spinosely toothed: lobes ovate, terminal one elongated: pedicels subumbellate: achænia compressed, subcylindric, not attenuated at the base, muriculate.

Hurdwar.

(4) S. WIGHTIANUS. *(Dec.)*

Ident. Dec. prod. VII. p. 187.
Engrav. Wight's Icon. L. 1142.

SPEC. CHAR. Root somewhat woody: stems ascending, erect, somewhat angular, glabrous: leaves stem-clasping, with roundish auricles, oblong-lanceolate, unequally and acutely toothed, glabrous, glaucescent, upper ones nearly linear: corymbs lax and with the pedicels and involucres glandulosely hairy: achænia oblong, striated, transversely wrinkled.

In shady places near Coimbatore, flowering in the cold season.

GENUS LXXXII. YOUNGIA.

Syn. Pol. Æqualis. *Ber; Syst;*

GEN. CHAR. Capitula about 12-flowered: corolla pilose at the apex of the tube: achænia oblong, compressed, subtrigonous, striated, beakless, attenuated at both ends: pappus pilose, scarcely denticulate.—Herbaceous plants; inferior leaves lyrate or pinnatifid: capitula paniculate: involucre cylindrical, 8-leaved, with about 5 calyculate accessory scales: receptacle naked.

(1) Y. RUNCINATA. *(Dec.)*

Ident. Dec. prod. VII. p. 192.
Syn. Chondrilla runcinata, *Wall. Cat. No.* 382.

SPEC. CHAR. Slightly glabrous: stems somewhat leafless at the apex, puberulous below together with the petioles and nerves of the leaves: leaves runcinately-pinnati-partite: lobes subtriangular, mid cronately toothed, acute, stem ones also petioled: panicle elongated, lax: pedicels naked filiform: capitula 15-16-flowered: involucral scales 7-8, submembranaceous at the margin, very minutely calyculate at the base.

Silhet.

(2) Y. NAPIFOLIA. *(Dec.)*

Ident. Dec. prod. VII. p. 193.

Syn. Prenanthes napifolia, *Wall. Cat. No.* 387.—Lactuca napifolia, *Dec. in Wight's Contrib.* p. 27.

Engrav. Wight's Icon. t. 1147.

SPEC. CHAR. Glabrous, subhirsute at the base: stem erect, loosely panicled at the apex, somewhat leafless; radical and lower stem leaves petioled, runcinately lyrate: lobes oval-oblong, obtusely sinuate, mucronately denticulate, the extreme ones confluent: involucre 8-leaved, very minutely calyculate: achænia attenuated at the apex.

Silhet. Coimbatore. Flowering in the rainy season.

(3) Y. AMBIGUA. *(Dec.)*

Ident. Dec. prod. VII. p. 193.

SPEC. CHAR. Stem angular, striated, somewhat naked, panicled at the apex, pubescent at the apex: radical leaves slightly hairy, petioled, runcinately pinnati-partite, lobes triangular, mucronately toothed, terminal one scarcely larger, cauline ones few, sessile, glabrous: involucre 8-leaved, very minutely calyculate; achænia subtrigonous, striated, attenuated at the apex.

North-West Provinces.

(4) Y. PROCUMBENS. *(Dec.)*

Ident. Dec. prod. VII. p. 193.

Syn. Prenanthes procumbens, *Roxb. fl. Ind.* III. p. 404.

SPEC. CHAR. Glabrous; stems procumbent, dichotomous: leaves mostly radical, linear, runcinate, lobes short, obtuse: racemes terminal: capitula remotely fascicled: flowers yellow.

Bengal, flowering in the cold season.

(5) Y. ACAULIS. *(Dec.)*

Ident. Dec. prod. VII. p. 193.

Syn. Prenanthes acaulis, *Roxb. fl. Ind.* III. p. 403.

SPEC. CHAR. Glabrous: stems many, scapose, 4-5-headed, shorter than the leaves; leaves radical, sessile, linear-lanceolate, entire, smooth: involucre 8-flowered: flowers yellow.

Dinajepore, flowering in March and April.

GENUS LXXXIII. PRENANTHES.

Syn. Pol. Æqualis. *Sex: Syst.*

Deriv. From *Prenes*, prostrate, and *Anthos*, a flower; alluding to the habit of the plant.

GEN. CHAR. Capitula 3–5-flowered: style exserted: achænia attenuated at the base, subcylindrical or subpentagonal, truncated: pappus many-series, pilose, rigid.—Herbs, with entire or dentate leaves: capitula drooping, racemose or paniculate; involucres cylindrical, 4–6-leaved, calyculate: flowers purple.

(1) P. ALLIARIÆFOLIA. (*Dec.*)

Ident. Dec. prod. VII. p. 195.

SPEC. CHAR. Stem simple, panicled at the apex, here and there together with the petioles particularly bristly: cauline leaves longpetioled, cordate, acuminate, coarsely toothed: teeth mucronate: leaves of the panicle subsessile, linear-lanceolate: capitula elongated, 3-flowered.

North-West Provinces.

(2) P. RAPHANIFOLIA. (*Dec.*)

Ident. Dec. prod. VII. p. 195.

SPEC. CHAR. Glabrous; stem erect, branched, panicled: cauline leaves sessile, attenuated at the base, pinnatifid, acuminate: lobes acute, toothed: teeth mucronate: capitula 3-flowered.

North-West Provinces.

(3) P. DISPIDELA. (*Dec.*)

Ident. Dec. prod. VII. p. 195.

SPEC. CHAR. Stem branched at the apex: petioles and leaves sparingly bristly; stem leaves some ovate, others 3-cleft, all acuminate, toothed; teeth long and callously mucronate: capitula glabrous, 4-flowered.

North-West Provinces.

GENUS LXXXIV. HIERACIUM.

Syn. Pol. Æqualis. *Sex: Syst.*

Deriv. From *Hierax*, a hawk; from a belief that birds of prey made use of the juice of these plants to strengthen their sight.

GEN. CHAR. Capitula many-flowered, homoocarpus: achænia beakless, striated or subprismatic; pappus bristly, 1-series, bound by the short annular margin of the achænium.—Perennial herbs, sprinkled all over with dentate glandulose or stellate hairs: involucre many-leaved; scales unequal, imbricated: receptacle naked, scrobiculate: flowers yellow.

(1) H. SILHETENSE. *(Dec.)*

Ident. Dec. prod. VII. p. 218.
Syn. Prenanthes Candolleana, *Wall. Cat. No.* 32.

SPEC. CHAR. Glabrous; root thick, cylindric: stems many, furrowed, angular, somewhat naked, paniculately corymbose at the apex: radical leaves acuminated at both ends, quite entire, stem leaves few, linear-subulate, elongated; involucre cylindric, subcalyculate, 10–12-flowered: achænia long striated.

Silhet mountains.

GENUS LXXXV. MULGEDIUM.

Syn. Pal. Æqualis. *Ser: Syst.*

GEN. CHAR. Capitula many-flowered: achænia glabrous, compressed, often nerved on both sides, attenuated upwards into a short thick beak, expanding at the apex into a cup-shaped disk; pappus one or several-series, setæ rigid, rough, greyish or white.—Erect ramous herbs, with pinnatifid leaves and racemose or panicled capitula: involucre calyculately imbricate, that is, the exterior scales are much shorter and sub-imbricate: receptacle naked, foviolate; flowers blue or purple.

(1) M. SAGITTATUM. *(Royle.)*

Ident. Dec. prod. VII. p. 250.
Engrav. Royle Ill. t. 61. f. 2.

SPEC. CHAR. Glabrous: stem erect: cauline leaves acutely sagittate, lanceolate-linear, acuminate, quite entire; panicle branched, many-headed; involucre subimbricated: flowers blue.

North-West Provinces.

(2) M. MACRORHIZUM. *(Royle.)*

Ident. Dec. prod. VII. p. 251.
Engrav. Royle Ill. t. 61. f. 1.

SPEC. CHAR. Glabrous: root thick: stems many, ascending: stem leaves amplexicaul, oblong, obtuse, sinuately toothed: capitula pedicellate, subcorymbose; involucre shortly calyculate: flowers blue.

North-West Provinces.

(3) M. NEILGHERRENSE. (*R. W.*)

Ideat. Wight's Icon. vol. III.—Spicil. II. t. 119.
Engrav. Wight's Icon. t. 1144.

SPEC. CHAR. Stem erect, glabrous, somewhat panicled at the apex: cauline leaves runcinately pinnatifid, doubly crenate, dilated and somewhat stem-clasping at the base: terminal lobe subrhomboid, attenuated upwards, mucronate, somewhat hairy on both sides: floral ones entire, lanceolate: pedicels hairy at the apex; capitula ovate: involucral scales imbricated: outer ones hairy on the back: achænia obovate, compressed, ending in a long beak: pappus double, outer short, paleaceous, inner long, slender, bristly; flowers purple.

Neilgherried, in jungles and by the road-side, flowering in the rainy and cold seasons.

ORDER LXXXIX. STYLIDIACEÆ.

ORD. CHAR. Herbs or undershrubs with a stem or scape: leaves occasionally in whorls or scattered, or the radical ones clustered, entire: stipules wanting: flowers racemose, spicate, or corymbose, terminal, rarely axillary, pedicels usually with 3 bracts: calyx adherent, limb 2-6-partite, regular or two-lipped, persistent: corolla monopetalous, slowly deciduous: limb rarely regular, 5-6-partite, imbricated in æstivation: stamens two, filaments connate with the style into a longitudinal column: anthers twin or simple, overlying the stigma: ovary 2-celled, or occasionally 1-celled from the construction of the dissepiment, many-seeded, often crowned with one, or two opposite glands: style one: stigma entire, or 2-cleft: ovules anatropal: capsules 2-valved, 2-celled, sometimes 1-celled by contraction: seeds small, erect, sometimes stalked, attached lengthwise to the axis of the dissepiment: albumen fleshy, somewhat oily: embryo minute, included.

GENUS. STYLIDIUM.

Gynandria Diandria. *Sex: Syst:*

Deriv. From *Stylos*, a column. The stamens and style are united.

Gen. Char. Limb of the calyx 2-lipped: corolla irregular, 5-cleft, the fifth segment dissimilar, smaller, more often deflexed, the rest open, sometimes cohering by pairs; staminal column reclinate, with a double flexure; anthers 2-lobed: lobes much divaricated: stigma obtuse, undivided: capsule 2-celled.

(1) S. Kunthii. *(Wall.)*

Ident. Dec. prod. VII. p. 335.—Wall. Cat. No. 3759.

Spec. Char. Herbaceous, glabrous, or here and there with glandulose hairs: scapes many, somewhat naked, cymosely corymbose: flowers sessile, few, spiked, sessile in the forks: leaves roseate a little above the middle, subrotund, petioled: one lip of the calyx somewhat 3-toothed, the other 2-cleft.

Silhet. Chittagong.

ORDER XC. GOODENIACEÆ.

Shrubby or herbaceous plants with alternate, exstipulate, simple, entire, dentate, or somewhat incised leaves: flowers distinct: tube of the calyx more or less adherent to the ovary; limb 4-5-lobed, entire or obsolete, persistent: corolla gamoepetalous, more or less irregular, tube split above, rarely 5-partible: limb 5-parted, two or rarely one-lipped, the middle lobes lanceolate, flat, the lateral ones thinner and more corolline; æstivation induplicate, rarely obsolete: stamina united with the corolla, not with the style, alternate with its lobes: filaments distinct: anthers united or oftener free, continuous with the filaments, 2-celled, bursting longitudinally: pollen simple or compound: ovary 1-2 or 4-celled; ovules few or numerous: style simple or rarely double: stigma fleshy, surrounded with a cup-shaped membranaceous indusium, entire or 2-lobed, ciliate or naked: fruit various, capsular, many-seeded with the septum, when present, usually parallel with the valves or drupaceous, or nucamentaceous with definite seed: seed erect, albuminous with a thick testa: embryo straight. foliaceous: radicle inferior.

GENUS. SCÆVOLA.

Pentandria Monogynia. *Sex. Syst.*

Deriv. From *Scæva*, the left hand; alluding to the form of the corolla.

GEN. CHAR. Shrubs, undershrubs or perennial herbs: leaves alternate, rarely opposite, quite entire or toothed and even sometimes cut; spikes or cymes dichotomous, springing from the axils: flowers bibracteate, sometimes solitary in the axils, blue, white, rarely yellowish: lobes of the corolla winged, often fimbriated; tube villous within, the throat bearing fringes at the apex: tube of the calyx adherent: limb 5-partite or 5-toothed, rarely entire: corolla cleft longitudinally on the upper side; segments winged, equal: anthers free: indusium of the stigma ciliated: drupe fleshy or juiceless, 1–4-celled: cells 1-seeded.

(1) S. KOENIGII. (*Vahl.*)

Ident. Vahl. Symb. III. p. 36.—Dec. prod. VII. p. 505.

Syn. S. Bela Mogadam, *R. & S.*—S. Lambertiana, *de Vriese*, —S. chlorantha, *de Vriese*.—S. Taccada, *Roxb. flor. Ind.* II. p. 146.

Engrav. Lam. Ill. t. 124. f. 2.—Wight's Ill. II. t. 137.—Hook. Bot. Mag. t. 2732.—Rheede Mal. IV. t. 59.

SPEC. CHAR. Fruticose: axils bearded: leaves obovate, subrepand at the apex, and with the branches and cymes glabrous: peduncles axillary, dichotomous: limb of the calyx 5-partite, equalling the length of the ovary: flowers and ripe fruit white.

Concans. Travancore. Chiefly on the sea-coast. Flowering all the year.

(2) S. PLUMIERI. (*Vahl.*)

Ident. Vahl. Symb. II. p. 36.—Dec. prod. VII. p. 506.

Syn. S. Thunbergii, *Eckl. & Zeyh.*—S. uvifera, *Stocks.*—S. Senegalensis, *Presl.*

Engrav. Wight's Icon. t. 1613.—Lam. Ill. t. 124. f. 1.

SPEC. CHAR. Fruticose: axils subbarbate: leaves obovate, quite entire, somewhat fleshy, glabrous: peduncles axillary, dichotomous: limb of the corolla cup-shaped, truncated, nearly quite entire; corolla tomentose within: flowers white: fruit purple.

Malabar. Scinde, on the sea-shore.

ORDER XCI. SPHENOCLEACEÆ.

Tube of the calyx adnate to the ovary, limb 5-parted, lobes round on the margin, inflexed, persistent, finally connivent over the ovary: corolla deciduous, 2-celled, dehiscing longitudinally: ovary 2-celled, many-ovuled: styles very short, stigma capitate, bilobate, glabrous: capsule membranaceous, 2-celled, cuneiform at the base, many-seeded, circumscissile: placentæ fungose, pendulous from the apex of the septum: seed tuberculate, minute, terete, sparingly furnished with fleshy albumen: embryo straight, terete, radicle about twice as long as the cotyledons.

GENUS. SPHENOCLEA.
Pentandria Monogynia. *Scr: Syst:*

Deriv. From *Sphen,* a wedge, and *Kleio,* to enclose; alluding to the wedge-shaped capsules.

GEN. CHAR. Same as that of the Order.

(1) S. PONGATIUM. *(Dec.)*

Ident. Dec. prod. VII. p. 548.
Syn. Pongatium Indicum, *Lam.*—S. Zeylanica, *Roxb fl. Ind.* I. p. 507.—Rapania herbacea, *Lour.*—Gaertnera pongati, *Retz.*
Engrav. Wight's Ill. II. t. 138.—Rheede Mal. II. t. 24.

SPEC. CHAR. Annual, herbaceous, erect, branched, glabrous: leaves alternate, exstipulate, lanceolate, entire, smooth; terminal or leaf-opposed, peduncled, cylindric: bracts 3 or 3-partite, under the flower: flowers very small, white.

Peninsula. Bengal. Hindostan. Common in rice fields, flowering nearly all the year.

ORDER XCII. CAMPANULACEÆ.

Herbaceous, rarely suffruticose, milky plants: leaves exstipulate, alternate, or rarely opposite, often dentate: inflorescence either definite, centrifugal or obscurely indefinite, in that case the flowers terminating, the lateral branches opening first;

flowers solitary or glomerate, generally pedicelled, seldom involucrate: corolla usually blue, sometimes yellow or purple: calyx usually 5-lobed, occasionally 3-6-8 or 10-lobed, adnate to the ovary, the lobes equal: petals united, regular, or rarely somewhat irregular, divisions alternate with the lobes of the calyx valvate in æstivation: stamens 3-10, usually equalling, never exceeding the lobes of the corolla, alternate with them, and not adhering to the tube: filaments usually dilated, membranaceous at the base: anthers for the most part free, the cells bursting longitudinally before dehiscence: ovary inferior, 2-10-celled, from the incomplete partitions: ovules numerous: styles more or less covered with caducous collecting hairs: stigma naked, sometimes bound with an indusium, usually branched, the branches equalling the cells of the ovary, erect in the flower-bud, hairy on the back, papillose within, diverging or recurved in the flower: capsule dehiscing at the apex or sides, the valves for the most part bearing the partitions, more rarely without valves, opening by pores or fissures: seeds numerous, small: embryo straight: albumen fleshy.

GENUS I. CEPHALOSTIGMA.

Pentandria Monogynia. *Sm. Syst.*

Deriv. From *Kephale*, a head, and *Stigma*; alluding to the capitate stigma.

GEN. CHAR. Calyx 5-cleft: corolla 5-parted; segments alternating with the calycine lobes, and longer than them: stamens free, filaments broader at the base; anthers 2-celled: style more or less hairy: stigma simple, capitate, pilose: capsule 2-3-celled, dehiscing by 2-3 short valves at the apex, which are septiferous in the middle: seeds numerous, small, ovoid, triquetrous.

(1) C. MINUTUM. (*Edgew.*)

Ident. Edgew. in Linn. Trans. XX. p. 81.—H. F. & T. in Jour. Proc. Linn. Soc. II. p. 9.

Syn. Wahlenbergia perotifolia, *W. & A. Dec. prod.* VII. p. 434.—Campanula anagalloides, *Royle Ill.* p. 254.

Engrav. Wight's Icon. t. 842.

Spec. Char. Stem erect, flexuose, pilose, angular, branched: leaves alternate, sessile, lanceolate, acuminate, attenuated at the base, glabrous, denticulate on the margin: peduncles terminal, pubescent, naked; tube of the calyx hairy; lobes linear-acuminate: capsule globose.

Concans. In cultivated sandy soils near the Coast. Flowering in September.

(2) C. FLEXUOSUM. *(H. F. & T.)*

Ident. H. F. & T. in Jour. Proc. Linn. Soc. II. p. 9.—Dalz. Bomb. Flor. p. 133.

Spec. Char. Stems hispidly hairy, very slender, flexuose, paniculately branched above: branches filiform: leaves sessile, broad ovate-oblong, obtuse, somewhat sinuately toothed, glabrous above: flowers on slender pedicels; tube of the calyx broad-hemispherical; corolla deeply 5-cleft; lobes linear-oblong: filaments ciliated.

Concans.

GENUS II. CAMPANUMŒA.

Pentandria Monogynia. *Ser ; Syst:*

Deris. Altered from Campanula.

Gen. Char. Calyx hemispherical, combined with the involucre at the base: sepals 5, adhering to the base of the ovary, patent: corolla shortly 5-lobed; stamens free; filaments filiform, somewhat dilated at the base: anthers oblong: disk epigynous, depressed, obscurely lobed: ovary depresso-globose, 5-celled, 5-ribbed at the base: cells many-ovuled; ovules in many rows, adnate to the thick axillary placentæ: stigma club-shaped, 5-lobed: lobes valvate, densely hairy outside: fruit membranaceous or somewhat baccate, indehiscent, irregularly broken: seeds minute, oblong.

(1) C. JAVANICA. *(Blume.)*

Ident. Blume Bijdr. p. 726.—Dec. prod. VII. p. 423.

Engrav. H. F. & T. in Ill. Illmal. plants t. 16. [B.]

Spec. Char. Stems slender, twining: leaves cordate or 2-lobed at the base, with the sinus broad and sometimes dilated at the insertion of the petiole, acute or subobtuse, crenated, rarely quite entire, membranaceous, pale green above; sepals adnate to the base of the ovary, ovate-lanceolate or ovate-oblong, obtuse or acute, increasing after flowering: corolla shortly tubular-campanulate, broad at the base and obscurely 5-angled: lobes scarcely patent, broad-ovate, with acute papillose apices: filaments glabrous, scarcely dilated at the base; anthers linear, apiculated with a connectivum: style puhescent at the apex; ovary broad-hemispherical, obscurely 10-ribbed, 5-celled: berry pulpy or slightly membranaceous, purple, very often crowned with the persistent calyx, 5-celled, without valves: flowers solitary, axillary, greenish.

Khasia mountains at 4—6000 feet.

GENUS III. CODONOPSIS.

Pentandria Monogynia. *Sex. Syst.*

Deriv. From *Kodon*, a bell, and *Opsis*, resemblance; alluding to the shape of the flowers.

GEN. CHAR. Twining or somewhat erect herbs, with milky or watery juice, very often fetid: calyx superior, 5-lobed: corolla tubular or campanulate, 4-6-lobed: stamens 4-6, free: filaments somewhat dilated at the base, filiform: anthers oblong: disk epigynous, fleshy, depressed, obscurely lobed: ovary ribbed, globose or obconical, upper part truncated or conical, attenuated into the straight style, 3-5-celled; cells many-ovuled; ovules in many rows, adhering to thick axillary placentæ: stigma club-shaped, 3-5-lobed: lobes valvate, densely hairy outside, recurved while flowering: fruit baccate below the corolla, indehiscent or irregularly broken, upper part conical, coriaceous or horny, 3-5-valved: seeds oblong: albumen copious, fleshy: embryo terete.

(1) C. VIRIDIS. *(Wall.)*

Ident. Wall. in Roxb. B. Ind. II. p. 103.—H. F. & T. in Jour. Proc. Linn. Soc. II. p. 12.

Syn. Wahlenbergia viridis, Dec. prod. VII. p. 425.—Campanula viridis, *Spreng.*

SPEC. CHAR. Twining; branches glabrous; branchlets and leaves, especially below, hoary: leaves opposite and alternate, ovate, oblong or ovate-lanceolate, acute or acuminated: pedicels axillary and leaf-opposed: calyx pubescent, with narrow lobes: ovary hemispherical: corolla broad-campanulate: berry depressed, globose, with a conical apex: valves 3-5, horny: flowers pale green, purplish at the base.

Khasia mountains at 5—6000 feet, flowering in September.

GENUS IV. CYCLOCODON.

Pentandria Monogynia. *Sex. Syst.*

Deriv. From *Kuklos*, a circle, and *Kodon*, a bell; alluding to the shape of the flowers.

Gen. Char. Herbs: calyx adherent to the base of the ovary or altogether free, 5-partite, the leaflets subserrate: corolla shortly campanulate, 4-5-lobed; style erect; stigma clavate, 4-5-lobed, lobes afterwards revolute: ovary 4-5-celled, cells opposite the sepals, ovules many, affixed in many rows to thick axillary placentæ: fruit baccate, irregularly broken; seeds numerous, subangled, compressed; testa smooth, coriaceous: embryo broadly clavate, cotyledons and radicle short.

(1) C. parviflorum. *(H. F. & T.)*

Ideal. H. F. & T. in Jour. Proc. Linn. Soc. II. p. 18.

Syn. Codonopsis parviflora, *Wall. Cat. No.* 1300.—*Dec. prod.* VII. p. 423.—Campanumæa Celebica, *Blume Bijdr.* p. 727.—*Dec. l. c.*

Spec. Char. Herbaceous, annual, erect, dichotomously branched: leaves opposite, ovate-lanceolate, long-acuminate, serrated, short-petioled, glaucous below: sepals free: flowers tetramerous, arranged in trichotomous cymes, small, white: peduncles curved, nodding: berry globose, 4-celled.

Assam. Khasia mountains at 2—4000 feet. Flowering in August and September.

GENUS V. WAHLENBERGIA.

Pentandria Monogynia. *Ser: Syst:*

Deriv. Named after George Wahlenberg, a celebrated German Botanical author.

Gen. Char. Herbs, rarely shrubs with alternate, rarely opposite, leaves: calyx 3-5-cleft: corolla 3-5-lobed, rarely divided to the middle: stamens 3-5, free, filaments broader at the base: style hairy, especially above: stigmas 2-5: ovary adherent; capsule 2-5-celled, opening by valves at the apex.

(1) W. agrestis. *(Dec.)*

Ideal. Dec. prod. VII. p. 434.—H. F. &. T. in Jour. Proc. Linn. Soc. II. p. 21.

Syn. W. dehiscens, *Dec. l. c.*—W. Indica, *Dec. l. c.*

Engrav. Wight's Icon. t. 1175 and 1176.—Spicil. II. t. 123 and 124.

Spec. Char. Stem erect, branched from the base, hairy below; lower leaves approximated, narrow-linear, nearly entire, undulated on the margin: peduncles usually dichotomous, with very short bracts: tube of the calyx glabrous, obovoid, shorter than the erect narrow linear lobes: corolla funnel-shaped, about twice the length of the lobes of the calyx: capsule ovoid: flowers pale blue.

Peninsula. Bengal. Khasia mountains. Neilgherries. Flowering all the year.

GENUS VI. PERACARPA.

Pentandria Monogynia. *Sm: Syst:*

Deriv. From *Pera*, a bag, and *Karpos*, fruit.

GEN. CHAR. Calyx-tube obconical, lobes of the limb triangular: corolla campanulate, deeply 5-cleft, lobes equal, linear, acuminate: stamens epigynous, filaments linear, free: anthers linear: style elongated, stigmas 3, revolute: ovary 3-celled: capsule oblong, pendulous, thinly membranaceous, contracted at the apex, few-seeded, irregularly broken: seeds large, oblong.

(1) P. CARNOSA. (*H. F. & T.*)

Ident. H. F. & T. in Jour. Proc. Linn. Soc. II. p. 28.
Syn. Campanula carnosa, *Wall. Cat. No.* 1282.—*Roxb. fl. Ind.* II. p. 102.—*Dec. prod.* VII. p. 474.

SPEC. CHAR. Slender herb, branched, slightly fleshy, prostrate or creeping, quite glabrous: leaves petioled, ovate, subacute, sinuately toothed: pedicels axillary, slender, erect: flowers small, white.

Khasia mountains at 4—6000 feet, flowering in July.

GENUS VII. CAMPANULA.

Pentandria Monogynia. *Sm: Syst:*

Deriv. A diminutive of *Campana*, a bell; alluding to the shape of the flowers.

GEN. CHAR. Herbs usually perennial: radical leaves usually different in form from the cauline ones and larger: flowers generally racemose, rarely spicate or glomerate, blue or white: calyx 5-cleft, the sinuses usually covered by appendages: corolla 5-lobed or 5-cleft, usually bell-shaped: stamens free, filaments broader at the base: style covered by fascicles of hairs, except at the base: stigmas 3–5, filiform: ovary inferior, 3–5-celled: capsule 3–5-valved, dehiscing laterally: seeds usually ovate, flattened, sometimes ovoid, small.

(1) C. CANESCENS. (*Wall.*)

Ident., Dec. prod. VII. p. 473.—Wall. Cat. No. 1289.

SPEC. CHAR. Hispid: stem erect: radical leaves clustered, lanceolate, crenulate: cauline ones remote, narrower, repandly denticulate: flowers racemose, often approximated, short-pedicelled, small: tube of the calyx spherical, lobes linear, entire: corolla tubular, hairy: capsule spherical.

Upper and Eastern Bengal. Khasia mountains. Flowering all the year.

(2) C. COLORATA. (*Wall.*)
Var. ramulosa.

Ident. Dec. prod. VII. p. 473.—Wall. in Roxb. flor. Ind. II. p. 100.
Engrav. Wight's Icon. t. 1178.—Spicil. II. t. 126.

SPEC. CHAR. Stem erect, hairy, branched: leaves lanceolate, sessile, crenately toothed: pedicels axillary and terminal: calyx hairy, lobes broad acute, subdentate, about half the length of the cylindrical villous corolla: capsule turbinate, drooping: flowers purplish.

Neilgherries. Khasia mountains.

(3) C. ALPHONSII. (*Wall.*)

Ident. Dec. prod. VII. p. 473.
Engrav. Wight's Icon. t. 1177.—Spicil. II. t. 125.

SPEC. CHAR. Decumbent, 1-flowered: stem pubescent: cauline leaves sessile, sublanceolate, acute, denticulate, hairy above, hoary beneath: calyx pubescent: segments acute, serrated or sometimes lobed, about half the length of the puberulous corolla: flowers purple.

Neilgherries, forming dense tufts in clefts of rocks. Flowering in June and July.

(4) C. FULGENS. (*Wall.*)

Ident. Dec. prod. VII. p. 477.—Wall. in Roxb. flor. Ind. II. p. 99.
Engrav. Wight's Icon. t. 1179.—Ill. t. 136.

SPEC. CHAR. Stem erect, about a foot high, hairy: leaves lanceolate, acuminated at both ends, short-petioled, serrated: lobes of the calyx subulate, erect, entire: corolla glabrous, infundibuliform: flowers subsessile, axillary, solitary or three together, approximated towards the apex, purple.

Khasia mountains. Neilgherries. Common on grassy slopes and pastures. Hills in Canara. Flowering in June and July.

(3) C. KHASIANA. *(H. F. & T.)*

Ident. H. F. & T. in Jour. Proc. Linn. Soc. II. p. 25.

SPEC. CHAR. Hispidly pubescent; stem simple, erect, straight, somewhat robust, furrowed: leaves suberect, sessile, obovate-oblong, acuminated, serrated, hispidly pubescent on both sides: raceme long, terminal, simple, or paniculately branched: tube of the calyx subglobose: lobes bristly-lanceolate; corolla glabrous, campanulate, shortly 3-lobed: style slender: stigmas 2: flowers middle-sized, nodding, short-pedicelled: pedicels bracteate.

Khasia mountains, flowering in July.

GENUS VIII. PIDDINGTONIA.

Pentandria Monogynia. *See: Syst:*

Deriv. Named after H. Piddington, author of an English Index to Indian Plants.

GEN. CHAR. Calyx-tube ovoid, narrowed below: lobes linear-acuminate, equal, 3 upper ones approximated, and less patent: corolla longitudinally cleft at the back; two upper lobes linear, erect, lower lip 3-cleft, patent: lobes ovate-acute; lips equal in length; anthers 2, terminated by solitary bristles: stigma 2-lobed: berry thick, ovoid-globose, 2-celled.

(1) P. NUMMULARIA. *(Lam.)*

Ident. Lam. Dict. II. p. 589.—Dec. prod. VII. p. 341.

Syn. Rapuntium nummularinm, *Presl.*—Lobelia begonifolia, *Wall. in Rosb. fl. Ind.* II. p. 115.

Engrav. Lindl. Bot. Reg. t. 1373. (under Pratia.)

SPEC. CHAR. Herbaceous, creeping: stems rooting, hairy: leaves pubescent, unequally reniform, cordate, toothed, mucronulate: pedicels quite glabrous, naked, in the axils of the middle leaves: flowers purplish.

Khasia mountains at 4—7000 feet, flowering in the rainy season.

GENUS IX. LOBELIA.

Pentandria Monogynia. *See: Syst:*

Deriv. Named after Matthew Lobel, author of various Botanical works; he was Physician to James I. and died in London in 1616.

GEN. CHAR. Herbs, rarely undershrubs: leaves alternate: pedicels axillary: flowers often racemosely-spiked, blue, white, violet, red or golden-coloured: calyx 5-lobed, tube obconical, ovoid or hemispherical: corolla longitudinally cleft above, 2-lipped, tube cylindric or infundibuliform, straight, upper lip often less and erect, lower often patent, broader, 3-cleft or rarely 3-toothed, two lower anthers, sometimes all, bearded at the apex: ovary inferior or half-superior.

(1) L. TRIGONA. (Roxb.)

Ident. Roxb. fl. Ind. II. p. 111.—Dec. prod. VII. p. 359.

Syn. L. Zeylanica, *Linn.*—L. trialata, *Ham.*—L. micrantha, *Hook.*—L. Subincisa, *Wall.*

Engrav. Wight's Icon. t. 1170.—Spicil. II. t. 120.

Spec. Char. Glabrous: branches diffuse, erect or ascending and with the stem 3-cornered: leaves subsessile, ovate, subcordate, repand-toothed, mucronulate: pedicels slender, longer than the leaf, bibracteolate at the base: lobes of the calyx linear-acuminate: corolla small, slightly longer than the calycine lobes: anthers included, all bearded at the apex: capsule obovoid: flowers purple.

Common everywhere, chiefly found in rice-fields and flowering all the year.

(2) L. AFFINIS. (*Wall.*)

Ident. Dec. prod. VII. p. 360.

Spec. Char. Pubescent: stem procumbent, branched: leaves ovate-rotundate, short petioled, sub-cordate, repandly mucronato-dentate: pedicels longer than the leaf: lobes of the calyx linear-acuminate, patent in the capsule: corolla slightly hairy above: anthers included, hairy at the apex.

Silhet mountains. Eastern Bengal. Flowering in the rainy season.

(3) L. COLORATA. (*Wall.*)

Ident. Dec. prod. VII. p. 380.—Wall. Pl. As. Rar. II. p. 42.

Syn. L. purpurascens, *Wall.*—Rapuntium coloratum, *Presl.* prod. Lob. p. 24.

Spec. Char. Stem erect, simple, glabrous: leaves linear-lanceolate, subdentate, glabrous: racemes spiciform, long, few-flowered: bracts glanduloso-dentate, longer than the pedicel: pedicels hairy: calyx-tube ovoid: lobes linear, toothed, hairy, 3 times shorter than the corolla: petals linear, narrow: anthers hairy, 2 lower ones bearded at the apex: flowers purplish.

Khasia mountains at 5—6000 feet, flowering in August.

(4) L. PYRAMIDALIS. *(Wall.)*

Ident. Wall. in As. Res. XIII. p. 376.—Dec. prod. VII. p. 381.

Syn. Rapuntium pyramidale, *Presl. prod. p.* 23.

Engrav. Bot. Mag. t. 2387.

SPEC. CHAR. Stem erect, branched, glabrous: branches angular: leaves sessile, linear-lanceolate, acute, serrated, glabrous: racemes many-flowered: bracts acuminate, leafy, longer than the pedicel: calyx-tube ovoid: lobes subulate, serrated, longer than the tube and equalling the corolla: petals all reflexed, lateral ones narrower: capsule 10-ribbed, nodding: flowers violet.

Silhet mountains, flowering in the rainy season.

(5) L. EXCELSA. *(Lesch.)*

Ident. Dec. prod. VII. p. 381.—Roxb. flor. Ind. II. p. 114.

Syn. Rapuntium Leschenaultianum, *Presl. prod. p.* 24.

Engrav. Wight's Icon. t. 1173-4.—Spicil. II. t. 129.

SPEC. CHAR. Stem herbaceous, erect: leaves lanceolate, short-petioled, narrowed at the base, acuminated, denticulate, puberulous above, tomentose below: raceme leafy, pubescent, many-flowered: bracts long-acuminate, glanduloso-dentate, twice as long as the pedicels: lobes of the calyx erect, linear-lanceolate, denticulate, 3 times longer than the hemispherical tube, and equalling the tube of the pubescent corolla: flowers pale yellow tinged with lilac.

Common on the Neilgherries, flowering during the rains from May to September.

(6) L. NICOTIANÆFOLIA. *(Heyne.)*

Ident. Roem & Schult. Syst. V. p. 47.—Dec. prod. VII. p. 381.

Syn. Rapuntium nicotianæfolium, *Presl. prod. p.* 24.

Engrav. Wight's Ill. t. 135.

SPEC. CHAR. Stem erect: leaves subsessile, oblong-lanceolate, denticulate, narrowed at the base, acuminated: racemes many-flowered: bracts leafy: pedicels puberulous, slightly longer than the bract, bibracteolate in the middle: calycine lobes lanceolate, serrated, patent, longer than the hemispherical tube: corolla pubescent, 4 times longer than the lobes of the calyx: lateral lobes long-linear, central ones lanceolate: 2 lower anthers penicillate at the apex: flowers purple.

Neilgherries. Canara.

(7) L. ROSEA. *(Wall.)*

Ident. Wall. in Roxb. fl. Ind. II. p. 115.—Dec. prod. VII. p. 361.

Syn. Rapuntium roseum, *Presl. prod.* p. 24.—L. trichandra, R. W.

Engrav. Wight's Icon. t. 1171.

SPEC. CHAR. Everywhere velvety: stem branched; leaves sessile, lanceolate, acuminated at both ends, serrulate: flowers racemosely spiked, secund: pedicels shorter than the entire lanceolate bract: calyx-tube ovoid: lobes linear, entire, twice as long as the tube: corolla velvety, a half longer than the calycine lobes; lateral divisions narrower, deflexed; centre ones ovate-acute; flowers numerous: petals rose-coloured at the apex.

Pundua, Silhet. Nellgherries. Flowering from January to April.

(8) L. LOBBIANA. *(H. F. & T.)*

Ident. H. F. & T. in Jour. Proc. Linn. Soc. II. p. 28.

SPEC. CHAR. Stem decumbent or prostrate, branched, quite glabrous: branchlets puberulous: leaves petioled, ovate, acute, sharply serrulate: flowers axillary, solitary, on long slender pedicels: calyx-tube puberulous, lobes linear-bristly, spreading or recurved: tube of the corolla hard at the base, cleft, shortly 2-lipped, lower lip 3-cleft, lobes oblong-lanceolate; anthers glabrous, two lower ones penicillate at the apex.

Khasia mountains.

(9) L. WALLICHIANA. *(H. F. & T.)*

Ident. Journ. Proc. Linn. Soc. II. p. 29.

Syn. L. pyramidalis, *var.* a. *Dec. prod.* VII. p. 361.—Rapuntium Wallichianum, *Presl. prod.* p. 24.

SPEC. CHAR. Stem erect: branches angular: leaves sessile, linear-lanceolate, acute, serrated: racemes many-flowered: bracts and calycine lobes entire: anthers hairy and ciliated: petals reflexed: capsule 10-ribbed: flowers violet.

Khasia mountains, flowering in the rainy season.

ORDER XCIII. SIPHONANDRACEÆ.

Trees, shrubs or undershrubs: buds scaly; corolla gamopetalous, regular, deciduous: leaves alternate: anthers at last introrse, cells separated at the apex, dehiscing from the apex in front by a pore or more or less elongated foramen: fruit inferior or superior, baccate, drupaceous or loculicidally capsular.

TRIBE I.   VACCINIEÆ.

Calyx adherent, limb epigynous, 4–7-partite, deciduous or persistent: corolla epigynous, gamopetalous, 4–7-divided, deciduous, imbricated in æstivation: stamens double the number of the segments of the corolla, epigynous, in one row, filaments connate into a tube or free: anthers 2-celled, introrse, affixed by the back, cells parallel, divided at the apex, tubulose: ovary inferior, or half-inferior, 4–10-celled: placentæ adnate to the central column, 1-many-ovuled: style single, stigma capitate.—Branched shrubs: branches and branchlets terete or irregularly angled: leaves alternate, simple, entire, deciduous or persistent: flowers solitary or racemose, rarely spiked.

GENUS I.   AGAPETES.

Decandria Monogynia. *Bar: Syst:*

Deriv. A Greek term in reference to the plants being showy.

GEN. CHAR. Calyx campanulate; limb 5-partite, segments acute: corolla tubulose incurved, hairy outside, lobes narrow, suberect: stamens slightly incurved towards the apex: filaments short, distinct: anthers 2-celled, hirsute, produced into two combined forked small tubes, dehiscing at the apex in front: style ascending, stigma depresso-capitate, 5-lobed: berry fleshy, crowned by the limb of the calyx and cup-shaped disc, 5-celled, many-seeded: seeds angular, affixed to central placentæ.—Erect shrubs: leaves alternate, coriaceous, large, evergreen: racemes axillary at the tops of the branchlets: flowers large, hirsute or pubescent.

(1) A. VARIEGATA. (D. Don.)

Ident. Dec. prod. VII. p. 584.

Syn. Ceratostema variegatum, *Roxb. fl. Ind.* III. p. 413.— Thibaudia variegata, *Royle.*

Engrav. Royle Ill. t. 79. f. 1.

Spec. Char. Branches terete, tubercled, somewhat angular at the apex: leaves very shortly petioled, oblong, obtuse, subspathulate, glabrous, minutely or sparingly serrated: petioles callous: racemes solitary, axillary: pedicels long, thickened at the apex, 5-angled: corollas tubulose, curved at the apex, 5-toothed, red.

Pundua, Silhet.

(2) A. SETIGERA. (*D. Don.*)

Ident. Dec. prod. VII. p. 554.

Syn. Thibaudia setigera, *Wall. Cat. No.* 752.

Spec. Char. Branches terete, here and there hairy: hairs short, darkish, glanduloso-tuberculate, glands whitish: leaves short-petioled, coriaceous, elliptic-lanceolate, attenuated at the base, obtuse, acuminated at the apex, racemes few-flowered, axillary, twice shorter than the leaf: pedicels and calyx hairy: flowers purple.

Pundua, Silhet.

(3) A. VERTICILLATA. (*D. Don.*)

Ident. Dec. prod. VII. p. 554.

Syn. Thibaudia verticillata, *Wall. Cat. No.* 753.—Vaccinium verticillatum, *R. W.*

Engrav. Wight's Icon. t. 1181.

Spec. Char. Stem fruticose: leaves verticillate, lanceolate, acuminate, minutely denticulate, acute at the base: flowers racemosely corymbose: peduncles and calyx hispid: corolla glabrous.

Pundua, Silhet. *A. vacciacea*, (Roxh. fl. Ind. II. p. 412)—appears to be a mere variety of this species.

(4) A. ACUMINATA. (*D. Don.*)

Ident. Dec. prod. VII. p. 554.

Syn. Thibaudia acuminata, *Wall. Cat. No.* 6297.

Spec. Char. Leaves petioled, lanceolate, long-acuminate, dentate: flowers corymbose: calyx and peduncles slightly tomentose: lobes of the calyx ovate, mucronate.

Silhet.

(5) A. WALLICHIANA. (*Klotsch.*)

Ident. Walp. Ann. II. p. 1089.—Klotsch in Linn. XXIV. p. 87.

Syn. Vaccinium Wallichianum, *R. W.*

Engrav. Wight's Icon. t. 1180.

Spec. Char. Leaves subsessile, lanceolate-acuminate, entire, glabrous, congested towards the ends of the branchlets: racemes axillary, erect, shorter than the leaves: flowers tubular, drooping, and with the calyx and pedicels sprinkled with longish hairs: pedicels dilated, cup-shaped at the apex: anthers rough, without bristles, ending in two long tubes cohering nearly half their length: stigma dilated: flowers dark pink.

Silhet ?

(6) A. HIRSUTA. *(Klotsch.)*

Idem. Walp. Ann. II. p. 1089.—Klotsch. in Linn. XXIV. p. 37.

Syn. Vaccinium hirsutum, *R. W.*

Engrav. Wight's Icon. t. 1182.

Spec. Char. Leaves elliptic-lanceolate, entire, glabrous or sub-pubescent: racemes erect, corymbose, many-flowered: flowers tubular, long-pedicelled: pedicels, calyx and corolla hairy: filaments short, anthers pubescent.

Silhet ?

GENUS II. CALIGULA.

Decandria Monogynia. *Ser; Syst;*

Gen. Char. Calyx urceolate, limb 5-cleft, segments lanceolate: corolla tubulose, 5-sided, limb 5-cleft, revolute: filaments distinct, short, ciliated: anthers very long, 2-celled, subulate, puberulous below, incurved at the base, long-tubulose, affixed above the base, 2-awned below the middle, alternately of unequal length, awns pendulous, pubescent: style cylindric, the length of the corolla: stigma obtuse: ovary fleshy, 10-celled.

(1) C. ODONTOCERA. *(Klotsch.)*

Idem. Walp. Ann. II. p. 1083.—Kl. in Linn. XXIV. p. 28.

Syn. Vaccinium odontocerum, *R. W.*

Engrav. Wight's Icon. t. 1187.

Spec. Char. Shrub: stem erect, diffuse, glabrous: branchlets furnished with scattered subulate scales: leaves coriaceous, oblong, narrow, acuminated, attenuated at the base, serrated, short-petioled: flowers corymbose, axillary: stigma capitate.

Khasia mountains.

GENUS III. EPIGYNIUM.

Decandria Monogynia. *See: Syst:*

GEN. CHAR. Calyx adherent, tube half-globose, smooth, limb free, 5-partite: corolla ovate, campanulate or urceolate: stamens distinct: filaments subulate: anthers 2-celled, mutic or awned, produced into two separate tubes, dehiscing by oval openings at the apex: ovary inferior, 5-celled, many-ovuled, disk 5-gibbous, and crowned with the limb of the calyx: berry pulpy, subglobose-obovate.

(1) E. SERRATUM. (*Klotsch*.)

Ident. Klotsch. in Linn. XXIV. p. 50.

Syn. Vaccinium serratum, *R. W.*—Gaylussacia serrata, *Lindl.* Dec. prod. VII. p. 558.—Agapetes serrata, *G. Don.*

Engrav. Wight's Icon. t. 1181.—Royle. Ill. t. 70. f. 2.

SPEC. CHAR. Stem fruticose: leaves narrow-lanceolate, serrated, acute, stiff, coriaceous, shining, short-petioled, approximated: bracts coloured, subulate: racemes axillary, few-flowered: flowers withering, long-pedicelled, whitish-green.

Khasia mountains.

(2) E. LESCHENAULTII. (*Klotsch*.)

Ident. Klotsch. l. c.

Syn. Vaccinium Leschenaultii, *R. W.*—Agapetes symplocifolia and A. arborea, *DNS. in Dec. prod.* VII. p. 555.

Engrav. Wight's Icon. t. 1188.—Spicil. II. t. 128.

SPEC. CHAR. Tree: older branches glabrous, greyish-white: branchlets pubescently villous: leaves short-petioled, ovate-elliptic, serrated, acute, paler below, hairy on the rib: racemes axillary and terminal, the length of the leaves: flowers red.

Neilgherries. Travancore mountains, flowering in March and April. The berries are agreeably acid and make excellent tarts.

(3) E. NEILGHERRENSE. (*R. W.*)

Ident. Klotsch. l. c.

Syn. Vaccinium Neilgherrense, *R. W.*

Engrav. Wight's Icon. t. 1189.—Spicil. II. t. 129.

SPEC. CHAR. Shrubby, glabrous, except the pubescent young shoots and leaves: leaves lanceolate, acute at the base: racemes longer than the leaves, axillary, usually towards the ends of the branchlets: corolla ovate, slightly pubescent: flowers whitish or rose-coloured, usually furnished with a large leafy bract.

Low banks of streams on the Neilgherries, flowering from February till April.

(4) E. AFFINE. *(Klotsch.)*

Ident. Klotsch. l. c.
Syn. Vaccinium affine, *R. W.*
Engrav. Wight's Icon. t. 1190.

SPEC. CHAR. Glabrous shrub: leaves short-petioled, ovate-lanceolate or elliptic-lanceolate, pointed at both ends, crenato-serrated towards the apex: racemes axillary, flowers secund, drooping, pedicels as long as the flowers: bracts leafy, lanceolate, caducous, with two subulate bracteoles at the base of the pedicels: filaments as long as the anthers, both hairy.

Khasia mountains.

(5) E. DONIANUM. *(Klotsch.)*

Ident. Klotsch. l. c. p. 51.
Syn. Vaccinium Donianum, *R. W.*
Engrav. Wight's Icon. t. 1191.

SPEC. CHAR. Branchlets virgate, glabrous, terete: leaves short-petioled, obovate-lanceolate, acuminate, coriaceous, crenato-serrated: racemes axillary, cernuous, about the length of the leaves, many-flowered: flowers drooping: corolla glabrous, villous within: filaments short, thick, covered with matted hair: anthers glabrous.

Khasia mountains.

(6) E. ROTUNDIFOLIUM. *(Walp. Mss.)*

Ident. Walp. Ann. II. p. 1095.
Syn. Vaccinium rotundifolium, *R. W.*
Engrav. Wight's Ill. II. t. 139.

SPEC. CHAR. Arboreous: leaves orbicular, coriaceous, entire of slightly crenato-serrate, glabrous: racemes axillary and terminal, longer than the leaves: flowers 5-lobed: filaments filiform, hairy: anther bristles minute or wanting: flowers pale-pink, often streaked with darker lines: lobes of the corolla hairy: berries purplish-red when ripe.

Ootacamund.

TRIBE II. ANDROMEDEÆ.

Fruit capsular, loculicidally dehiscing, valves septiferous: corolla deciduous.—Shrubs; leaves evergreen or deciduous: leaf-buds almost always scaly.

GENUS IV. GUALTHIERIA.

Decandria Monogynia. *Ear: Syst:*

Deriv. Named after Gualthier, a Physician and Botanist of Canada.

GEN. CHAR. Calyx 5-lobed, afterwards enlarged, more or less baccate and surrounding the capsule: corolla ovate, often contracted at the mouth, 5-toothed: stamens included, filaments often villous, anthers 4-awned, cells 2-awned, very rarely mutic: style filiform: stigma obtuse: scales hypogynous, 10, distinct or united: capsule depresso-globose, 5-celled, 5-furrowed, 5-valved, valves septiferous, loculicidally dehiscing: placentas adnate to the axis: seeds numerous, small.

(1) G. LESCHENAULTII. *(Dec.)*

Ident. Dec. prod. VII. p. 593.

Syn. Andromeda Kotagherrensis, *Hook. Icon. t.* 246.—Leucothoe Kotagherrensis, *Dec. prod.* VII. p. 606.

Engrav. Wight's Icon. t. 1195.—Spiel. II. t. 130.

SPEC. CHAR. Glabrous, branches somewhat 3-cornered: leaves petioled, ovate or obovate, terminating in a gland, crenulate, punctuate beneath: racemes axillary or lateral, pubescent, shorter than the leaves, erect: bracts concave, acute, glabrous, one under the pedicel, two near the flower: flowers pure white: berries blue.

Neilgherries, flowering all the year.

GENUS V. ANDROMEDA.

Decandria Monogynia. *Ear: Syst:*

GEN. CHAR. Calyx 5-partite: segments acute, not imbricated: corolla globose-urceolate, contracted at the mouth, 5-toothed: stamens included: filaments bearded: anthers short: cells 1-awned: stigma truncate: capsules 5-celled, 5-valved, loculicidally dehiscing: placentas 3-lobed: seeds elliptic, compressed.—Undershrubs: leaves alternate, quite entire, revolute at the margin, glaucous-white below, short-petioled: flowers subterminal, almost umbellate, pedicelled, erect: bracts ovate: corolla white or rose-coloured.

(1) A. LANCEOLATA. *(Wall.)*

Ident. Wall. in As. Res. XIII. p. 390.
Syn. Pieris lanceolata, *Don. in Don. prod.* VII. p. 509.—A. squamulosa, *Don. prod. flor. Nep.* p. 149.
Engrav. Wight's Icon. t. 1198.
Spec. Char. Leaves lanceolate, acute at the base, acuminated, entire on the margin: racemes simple, secund: corolla oval, cylindrical, pubescent: flowers white.

Khasia mountains.

ORDER XCIV. RHODORACEÆ.

Trees, shrubs or undershrubs: buds cone-shaped, coverings large: leaves alternate: corolla gamopetalous or pleiopetalous, slightly irregular, deciduous: anthers mutic: cells joined together up to the apex, dehiscing by a pore at the top: fruit septicidally capsular.

GENUS I. RHODODENDRON.

Decandria Monogynia. *Ber: Syst.*

Deriv. From *Rhodon*, a rose, and *Dendron*, a tree, from the appearance of the terminal bunches of flowers.

Gen. Char. Calyx 5-partite: corolla infundibuliform, seldom campanulate or rotate, sometimes regular, sometimes more or less irregular, always 5-lobed: stamens 10, seldom by abortion 6-9, not adhering to the corolla, but placed before and between the lobes, often declinate, exserted: anthers dehiscing by two terminal pores: capsules 8-celled, 5-valved: seeds adnate to the columnar angled axis, compresso-scobiform, subulate.

(1) R. ARBOREUM. *(Sm.)*

Ident. Smith's Exot. Bot. No. 6.—Dec. prod. VII. p. 720.
Engrav. Wight's Ill. II. t. 140.—Spicil. II. t. 131.—Bot. Reg. t. 890.
Spec. Char. Tree: leaves very coriaceous, lanceolate, acute, cordate at the base, or attenuated into the thick petiole, shining-green above, below glabrous, silvery, or rusty-pubescent: flowers densely capitate: calyx none: corolla campanulate, white, rose or blood-coloured: ovary 7-10-celled.

Neilgherries and other lofty mountain ranges, flowering in March and April. The varieties of this species are very numerous.

(2) R. formosum. *(Wall.)*
Ident. Wall. Pl. As. Rar. III. p. 3.—Dec. prod. VII. p. 721.
Engrav. Wall. l. c. t. 207.

Spec. Char. Leaves lanceolate, obtuse, shining above, below and with the corollas scurfy: flowers few, terminal: calyx small, scarcely lobed: corolla subcampanulate, with an angular tube: filaments villous: flowers white, tinged with purple and yellow.

Silhet mountains.

ORDER XCV. LENTIBULACEÆ.

Herbaceous aquatic or marshy plants, with entire or compound radical leaves; in the latter case usually floating, the leaves resembling roots and bearing numerous small bladders: scapes slender, erect or twining, naked or furnished with scales; flowers either solitary or several, forming a raceme towards the apex: calyx 2-parted or more or less bilabiate, the inferior one often larger, often emarginate or bidentate: corolla personate or bilabiate, rarely regular; the upper lip 2-lobed or entire, the under larger, spurred at the base, 3-lobed or entire: stamens two, inserted on the base of the corolla, between the spur and ovary: filaments somewhat flattened, often bent, approximated at the base and apex: anthers terminal, ellipsoid, one-celled, often contracted in the middle, as if two-celled, dehiscing above: pollen (dry) broadly elliptical; ovary free, ovoid, one-celled: placentæ central, free, ovoid or globose, shortly stipitate at the base: ovules numerous, anatropous or peltate: stigma 2-or 1-lipped (the upper one being obsolete) the lower one larger, often dilated and revolute over the anthers: capsule globose or ovoid, bursting laterally or irregularly: seeds numerous, minute: testa often wrinkled: albumen wanting: embryo orthotropous, sub-cylindrical, sometimes undivided, sometimes with two short cotyledons.

GENUS I. UTRICULARIA.

Decandria Monogynia. *Ses: Syst:*

Deriv. From *Utriculus*, a little bladder; alluding to the small inflated appendages of the roots.

Gen. Char. Calyx bipartite: upper lobe entire, lower often emarginate or 2-toothed: corolla personate, spurred below the lower lip, the upper lip erect, entire or emarginate, lower one usually longer, often 3-lobed, plain, or reflexed below at the margin: palate a little prominent: stamens arched, approximated at the base and apex: anthers 1-celled, middle one sometimes constricted and then as if 2-celled: ovary ovoid: style none or filiform, persistent: stigma unequally 2-lipped, upper lip sometimes abortive; lips plain, lower one greater, rotund: placentæ shortly pedicelled: pedicel concealed above in a hollow: capsule ovoid or globose, often many-seeded, variously and irregularly dehiscent: seeds many, lenticularly angled, small: embryo acotyledonous.——Cosmopolite plants; some aquatic, freely floating: leaves radical, demersed, multisect, the axis inflated into a vesicle, or more often with the lateral segments utriculiferous, the utricles usually terminating in ramous bristles only aeriferous during flowering: others marshy, affixed by white fibrous roots, sometimes inflated, more often laterally glandulosely utriculiferous, the lateral leaves are then erect, entire, rarely utriculiferous on the margin: all with scapes more or less erect, with remote stipate scales: flowers racemose or solitary, yellow, purple or blue, very rarely white.

(1) U. STELLARIS. (*Linn.*)

Ident. Dec. prod. VIII. p. 3.—Oliver in Jour. Proc. Linn. Soc. III. p. 174.—Roxb. flor. Ind. I. p. 143.

Engrav. Wight's Icon. t. 1567.—Roxb. Cor. II. t. 180.

Spec. Char. Scapes furnished above or below the middle with about 3–5 lanceolate, oblong, or ovate-oblong vesicles with more or less branched capillary filaments towards the apex, very often arranged in a verticil: pedicels often thickened at the apex, fruit-bearing, patent or deflexed, equalling or exceeding the capsule: upper lip of the corolla ovate or round, obtuse, often twice the length of the calyx: spur short, saccate, obtuse, curved towards the lower lip of the corolla and almost equalling the same: capsules globose, more or less covered by the ovate, round or often unequal lobes of the calyx: seeds peltate, 5–6-angled: flowers yellow.

All tropical India. Moradabad. Rohilcund. Concan. Tanjore. Orissa. Khasia hills. Flowering in the rains.

(2) U. FLEXUOSA. (*Vahl.*)

Ident. Vahl. En. Pl. I. p. 199.—Oliver in Jour. Proc. Linn. Soc. III. p. 175.

Syn. U. fasciculata, *Roxb.*—U. inaequalis, *Benjamin in Linnaea* XX. p. 304.

Engrav. Wight's Icon. t. 1568.

SPEC. CHAR. Vesicles floating: here and there furnished with capillary multisect leaflets not unfrequently arranged in a verticil at the base of the scape: pedicels at first erect, fructiferous, more often deflexed: lobes of the calyx fructiferous, accrescent, diverging, often more or less unequal: spur short curved or stretched out, slightly obtuse, lower lip of the corolla broad-ovate, palate very prominent marked with an orange spot: seeds peltate, usually 2-6-angled: flowers yellow.

Assam and Silhet. Malabar. Bengal. Travancore. Chittagong. Flowering in the rainy season.

(3) U. DIANTHA. (*Roem. & Sch.*)

Ident. Roem. & Sch. Syst. Veg. I. p. 169. (not Dec.)—Oliver in Jour. Proc. Linn. Soc. III. p. 176.

Syn. U. biflora, *Roxb. flor. Ind.* I. p. 43.—U. Roxburghii, *Sprng.*—U. elegans, *Wall.*—U. pterosperma, *Edgew. Proc. Linn. Soc.* I. p. 352.

Engrav. Wight's Icon. t. 1569.

SPEC. CHAR. Floating or terrestrial: axis demersed, somewhat branched, segments leafy, capillaceous, utriculiferous: scape slender, naked or furnished with 1-2 minute scales, 1-2 (rarely 3)-flowered, in the single-flowered scapes opposite the bract is occasionally an apparently abortive axis or flower bud: bract small, stem-clasping: pedicels erect: calyx-lobes slightly unequal, much shorter than the mature capsule: lower lip of the corolla entire embracing the base of the spur, and often shorter than it: capsule subglobose, stigma minute, sessile: seeds peltate, winged, 5-6-angled: flowers yellow with an orange-streaked scale.

North-West Provinces. Khasia mountains. Bengal. Coromandel. Quilon. Flowering in the rainy season.

(4) U. ALBO-CÆRULEA. (*Dalz.*)

Ident. Dalz. in Kew Journ. Bot. VIII. p. 279.—Oliver in Journ. Proc. Linn. Soc. III. p. 177.

SPEC. CHAR. Fragrant: roots sparingly branched, bladders few: scape terete, erect, firm or slender, furnished with few small ovate scales, 1-2-flowered, sometimes elongated and 4-flowered:

bracts ovate, acute: upper lobe of the calyx broad-ovate, very acute, shorter than the upper white suborbiculate entire lip of the corolla, margins reflexed, lower lobe often much shorter than the subulate acute spur: lower lip of the corolla quadrate-orbiculate, emarginate, the palate marked in front with a pale, 3-lobed, veined spot ; capsule globose or sacciform, covered by the enlarged lobes of the calyx : seeds minute, oblong or ovate, obtuse or subspherical at both ends, scrobiculately reticulate: flowers few, blue and white.

Concan. Rocky places near Vingorla, flowering in the rainy season.

(5) U. ARCUATA. (R. W.)

Ident. Oliver in Journ. Proc. Linn. Soc. III. p. 177.

Engrav. Wight's Icon. t. 1570-1.

SPEC. CHAR. Roots branched, utriculiferous: leaves short, linear-spathulate, obtuse, sometimes furnished with dark bladders, few or none while flowering: scapes erect, sometimes 2-cleft, scales few, small, ovate ; lobes of the calyx somewhat equal, upper one larger, broad-ovate, or cordate-ovate, shorter than the suborbiculate, or obcordate, entire, or more or less deeply emarginate upper lip of the corolla, lower lobe ovate, 3–4 times shorter than the spur ; lower lip of the corolla entire, rarely emarginate, somewhat equalling the long, slender, linear-subulate, dependent or falcately-curved spur: pedicels slightly margined towards the apex: capsule ovate, or elliptic : flowers few, blue or pale violet.

Var. 1. Upper lip of the corolla 2-lobed, spur dependent, pedicels scarcely exceeding the spur. Bombay.

Var. 2. Upper lip of the corolla suborbiculate, entire, or subentire, spur often arched, pedicels equalling or scarcely exceeding the flower. Malabar. Concan. Belgaum.

(6) U. AFFINIS. (R. W.)

Ident. Oliver in Journ. Proc. Linn. Soc. III. p. 178.

Engrav. Wight's Icon. t. 1580. fig. 1.

SPEC. CHAR. Roots sparingly utriculiferous: leaves usually deciduous before flowering : scapes often erect, terete, few or many-flowered : scales few and with the bracts ovate or ovate-lanceolate : pedicels short, more or less arched, ascending, scarcely patent, never deflexed: lobes of the calyx somewhat equal, upper one broad-ovate or orbiculate-ovate, very acute or cuspidate, lower one ovate, much shorter than the subfalcate acutish spur, fructiferous lobes more or less round apiculated, larger lobe often suborbiculate ; upper lip of the corolla obovate-cuneate, subovate, or broad-oblong, entire or emarginate, lower lip often nearly equalling the spur: seeds more or less deeply scrobiculate: flowers blue or violet.

Var. 1. Scape at first few, afterwards 5-7-flowered; pedicels nearly equalling the calyx: upper lip of the corolla often more or less obovate, emarginate or entire.—U. deciptens, *Dals. in Kew Journ. Bot.* III. p. 279.

Var. 2. Scape 1–4-flowered: pedicels very short, scarcely exceeding the bract: upper lip of the corolla obovate, emarginate.— U. brachypoda, *Wight's Icon.* t. 1578-1.

(1) Concan. Neilgherries.——(2) Quilon.

(7) U. CÆRULEA. (*Linn.*)

Ident. Oliver in Journ. Proc. Linn. Soc. III. p. 179.—Linn. Herb. (not Dec.)

SPEC. CHAR. Leaves spathulate or linear-spathulate, usually evanescent before flowering: scape firm or slender, erect or somewhat twining, furnished with few ovate acute scales: fructiferous pedicels slender, erect, ascending, never deflexed, at last equalling or exceeding the calyx: upper lobe of the calyx ovate-acuminate, shorter than the upper obovate, or orbiculate-obovate, entire lips of the corolla; lobes of the fructiferous calyx ovate, acute or acuminate: seeds minute, reticulately striated: flowers blue.

Var. 1. Scape firm or slender: racemes finally elongated, flowers remote: fructiferous pedicels equalling or exceeding the calyx, erect or erecto-patent: upper lip of the corolla very often obovate, lower usually more or less ascending.—U. pedicellata, *Wight's Icon.* t. 1578. *f.* 2.—U. conferta, *Wight l. c. t.* 1575.

Var. 2. Scape often firmer furnished with numerous ovate, acute scales: flowers few, subterminal: pedicels straight, nearly equalling the calyx.—U. squamosa, *Wight l. c.*

Var. 3. Elongated: scape slender or twining: pedicels more slender, loosely-ascending or spreading archwise: upper lip of the corolla narrower.—U. uliginoides, *Wight l. c. t.* 1579.

(1) Mountains of the Peninsula. Concan. Neilgherries. Pulney Hills. Courtallum. (2) Sitpara. (3) Courtallum. Flowering in the rainy season.

(8) U. RETICULATA. (*Sm.*)

Ident. Oliver in Journ. Proc. Linn. Soc. III. p. 180.—Dec. prod. VIII. p. 19.—Wall. Cat. No. 1493.

Engrav. Sm. Exot. Bot. t. 119.

SPEC. CHAR. Scape twining, or firm and erect: leaves linear, often deciduous before flowering: upper lobe of the calyx ovate, very acute or cuspidate, shorter than the entire ovate, very obtuse upper lip of the corolla: spur subulate-dependent: calyx fructiferous, accrescent, margins often decurrent with the deflexed pedicel,

Var. 1. Scape very often twining, furnished with scales and ovate, acute or acuminated bracts: pedicels fructiferous, patent or more often deflexed, margined towards the apex, equally or much exceeding the calyx: flowers remote: lobes of the calyx fructiferous, much increasing, broad-ovate, acute: upper lip of the corolla ovate, round at the apex or ovate-orbiculate, entire, exceeding the upper lobe of the calyx and almost covering it with reflexed margins: lower lip large, galeate, entire or marginate, equally or often much exceeding the straight or scarcely curved subulate, acute spur: capsule ovate or elliptic covered by the enlarged calyx : seeds ovate or oblong, striated: flowers blue or violet, lower lip of the corolla towards the throat pale or whitish longitudinally and transversely streaked with coloured nerves: spur pale or whitish.—*Wight's Ill.* II. t. 143.—*Rheede Mal.* IX. t. 70.—U. uliginosa, *Wight's Icon.* t. 1574. *(partly.)*

Var. 2. Scape erect, firm : scales more or less acute ; fructiferous pedicels marginate, equalling or shorter than the enlarged calyx, a little exceeding the bract, ascending or deflexo-patent : lobes of the calyx finally elliptic, narrowed downwards : corolla smaller, the upper lip equalling or shorter than the calyx, lower one nearly equalling the spur, which itself is scarcely longer than the calyx.— U. uliginosa, *Vahl.*—U. humilis, *Heyne.*—U. polygaloides, *Edgw.*

(1) Neilgherries. Malabar. Mysore. Concans. Mangalore. Quilon. (2) Coromandel. Mysore. Bengal. Flowering in the rainy season:

(9) U. SCANDENS. *(Benj.)*

Ident. Benj. Mss. in Herb. Hook. (partly.)—Oliver in Journ. Proc. Linn. Soc. III. p. 181.

SPEC. CHAR. Very slender: scape twining, very thin, furnished with few, very minute, ovate scales: pedicels remote, often very short or equalling the calyx ; fructiferous ones usually deflexed: lobes of the calyx somewhat unequal, upper one shorter, ovate-orbiculate, almost equalling the very obtuse, entire or submarginate upper lip of the corolla: lower lobe ovate, shorter than the conical, subulate, dependent spur: lobes of the fructiferous calyx obtuse, nearly equalling the ovate or elliptic capsule: flowers blue, usually remote: seeds reticulately striated, scrobiculate.

Near the Mádura hills.

(10) U. BIFIDA. *(Linn.)*

Ident. Oliver in Journ. Proc. Linn. Soc. III. p. 182.

Syn. U. biflora, *Wall.*—U. diantha, *Dec. prod.* VIII. p. 21.—U. humilis, *R. W.*

Engrav. Wight's Icon. t. 1572. f. 2.

Spec. Char. Scape often erect, 2 or many-flowered, furnished with ovate, more or less acute scales and bracts: roots fibrous, sparingly utriculiferous: leaves linear-spathulate, none or almost none while flowering: pedicels short, marginate, fructiferous, nodding, shorter than the bract: upper lobe of the calyx ovate, slightly obtuse, a little shorter than the quadrate-oblong, ovate or subovate, entire upper lip of the corolla; lower lobe ovate, obtuse or 2-toothed, shorter than the spur; lower lip of the corolla entire or emarginate, shorter or nearly equalling the slightly falcately curved dependent spur: capsule globose or ovate, covered by the increasing lobes of the calyx: seeds usually ovate, obliquely striated: flowers yellow.

Malabar and Mysore. Chittagong. Silhet.

(11) U. WALLICHIANA. *(R. W.)*

Ident. Oliver in Journ. Proc. Linn. Soc. III. p. 182.

Engrav. Wight's Icon. t. 1572. f. 1.

Spec. Char. Roots utriculiferous: leaves almost none while flowering, linear or linear-spathulate: scape straight or twining, firm or slender, 1-2 or many-flowered, furnished with 1-4 minute, ovate or ovate-lanceolate, acute scales: pedicels erect or ascending; fructiferous ones more or less marginate towards the apex with the decurrent calyx: floriferous ones equalling or exceeding the spur: upper lobe of the calyx broader, very acute or cuspidate, lower one acute or minutely 2-toothed, often much shorter than the spur: upper lip of the corolla entire: capsule ovate or elliptic: stigma sessile: seeds minute, reticulately striated: flowers golden-coloured.

Var. 1. Scape slender or capillaceous, often twining: flowers remote: pedicels ascending or patent.—U. scandens, *Benj. in Linn.* XX. p. 309.

Var. 2. Scape 1-3-flowered: upper lip of the corolla entire or submarginate: spur often slightly exceeding the calyx.

(1) Courtallum Hills. Khasia mountains. (2) Neilgherries. Arcot. Dindigul Hills.

(12) U. MIXTA. *(Klein.)*

Ident. Oliver in Journ. Proc. Linn. Soc. III. p. 183.

Syn. U. setacea, *Wall. Cat. No.* 6398. *(not Michx.)*

Spec. Char. Scape slender, erect, 1-2 rarely 3-flowered: scales few and with the bracts and bracteoles narrow, linear-subulate: lobes of the calyx equal, ovate: pedicels bracts and calyx loosely subpatent, hairy or slightly so: pedicels very short scarcely exceeding the bract: upper lip of the corolla obovate, or quadrate-oblong, entire, very obtuse, often exceeding the calyx: spur conical, ascending, or stretched out, more or less long than the lower lip of the corolla: ripe capsule nearly equalling the calyx: stigma sessile: seeds reticulated: flowers blue or purple, veined, whitish or yellowish towards the throat and in the spur.

Khasia.

(13) U. ROSEA. *(Edgew.)*

Ident. Edgew. in Proc. Linn. Soc. I. p. 352.—Oliver in Journ. Proc. Linn. Soc. III. p. 184.

Syn. U. racemosa, *R. W.*—*Benj. in Linnæa* XX. p. 307.— U. cærulea, *Dec. prod.* VIII. p. 19.

Engrav. Wight's Icon. t. 1584. f. 1.

SPEC. CHAR. Roots fibrous, bladders few or none: leaves spathulate or linear-spathulate, often disappearing before flowering: scape sometimes 2-cleft, erect, slender or firm, 2-many-flowered: scales few or many: pedicels very short with minute lanceolate bracteoles more or less volute from the base: lobes of the purple calyx ovate-round, upper one more or less short than the upper round very obtuse entire lip of the corolla: spur stretched out, thick, cylindric-conical, very obtuse, lower lip shorter; lower lip of the corolla hooded, marked with four broad handsome streaks, covering the spur: capsule globose: seeds minute: flowers purple or rose-coloured?

Bengal. Pulney mountains.

(14) U. RACEMOSA. *(Wall.)*

Ident. Wall. Cat. No. 1496.—Oliver in Journ. Proc. Linn. Soc. III. p. 186.

SPEC. CHAR. Leaves linear-spathulate or spathulate: scape erect or sometimes weak: 1-2-many-flowered: upper lobe of the calyx slightly longer: upper lip of the corolla obovate ovate-oblong or quadrate-ovate, more or less abruptly obtuse, entire or submarginate, often twice as long as the upper lobe of the calyx: lower lip of the corolla hooded: spur stretched out, thick, conical-cylindric or conical: capsule globose: flowers variable in size, purple.

Chittagong. Silhet.

(15) U. NIVEA. *(Vahl?)*

Ident. Oliver in Journ. Proc. Linn. Soc. III. p. 186.—Wall. in Roxb. flor. Ind. I. p. 144.

SPEC. CHAR. Leaves linear-spathulate or spathulate: scape suberect, filiform, sometimes 2-cleft, fructiferous one elongated, twining: flowers small, 4–8: upper lip of the corolla short, linear, emarginate, lower one hooded, entire, almost twice as short as the conical ascendent spur: capsule globose: corolla white with a yellowish palate.

In rice fields near Serampore.

(16) U. ORBICULATA. *(Wall.)*

Idem. Wall. Cat. No. 1500.—Oliver in Journ. Proc. Linn. Soc. III. p. 187.
Syn. U. glochidiata, *R. W.*
Engrav. Wight's Icon. t. 1581.

SPEC. CHAR. Leaves orbiculate, petioled: scape very thin, 2–4-flowered, sometimes longer and many-flowered, naked or furnished with 1–2 scales: upper lip of the corolla very short, minutely 2-lobed or 2-toothed, shorter than the calyx, lower one 5-lobed; lobes obtuse, equal or sometimes with 2 upper small lobes and 3 lower smaller ones: spur dependent, linear-subulate, equalling the lower lip: seeds ovate or ovate-oblong, more or less armed with the lengthened or glochidiate cells of the testa: corolla lilac, with a yellow throat.

Concans. Cochin. Khasia.

(17) U. FURCELLATA. *(H. F. & T.)*

Idem. Oliver in Journ. Proc. Linn. Soc. III. p. 189.

SPEC. CHAR. Leaves orbiculate, petioled: scape very slender, 1–2, rarely 3-flowered: lower lip of the corolla often 4-lobed, 2 lower divisions equal, obtuse or submarginate; upper and lateral ones short, sometimes minute or almost wanting: spur dependent, subulate, 3–5 times longer than the lower lobe of the calyx, equalling or exceeding the lower lip of the corolla: seeds ovate, furnished at the thick end with the glochidiate or capitate cells of the testa: flowers white.

Khasia mountains.

ORDER XCVI. PRIMULACEÆ.

Annual or perennial herbaceous plants, sometimes almost shrubby: leaves usually radical, when cauline, whorled, opposite or alternate: stipules none: flowers either on radical scapes umbelled, or variously arranged in the axils of the leaves, or forming terminal racemes or spikes; calyx 5-seldom 4-cleft, inferior or half superior, regular, persistent; corolla monopetalous, hypogynous, regular, the limb 5-seldom 4-cleft: stamens inserted on the corolla, equal in number to its segments and opposite to them: ovary 1-celled; style 1; stigma capitate; ovules usually amphitropal, rarely anatropal: capsule opening with valves: placentæ central, distinct, seeds numerous, peltate; embryo included within a fleshy albumen, the radicle indeterminate or across the hilum.

GENUS I. PRIMULA.

Pentandria Monogynia. *Ser. Syst.*

Deriv. From *Primus*, first, because of the early appearance of the flowers.

GEN. CHAR. Herbs: leaves usually radical: scape simple: flowers umbellate, involucrate, rarely verticilled: calyx subcampanulate or tubulose, more or less deeply 5-toothed or even 5-cleft: corolla cup-shaped or funnel-shaped, limb 5-cleft, lobes usually emarginate, throat dilated at the limb, tube terete, equalling or exceeding the calyx: stamens included: filaments very short: anthers often acuminated: ovary globose or ovate-globose: ovules numerous peltately amphitropous: capsules ovate, 5-valved, valves entire or 2-cleft, only dehiscing at the apex: seeds minute, numerous.

(1) P. PROLIFERA. *(Wall.)*

Ident. Wall. in As. Res. 13. p. 372. in Roxb. flor. Ind. II. p. 18.—Dec. prod. VIII. p. 34.

SPEC. CHAR. Quite glabrous, not farinaceous: lower leaves rosulate, oblong-spathulate, obtuse, serrated, contracted into the petiole: scape long: flowers verticilled: leaflets of the involucre in the lower verticels somewhat of the same form as the leaves, in the upper ones linear-subulate: segments of the calyx lanceolate-acuminate, 2–3 times shorter than the tubes: corolla cup-shaped, lobes obcordate, crenulate, tube 10-striated: throat contracted and marked with five minute 2-lobed tubercles.

Silhet mountains, flowering from February to April.

GENUS II. MICROPYXIS.

Pentandria Monogynia. *Ser. Syst.*

Deriv. From *Micros*, small, and *Pyxis*, a capsule.

GEN. CHAR. Calyx 5-parted: corolla funnel-shaped, usually shorter than the calyx, persistent even to the ripening of the capsule, tube short, limb 5-parted, lobes narrow, acute, connivent after flowering: stamens equal: filaments slightly bearded, dilated at the base and combining into a connivent tube covering the ovary and inserted at the throat of the tube: anthers affixed at the base, erect, introrse: capsule globose, cut circularly round the sides, membranaceous; seeds numerous, very small.

(1) M. TENELLUS. (R. W.)

Ident. Wight's Icon. vol. IV.
Syn. Centunculus tenellus, *Dec. prod.* VIII. p. 72.—Lysimachia tenella, *Wall. Cat. No.* 1401. *(not Linn.)*
Engrav. Wight's Icon. t. 1585.

SPEC. CHAR. Small, erect, simple or branched from the base: branches erect; leaves broad-ovate, subacute, entire, subsessile or contracting into the petiole: flowers axillary; peduncles slender, shorter than the leaves: calycine lobes linear-lanceolate, acuminate-subulate, short, equalling the corolla: corolla deciduous, urceolate at the base: capsule equalling the calyx.

Pulney mountains, flowering in September.

GENUS III. ANAGALLIS.

Pentandria Monogynia. *Ser: Syst.*

Deriv. From *Anagelao*, to laugh. The power of removing despondency is attributed to this genus.

GEN. CHAR. Calyx 5-parted: corolla rotate, deciduous, deeply 5-parted; lobes broad, obtuse: stamens 5, inserted into the bottom of the corolla, free or rarely more or less united at the base: filaments bearded; anthers attached by the back near the base, more or less nodding, introrse: capsule globose, cut circularly round the sides, membranaceous: seeds numerous, angular, immersed in a central placenta.

(1) A. ARVENSIS. *(Linn.)*
Var. CAERULEA.

Ident. Roxb. flor. Ind. [Ed. Car.] II. p. 24.

SPEC. CHAR. Stem smooth, with procumbent, sharply 4-cornered branches: upper leaves frequently ternate, sessile, broad-ovate, acute, obscurely 3-nerved; lower ones remote: peduncles opposite, axillary, filiform, longer than the leaves, slightly thicker at the apex and there marked with 5 angles which run into the keel of the subulate, acuminate segments of the calyx: segments of the corolla obovate, strongly and unequally gland-crenulate: filaments covered with glandular hairs: capsule globular: flowers dark blue.

Common in all the northern parts of Hindostan. Neilgherries. Flowering in the cold season. The A. latifolia, *Linn.* *(Wight's Spicil. and Icon. t. 1205.)* is evidently a mere variety of the above.

GENUS IV. LYSIMACHIA.

Pentandria Monogynia. *Sex: Syst.*

Derив. From *Lysis,* dissolution, and *Mache,* strife.

GEN. CHAR. Herbs, usually perennial: leaves alternate, opposite or verticelled, entire: flowers axillary, racemose, spicate or panicled: calyx 5-parted: corolla 5-parted, subrotate or campanulate, longer than the calyx: stamens 5, inserted into the base of the corolla: filaments sometimes united at the base, sometimes as many sterile filaments as fertile ones: anthers oblong: capsula globose, 5–10-valved, dehiscing at the apex, many-seeded,

(1) L. LESCHENAULTII. *(Dec.)*

Ident. Dec. prod. VIII. p. 61.
Engrav. Wight's Icon. t. 1204.—Spicil. II. t. 132.

SPEC. CHAR. Erect, branched; leaves opposite or ternate, lanceolate, entire, acuminate, glabrous: flowers racemose, crowded: bracteoles linear-subulate, acuminated, much shorter than the pedicels: calyx much shorter than the campanulate corolla; divisions linear-lanceolate, acuminate: lobes of the corolla obovate, obtuse, entire: stamens equal, exserted: style filiform.

Neilgherries. In low marshy soils, flowering nearly all the year. The flowers on first opening are reddish white, streaked with darker lines, and afterwards acquire a rather deep lilac tinge.

(2) L. DELTOIDEA. *(R. W.)*

Ident. Wight's Ill. vol. II. p. 137.
Engrav. Wight l. c. t. 144.

SPEC. CHAR. Procumbent: extremities of the branches ascending, densely pubescent: hairs jointed: leaves subsessile, opposite or whorled, ovate, obtuse, sparingly sprinkled on both sides with jointed pubescence: peduncles axillary, solitary, about the length of, or exceeding the leaves: calyx-lobes lanceolate, perforated with numerous translucent orange-coloured glands: corolla rotate: filaments short, monadelphous at the base: anthers subtriangular: seed hispid: flowers yellow.

Neilgherries. Pulney mountains.

(3) L. RAMOSA. *(Wall.)*

Ident. Wall. Cat. No, 1490.—Dec. prod. VIII. p. 65.

SPEC. CHAR. Stem elongated, tetragonal, furrowed, branched: leaves alternate, lanceolate, entire, acuminated, short-petioled. lax: peduncles axillary, recurved, nearly equalling the leaves: segments of the calyx ovate-rotund, quite entire, abruptly acuminated: lobes of the corolla ovate-obtuse: stamens inserted at the base of the corolla, not monadelphous: anthers large, subsessile, erect: style filiform, elongated.

Pundua, Silhet.

ORDER XCVII. MYRSINACEÆ.

Trees or shrubs: leaves alternate, undivided, serrated or entire, coriaceous, smooth, stipules none, sometimes under-shrubs with opposite or ternate leaves; inflorescence in umbels, corymbs, or panicles, axillary or terminal: flowers bisexual or sometimes unisexual: calyx 4–5-cleft, persistent; corolla usually deeply 4–5-cleft, rarely 4-petaled, equal: stamens 4–5, opposite the lobes of the corolla, into the base of which they are inserted; filaments distinct, rarely connate, sometimes wanting; anthers attached by their emarginate base, two-celled, dehiscing longitudinally: ovary free, or partially adherent with a single cell and a free central placenta, in which is immersed the campolitropal ovules: style 1, short: stigma lobed or undivided; fruit drupaceous or baccate, usually 1-seeded sometimes with two or more; seeds angular or subglobose with a hollow hilum and simple integument: albumen copious horny, of the same shape as the seed: embryo taper, usually curved, lying across the hilum when the seed is solitary or inferior and touching the hilum when the seeds are numerous and lateral; cotyledons short.

GENUS I. MÆSA.

Pentandria Monogynia. *Ser: Syst:*

Deriv. From *Mæss,* the Arabic name of one of the species.

GEN. CHAR. Shrubs or trees, often hermaphrodite: leaves alternate: racemes axillary or rarely terminal, simple or composite at the base, many-flowered: flowers small, whitish: bracts persistent, small: bracteoles narrower, adpressed to the flower: calyx bibracteolate at the base, 5-lobed; lobes 2 outer, 3 inner: corolla 5-cleft, subcampanulate: lobes obtuse, in æstivation one lobe exterior, another interior, 3 middle ones, imbricately-convolute at the margin, all obtuse, inflexed at the apex: stamens included, free: filaments filiform: anthers ovoid-spherical, cordate, 2-celled: ovary adnate to the calyx, half superior in flower: placenta basilar within the calyx-tube; style short: stigma capitate, often obsoletely 3-5-lobed, lobes distinct, opposite the lobes of the calyx: berry covered by the calyx, ovoid.

(1) M. RAMENTACEA. *(Wall.)*

Ident. Wall. in Roxb. flor. Ind. II. p. 230.—Dec. prod. VIII. p. 77.

Syn. Bæobotrys ramentacea, *Roxb.*

SPEC. CHAR. Leaves ovate-lanceolate, acuminated, membranaceous: panicles axillary, somewhat shorter than the leaf, branched from the base, slightly glabrous: bracts lanceolate, acuminate, ciliated, shorter than the pedicel: bracteoles close to the flower, ovate: calycine lobes ovate; corolla a half longer than the calyx: flowers small, greenish yellow.

Silhet, flowering in March and April.

(2) M. MONTANA. *(Dec.)*

Ident. Dec. prod. VIII. p. 70.

Syn. Bæobotrys nemoralis, *Roxb. fl. Ind.* II. p. 232.—B. Indica, *Wall.*

SPEC. CHAR. Branches slightly glabrous: leaves ovate-oblong, acuminate, obtuse at the base, subrevolute at the margin, remotely denticulate: racemes axillary, twice as long as the petiole, puberulous, simple or branched at the base: bracts lanceolate-acuminate: bracteoles lanceolate, and with the ovate-acute calycine lobes puberulous: corolla 3 times larger than the calyx; lobes obovate, subciliated: ovary half-superior, hemispherical above: stigma somewhat lobed: flowers small, white.

Silhet, flowering in March.

(3) M. INDICA. *(Dec.)*

Ident. Dec. prod. VIII. p. 80.

Syn. Bæobotrys Indica, *Roxb. fl. Ind.* II. p. 230.—M. Perottetiana, *Dec. prod.* VIII. p. 80.

Engrav. Wight's Icon. t. 1200.—Spicil. II. t. 134.

SPEC. CHAR. Leaves ovate-elliptic, acuminated, coarsely toothed, membranaceous, subrevolute at the margin: racemes axillary and terminal, simple or branched at the base, twice as long as the petiole: bracts lanceolate-acuminate: bracteoles ovate-acute: calycine-lobes ovate, subciliated: corolla 5-cleft, 3 times larger than the calyx: lobes obovate, subciliated, patent: ovary hemispherical above: stigma capitate, somewhat lobed: corolla white, with purplish veins.

Neilgherries. Silhet. Chittagong. Flowering in March and April.

(4) M. DUBIA. *(Wall.)*

Ident. Wall. in Roxb. fl. Ind. II. p. 235.—Dec. prod. VIII. p. 81.

SPEC. CHAR. Branches, petioles, and nerves of the younger leaves slightly hirsute: leaves ovate-acute, coarsely toothed, obtuse at the base, membranaceous, glabrous above: racemes axillary, composite, somewhat shorter than the leaf: bracts lanceolate-acuminate: bracteoles ovate-acute: calycine lobes ovate, round, subciliated: corolla twice as large as the calyx: lobes obovate, patent: ovary half-superior; style very short: stigma obtuse.

Wynaad and Coorg.

(5) M. GLABRA. *(Roxb.)*

Ident. Roxb. fl. Ind. II. p. 233.—Dec. prod. VIII. p. 82.

SPEC. CHAR. Small tree: branches straight, smooth: leaves ovate-lanceolate, entire: panicles axillary.

Chittagong, flowering in March.

GENUS II. EMBELIA.

Pentandria Monogynia. *Sor: Syst:*

Deriv. The Cinghalese name latinised.

GEN. CHAR. Climbing shrubs or small trees: leaves alternate, often entire, petioles frequently marginate and denticulate: panicles or racemes many-flowered: flowers often almost diœcious: calyx 5-parted or deeply 5-cleft, lobes convolute to the left in æstivation: petals 5, patent or reflexed often quincuncial in æstivation, 2 outer, 3 inner, rarely convolute at the left: filaments connate at the base with the opposite petal, filiform above: anthers much shorter than the filament, ovoid, at the base and sometimes at the apex emarginate, 2-celled, cells ovate, longitudinally dehiscing, subdivided with a longitudinal membranaceous nerve: ovary ovoid, often very small; style short: stigma included, capitellate, sublobed: ovules 4-1, often abortive, inserted on a central (often very small or almost wanting) placenta: drupe globose: seed single, not filling the cavity of the pericarp.

(1) E. RIBES. (Burm.)

Ident. Burm. flor. Ind. p. 62.—Dec. prod. VIII. p. 85.
Syn. E. Ribesoides, *Linn. flor. Zeyl. No.* 403.
Engrav. Burm. l. c. t. 23.

SPEC. CHAR. Climbing shrub: branches glabrous; leaves ovate or obtusely acuminate, glabrous, entire, coriaceous, short-petioled: panicle much branched, many-flowered, puberulous, much longer than the leaves: pedicels longer than the calyx: lobes of the calyx ovate-acute: petiole elliptic, puberulous, ciliated: flowers small, green-yellowish: drupes white.

Peninsula: Western Ghauts. Flowering in February and March.

Var. Silhetensis, *Dec. l. c.*—E. Ribes, *Wall. in Roxb. for. Ind.* II. p. 285.—Leaves membranaceous, ovate-lanceolate, pellucid-dotted, dots reddish; petioles often winged-denticulate; drupes black. Eastern Bengal towards Silhet. With the fruit of this species the natives adulterate black pepper. *(Roxb.)*

(2) E. FLORIBUNDA. *(Wall.)*

Ident. Wall. in Roxb. flor. Ind. II. p. 291.—Dec. prod. VIII. p. 85.—E. esculenta, *Don. prod. for. Nep.* p. 147.

SPEC. CHAR. Climbing: branches glabrous: leaves lanceolate, acuminated, obtuse at the base, glabrous, membranaceous, pellucid-dotted and with the larger dots situated at the margin: petioles crisp-margined: panicles many-flowered, slightly hairy, somewhat shorter than the leaf: calycine lobes ovate-acute: petals spreading, lanceolate, four times longer than the calyx: flowers yellowish, green, fragrant: berries red.

Silhet, flowering from December to February.

(3) E. VILLOSA. *(Wall.)*

Ident. Wall. in Roxb. flor. Ind. II. p. 290.—Dec. prod. VIII. p. 85.

SPEC. CHAR. Climbing: branches covered with cinnamon pubescence: leaves elliptic-obovate, acute at the base, slightly glabrous above, below and with the petioles velvety: racemes slender, the length of the leaves: lobes of the calyx lanceolate, acute: petals lanceolate, reflexed, four times longer than the calyx, papillosely puberulous; flowers small, whitish, very numerous.

Rajmahal, flowering in the hot season.

(4) E. PICTA. *(Alph. Dec.)*

Ident. Alph. Dec. in trans. Linn. Soc. 17. p. 130.—Dec. prod. VIII. p. 86.

SPEC. CHAR. Small tree: branches and peduncles covered with

velvety cinnamon pubescence: leaves oval, membranaceous, remotely subdenticulate: racemes solitary or twin, slightly longer than the leaf: calycine teeth ovate-acute; petals reflexed, oblong, velvety at the margin, four times longer than the calyx.

Southern parts of the Peninsula. Goalpara.

(5) E. ROBUSTA. *(Roxb.)*

Ident. Roxb. flor. Ind. II. p. 267.—Dec. prod. VIII. p. 86.

SPEC. CHAR. Shrub: younger branches puberulous: leaves ovate-oblong, entire, pubescent below, short-petioled: racemes many times shorter than the leaf: calycine lobes ovate, hairy; petals ovate, revolute, hairy, three times longer than the calyx: flowers greenish-white.

Rajmahal Hills, flowering in the rainy season.

(6) E. PARVIFLORA. *(Wall.)*

Ident. Dec. prod. VIII. p. 86.

SPEC. CHAR. Small tree: branches velvety-cinnamon: leaves approximated, bifarious, ovate-acute, entire, short-petioled: nerves and petioles puberulous below: racemes velvety, 4 times shorter than the leaf: flowers approximated, shorter than the pedicel: lobes of the calyx ovate-acute, puberulous: petals oblong, 5 times larger than the calyx.

Silhet.

(7) E. VESTITA. *(Roxb.)*

Ident. Roxb. fl. Ind. II. p. 288.—Dec. prod. VIII. p. 86.

Engrav. Delesa. Icon. L. 30.

SPEC. CHAR. Climbing shrub: branches glabrous: leaves ovate-lanceolate, acute, subserrate, glabrous, obtuse at the base: petioles denticulate: racemes puberulous, a half shorter than the leaf: bracteoles 4 times shorter than the pedicel: lobes of the calyx ovate-acute; petals lanceolate, patent, glabrous, much longer than the calyx: berries red.

Silhet, flowering in January.

(8) E. NUTANS. *(Wall.)*

Ident. Wall. in Roxb. fl. Ind. II. p. 290.—Dec. prod. VIII. p. 87.

SPEC. CHAR. Shrubby: branches pendulous, diffuse, velvety at the apex: leaves ovate-acuminate, approximated, entire, short-petioled, glabrous above, nerves below and petioles puberulous: racemes many times shorter than the leaf, pubescent: pedicels somewhat longer than the flower: calycine lobes acute, puberulous: petals obovate, spreading: flowers greenish-white.

Silhet, flowering in December and January.

(9) E. BADAAL. (*Alph. Dec.*)

Ident. Alph. Dec. in trans. Linn. Soc. XVII. p. 131.—Dec. prod. VIII. p. 87.

Engrav. Rheede Mal. V. t. 12.

Spec. Char. Shrub: leaves approximated at the apex of the branches, ovate, acute, entire: racemes lateral, 3 times shorter than the leaf: petals expanded, acuminate: flowers very small, greenish-yellow: berries globose, red when ripe, sweet.

Malabar. Travancore. Vingorla. Flowering in the hot season.

(10) E. TSJERIAM-COTTAM. (*Alph. Dec.*)

Ident. Dec. prod. VIII. p. 87.

Engrav. Rheede Mal. V. t. 11.—Wight's. Icon. t. 1200.

Spec. Char. Shrub; leaves oval, entire, coriaceous, short-petioled: racemes terminal and axillary, solitary twin or in threes, a half shorter than the leaf: petals subrotund; flowers fragrant, greenish-brown: drupes fleshy, acid, red.

Malabar.

(11) E. GLANDULIFERA. (*R. W.*)

Ident. Wight's Icon. vol. IV.—Dals. Bomb. flor. p. 137.

Engrav. Wight's Icon. t. 1207.

Spec. Char. Shrub, glabrous; leaves ovate-lanceolate, obtusely acuminate, entire, furnished with numerous glands on either side of the midrib: panicles axillary, sometimes reduced to a simple raceme: lobes of the calyx ovate-acute: petals elliptic, puberulous: fruit small, globose.

Neilgherries. Belgaum. Western Ghauts.

(12) E. GARDNERIANA. (*R. W.*)

Ident. Wight's Icon. vol. IV.

Engrav. Wight's Icon. t. 1208.

Spec. Char. Diffuse shrubs: young branches and petioles covered with rusty-coloured hairs: leaves ovate, rounded at the base, crenulato-serrated, coriaceous, glabrous; peduncles axillary, short, rusty-tomentose: racemes capitulate: pedicels about as long as the peduncles, glabrous: calyx much shorter than the corolla: petals obovate, obtuse, yellowish, sprinkled with purplish coloured spots.

Sisparah jungles, rare, flowering in February and March.

GENUS III. SAMARA.

Tetrandria Monogynia. *Sex. Syst.*

GEN. CHAR. Calyx minute, 4-parted, acute, persistent: petals 4, ovate, sessile, with a longitudinal furrow at the base: filaments long-subulate, immersed in the furrow: anthers subcordate: stigma funnel-shaped: drupe round: seed solitary.

(1) S. AURANTIACA. (*R. W.*)

Ident. Wight's Spicil. II. p. 32.

Syn. Choripetalum aurantiacum, *Dec. prod.* VIII. p. 88.—Myrsine aurantiaca, *Wall. in Roxb. flor. Ind.* II. p. 300.

Engrav. Wight's Spicil. II. t. 136.—Icon. t. 1210.

SPEC. CHAR. Climbing: leaves ovate-lanceolate, subacute at both ends, entire, coriaceous, long-petioled: racemes much shorter than the leaves, longer than the petioles: bracts acuminated, as long as the pedicels: petals linear-lanceolate, reflexed: filaments longer than the petals, much larger than the anthers: flowers bright orange.

Neilgherries. Malabar. Flowering in the dry season.

(2) S. PANICULATA. (*Roxb.*)

Ident. Roxb. flor. Ind. I. Ed. Car. p. 435.—Wight's Spicil. II. p. 32.

Syn. Ardisia paniculata, *Dec. prod.* VIII. p. 139.

SPEC. CHAR. Tree: leaves opposite, petioled, broad-lanceolate, acuminated, entire, glabrous: corymbs axillary, small, peduncled, shorter than the leaves, divided by three forked divisions: drupe dry, oblong.

Circar mountains.

(3) S. RHEEDII. (*R. W.*)

Ident. Wight's Icon. vol. IV.

Engrav. Wight l. c. t. 1591.—Rheede Mal. 7. t. 42.

SPEC. CHAR. Scandent shrub: floriferous branchlets sub-bifarious, ascending: leaves petioled, ovate-elliptic, entire, subacuminate, glabrous, coriaceous: spikes axillary, usually solitary, numerous towards the extremities of the branchlets: flowers short pedicelled, each furnished with a small ovate bract: calyx lobes broad-ovate, dentate on the margin: lobes of the corolla scarcely cohering at the base, ovate-obtuse, ciliated towards the apex: stamens scarcely exceeding the petioles: anthers glanduloso-cuspidate: drupe about the size of a pea, 1-seeded: flowers greenish-white.

Malabar. Neilgherries. Flowering in the rainy season.

GENUS IV. AMBLYANTHUS.

Pentandria Monogynia. *Sex. Syst.*

Derio. From *Amblys*, obtuse, and *Anthos*, a flower, alluding to the very obtuse flower-bud.

GEN. CHAR. Calyx 5-cleft, tube funnel-shaped, lobes ovate-acute; corolla 5-cleft, tube cylindric, lobes round, reflexed, twisted to the left in aestivation: filaments very short, inserted at the lowest base of the corolla: anthers many times longer than the filament, connate above into a convex tube, free at the base and apex, included, longitudinally dehiscing inwards by two clefts: ovary ovoid, 1-celled: placenta central; ovules few (5–3) immersed: style cylindric: stigma obtuse, depressed in the middle, obscurely 4–5-cornered, not exceeding the anthers.

(1) A. GLANDULOSUS. *(Dec.)*

Icon. Dec. prod. VIII. p. 91.

Syn. Ardisia glandulosa, Roxb. *flor. Ind.* II. p. 282.—A. Roxburghiana, *Dietr. Syn.* I. p. 617.

SPEC. CHAR. Shrub, quite glabrous: branches diffuse: leaves lanceolate, acuminated at both ends, glanduloso-crenate, petioled, peduncles 7–8, subumbellate at the apex of the branches, unequally shorter than the leaf: pedicels 4–6 at the apex of the peduncles; umbellate, somewhat longer than the flower: flower-buds pear-shaped, obtuse: flowers small, white.

Silhet, flowering in the rainy season.

GENUS V. HYMENANDRA.

Pentandria Monogynia. *Sex. Syst.*

Derio. From *Hymen*, a membrane, and *Aner, andros*, male, alluding to the anthers.

GEN. CHAR. Calyx 5-parted, lobes ovate-acute, not valvate in aestivation, quincuncial; corolla 5-parted, rotate, lobes lanceolate-acuminate, elongated, twisted to the left in aestivation: anthers sessile, produced laterally and upwards into connate membranes, free, obtuse at the apex: membranaceous tube of the anthers twisted to the left, twice as long as the intorse cells, equalling the corolla: ovary ovoid, angular: style filiform: stigma slightly obtuse: ovules numerous.

(1) H. Wallichii.

Ident. Dec. prod. VIII. p. 91.
Syn. Ardisia hymenandra, *Wall. in Roxb. flor. Ind.* II. p. 288.

Spec. Char. Glabrous shrub: leaves obovate, large, sessile, dotted, crenated, entire at the base: peduncles lateral, a half shorter than the leaf: bracts leafy, 2–3, oblong, toothed, verticellate at the apex of the peduncles: branches of the panicle beyond the bracts shorter than the common peduncle, compressed: pedicels umbellate, somewhat longer than the flower, angular: flowers largish, pink.

Juntipoor mountains. Silhet. Flowering in March.

GENUS VI. ANTISTROPHE.

Pentandria Monogynia. *Sex : Syst :*

Deriv. From *Anti*, contrary, and *Strophe*, turning, in allusion to the æstivation.

Gen. Char. Calyx 5-partite: lobes (æstivation unknown) already spreading before the corolla has opened, lanceolate, acuminate, subciliated: corolla 5-parted, much longer than the calyx, glabrous, narrow-lanceolate, acuminate, imbricato-convolute to the right in æstivation: stamens as long as the corolla: filaments very short: anthers subsessile, free, produced at the apex and beyond the cells and laterally into thin acuminated membranes which are somewhat longer than the cells: ovary free, ovoid: style filiform, the length of the corolla: stigma spherical, minute: ovules and fruit unknown.

(1) A. oxyantha. *(Dec.)*

Ident. Dec. prod. VIII. p. 92.

Spec. Char. Small tree (?) leaves lanceolate, acuminate at both ends, thin, glabrous, entire: leaf-buds of the bracts floriferous, axillary, 1–2-flowered: bracts imbricated, somewhat velvety, subulate: pedicels slender, compressed, smooth, somewhat equalling the petiole; flower-buds acuminate.

Silhet mountains.

GENUS VII. MYRSINE.

Polygamia Diœcia. *Sex : Syst :*

Deriv. The Greek term for myrrh.

Gen. Char. Shrubs or small trees, with alternate, entire, rarely-toothed leaves: fascicles of flowers axillary: flowers often 4–5-

androus is the same plant, the males larger: calycine lobes often unequal: segments of the corolla inflexed in the flower-bud: stigma in the female flowers sometimes large and coloured: calyx 4–5-cleft: corolla 4–5-parted, lobes with imbricated subquincuncial æstivation, 2 outer ones, 2–3 inner ones or rarely valvular: stamens free, sometimes longer than the corolla: filaments very short, inserted at the base of the corolla: anthers 2-celled, erect, lanceolate, glanduloso-acute, somewhat 2-lobed at the base, much longer than the filaments, the cells dehiscing lengthwise from the base towards the apex: ovary globose: style cylindric, short, caducous: stigma capitate, papillose, irregularly lobed or fimbriated: placenta spherical, frequently depressed at the apex: ovules 4–5, peltately amphitropal round the vertex of the placenta: drupe pea-shaped, with crustaceous putamen, 1-seeded by abortion.

(1) M. CAPITELLATA. *(Wall.)*

Ident. Wall. in Roxb. flor. Ind. II. p. 295.—Dec. prod. VIII. p. 94.

Engrav. Hook. Bot. Mag. t. 3222.

SPEC. CHAR. Tree: leaves elliptic-obovate, entire, coriaceous, smooth, narrowing into the petiole; fascicles numerous, 5–8-flowered, bracteate; bracts imbricated, ovate: teeth of the calyx ciliated: lobes of the corolla lanceolate, acute, 2–3-times longer than the calyx, exceeding the stamens: flowers short-pedicelled.

Var. 1. Parvifolia, *Alph. Dec.*—Leaves smaller; fascicles few-flowered. Silhet.

Var. 2. Grandiflora, *Dec. l. c. Wight's Icon. t.* 1211. *Spicil. t.* 137.—Leaves smaller: lobes of the corolla 4 times longer than the teeth of the calyx: flowers white. Ootacamund, flowering in February and March.

GENUS VIII. ARDISIA.

Pentandria Monogynia. *Bur: Syst:*

Deriv. From *Ardis*, a spear-point, alluding to the acute segments of the corolla.

GEN. CHAR. Trees, shrubs or undershrubs: leaves alternate, rarely opposite or ternate, punctuate, entire or serrated: flowers panicled, rarely racemose, peduncles terminal or axillary: corolla white or rose-coloured: drupes usually purple: calyx 5-parted: corolla 5-parted or cleft, lobes spreading or reflexed: æstivation of both tending to the left: stamens inserted into the base of the tube of the corolla: filaments free, usually short: anthers free, erect, emarginate or bifid at the base, often triangular, acuminate: cells dehiscing longitudinally: ovary rounded, 1-celled: style filiform, subulate at the apex: placenta central, spherical: ovules numerous,

8–12, peltate: drupe globose, usually fleshy externally, usually smooth, coriaceous, hard within: seed single.

(1) A. HUMILIS. *(Vahl.)*

Ident. Vahl. Symb. III. p. 40.—Dec. prod. VIII. p. 129.

Syn. A. solanacea, *Roxb. Cor.* I. p. 27.—A. litoralis, *Andr.*— A. umbellata, *Roxb. flor. Ind.* I. p. 582.—Anguillaria Zeylanica, *Gærtn. fr.* I. p. 373.

Engrav. Roxb. l. c. t. 27.—Andr. Bot. Rep. X. t. 630.—Bot. Mag. XL. t. 1677.—Burm. Zeyl. t. 103.—Wight's Icon. t. 1212.—Spicil. II. t. 138.

SPEC. CHAR. Shrub or small tree, sometimes 20 feet high: leaves obovate-lanceolate, obtuse, somewhat entire, coriaceous, contracted at the base into the petiole: racemes umbelliform, axillary and terminal, reflexed, shorter than the leaves: lobes of the calyx orbiculate, subciliate: lobes of the corolla lanceolata, subacute, twice the length of the calyx: flowers rose-coloured or light purplish.

Subalpine forests on the Eastern slopes of the Neilgherries near the banks of streams. Bengal. Western Ghauts. Coromandel Coast. Flowering nearly all the year.

Var. Leaves acute, membranaceous: peduncle and pedicels elongated, few-flowered. Silhet.

(2) A. PANICULATA. *(Roxb.)*

Ident. Roxb. flor. Ind. II. p. 270.—Dec. prod. VIII. p. 126.

Engrav. Bot. Mag. 50. t. 2364.—Bot. Reg. 8. t. 638.

SPEC. CHAR. Shrub or small tree: leaves obovate-oblong, subsessile, coriaceous; panicles terminal, ovoid, somewhat longer than the leaf, many-flowered: peduncles spreading, cylindric: pedicels reflexed, longer than the flower: calycine lobes ovate-obtuse: lobes of the corolla ovate-acuminate, five times longer than the calyx: drupe red: flowers smallish, rose-coloured.

Chittagong. Silhet. Flowering in April.

(3) A. ANCEPS. *(Wall.)*

Ident. Wall. in Roxb. fl. Ind. II. p. 280. (not Blume.)—Dec. prod. VIII. p. 126.

SPEC. CHAR. Shrub: branches 2-edged at the apex, climbing: leaves oblong-lanceolate, acute at both ends, narrowing at the base into the petiole: panicles terminal, somewhat shorter than the leaf, elongated: rachis and peduncles compressed, smooth: pedicels shorter than the flower: lobes of the calyx ovate-acute, somewhat pubescent, of the corolla ovate-acuminate, much longer than the calyx: flowers rose-coloured.

Silhet mountains.

(4) A. QUINQUANGULARIS. *(Alph. Dec.)*

Ident. Dec. prod. VIII. p. 127.

SPEC. CHAR. Shrub: leaves lanceolate-acute, narrowed into the petiole; panicle terminal, a half shorter than the leaf: peduncles erect, somewhat compressed: pedicels alternate; lobes of the calyx ovate-acute: drupe globosely 5-sided.

Banks of the Ganges.

(5) A. PAUCIFLORA. *(Heyne.)*

Ident. Dec. prod. VIII. p. 127.—Roxb. flor. Ind. II. p. 270.

SPEC. CHAR. Shrub: leaves long-elliptic, narrowed at both ends, broken at the extremity, entire, thin; racemes axillary, few, few-flowered, smooth, 3 times shorter than the leaf: pedicels umbellate, much longer than the flower: lobes of the calyx ovate-acute, subciliated: corolla 4 times longer than the calyx.

Neilgherries.

(6) A. NERIIFOLIA. *(Wall.)*

Ident. Dec. prod. VIII. p. 127.

Engrav. Alph. Dec. trans. Linn. Soc. XVII. t. 6.

SPEC. CHAR. Shrub: leaves oblong-lanceolate, acuminated, long-petioled: panicles axillary and terminal, loose, shorter than the leaf, somewhat velvety: calycine lobes ovate-acute, ciliated, spreading: segments of the corolla lanceolate-acuminate, much longer than the calyx.

Silhet mountains.

(7) A. PEDUNCULOSA. *(Wall.)*

Ident. Wall. in Roxb. fl. Ind. II. p. 279.—Dec. prod. VIII. p. 128.

SPEC. CHAR. Glabrous: leaves frequent at the apex of the branches, lanceolate, acuminated at both ends: panicles terminal and axillary, shorter than the leaf: pedicels alternate or umbellate, much longer than the flower: calycine lobes ovate-acuminate, spreading: segments of the corolla acuminate, much longer than the calyx: flowers small, red.

Silhet.

(8) A. BUGENIÆFOLIA. *(Wall.)*

Ident. Dec. prod. VIII. p. 130.

SPEC. CHAR. Shrub: leaves oblong-lanceolate, acute at the base, acuminate at the apex, coriaceous, lateral nerves near the margin arched and connected: panicles axillary, longer than the petioles ; pedicels the length of the flowers : peduncles and calyx velvety : calycine lobes ovate-acute ; corolla campanulate, 5-cleft.

Silhet mountains.

(9) A. MEMBRANACEA. *(Wall.)*

Ident. Dec. prod. VIII. p. 134.—Alph. Dec. trans. Linn. Soc. 17. p. 123.

SPEC. CHAR. Undershrub: leaves approximated, oblong-lanceolate, acuminated, narrowed at the base into the petiole, membranaceous, irregularly crenated; peduncles axillary, much shorter than the leaf; flower-buds obtuse; calycine lobes ovate-acute, almost equal to the segments of the corolla.

Silhet mountains.

(10) A. ODONTOPHYLLA. *(Wall.)*

Ident. Dec. prod. VIII. p. 135.

Engrav. Alph. Dec. trans. Linn. Soc. 17. t. 6.

SPEC. CHAR. Perennial, shrubby: leaves lanceolate-oblong, acute at both ends, long-petioled, sharply and frequently toothed, puberulous; racemes axillary, much shorter than the leaf; pedicels short, alternate and with the peduncles velvety; lobes of the calyx ovate-acute, ciliated, puberulous, of the corolla deeply parted, segments ovate-acute, much longer than the calyx.

Eastern Bengal towards Silhet. Goalpara.

(11) A. ICANA. *(Wall.)*

Ident. Dec. prod. VIII. p. 136.

Engrav. Alph. Dec. trans. Linn. Soc. 17. t. 7.

SPEC. CHAR. Shrub: leaves oblong-lanceolate, acuminate at the base, acute at the extremity, denticulate, slightly hairy above, smooth below; panicles terminal or axillary, somewhat equalling the leaf, very slightly puberulous: lobes of the calyx subulate, pubescent, of the corolla ovate-acute, longer than the calyx.

Bengal.

(12) A. RHOMBOIDEA. *(R. W.)*

Ident. Wight's Icon. vol. IV.

Engrav. Wight l. c. t. 1213.

SPEC. CHAR. Shrub: leaves rhomboidal, acuminated, narrowed into the petiole, glabrous, slightly crenulately undulated on the margin; racemes axillary, much shorter than the leaves, few-flowered; pedicels umbellate; lobes of the calyx ovate, subciliated, shorter than the corolla; lobes of the corolla broad-ovate, acute; fruit globose.

Shevagherry mountains, flowering in August.

(13) A. COURTALLENSIS.

Icon. Wight's Icon. vol. IV.
Engrav. Wight l. c. t. 1215.

Spec. Char. Shrub: leaves obovate-cuneate, bluntly acuminate, entire, subsessile: panicles longer than the leaves, terminal, lax, branches umbellate, few-flowered: pedicels umbellate, 4-5 times longer than the flowers, spreading: lobes of the calyx ovate, pointed, ciliated, of the corolla ovate, subcuspidate.

Courtallum, flowering in August and September. Very closely allied to *A. paniculata*.

(14) A. SERRATIFOLIA. (R. H. B.)

Ident. Beddome in Madr. Journ. Lit. Ser. III. No. I. p. 51.

Spec. Char. Shrubby: leaves short-petioled, narrow-lanceolate, attenuated at both ends, long-acuminated, very sharply serrated, nearly glabrous above; young branches petioles and under surface of the leaves rufo-tomentose: peduncles axillary: pedicels 2-5, slender, glabrous: calyx pubescent: petals gland-dotted: style long, slender.

Moist woods on the Annamullay Hills.

ORDER XCVIII. ÆGICERACEÆ.

Small trees or shrubs inhabiting salt marshes near the sea coast; leaves alternate, exstipulate, coriaceous, entire, obtuse, marked with minute depressed points; flowers usually umbelled, white, fragrant; calyx and corolla 5-parted, imbricato-convolute to the left in æstivation; stamens 5, attached to the base of the corolla; filaments united at the base into a short tube; anthers versatile, 2-lobed at the base, opening longitudinally, pollen lodged in a double row of cells or alveolæ; ovary free, fusiform, 1-celled; central placenta ovoid, shortly stipulate; style filiform, persistent; stigma acute; ovules numerous, not immersed in the placenta, all except 1, aborting but persistent, the fertile ovule rapidly growing after the fall of the corolla, long-ellipsoid, exceeding the dry compressed placenta; shortly after the ovule begins to germinate at the base, the elongation of the radicle stretches the stipe of the placenta, which finally becomes the apparent, but false funiculus, while the persistent placenta forms the hood covering the cotyledons at the apex of the mature seed; seed exalbuminous, curved: embryo cylindrical, curved, thick; radicle inferior, sulcated; pericarp elongated, arched, without nerve or suture, longitudinally striated, somewhat fleshy within, punctuate with resinous matter, at length splitting lengthwise on one or both sides.

GENUS ÆGICERAS.

Pentandria Monogynia. *Sex: Syst:*

Deriv. From *Aix, Aigos*, a goat, and *Keras*, a horn, alluding to the form of the fruit.

Gen. Char. Same as that of the Order.

(1) Æ. MAJUS. (*Gærtn.*)

Ident. Dec. prod. VIII. p. 142.—Gærtn. ft. I. p. 216.—Roxb. fl. Ind. III. p. 130.

Syn. Æ. fragrans, *Koen.*—Æ. obovatum, *Blume.*—Æ. floridum, *Roem. & Schult.*—Rhizophora corniculata, *Linn.*

Engrav. Hook. Bot. Misc. III. t. 21.—Rheede Mal. 6. t. 36.—Koen. Ann. Bot. I. t. 3.—Wight's Ill. II. t. 146.—Rumph. Amb. III. t. 77.

Spec. Char. Milky shrub: leaves obovate, rounded, obtuse, often retuse; flowers in terminal umbels, pure white, fragrant: fruit elongated, falcate, 3-4 times longer than the pedicels.

Common in salt marshes in Malabar and Eastern Coast, Soonderbuns. Flowering in the hot season.

ORDER XCIX. SAPOTACEÆ.

Trees or shrubs often abounding in milky juice; leaves alternate or almost whorled, entire, coriaceous: inflorescence axillary: flowers hermaphrodite: calyx 5-or occasionally 4-8-lobed, valvate or imbricate in æstivation: corolla monopetalous, hypogynous, regular, deciduous, its segments usually equal in number to those of the calyx, seldom twice or thrice as many, imbricated in æstivation; stamens inserted on the corolla, distinct; usually partly fertile, partly sterile, the former equalling the number of lobes and opposite them, the latter alternate, sometimes twice as many, rarely all fertile; anthers extrorse, ovary superior, several-celled, cells usually opposite the lobes of the calyx, with a single erect or suspended ovule in each; style one; stigma undivided or occasionally lobed; fruit fleshy with several one-seeded cells, or, by abortion, only one; seeds nut-like, sometimes adhering to a several-celled putamen; testa bony, shining, with a long scar on the inner face; embryo erect, large, white, usually enclosed in fleshy albumen; cotyledons, when albumen is present, foliaceous, when absent, fleshy and sometimes connate; radicle short, straight, or a little curved, turned towards the hilum.

GENUS I. CHRYSOPHYLLUM.

Pentandria Monogynia. *Sex. Syst.*

Deriv. From *Chrysos*, gold, and *Phyllon*, a leaf, alluding to the under surface of the leaves being covered with bright yellow hairs.

GEN. CHAR. Trees with entire alternate leaves, usually silky-tomentose below: pedicels axillary, umbellately fascicled, shorter than the petiole, and with the calyx usually silky-ferrugineous: calyx 5-rarely 6-parted: lobes obtuse, pubescent, imbricated in aestivation, 2 exterior: corolla tubulose or campanulato-rotate, 5-rarely 6–7-lobed, imbricated in aestivation: stamens 5-rarely 6–7, opposite the lobes of the corolla, inserted on the tube; filaments slender: anthers 2-celled, often laterally dehiscing, subintrorse, included, equalling the filament in length, ovoid or lanceolate: ovary 5–9-celled, ovoid, hirsute: style usually shorter than the ovary: stigma obtuse or rarely acute: ovules solitary, in the inner angle of each cell, ascending: berry (by abortion) few or 1-celled, few or 1-seeded: seeds bony, erect, ovoid-acute, subcompressed: albumen small: embryo large, erect: cotyledons thick, smooth, ovate: radicle short.

(1) C. ROXBURGHII. (*G. Don.*)

Ident. Don. Syst. gard. IV. p. 33.—Dec. prod. VIII. p. 162.

Syn. C. acuminatum, *Roxb. fl. Ind.* II. p. 345.

SPEC. CHAR. Tree: leaves lanceolate, long acuminated, smooth: pedicels axillary, fascicled, recurved, the length of the petiole, younger ones tomentose: lobes of the calyx ovate, ciliated: corolla 5-cleft, the length of the calyx, smooth outside: lobes ovate, ciliolate: ovary very hispid: berry yellow, spherical, the size of an apple: flowers small, pale-yellow.

Khasia mountains. Assam. Chovia Ohaut and Soonda jungles, (Western coast.) Flowering in April and May.

GENUS II. SAPOTA.

Pentandria Monogynia. *Sex, Syst.*

Deriv. The name of the fruit *Sapotiller, Seppodilla, Zapota,* and hence *Sapota.*

GEN. CHAR. Milky trees: branches sometimes spinous: leaves alternate, entire, coriaceous: flowers axillary: berry like an apple, fleshy: sepals 5–6, obtuse, imbricated: corolla tubuloso-campanulate, 5–6-lobed, with as many epipetalous scales (sterile stamens) inserted on the tube, alternate with its lobes: stamens 5–6 opposite the lobes of the corolla below the scales: anthers extrorse, 2-celled, dehiscing lengthwise: ovary ovoid, hairy, 5–12-celled: style cylindrical, glabrous: stigma undivided, obtuse: ovules solitary, ascending, anatropous: berry (by abortion) few or 1-seeded, seed nut-like, compressed, elongated, the inner angle furrowed: testa shining: albumen fleshy: embryo central: radicle inferior: cotyledon leafy.

(1) S. ? TOMENTOSA. *(Dec.)*

Ident. Dec. prod. VIII. p. 175.
Syn. Sideroxylon tomentosum, *Roxb. fl. Ind.* II. p. 348.
Engrav. Roxb. Cor. I. t. 28.

SPEC. CHAR. Small tree, unarmed: leaves oval, undulated, obtusely narrowed at the apex, younger ones rufous-tomentose: fascicles of flowers axillary, many-flowered: pedicels shorter than the flower and with the calyx silky-pubescent: lobes of the calyx ovate-acute, outer ones broader, corolla twice as long as the calyx: sterile stamens lanceolate-subulate, bispid outside: berry broad ellipsoid, size of an olive, yellow: flowers small, dull white.

Mountains of Coromandel and Western Ghauts, flowering in the hot season.

(2) S. ELENGOIDES. *(Dec.)*

Ident. Dec. prod. VIII. p. 176.
Engrav. Wight's Icon. t. 1218.—Spicil. t. 141.

SPEC. CHAR. Large tree: branches often spinous, branchlets rusty-tomentose: leaves acute at both ends, glabrescent, entire: pedicels the length of the petiole and with the calyx clothed with rusty coloured pubescence: lobes of the calyx ovate-acute, the three exterior ones broader: corolla about twice the length of the calyx, 5-cleft, lobes erect, ovate-acute, tube externally hairy: anthers apiculate: sterile stamens oblong-subulate, the length of the stamens, the back and margins hairy: flowers axillary, few, white.

Common on the Neilgherries, flowering all the year.

GENUS III. SIDEROXYLON.

Pentandria Monogynia. *Ser: Syst:*

Deriv. From *Sideros*, iron, and *Xylon*, wood, because of the hardness of the timber.

GEN. CHAR. Trees: branches often unarmed: leaves alternate, entire: flowers fascicled, usually whitish: very small, often cherry-shaped: calyx 5-parted or half 5-cleft, lobes imbricated, two exterior: corolla half-5-cleft, 5-left, or almost 5-parted, lobes often spreading, imbricated in aestivation, one exterior: stamens inserted on the tube of the corolla, five sterile alternating with the lobes, petaloid, included, often toothed and cut: fertile ones five, opposite the lobes, anthers ovate, obtuse, extrorse, often affixed to the middle of the filament, at last oscillating, shorter than the filament: ovary hirsute, often 5-celled, rarely 4-or 2-celled: style longer than the ovary, acutish at the apex, at least not lobed; fruit baccate, ovoid or globose: seeds by abortion 3–1, globose or rarely ovoid, bony, shining, ribs obtuse, 4–2, umbilicus rotund, small, depressed; albumen copious: cotyledons large, smooth, leafy.

(1) S. GRANDIFOLIUM. (*Wall.*)

Ident. Dec. prod. VIII. p. 178.—Wall. in Roxb. fl. Ind. II. p. 349.

SPEC. CHAR. Leaves obovate-elliptic, cuneate at the base, obtusely acuminated at the apex, entire, smooth: fascicles axillary, 6–10 flowered: pedicels the length of the flower, much shorter than the petiole and with the calyx puberulous: corolla scarcely exceeding the calyx, globose: flowers small, pea-green.

Silhet forests, flowering in April.

GENUS IV. ISONANDRA.

Octandria Monogynia. *Ser: Syst:*

Deriv. From *Isos*, equal, and *Aner, Andros*, male, because the stamens are all perfect.

GEN. CHAR. Trees with alternate, entire leaves: flowers axillary, aggregated: petioles short or wanting: calyx 4-parted, the two exterior lobes larger: corolla 4-cleft or 4-parted, lobes in aestivation twisted to the left, no scales: stamens 8 in a single series, all equal, cohering at the base with the tube of the corolla: anthers hastate, erect, 2-celled, extrorse, dehiscing lengthwise, four longer opposite the lobes of the corolla: ovary free, hispid, 4-celled: ovules 4, ascending: style exserted, smooth: berry fleshy, 1-

seeded by abortion: seed obovoid, erect: testa cartilaginous: albumen copious: cotyledons leafy, longer than the radicle.

(1) I. LANCEOLATA. *(R. W.)*

Ident. Wight's Icon. vol. II.—Dec. prod. VIII. p. 187.
Engrav. Wight l. c. t. 350.

SPEC. CHAR. Tree: leaves lanceolate, acute at both ends, smooth: flowers pedicelled; pedicels the length of the calyx, shorter than the petiole and, with the ovate, subacute lobes of the calyx, glabrous: corolla 4-cleft: drupes obovoid-elliptic, mucronulate.

Peninsula.

(2) I. PERROTTETIANA. *(DeC.)*

Ident. Dec. prod. VIII. p. 188.
Engrav. Wight's Icon. t. 1219.—Spicil. II. t. 142.

SPEC. CHAR. Branchlets clothed with rusty-coloured silky hairs: leaves elliptic, narrowing at both ends, acute at the base, glabrous above, slightly pilose beneath: lobes of the calyx ovate-rotundate, silky: flowers small, sessile, forming dense capitula on the leafless branches, white.

Neilgherries, flowering in February and March.

(3) I. VILLOSA. *(R. W.)*

Ident. Wight's Icon. vol. II.—Dec. prod. VIII. p. 188.
Engrav. Wight l. c. t. 360.

SPEC. CHAR. Tree: branchlets, petioles and leaves beneath rusty-villous: leaves coriaceous, elliptic or subrotund: fascicles of flowers shorter than the petiole: pedicels shorter than the calyx: lobes of the calyx ovate, obtuse, ciliated, corolla deeply 5-cleft: lobes emarginate: flowers orange-coloured.

Peninsula.

(4) I. CANDOLLEANA. *(R. W.)*

Ident. Wight's Icon. vol. IV.
Engrav. Wight l. c. t. 1220.

SPEC. CHAR. Leaves obovate-oblong, bluntly acuminate, tapering at the base, glabrous beneath: lobes of the calyx very unequal: exterior ones much larger and hairy: corolla deeply 4-cleft: lobes emarginate, much longer than the stamens: anthers pubescent at the apex: flowers sessile, dullish white.

Neilgherries, flowering in March and April.

GENUS V. BASSIA.

Dodecandria Monogynia. *Lin : Syst.*

Deriv. Named in honor of Signor Bassi, Curator of the Botanic gardens at Bologna.

GEN. CHAR. Milky trees: leaves alternate, entire: pedicels axillary, usually fascicled; seeds butyraceous or oily: calyx 6-8-parted, in two rows: exterior lobes sub-valvate, interior sub-imbricated, one before expansion with the margin not covered: corolla tubuloso-campanulate, nearly equalling the calyx in length: lobes erect, 6-14: stamens all fertile, about twice the number of the lobes of the corolla, biserial, or (rarely) almost 1 or triserial, inserted on the tube: filaments usually very short: anthers lanceolate-hastate, cordate at the base, 1-3-cuspidate at the apex, 2-celled, dehiscing lengthwise: ovary free, hirsute, 6-8-celled: style exserted, compressed, linear: stigma undivided: ovules solitary, hanging from the apex of the inner angle, half-anatropous: berry oblong or globose, 5-1-seeded: seeds obovoid-oblong: testa shining, crustaceous: albumen none: embryo erect, white: cotyledons fleshy, oily, contrary to the hilum: radicle inferior, short.

(1) B. LONGIFOLIA. (*Linn.*)

Ident. Linn. Mant. p. 563.—Dec. prod. VIII. p. 197.—Roxb. fl. Ind. II. p. 523.

Engrav. Gaertn. fr. II. t. 101. f. 2.—Lam. Ill. t. 308.—Wight's Ill. II. t. 147.

SPEC. CHAR. Tree: leaves lanceolate, acuminate at both ends: petioles slightly villous: pedicels a half shorter than the leaf, suberect: corolla 8-9-cleft: stamens 16-20 in two rows: calyx rufous-pubescent: berry oblong, villous, yellow, size of a large plum, 3-1-seeded: flowers whitish.

Southern parts of the Coromandel coast. Malabar. Common in Canara. Flowering in May. The wood is hard and durable. Almost every part of this valuable tree is applied to some use.

(2) B. LATIFOLIA. (*Roxb.*)

Ident. Roxb. flor. Ind. II. p. 526.—Dec. prod. VIII. p. 198.

Engrav. Roxb. Cor. I. t. 19.

SPEC. CHAR. Tree: leaves elliptic-oblong or oval: petioles somewhat villous: pedicels subumbellate at the apex of the branches, reflexed, 5-6 lines shorter than the leaf, rufous-tomentose: corolla 7-14-cleft: stamens 16-30, somewhat in three rows: calyx rufous-pubescent: berry oblong, size of a small apple, 1-4-seeded; flowers white with a tinge of green and cream-colour.

Circar mountains. Bengal. Bombay. Concans. Guzerat. Flowering in March and April.

(3) B. ELLIPTICA. (Dalz.)

Ident. Dalz. in Hook. Journ. Bot. III. p. 36.—Bomb. for. p. 139.

SPEC. CHAR. Tree: leaves elliptic or elliptic-obovate, shortly and obtusely acuminated, coriaceous, smooth on both sides: pedicels axillary, twin or tern; 3-4 times longer than the petiole, erect in fruit: filaments in one series, those opposite the lobes of the corolla in pairs, those alternate with them single: fruit oblong, smooth.

Canara. Travancore forests. Wynaad. Flowering in February: This tree yields the Pauchontee gum, which at one time was supposed to be a good substitute for Gutta Percha.

(4) B. POLYANTHA. (Wall.)

Ident. Dec. prod. VIII. p. 198.

SPEC. CHAR. Tree: leaves obovate, cuneate at the base, slightly obtuse at the apex, quite glabrous above, below glaucescent with inconspicuous pubescence: pedicels fascicled, shorter than the petiole, rusty-pubescent: corolla 6-cleft; stamens 12: filaments very hispid.

Silhet.

GENUS VI. MIMUSOPS.

Octandria Monogynia. *Ser: Syst.*

Deriv. From *Mimo*, an ape, and *Ops*, a face, from a fancied resemblance in the flower.

GEN. CHAR. Milky trees or shrubs, with alternate, entire leaves: pedicels axillary, usually closely packed: calyx 6-8-parted; lobes in 2 series: corolla 18-24-parted; tube short: partitions 3 times the number of the lobes of the calyx, in 2 series, the outer consisting of 12-16 linear spreading lobes, almost distinct from the base, two being before each lobe of the calyx; the interior of 6-8 lobes, exactly opposite the lobe of the calyx, linear, erect, distinct before flowering from the exterior row: æstivation of either series imbricated: stamens inserted on the tube of the corolla; fertile ones the number of the lobes of the calyx, opposite to them and at the same time to the inner lobes of the corolla: filaments slender, short: anthers lanceolate-sagittate, longer than the filament, extrorse, 2-celled, dehiscing lengthwise; sterile ones alternating with the fertile ones, ovate-acute, toothed, hairy at the back, sometimes 2-lobed or 2-parted: ovary free, hirsute, often angular; cells the number of the stamens or lobes of the calyx, opposite them 1-seeded: ovules ascending: style cylindric: berry 1-2-celled by abortion, globose, or ellipsoid: seeds 1-2, erect: testa coriaceous, smooth: albumen fleshy: embryo erect: cotyledons leafy and fleshy: radicle inferior.

(1) M. ELENGI. (*Linn.*)

Ident. Linn. Sp. p. 497.—Dec. prod. VIII. p. 202.—Roxb. fl. Ind. II. p. 236.

Engrav. Roxb. Cor. I. t. 14.—Rheede Mal. I. t. 20.—Lam. Ill. t. 300.—Wight's Icon. t. 1586.

Spec. Char. Tree: leaves elliptic-oblong, obtusely acuminated, smooth: fascicles of pedicels axillary, shorter than the petiole, 3-6-flowered: pedicels rusty-pubescent: lobes of the calyx lanceolate-acuminate, equal, 4 outer ones rusty-velvety outside, smooth within, 4 inner ones narrower, whitish-velvety outside, smooth within: flowers middle-sized, white, fragrant.

Circars. Western Ghauts. Peninsula. Silhet. Flowering in the hot season.

(2) M. HEXANDRA. (*Roxb.*)

Ident. Roxb. flor. Ind. II. p. 238.—Dec. prod. VIII. p. 204.

Syn. M. Indica, *Dec. prod. l. c. p.* 205.—*Wight's Icon. Fol.* IV.

Engrav. Wight's Icon. t. 1587.—Roxb. Cor. t. 15.

Spec. Char. Tree: leaves obovate-elliptic, emarginate, glabrous: axils 1-6-flowered: pedicels shorter than the petiole: lobes of the calyx ovate-acute, spreading, 3 outer ones coriaceous: 12 outer lobes of the corolla white, linear-lanceolate, 6 inner ones yellow, all spreading: berry olive-shaped, yellow.

Mountainous parts of the Circars. Concans. Guzerat. Flowering in the hot season.

(3) M. ROXBURGHIANA. (*H. W.*)

Ident. Wight's Icon. vol. IV.

Engrav. Wight l. c. t. 1588.

Spec. Char. Tree: leaves obovate, oval, obtuse at both ends: fascicles 2-3-flowered: pedicels about thrice the length of the petioles: lobes of the calyx ovate-acute, about the length of the corolla, rusty-velvety: sterile stamens about the length of the filaments, broad-obovate, fimbriated on the margin: fruit globose, depressed above, about 6-seeded.

Coimbatore district, flowering in March and April.

ORDER C. EBENACEÆ.

Trees, shrubs or under-shrubs without milk; leaves alternate or sub-opposite, entire, exstipulate, short petioled; cymes axillary, rarely terminal, in the males few or many-flowered, one flower evidently terminal; in the female, one-flowered by the lateral flowers aborting; pedicels articulated at the apex; females usually larger, with the calyx growing with the fruit; flowers, by abortion, usually unisexual, rarely bisexual, the male with a rudimentary ovary, the female with sterile stamens; calyx 3-7-lobed, nearly equal, persistent; corolla monopetalous, somewhat coriaceous, usually pubescent externally and glabrous within; limb 3-7-lobed, imbricated in æstivation; stamens definite, either arising from the corolla or hypogynous, twice as many as the segments of the corolla, sometimes four times as many, or the same number and then alternate with them, often inserted in pairs near the bottom of the tube and those neither opposite nor alternate; filaments simple in the hermaphrodite species, generally doubled in the polygamous and diœcious ones, both their divisions bearing anthers but the inner generally smaller; anthers attached by their base, generally lanceolate, 2-celled, dehiscing lengthwise, sometimes bearded; pollen round, smooth: ovary sessile without any disk, several-celled, the cells each having one or two ovules, pendulous from the apex; style divided, seldom simple; stigmas bifid or simple; fruit fleshy round, or oval by abortion, often few-seeded, its pericarp often opening in a regular manner; seed with a membranaceous testa, of the same figure as the albumen, which is cartilaginous and white; embryo in the axis or but little out of it, straight, white, generally more than half as long as the albumen; cotyledons foliaceous, generally somewhat veiny, lying close together, or occasionally slightly separate; radicle taper, of middling length or long, superior, turned towards the hilum.

GENUS I. DIOSPYROS.

Polygamia Diœcia. *Linn. Syst.*

Derio. From *Dius*, noble, and *Pyrus*, a pear-tree, alluding to the general appearance of the plants.

GEN. CHAR. Trees, or rarely shrubs, with entire, alternate leaves; male flowers smaller, often racemose or rather arranged in lateral cymes; females often solitary: calyx 4–6-lobed: corolla tubulose or campanulate, 4–6-cleft, convolute to the left side in æstivation: male stamens 8–30, frequently 16, inserted at the base of the corolla or on the disk, or partly on the corolla and partly on the disk: filaments shorter than the anther, distinct or twin-connate at the base, in which case one is interior, the other exterior and longer: anthers linear-lanceolate, laterally dehiscing on both sides: female stamens fewer, often 8: anthers sterile: ovary in the males almost abortive, in the females 4 or often 8-celled, occasionally 10–12-celled: styles 2–4, more or less connate at the base, usually 2-lobed at the apex: stigmas punctiform: ovules solitary in the cells, pendent, anatropal: berry globose or ovoid, covered at the base by the increasing calyx, 4–8-celled: seeds oblong, convex at the back, more or less compressed at the sides: albumen cartilaginous: embryo axile, erect: cotyledons leafy.

(1) D. EXSCULPTA. (*Ham.*)

Ident. Ham. in trans. Linn. Soc. XV. p. 110.—Dec. prod. VIII. p. 223.

Syn. T. tomentosa, Roxb. *flor. Ind.* II. p. 532. (*not Poir.*)

Engrav. Wight's Icon. t. 182 & 3.

SPEC. CHAR. Branchlets, peduncles and flowers rusty-tomentose; leaves alternate and opposite, broad-elliptic, especially tomentose below: peduncles of the male flowers the length of the petiole, 3-flowered at the apex: calyx campanulate, acute at the base, 4–6-lobed at the apex: lobes acute, erect: female flowers solitary, short-pedicelled: calyx deeply 4–6-cleft: lobes ovate-acute, reflexed at the margin outwards, somewhat winged: tube of the corolla hairy outside: flowers small, whitish.

Northern Bengal. Oude. Flowering in March and April.

(2) D. DUBIA. (*Wall.*)

Ident. Dec. prod. VIII. p. 223.

Engrav. Wight's Icon. t. 1223.

SPEC. CHAR. Branchlets tomentose: leaves ovate-elliptic, obtuse at both ends, slightly hairy above, beneath with the petioles pubescent; male flowers short-peduncled, ternate, sessile: calyx 4–

5-cleft, tomentose on both sides: lobes acute, erect: corolla twice the length of the calyx, externally pubescent: flowers pale, tomentose.

Neilgherries. Dindigul hills.

(3) D. INSCULPTA. *(Ham.)*

Ident. Ham. in trans. Linn. Soc. XV. p. 112.—Dec. prod. VIII. p. 223.

SPEC. CHAR. Branchlets glabrous: leaves alternate, oblong, broader towards the base, acute at the base, acuminated at the apex, entire, shining: berry solitary, very shortly peduncled, 4–3-celled: calyx 4-cleft.

Mountains of Bengal, in the Province of Camrup.

(4) D. MELANOXYLON. *(Roxb.)*

Ident. Dec. prod. VIII. p. 224.—Roxb. fl. Ind. II. p. 530.

Engrav. Roxb. Cor. I. t. 46.

SPEC. CHAR. Large tree: leaf-buds hirsute; leaves sub-opposite, oblong, obtuse, younger ones pubescent: peduncles of the male flowers equalling the petiole, 3–5-flowered: calyx 4-cleft: lobes acute: corolla campanulate, 4 times longer than the calyx, 4-lobed at the apex: lobes spreading, ovate: female flowers solitary: pedicels shorter than the petiole: calyx 5-parted, tomentose: lobes acute: corolla twice as long as the calyx, 5-lobed at the apex: flowers white: fruit globose, yellowish, 8-celled, 6–8-seeded.

Malabar. Coromandel. Orissa. Flowering in April and May. The Coromandel Ebony tree.

(5) D. MONTANA. *(Roxb.)*

Ident. Dec. prod. VIII. p. 230.

Engrav. Roxb. Cor. I. t. 48.

SPEC. CHAR. Leaves ovate-acute, obtuse at the base, smooth, membranaceous: racemes reflexed, nearly twice as long as the petiole; males 5–6-flowered, hermaphrodites 1-flowered; bracts and lobes of the calyx ovate-acute, ciliated: corolla in the males twice as long as the calyx: stamens the length of the tube: ovary globose, smooth: flowers small, green, fragrant.

Circar mountains. Hills eastward of Panwell. Flowering in March and April.

(6) D. TOMENTOSA. *(Poir.)*

Ident. Poir. dict. Enc. V. p. 436. (not Roxb.)—Dec. prod. VIII. p. 229.

SPEC. CHAR. Branches smooth, shining, ashy, spinescent: spines lateral and terminal: leaves alternate, elliptic, on both sides together with the branchlets and petioles yellow-tomentose : flowers axillary, solitary or ternate, subsessile, hispid: bracts ovate, adpressed to the flowers: corolla hairy outside.

Tranquebar.

(7) D. OBIZENSIS. *(Willd.)*

Ident. Willd. Sp. IV. p. 1110.—Dec. prod. VIII. p. 230.

SPEC. CHAR. Branchlets pubescent: leaves oblong, obtuse at the base, slightly acute at the apex, smooth above, below spread with thin and very soft down.

Orissa.

(8) D. CORDIFOLIA. *(Roxb.)*

Ident. Dec. prod. VIII. p. 230.—Roxb. flor. Ind. II. p. 538.
Engrav. Roxb. Cor. t. 50.

SPEC. CHAR. Trunk and older branches spinous: leaves ovate-lanceolate, acute or obtuse, subcordate at the base, smooth above, pubescent below: male racemes 3-flowered; female 1-flowered: peduncle equalling the petiole: lobes of the calyx ovate, pubescent: corolla longer than the tube of the calyx: stamens of the male flowers longer than the tube of the corolla: drupe globose, yellow: flowers small, greenish-white.

Circars. Pondicherry. Bengal. Flowering in March and April.

(9) D. CHLOROXYLON. *(Roxb.)*

Ident. Roxb. flor. Ind. II. p. 588.—Dec. prod. VIII. p. 230.
Engrav. Roxb. Cor. I. t. 49.—Wight's Icon. t. 1224 & 1388.

SPEC. CHAR. Tree: branchlets tomentose: leaves elliptic-obovate, obtuse, below with the petiole tomentose; flowers axillary, fascicled, subsessile, white: calyx 4-parted: calycine lobes ovate, silky outside, smooth within: corolla of male flowers 4-lobed, lobes hairy at the apex outside: fruit sessile, globose, size of a large pea, and resembling a cherry.

Orissa mountains. Circars. Nassick districts. Surat. Flowering in March and April. This appears to be identical with *D. capitulata, R. W.* (Icon. l. c.)

(10) D. SYLVATICA. *(Roxb.)*

Ident. Roxb. flor. Ind. II. p. 537.—Dec. prod. VIII. p. 231.
Engrav. Roxb. Cor. I. t. 47.

SPEC. CHAR. Leaves elliptic-oblong, acute at both ends; peduncles of male flowers many-flowered, twice as long as the petiole: pedicels much shorter than the flower; calyx 4-cleft: lobes ovate:

corolla campanulate, half 4-cleft, three times longer than the calyx : female flowers solitary : pedicel shorter than the petiole.

Circars.

(11) D. CANDOLLEANA. *(R. W.)*

Ident. Wight's Icon. vol. IV.
Engrav. Wight l. c. t. 1221.

SPEC. CHAR. Glabrous tree: leaves elliptic-oblong, obtusely acuminate, veinless: calyx 4–5-cleft: lobes of the female revolute on the margin, clothed with rusty-coloured hair; corolla tubular, 4–5-cleft: stamens of the male 10; filaments united by pairs at the base; of the female 4–5: fruit ovoid, hard, about the size of a nutmeg.

Malabar. Western Ghauts. Flowering in June.

(12) D. AMŒNA. *(Wall.)*

Ident. Dec. prod. VIII. p. 231.—Wall. Cat. No. 4139.

SPEC. CHAR. Leaves lanceolate, smooth, with a lengthened acumen, slightly obtuse: fascicles of male flowers 4–7-flowered, rusty-tomentose, shorter than the petiole: pedicels very short: calyx somewhat 4-cleft, tomentose on both sides: lobes ovate-acute: corolla tubular, 3 times longer than the calyx, somewhat 4-cleft, silky without, smooth within: female flowers solitary, with ovate bracts.

Silhet.

(13) D. MULTIFLORA. *(Wall.)*

Ident. Wall. Cat. No. 2144.—Dec. prod. VIII. p. 231.

SPEC. CHAR. Leaves lanceolate or ovate, acute at the base, quite glabrous above, younger ones hairy below: male flowers fascicled, subsessile: fascicles 3–5-flowered, axillary, everywhere tomentose: calyx 4-cleft: lobes ovate-acute, silky-tomentose on both sides: corolla twice as long as the calyx, ventricose at the base, 4-cleft, smooth within: filaments hairy.

Silhet.

(14) D. LANCEÆFOLIA. *(Roxb.)*

Ident. Roxb. flor. Ind. II. p. 537.—Dec. prod. VIII. p. 232.

SPEC. CHAR. Leaves lanceolate, bifarious, coriaceous, shining: male flowers sessile, aggregated in the axils: calyx 4-toothed, pubescent: tube of the corolla inflated: females solitary, sessile, drooping: calyx 4–5-toothed, pubescent, corolla pubescent outside, tube inflated.

Silhet, flowering in April.

(15) D. STRICTA. *(Roxb.)*

Ident. Roxb. for. Ind. II. p. 539.—Dec. prod. VIII. p. 232.

Spec. Char. Tree with a straight trunk, branching only at the top: branches tomentose: leaves lanceolate, long acuminated; petioles on the central nerve and lower surface pubescent: racemes lateral, males 3-5-flowered, hirsute: peduncle the length of the petiole, shorter than the flower: bracts ovate-acute: pedicels almost none: calycine lobes ovate-acute, smooth inside: corolla half 5-cleft, hirsute externally, tube ventricose below, lobes ovate, spreading: stamens twice as short as the tube.

Tipperah, flowering in March.

(16) D. RAMIFLORA. *(Roxb.)*

Ident. Roxb. fl. Ind. II. p. 535.—Dec. prod. VIII. p. 233.

Engrav. Wight's Icon. t. 189.

Spec. Char. Branches and leaves smooth: leaves bifarious, lanceolate, acute, stiff: flowers fascicled towards the older branches: peduncles subdivided, bracteated at the base, pedicels and calyx, dark-tomentose: calyx of female flower campanulate, narrower at the apex, half 4-6-cleft; lobes acute, tomentose on both sides: corolla 4-6-cleft, cylindric, rusty-tomentose, smooth within: flowers small, white.

Silhet mountains, flowering in March and April.

(17) D. FOLIOLOSA. *(Wall.)*

Ident. Wall. Cat. No. 4143.—Dec. prod. VIII. p. 234.

Spec. Char. Glabrous: branches slender: leaves lanceolate, obtusely acuminated; peduncles of male flowers thin, 3-5-flowered, twice as long as the petiole: pedicels the length of the flowers: calyx 4-parted: lobes ovate; corolla 6 times longer than the calyx.

Courtallum.

(18) D. EMBRYOPTERIS. *(Pers.)*

Ident. Pers. ench. II. p. 624.—Dec. prod. VIII. p. 235.

Syn. D. glutinosa, *Roxb. for. Ind.* I. p. 533.—E. peregrina, *Garis.*—E. glutinifera, *Roxb. Cor.*

Engrav. Wight's Icon. t. 844.—Bot. Reg. t. 499.—Rheede Mal. III. t. 41.—Roxb. Cor. I. t. 70.

Spec. Char. Tree, with smooth branches: leaf-buds silky: leaves lanceolate or elliptic, coriaceous, quite glabrous: peduncles of male flowers 3-5-flowered, the length of the petiole, and with the reflexed pedicels pubescent: calyx spreading, shortly 4-lobed, hairy outside: corolla campanulate, 3 times longer than the calyx, half 4-cleft: lobes ovate, ciliated; female flowers solitary; pedicels and calyx pubescent: corolla somewhat longer than the calyx: ovary hirsute; fruit 12-celled: flowers white.

Peninsula. Bengal. Silhet. Assam. Hurdwar. Flowering in March and April.

(19) D. GLAUCA. *(Rottl.)*

Ident. Rottl. et. Willd. Nov. Act. Nat. Cur. IV. *(Ann.* 1803.) —Dec. prod. VIII. p. 238.

SPEC. CHAR. Stem arboreous, prickly: leaves alternate, short-petioled, oval, acute, tomentose below, coriaceous; glaucous; flowers axillary, sessile, solitary.

Wandewash.

(20) D. RACEMOSA. *(Roxb.)*

Ident. Roxb. flor. Ind. II. p. 536.—Dec. prod. VIII. p. 239.
Engrav. Wight's Icon. t. 416.

SPEC. CHAR. Leaves oblong-lanceolate, obtuse, silky: male and female flowers racemose: racemes axillary, comose: stamens 20-30: ovary 4-celled: style wanting: stigma 4-partite: berry globose, smooth, 4-seeded.

Silhet. Tipperah. Flowering in March and April.

(21) D. PANICULATA. *(Dals.)*

Ident. Dals. in Hook. Journ. Bot. IV. p. 109.—Bomb. flor. p. 141.

SPEC. CHAR. Tree with glabrous branches: leaves lanceolate-oblong, obtusely acuminated, short-petioled, coriaceous, smooth: male flowers panicled in the axils of the fallen leaves: panicles shorter than the leaf, with the buds and pedicels sooty-velvety: calyx 5-parted: segments leafy, broadly oval-obtuse: corolla sooty and velvety outside, twice as long as the calyx: segments 5, oblong-obtuse: female flowers lateral, solitary: pedicels as long as the petiole: fruit ovoid, densely tomentose, enclosed in the enlarged calyx.

Chorla Ghaut and Baighur. Flowering in the cold season.

(22) D. PRURIENS. *(Dals.)*

Ident. Dals. l. c.

SPEC. CHAR. Branchlets softly hairy: leaves narrow-oblong, acuminate, subsessile, hirsute on both sides: male flowers twin, on axillary peduncle, 3 times longer than the petiole: female flow-

are axillary and lateral, solitary, subsessile: fruit ovoid conical, densely clothed with fulvous stinging hairs, size of a large cherry.

Chorla Ghaut.

(23) D. NIGRICANS. (*Dals.*)

Ident. Dals. l. c.

SPEC. CHAR. Arboreous: leaves oblong or lanceolate, acuminated, membranaceous, glabrous: male flowers in threes, sessile on the apex of a very short peduncle: calyx villous, turbinate, 4-partite: lobes ovate-acute, ciliated, spreading: corolla glabrous, with a short tube; segments narrow-linear; stamens 26, in twos, threes or fours.

Chorla Ghaot.

(24) D. GOINDU. (*Dals.*)

Ident. Dals. l. c.
Syn. D. cordifolia, *R. W.*
Engrav. Wight's Ill. t. 148.—Icon. t. 1225.

SPEC. CHAR. Arboreous: leaves ovate-oblong, rounded or truncated at the base, obtuse at the apex, glabrous: male flowers in threes, on an axillary peduncle, as long as the petiole: female flowers axillary, solitary: calyx 4-partite: lobes short, rounded, glabrous: corolla urceolate: segments 4, rounded: stamens 16: fruit globose, size of a cherry, yellow when ripe.

Common on the Ghauts, flowering from April to June.

GENUS II. HOLOCHILUS.

Polygamia Dioecia. *Bar. Syst.*

Deriv. From *Holos*, entire, and *Cheilos*, a lip, alluding to the calyx.

GEN. CHAR. Calyx tubular, entire, truncated, seated on few bifarious imbricated scales: corolla tubular, 3-cleft almost to the middle, three times longer than the calyx, lobes ovate, obtuse, spreading: stamens in female flowers 6, sterile, inserted at the base of the corolla, free between themselves, anthers twice shorter than the filaments: ovary in females hemispherical, smooth, 6-celled: styles 3, erect, thickish, obtuse at the apex: ovules in the cells solitary, pendent; male flowers unknown.

(1) H. MICRANTHUS. (*Dals.*)

Ident. Dals. in Hook. Journ. Bot. IV. p. 200.—Bomb. flor. p. 142.

SPEC. CHAR. Middling sized tree: leaves elliptic or oblong, ob-

tenuated at the base, obtusely acuminated at the apex, short-petioled, coriaceous, smooth: flowers white, minute, axillary, solitary, sessile: fruit cylindric-oblong, supported at the base by the enlarged truncated calyx, dry, hard, 1-inch long.

Southern Ghauts, flowering in February and March.

GENUS III. MABA.

Diœcia Octandria. *Sm: Syst.*

Deriv. The name applied to the plant in Tongataboo.

GEN. CHAR. Flowers diœcious: calyx half 3-cleft or 3-cleft, cup-shaped: corolla urceolate or campanulate, 3-cleft: lobes convolute to the left in æstivation. *Males.* Stamens 3 or 6, sometimes 9, or 12 in pairs connate at the base, hypogynous, accrete at the base with the hairy rudiment of a pistil: filaments slender: anthers linear, often apiculated, dehiscing laterally. *Females.* Stamens none: ovary 8-celled, hirsute: cells 2-seeded: stigma 3-parted: berry ellipsoid, seldom globose, smooth, 3–2-celled: seeds often solitary in the cells, pendent, oblong, transversely furrowed near the base, blackish.—Trees, or shrubs (?) leaves alternate, small, entire: flowers solitary or twin, axillary, small, subsessile: corolla usually hairy outside; hairs long, whitish, specially thick at the middle of the lobes: filaments somewhat equalling the anthers in length.

(1) M. BUXIFOLIA. *(Pers.)*

Ident. Pers. Ench. 2. p. 606.—Dec. prod. VIII. p. 240.

Syn. Ferreola buxifolia, *Roxb. for. Ind.* III. *p.* 790.

Engrav. Wight's Icon. t. 763.—Roxb. Cor. 1. t. 45.

SPEC. CHAR. Small tree: branchlets and young parts pubescent: leaves obovate, cuneate at the base, coriaceous, thickened at the margin, glabrous: flowers subsessile, small, yellowish: males often ternate, females solitary: calyx campanulate, somewhat 3-cleft, hairy outside, lobes ovate, acute: corolla cylindric, 3-cleft, hairy outside, twice as long as the calyx, lobes ovate-oblong: stamens 6: fruit ovoid-globose.

Circar mountains, flowering in the hot season.

(2) M. NEILGHERRENSIS. *(R. W.)*

Ident. Wight's Icon. vol. IV.

Engrav. Wight l. c. t. 1228-9.

SPEC. CHAR. Branchlets slender, glabrous: leaves elliptic-lanceolate, obtusely acuminate, membranaceous, glabrous: flowers axillary, males several, females solitary: calyx campanulate, 5-lobed, hairy on both sides: corolla tubular, 5-lobed, about twice the length of the calyx: stamens 6, unequally hairy at the base: ovary 3-celled, ovules paired: stigma 3-lobed: berry 3-seeded.

Woods about Concor on the Neilgherries.

(3) M. NIGRESCENS. (Dalz.)

Ident. Dalz. Bomb. flor. p. 142.

SPEC. CHAR. Small tree: branches rigid, erect: branchlets covered with rusty-pubescence: leaves small, subsessile, ovate or ovate-acute, margins slightly narrowed and undulated: tawny with adpressed hairs beneath especially on the margin and midrib.

Forests of the Western Ghauts.

ORDER CI. AQUIFOLIACEÆ.

Evergreen trees or shrubs whose branches are often angular; leaves alternate or opposite, simple, coriaceous, without stipules; flowers small, white or greenish, axillary, solitary or clustered, sometimes unisexual by abortion, sepals 4-6-imbricated in æstivation, corolla 4-6-parted, hypogynous, imbricated in æstivation; stamens inserted in the corolla, alternate with its segments; filaments erect; anthers adnate, two-celled, opening longitudinally; disk none: ovary fleshy, superior, somewhat truncate, with from 2 to 6 or more cells; ovules solitary, anatropal, pendulous, and often hanging from a cup-shaped funiculus; stigma subsessile, lobed; fruit fleshy, indehiscent, with from 2 to 6 or more stones; seed suspended; nearly sessile; albumen large, fleshy; embryo small, 2-lobed, lying next the hilum, with minute cotyledons, and a superior radicle.

GENUS I. ILEX.

Tetrandria Tetragynia. *Sex: Syst:*

Derio. Originally from *Oc*, or *Ac*, a Celtic term having allusion to the prickly leaves.

GEN. CHAR. Calyx inferior, 4-6-lobed, permanent: corolla rotate in 4-6 deep elliptical, spreading, concave lobes, or as many petals, slightly cohering by their bases, much larger than the calyx: filaments oval-shaped: shorter than the corolla and alternate with its lobes: anthers 2-lobed: ovary roundish: styles none: stigmas 4-fid, obtuse: berry globular, 4-6-celled: seeds solitary in each cell, oblong, pointed, angular inside, rounded externally.

(1) I. WIGHTIANA. (*Wall.*)

Ident. Wight's Spicil. II. p. 35.—Dals, Bomb. flor. p. 143.
Engrav. Wight's Spicil. II. t. 139.—Icon. t. 1216.

SPEC. CHAR. Large tree: leaves alternate, ovate-elliptic or elliptic-acuminated, entire, coriaceous: umbels numerous, axillary, or from the scars of fallen leaves; pedicels about the length of the peduncles, often longer: flowers often polygamous by abortion: corolla 5-6-cleft: berry 5-0-seeded: flowers white.

Neilgherries, flowering nearly all the year.

(2) I. GARDNERIANA. (*R. W.*)

Ident. Wight's Spicil. II. p. 35.
Engrav, Wight l. c. t. 140.—Icon. t. 1217.

SPEC. CHAR. Small tree or large shrub: leaves ovate-lanceolate or subcordate, long acuminated : umbels axillary, or aggregated on the naked branches: pedicels often shorter than the peduncles, sparingly hairy : calyx and corolla 5-lobed, the former sprinkled with short hairs : flowers white.

Sisparah jungles. Neilgherries. Flowering in February. A third species, *I. denticulata*, (*Wall.*) grows on the Neilgherries, Wight's Ill. II. t. 149, of which no description is given.

ORDER CII. STYRACACEÆ.

Calyx adherent to the ovary, persistent, 5-(or 4-) lobed; lobes imbricating in æstivation; corolla monopetalous, the number of its lobes frequently different from those of the calyx (in the Indian species both usually 5) with imbricated æstivation: stamens definite or indefinite, arising from the tube of the corolla, of unequal length, cohering in various ways, but generally in a slight degree only, round the throat of the tube; anthers 2-celled, bursting inwardly; pollen broadly elliptical, smooth; ovary adhering to the calyx, rarely free, from 2-to 5-celled, cells opposite the lobes of the calyx, when the same number, the partitions sometimes scarcely adhering in the centre; ovules anatropal, 2 or several in each cell, either all pendulous, or the upper one ascending; style simple; stigma somewhat capitate; fruit drupaceous, enclosed in the persistent calyx, generally with all the cells, except one, abortive; seeds ascending or suspended, with the slender embryo lying in the midst of the fleshy albumen; radicle long, directed to the hilum; cotyledons flat.

GENUS I. SYMPLOCOS.

Polyadelphia Icosandria. *Lin. Syst.*

Deriv. From *Symploke*, a connexion ; the stamens are united at the base.

Gen. Char. Calyx 5-cleft, often ciliated : corolla of 5–8–10 petals in one or two series scarcely united at the base, but cohering by means of the adnate stamens: stamens inserted into the extreme base of the corolla, often monadelphous, the tube of the stamens often more or less extensively united to the corolla ; filaments filiform or ligulate, contracted at the apex: anthers ovoid, globose, 2-celled, ovary inferior, 2–4–5-celled : ovules 2–4 pendulous from the apex of the cells : style filiform : stigma capitulate, simple or 3-sided : berry crowned by the calyx, often by abortion reduced to one or two cells : seeds solitary in each cell : albumen copious : embryo axile : cotyledons very short.—Trees or shrubs with alternate, serrated or crenulated leaves : racemes axillary, many-flowered, bracteate : flowers sessile or pedicelled, white or red.

(1) S. POLYSTACHYA. *(Wall.)*

Ident. Dec. prod. VIII. p. 254.—Wall. Cat. No. 4128.

Spec. Char. Tree or shrub : leaves oblong-lanceolate, acuminate, acute at the base, serrated, glabrous : racemes axillary, branched from the base, nearly equalling the leaf, velvety, somewhat shorter than the calyx : flowers sessile : calycine lobes ovate, puberulous.

Silhet mountains.

(2) S. SPICATA. *(Roxb.)*

Ident. Roxb. flor. Ind. II. p. 541.—Dec. prod. VIII. p. 254.

Syn. Eugenia laurina, *Willd.*—Bobua laurina, *Dec. prod.* III. p. 24.—Myrtus laurina, *Retz.*

Spec. Char. Small tree : leaves oblong-lanceolate, acute, serrated, quite glabrous : racemes axillary, composite, 4–6 times longer than the petiole, slightly glabrous : bracts ovate, obtuse, very short : flowers sessile, yellowish : calyx spreading, 5-cleft : lobes ovate, obtuse, glabrous : drupe areolate, size of a pea.

Neilgherries. Assam. Khasia Hills. Flowering in the rainy season.

(3) S. HAMILTONIANA. *(Wall.)*

Ident. Wall. Cat. No. 4420.—Dec. prod. VIII. p. 254.

Spec. Char. Shrub or tree : leaves oblong, acute or acuminated, obtuse at the base, subserrated, nerve hairy below : racemes axillary, simple, 5 times longer than the petiole, pubescent : bracts ovate, hairy outside : pedicels shorter than the calyx, pubescent : calycine lobes obtuse, ovate, ciliolate.

Peninsula.

(4) S. LUCIDA. *(Wall.)*

Ident. Wall. Cat. No. 4414.—Dec. prod. VIII. p. 255.

SPEC. CHAR. Small tree: leaves lanceolate, long-acuminate, acute at the base, glabrous, shining on both sides, serrulate: racemes composite, many-flowered, axillary and terminal, the length of the petiole, puberulous: calycine lobes ovate, obtuse, ciliated.

Silhet mountains.

(5) S. OBTUSA. *(Wall.)*

Ident. Dec. prod. VIII. p. 255.

Engrav. Wight's Spicil. II. t. 146.—Icon. t. 1233.

SPEC. CHAR. Tree: leaves elliptic, obovate-orbicular, tapering towards the base, subdenticulate: racemes axillary, twice the length of the petioles, simple, glabrous: calycine lobes roundish, ciliolate: flowers subsessile, white.

Neilgherries, flowering in April and May.

(6) S. RACEMOSA. *(Roxb.)*

Ident. Roxb. flor. Ind. II. p. 539.—Dec. prod. VIII. p. 255.

Syn. S. thexifolia, Don. prod. flor. Nep. p. 145.

SPEC. CHAR. Tree: leaves oblong-lanceolate, acuminate, acute at the base, quite glabrous, subdenticulate, shining above: racemes simple, axillary, nearly equalling the petiole, hairy: calycine lobes and bracteoles ovate, obtuse, ciliated: ovary free at the apex: flowers small, yellow.

Burdwan. Midnapore. Western Ghauts. Flowering in December.

(7) S. ATTENUATA. *(Wall.)*

Ident. Wall. Cat. No. 4426.—Dec. prod. VIII. p. 256.

SPEC. CHAR. Tree: leaves elliptic, acuminated at both ends, quite glabrous, almost entire: racemes axillary, simple or 3-forked from the base, pubescent, 2-3 times longer than the petiole: calyx spreading, 5-cleft, lobes ovate-acute.

Silhet.

(8) S. CAUDATA. *(Wall.)*

Ident. Wall. Cat. No. 4415.—Dec. prod. VIII. p. 256.

SPEC. CHAR. Tree: leaves lanceolate, long-acuminated, acute at the base, slightly toothed at the margin, quite glabrous: racemes axillary, simple, twice the length of the petiole, hairy: pedicels longer than the calyx: tube of the calyx obconical, glabrous: lobes ovate-lanceolate, slightly hairy.

Khasia Hills.

(9) S. OXYPHYLLA. *(Wall.)*

Ident. Wall. Cat. No. 4430.—Dec. prod. VIII. p. 256.

SPEC. CHAR. Tree: leaves oblong-lanceolate, long-acuminated, acute at the base, slightly toothed at the margin, glabrous: racemes axillary, composite, slightly glabrous, shorter than the leaf, branches alternate, subdivided: bracts ovate, obtuse, puberulous: flowers sessile.

Khasia Hills.

(10) S. TRIFOLIA. *(Wall.)*

Ident. Wall. Cat. No. 4415.—Dec. prod. VIII. p. 256.

SPEC. CHAR. Tree: leaves ovate-oblong, acuminate, subacute at the base, glabrous, denticulate from the middle to the apex: racemes axillary, twice as long as the petiole, simple, glabrous; flowers alternate, subsessile: calyx 5-cleft: tube glabrous, lobes ovate-acute, hairy outside.

Khasia Hills.

(11) S. GRANDIFLORA. *(Wall.)*

Ident. Wall. Cat. No. 4421.—Dec. prod. VIII. p. 257.

SPEC. CHAR. Tree: leaves lanceolate, acuminated at both ends, somewhat entire, glabrous: racemes axillary, 4 times longer than the petiole, simple: pedicels the length of the calyx: bracteoles and calycine lobes ovate-obtuse, glabrous.

Khasia Hills.

(12) S. MACROPHYLLA. *(Wall.)*

Ident. Wall. Cat. No. 4431.—Dec. prod. VIII. p. 257.

SPEC. CHAR. Tree: branches at the apex, racemes, petioles and nerves of the leaves densely hispid-ferrugineous: leaves obovate-elliptic, acute, entire, glabrous above, hairy below: racemes axillary, simple, somewhat longer than the petiole: bracts ovate-acute, silky outside: flowers sessile: calycine lobes lanceolate, hairy outside.

Khasia Hills.

(13) S. FERRUGINEA. *(Roxb.)*

Ident. Roxb. fl. Ind. II. p. 342.—Dec. prod. VIII. p. 257.

SPEC. CHAR. Small tree: leaves elliptic-lanceolate, acuminate, acute at the base, denticulate, glabrous above, below, with the petioles branches and racemes, tawny-pubescent; racemes axillary, composite, 2-3 times longer than the petiole: flowers sessile: calycine lobes ovate-acute, pubescent: flowers middle-sized, yellow.

Khasia Hills, flowering in September and October.

(14) S. PULCHRA. (R. W.)

Ident. Wight's Icon. vol. IV.
Engrav. Wight l. c. t. 1230.—Spicil. t. 143.

SPEC. CHAR. Diffuse shrub: branchlets, leaves peduncles and bracts clothed with long brownish hair: leaves ovate-oblong, acuminate, slightly cordate, bristle-serrated: peduncles axillary, filiform, 3-4-flowered: calycine lobes ciliated: corolla glabrous, white: ovary pubescent, 3-celled.

Sisparah on the Western slopes of the Neilgherries, flowering in February.

(15) S. GARDNERIANA. (R. W.)

Ident. Wight's Icon. vol. IV.
Engrav. Wight l. c. t. 1231.—Spicil. t. 144.

SPEC. CHAR. Tree: branches rusty-tomentose: leaves elliptic-acuminate, denticulate, glabrous above, tomentose on the midrib beneath, pubescent on the lamina: racemes axillary, about half the length of the leaves: flowers white, crowded: bracts, bracteoles and calyx tomentose.

Neilgherries, flowering in February.

(16) S. MICROPHYLLA. (R. W.)

Ident. Wight's Icon. vol. IV.
Engrav. Wight l. c. t. 1232.—Spicil. t. 145.

SPEC. CHAR. Fruticose, branched, glabrous: leaves elliptic, obtuse, serrated, coriaceous, glabrous, or with a few hairs on the midrib beneath: racemes axillary, about twice the length of the petioles, hairy: bracts ovate, obtuse, and like the calyx pubescent: calycine lobes suborbicular, ciliate: flowers white.

Neilgherries on the banks of streams, flowering in February.

(17) S. FOLIOSA. (R. W.)

Ident. Wight's Icon. vol. IV.
Engrav. Wight's Icon. t. 1234.

SPEC. CHAR. Tree, much branched, very leafy towards the extremities: leaves ovate-lanceolate, acute or acuminate, coriaceous, serrato-dentate, glabrous or with a few scattered hairs on the midrib: racemes axillary, several congested on the ends of the branches, about twice the length of the petioles, hairy: flowers crowded, sessile, white: calycine lobes unequal, hairy on the back: ovary hairy, 3-celled.

Neilgherries, flowering in the hot season.

(18) S. NERVOSA. *(Dec.)*

Ident. Dec. prod. VIII. p. 256.
Engrav. Wight l. c. t. 1235.

Spec. Char. Leaves oblong-lanceolate, acuminate at both ends, crenately denticulate, very glabrous, shining above, beneath the veins areolate: racemes axillary, simple, twice the length of the petiole, and with the ovate acute bracts, hairy: calycine lobes ovate-oblong, acute, hairy on the back: flowers white.

Neilgherries, flowering in the dry season.

(19) S. MONANTHA. *(R. W.)*

Ident. Wight's Icon. vol. IV.
Engrav. Wight l. c. t. 1236.

Spec. Char. Fruticose, much branched, glabrous: leaves short-petioled, elliptic-lanceolate, acuminate, serrated: flowers axillary, solitary, sessile: calyx glabrous: lobes ovate, pointed, much shorter than the corolla: lobes of the corolla roundish-obovate: stigma capitate.

Shevagherry Hills, near Courtallum, flowering in August.

(20) S. PENDULA. *(R. W.)*

Ident. Wight's Icon. vol. IV.
Engrav. Wight l. c. t. 1237.

Spec. Char. Tree: leaves oval-obtuse or obovate, entire, coriaceous: peduncles axillary, short, 2–4-flowered: flowers pendulous, tubular: calyx ciliate: ovary 2-celled, fruit oblong.

Pulney mountains, flowering in September.

(21) S. UNIFLORA. *(R. H. B.)*

Ident. Madr. Journ. lit. Ser. III. No. I. p. 51.

Spec. Char. Tree: leaves glabrous, short-petioled, ovate-lanceolate, coriaceous, serrulate: peduncles axillary, solitary, nearly as long as the leaves, slender: berry cylindric, 3-celled.

Annamullay Hills.

(22) S. ROSEA. *(R. H. B.)*

Ident. Madr. Journ. lit. Ser. III. No. I. p. 51.

Spec. Char. Shrub: leaves from oblong to lanceolate, slightly attenuated at the base, and with a longish sharp acumen; mucronately serrulate, glabrous: young branches, petioles and inflorescence puberulous: racemes axillary, longer than the petioles: bracts, calyx and fruit puberulous: berry subcylindric, 3-celled: flowers rose-coloured.

Annamullay Hills.

GENUS II. STYRAX.

Decandria Monogynia. *Ser : Syst :*

Deriv. An alteration of *Astarak*, the Arabic name of *S. officinale*.

GEN. CHAR. Trees or shrubs, usually clothed with stellate tomentum: leaves entire, alternate, exstipulate: peduncles axillary or terminal, one or many-flowered: flowers racemose, bracteate, white or cream-coloured: calyx persistent, campanulate, 5-toothed: corolla funnel-shaped, deeply 3-7-cleft, usually 5-6-cleft, valvate in æstivation: stamens exserted: filaments monadelphous at the base, adnate to the tube of the corolla: anthers linear, 2-celled, dehiscing lengthwise inside: ovary superior, 3-celled, many-ovuled, erect: stigma obsoletely 3-lobed: drupe almost dry, containing a 1-celled, 1-3-seeded nut: testa double, inner cobwebbed: embryo inverted, with elliptic cotyledons: radicle thick: albumen fleshy.

(1) S. SERRULATUM. (*Roxb.*)

Ident. Roxb. flor. Ind. II. p. 415.—Dec. prod. VIII. p. 267.

SPEC. CHAR. Small tree: branchlets, petioles, racemes and calyxes tawny-tomentose: leaves oblong, acuminate, acute at the base, serrulate, glabrescent: racemes terminating the lateral branches, simple, shorter than the leaf: calyx campanulate, acute at the base, 5-toothed at the apex: capsule ovoid, pubescent, splitting irregularly from the base in 3-4-valves: seeds 1-4.

Chittagong, flowering in March.

(2) S. VIRGATUM. (*Wall.*)

Ident. Wall. Cat. No. 4400.—Dec. prod. VIII. p. 267.

SPEC. CHAR. Tree: branches velvety at the apex: leaves ovate, long-acuminate at the base, obtuse, serrulate, glabrous: racemes shorter than the leaf, few-flowered: pedicels somewhat longer than the calyx and with the calyx whitish-puberulous: calyx hemispherical, 5-toothed.

Khasia Hills.

ORDER CIII. OLEACEÆ.

Trees or shrubs with opposite, simple or unequally pinnate leaves; racemes or panicles axillary, or terminal, one bracteate; flowers often fragrant, white or lilac coloured, hermaphrodite or diœcious; calyx monophyllous, persistent, 4-cleft or 4-toothed;

corolla monopetalous, hypogynous, 4-cleft or sometimes 4-petalled, with the petals united by pairs by the filaments, sometimes wanting, by abortion, in the female flowers, sub-valvate in æstivation; stamens 2, attached to the base, alternate with the lobes or petals; anthers 2-celled, dehiscing longitudinally; ovary simple, free, without a hypogynous disk, 2-celled, with 2 collateral, pendulous or amphitropous ovules in each; style 1 or none; stigma bifid or undivided; fruit drupaceous, or baccate, or capsular, often 1-seeded by abortion; seed usually pendulous: albumen generally copious, dense, fleshy; sometimes sparing or wanting: embryo, when albumen is copious, straight, about half the length of the seed; cotyledons foliaceous, when wanting, amygdaloid.

GENUS I. FRAXINUS.

Polygamia Diœcia. *Sex: Syst:*

Deriv. From *Pùrasso*, to enclose or fence in; the ash was used formerly for making hedges.

GEN. CHAR. Trees or shrubs: leaves opposite, petioled, unequally pinnate, 2–7 pair, leaflets sessile or petioled, toothed, rarely quite entire: flowers racemose or panicled: petals (when present) white: calyx 4-cleft or wanting: petals sometimes none, sometimes 4, frequently cohering in pairs at the base, oblong or linear: stamens 2: stigma 2-cleft: samara 2-celled, compressed, winged at the apex, cells 2-ovuled, by abortion 1-seeded: seeds pendulous, compressed: albumen fleshy, thin: embryo the length of the albumen: cotyledons elliptic: radicle linear, superior.

(1) F. UROPHYLLA. *(Wall.)*

Ident. Wall. Cat. No. 2835.—Dec. prod. VIII. p. 275.
Syn. Ornus urophylla, Don's Mill. dict. IV. p. 57.

SPEC. CHAR. Tree: leaves long-petioled: leaflets 2–3 pair, long-petiolate, membranaceous, ovate-oblong, long-acuminated, serrated; panicles axillary.

Pundua. Silhet.

GENUS II. OLEA.

Diandria Monogynia. *Sex: Syst:*

Deriv. From *Elaia*, the Olive.

GEN. CHAR. Trees or shrubs: leaves opposite, quite entire

rarely toothed: flowers racemose, panicled or subcorymbose, usually fragrant, white: calyx campanulate, 4-toothed: limb of the corolla 4-partite, spreading: stamens, where there is a corolla, inserted at the bottom of the tube, opposite, exserted, in apetalous flowers, hypogynous: ovary 2-celled: ovules twin in the cells, pendulous from the apex: style short, bearing a 2-cleft or subcapitate stigma: drupe baccate, flesh oily: putamen bony, by abortion 2, often 1-seeded: seeds inverted: albumen fleshy: embryo inverted, straight, with leafy cotyledons.

(1) O. ROXBURGHIANA. *(Roem. & Schult.)*

Ident. Roem. & Schult. Syst. I. p. 77.—Dec. prod. VIII. p. 280.

Syn. O. paniculata, *Roxb. fl. Ind.* I. p. 105.—O. Roxburghii, *Spr.*

Engrav. Wight's Icon. t. 735.

SPEC. CHAR. Small tree: leaves oblong, attenuated at the base, quite entire, glabrous, waved: panicles axillary, and springing below the leaves: bracts deciduous: lobes of the stigma divaricate: flowers small, white: fruit small, purple.

Western Ghauts. Circar mountains. Flowering in the hot season.

(2) O. SALICIFOLIA. *(Wall.)*

Ident. Wall. Cat. No. 2821.—Dec. prod. VIII. p. 286.

SPEC. CHAR. Tree: leaves broad-lanceolate, subacute at the base, long-acuminated at the apex, remotely and sharply serrated: panicles axillary, nearly the length of the leaf, equal: bracts very minute: corolls campanulate, 4-toothed.

Khasia Hills.

(3) O. DIOICA. *(Roxb.)*

Ident. Roxb. flor. Ind. I. p. 106.—Dec. prod. VIII. p. 286.

Engrav. Rheede Mal. IV. t. 54.—Wight's Ill. II. t. 151.

SPEC. CHAR. Largish tree: leaves oblong, acuminated at both ends, remotely and sharply serrated, glabrous: panicles opposite, springing in branchlets below the leaves: flowers polygamo-dioicous: drupe subrotund, purplish: flowers small, white.

Chittagong. Silhet. Flowering in March. Khandalla. Mahableshwar. Vingorla. Flowering in July.

(4) O. LINDLEYI. *(Wall.)*

Ident. Wall. Cat. No. 6303.—Dec. prod. VIII. p. 288.

SPEC. CHAR. Tree, glabrous: branches warty: leaves lanceolate, attenuated at both ends: panicles thyrsoid, terminal, pubes-

ment: calyx obtusely toothed: corolla funnel-shaped, tube elongated.
Silhet.

(5) O. ROBUSTA. *(Wall.)*

Ident. Wall. Cat. No. 2822.
Syn. Phillyrea robusta, *Roxb. flor. Ind.* I. p. 101.—Visiania robusta, *Dec. prod.* VIII. p. 289.
Engrav. Wight's Spicil. II. t. 147.—Icon. t. 1242.
SPEC. CHAR. Small tree: leaves elliptic-oblong, acute at the base, acuminated at the apex, entire: panicles terminal, large, diffuse; rachis and pedicels pubescent: style clavate: fruit subcylindrical: flowers white, fragrant.
Eastern slopes of the Neilgherries. Silhet. Flowering nearly all the year.

(6) O. GLANDULIFERA. *(Wall.)*

Ident. Dec. prod. VIII. p. 285.
Engrav. Wight's Icon. t. 1238.
SPEC. CHAR. Low tree: leaves elliptic, acute at the base, acuminate at the apex, entire, glabrous, glandulose in the axils of the nerves: panicles axillary, shorter than the leaves: calyx 4-toothed: stigma capitate.
Neilgherries, flowering in March and April.

(7) O. POLYGAMA. *(R. W.)*

Ident. Wight's Icon. vol. IV.
Engrav. Wight l. c. t. 1239-40.
SPEC. CHAR. Polygamous: small tree: leaves obovate, cuspidate, tapering at the base, short-petioled, entire, coriaceous, those of the male plant smaller, tending to lanceolate; panicles axillary, cymose, each division terminating in a cluster of 8—10 flowers, those of the male larger and more diffuse: corolla 4-cleft: ovary ovate, with a distinct style and capitate stigma.
Western slopes of the Neilgherries, flowering in February and March.

(8) O. LINOCIERIODES. *(R. W.)*

Ident. Wight's Icon. vol. IV.
Engrav. Wight l. c. t. 1241.
SPEC. CHAR. Small tree or large shrub: leaves short-petioled, elliptic-oblong, abruptly acuminated, entire, glabrous: peduncles axillary, much shorter than the leaves, trichotomous, each division terminating in a head of flowers: calyx 4-lobed, ciliate: corolla

deeply 4-parted, divisions long-subulate, united by pairs to the filaments, 3–4 times the length of the stamens; ovary ovate: style short, 2-cleft; drupe oblong, bony, 1-seeded: flowers sessile, white.

Courtallum in dense forests, flowering in August.

GENUS III. LIGUSTRUM.

Diandria Monogynia. *Sex: Syst:*

Deriv. From *Ligare*, to tie, in allusion to the very flexible branches.

GEN. CHAR. Trees or shrubs with opposite leaves: flowers in terminal panicles or thyrses: calyx 4-toothed, deciduous: corolla funnel-shaped, tube longer than the calyx, limb 4-parted: stamens inserted within the tube: ovary 2-celled, with 2 ovules pendulous from the apex in each; style very short: stigma 2-cleft, obtuse: berry globose, 2-celled: seeds inverse: embryo straight: albumen subcartilaginous.

(1) L. PEROTTETII. *(Dec.)*

Ident. Dec. prod. VIII. p. 294.
Engrav. Wight's Icon. t. 1244.

SPEC. CHAR. Erect shrub: branches puberulous at the apex: leaves elliptic, obtuse or subacute, glabrous, coriaceous, somewhat fleshy: thyrses terminal, composite, contracted: flowers white, fragrant.

Neilgherries on hilly pastures and banks of rivulets.

(2) L. NEILGHERRENSE. *(R. W.)*

Ident. Wight's Icon. vol. IV.—Dalz. Bomb. flor. p. 159.
Engrav. Wight l. o. t. 1243.—Spicil. II. t. 148.

SPEC. CHAR. Small tree: leaves ovate, elliptic, acute or cuspidately acuminate, coriaceous: thyrses terminal, lax: flowers numerous, white, fragrant: fruit black, linear-oblong.

Neilgherries. Western Ghauts. Flowering in May and June.

GENUS IV. LINOCIERA.

Diandria Monogynia. *Sex: Syst:*

Deriv. Named after G. Linocer, a French Physician.

GEN. CHAR. Calyx minute, 4-cleft: petals 4, linear, or oblong, elongated, united by pairs at the base, through the medium of the stamens; stamens two, uniting the base of the petals, inclusive: ovary 2-celled, 4-ovulate, style very short, stigma emarginately

2-lobed: drupe baccate, 1-celled by abortion, 1-seeded: putamen, thin, sulcately striated: seed inverse, exalbuminous: cotyledons plano-convex, thick: radicle very short, superior.—Glabrous shrubs or rarely trees, with opposite, simple, entire leaves: peduncles axillary or terminal, racemose or panicled: corolla white, yellow or purple.

(1) L. INTERMEDIA. *(R. W.)*

Ident. Wight's Icon. vol. IV.
Engrav. Wight I. c. t. 1245.—Spicil. II. t. 149.
Spec. Char. Tree: leaves elliptic, acuminate at both ends, long petioled; panicles axillary, diffuse, about as long as the leaves; flowers aggregated on the points of the branchlets, sessile, often male by abortion: ovules ascending: stigma capitate, 2-lobed: fruit oval, 1-seeded.

Eastern slopes of the Neilgherries, flowering in the rainy season.

(2) L. MALABARICA. *(Wall.)*

Ident. Dec. prod. VIII. p. 297.—Dalz. Bomb. flor. p. 159.
Syn. Chionanthus Malabaricus, *Heyne. Herb.*
Engrav. Wight's Icon. t. 1246.
Spec. Char. Rambling shrub, or small tree: leaves elliptic-obtuse, attenuated towards the base, smooth on both sides: racemes axillary, much shorter than the leaf, few-flowered; pedicels bearing 1 to 3 sessile flowers at the top; pedicels and calyx pubescent: petals linear-channelled.

Courtallum. Western slopes of the Neilgherries. Khandalla. Ram Ghaut. Flowering from November till March.

(3) L. DICHOTOMA. *(Wall.)*

Ident. Wall. Cat. No. 2825.—Dec. prod. VIII. p. 297.
Syn. Chionanthus dichotomus, *Roxb. fl. Ind. I. p. 107.*
Spec. Char. Low branching shrub, dichotomous: lanceolate, acuminated at both ends, cuneate at the base, glabrous, recurved at the apex: racemes axillary, a little shorter than the leaf: branchlets 3-flowered at the apex: fruit ovate, purple: flowers white.

Coromandel Coast.

(4) L. MACROPHYLLA. *(Wall.)*

Ident. Wall. Cat. No. 2826.—Dec. prod. VIII. p. 297.
Spec. Char. Small tree, glabrous: leaves elliptic, acuminated at both ends, glabrous, stiffish, long-petioled: racemes composite, axillary, a little longer than the petiole: petals oblong, obtuse.

Silhet.

ORDER CIV. AZIMACEÆ.

Flowers dioicous. MALE: calyx urceolate, 4-cleft; petals 4, hypogynous, equal, æstivation valvate; stamens 4, hypogynous; anthers 2-celled, introrse, dehiscing longitudinally, connective shortly produced, apiculate; ovary abortive, conical. FEMALE: calyx irregularly 2–4-cleft; corolla as in the male; stamens rudimentary; ovary hypogynous, turgid, 2-celled, with a single ovule in each cell; style none; stigma sessile, peltate, somewhat 2-lobed; fruit a globose berry, 2-celled or rarely, by abortion, 1-celled; cells 1-seeded; seeds erect, plano-convex; testa coriaceous, rugose; albumen none; embryo lenticular; cotyledons fleshy, cordate-auriculate at the base; radicle inferior.

GENUS I. AZIMA.

Tetrandria Monogynia. *Ser. Syst.*

GEN. CHAR. Same as that of the Order.

(1) A. TETRACANTHA. *(Lam.)*

Ident. Lam. Ill. Gen.—Wight's Ill. p. 156.
Syn. Monetia tetracantha, *G. Don.*—M. barlerioides, *L'Herit.*, *Roxb. flor. Ind.* III. p. 765.
Engrav. L'Herit. Stirp. t. 1.—Lam. l. c. t. 807.—Wight l. c. t. 153.

SPEC. CHAR. Dioecious, shrubby: branches opposite, tetragonal, spreading, thorny: leaves opposite, petioled, acute, entire, smooth: flowers axillary, clustered, very shortly peduncled, yellow: berries white.

Peninsula. Dharwar. Flowering all the year.

ORDER CV. JASMINACEÆ.

Flowers hermaphrodite, unsymmetrical; calyx persistent, 4–8-toothed or lobed; corolla hypogynous, 1-petaled, 5–8-lobed, salver-shaped; lobes imbricated in æstivation, the two exterior ones twisted or valvate; stamens two, attached to the tube, inclose; anthers 2-celled, introrse, bursting longitudinally; ovary destitute of a hypogynous disk, 2-celled, 2-lobed at the apex; ovules ascending or amphitropous, 1-2, rarely more, in each cell; style simple; stigma 2-lobed; fruit bibaccate or capsular; capsules 2-celled, bipartible, cells indehiscent; seed exalbuminous or with sparing albumen, testa often tumid; embryo straight; radicle inferior.—Erect or scandent shrubs: leaves opposite, rarely alternate or often unequally pinnate, leaflets 3-5-7, or sometimes, by abortion of the lateral leaflets, reduced to one, but then on a jointed petiole, indicating its compound nature; flowers corymbose or panicled; pedicels opposite, 1-bracteate: corolla white or yellow, often fragrant.

GENUS I. CHONDROSPERMUM.

Diandria Monogynia. *Ser. Syst.*

Deriv. From *Chondros*, a lump, and *Sperma*, seed: the form of the seeds.

GEN. CHAR. Calyx urceolate, 4-toothed; teeth lanceolate, acute, with purple edges: corolla funnel-shaped, rather fleshy, having a large obscurely 4-sided tube which is longer than the calyx and a 4-parted spreading limb, which is longer than the tube; segments linear-clavate, blunt, vertical, and thickened at the apices: throat closed by the anthers; stamens 2, inserted above the base of the tube; anthers fleshy, with 2 marginal cells situated between the 2 opposite fissures of the border, hardly elevated above the tube; ovary ovate, 2-celled: ovule solitary.

(1) C. LAURIFOLIUM. (*Voight.*)

Ident. Voight. Hort. Calc. p. 348.

Syn. C. smilacifolium, *Wall.*—Chionanthus smilacifolius, *Wall.* Roxb. *flor. Ind. Ed. Car.* 1. p. 108.

SPEC. CHAR. Rambling shrub: leaves opposite, coriaceous, paler beneath, waved: panicles axillary and terminal, brachiate: flowers small, greenish-yellow.

Chittagong, flowering in April and May.

GENUS II. JASMINUM.

Diandria Monogynia. *Lin. Syst.*

Deriv. According to Linnæus, from *Ios*, a violet, and *Osme*, smell, but more probably from the Arabic name *Ysmyn*.

GEN. CHAR. Erect or scandent shrubs: leaves opposite, rarely alternate, all compound, or occasionally the petiole jointed in the middle and having one leaflet, or sometimes 3-7 leaflets, and then the leaves are trifoliolate or unequally pinnate: panicles few or many-flowered: corolla yellow or white, sometimes reddish externally: calyx campanulate, 5-8-lobed, teeth sometimes subulate, sometimes short: corolla salver-shaped: tube terete, limb flat, 5-8-parted, lobes oblique, twisted in æstivation: stamens adnate to the tube of the corolla, included: ovary 2-celled, 1-2-ovuled: ovules erect, ascending, lateral, or sometimes pendulous: style simple, 2-lobed at the apex: berry didymous: cells 1-(rarely 2)-seeded: seed erect, exalbuminous.

(1) J. ERECTIFLORUM. (*Dec.*)

Ident. Dec. prod. VIII. p. 308.

Engrav. Wight's Icon. t. 1251.—Spicil. II. t. 150.

SPEC. CHAR. Scandent: leaves ovate-lanceolate, subcordate, long-acuminate: peduncles on the ends of the branches, ternate, with from 5-7 erect, condensed flowers on the apex: bracts linear-subulate, somewhat longer than the pedicels: calycine lobes 6, linear-subulate: tube of the corolla 3 times longer than the calyx, lobes 6-7, oblong, acuminate, half the length of the tube: flowers white, fragrant.

Neilgherries, flowering in the hot season.

(2) J. REVOLUTUM. (*Sims.*)
Var. PENINSULARE. (*Dec.*)

Ident. Dec. prod. VIII. p. 312.—Don's prod. p. 106.

Syn. J. Chrysanthemum, Roxb. flor. Ind. I. p. 98.—J. Dignoniaceum, *Wall.*

Engrav. Wight's Icon. t. 1258.—Spicil. II. t. 151.—Bot. Reg. t. 178.

SPEC. CHAR. Erect, not scandent: leaves alternate, pinnated: branches angled: leaflets 3, 5, 7, 11-obovate, oblong, narrowed at the base, subacute at the apex: panicles terminal, opposite the leaves, corymbose: calyx acute and acutely denticulate: flowers few, yellow, fragrant.

Mountains of Northern India. Neilgherries, flowering all the year.

(3) J. SAMBAC. (*Ait.*)

Ident. Dec. prod. VIII. p. 301.—Ait. Hort. Kew. I. p. 8.

Syn. Nyctanthes Sambac, *Linn.*—Mogorium Sambac, *Lam.*— J. fragrans, *Salisb.* prod. p. 12.—Roxb. *flor. Ind.* I. p. 88.

Engrav. Wight's Icon. t. 704.—Bot. Reg. l. t. 1.—Rheede Mal. VI. t. 55. (var. simplex.)—Rheede l. c. t. 50.—Rumph. Amb. 5. t. 30.—Burm. Zeyl. t. 58. f. 2.—Bot. Repos. t. 497. (var duplex.)—Rheede l. c. t. 51.—Bot. Mag. 43. t. 1785.—Rumph. Amb. 5. t. 30. a. (var. plenum.)

SPEC. CHAR. Shrubby, somewhat scandent: branches and petioles pubescently hairy: leaves simple, short-petioled, ovate or subcordate, often acute: racemes terminal, ebracteate, few-flowered: calycine lobes about eight, subulate: flowers white, fragrant.

All over India, flowering in the hot season.

(4) J. QUINQUEFLORUM. (*Heyne.*)

Ident. Dec. prod. VIII. p. 302.

SPEC. CHAR. Scandent: branches, peduncles, calyx and petioles pubescent: leaves oblong, sometimes short-acuminated, always mucronate; peduncles terminal, 3-5-flowered; calycine lobes 8, long-subulate: lobes of the corolla lanceolate.

Gongachora. Patna.

(5) J. PUBESCENS. (*Willd.*)

Ident. Dec. prod. VIII. p. 302.

Syn. Nyctanthes pubescens, *Retz.*—N. hirsuta, *Linn.*—J. hirsutum, *Willd.*—J. multiflorum, *Andr.* Bot. Rep. t. 496.—N. multiflora, *Burm.*—Mogorium pubescens, *Lam.*

Engrav. Smith's Exot. Bot. II. t. 118.—Sim's Bot. Mag. t. 1931.—Bot. Reg. t. 15.—Burm. flor. Ind. t. 3. f. 1.

SPEC. CHAR. Shrubby: branchlets hirsute: leaves opposite, short-petioled, cordate, mucronate, below and with the petioles tomentose, above afterwards glabrous: flowers congested in terminal umbels, subsessile: calycine lobes 6-9, filiform, hirsute; tube of the corolla a little longer than the calyx: lobes oval, mucronate: flowers white, fragrant.

Coromandel. South Concans. Monghyr. Silhet. Assam. Flowering nearly all the year.

(6) J. ELONGATUM. (*Willd.*)

Ident. Willd. Sp. 1. p. 37.—Dec. prod. VIII. p. 302.—Roxb. Fl. Ind. I. p. 89.

Syn. Nyctanthes elongata, *Linn. f. suppl.* p. 82.

Engrav. Wight's Icon. t. 701.

Spec. Char. Scandent, whole plant velvety-hirsute; leaves opposite and alternate, ovate-lanceolate, attenuated at both ends; petioles jointed below the middle; corymbs terminal; calycine lobes 5-6, short; tube of the corolla elongated; lobes 10-12, linear-lanceolate, acute, the length of the tube; stigma 2-lobed; flowers white, fragrant.

Soonderbunds, flowering in March and April.

(7) J. PUNCTATUM. (*Wall.*)

Ident. Wall. Cat. No. 2877.—Dec. prod. VIII. p. 303.

Spec. Char. Shrub: puberulous when young: leaves ovate-lanceolate, acuminate, afterwards glabrous; peduncles terminal, trichotomous: calyx pubescent; lobes linear; segments of the corolla 8-10, acute; flowers white.

Near Patna.

(8) J. ARBORESCENS. (*Roxb.*)

Ident. Roxb. fl. Ind. I. p. 91.—Dec. prod. VIII. p. 303.

Syn. J. arboreum, Roem. & Schult.—Nyctanthes grandiflora, *Lour.*

Engrav. Wight's Icon. t. 699.

Spec. Char. Shrub or small tree: branches sub-erect, younger ones with the petioles, peduncles, bracteoles and calyx pubescently hirsute; leaves opposite or ternately verticilled, ovate, subcordate, acute, glabrous above, puberulous below: petioles jointed in the middle; calycine lobes 5-6, subulate, short: lobes of the corolla 10-12, linear-oblong: stigma 2-lobed: flowers numerous, arranged in terminal corymbs, large, snow-white, very fragrant.

Upper Bengal. Courtallum. Flowering in March and April.

(9) J. RETICULATUM. (*Wall.*)

Ident. Dec. prod. VIII. p. 303.

Spec. Char. Scandent; leaves oblong-lanceolate, acute at the base, acuminated at the apex; flowers terminal, capitate or corymbose; branchlets and calyx pubescent: calycine lobes subulate; tube of corolla long; lobes 8, linear.

Pundua, Silhet.

(10) J. LAURIFOLIUM. (*Roxb.*)

Ident. Roxb. flor. Ind. I. p. 91.—Dec. prod. VIII. p. 303.

Spec. Char. Shrubby, scandent; branchlets terete: petioles geniculate; leaves oval-lanceolate or lanceolate, acuminate, 3-nerved: pedicels 3 5, terminal, (rarely axillary) elongated: lobes of the calyx 9-12, linear, equalling the tube: flowers whitish with a pale green tube, fragrant.

Khasia Hills, flowering in November.

(11) J. ANGUSTIFOLIUM. *(Vahl.)*

Ident. Vahl. Enum. I. p. 29.—Dec. prod. VIII. p. 303.—Roxb. flor. Ind. I. p. 96.

Syn. J. vimineum and angustifolium, *Willd.*—J. triflorum, *Pers.*—Nyctanthes angustifolia, *Linn.*—N. viminea, *Retz.*—N. triflora, *Burm. Ind. t. 2.*—Mogorium triflorum, *Lam. Ill. t. 0.*

Engrav. Wight's Icon. t. 008.—Rheede Mal. 0. t. 53.

SPEC. CHAR. Shrub: branchlets obtusely sub-tetragonal, sub-pubescent, soon glabrous: leaves short-petioled, glabrous, ovate-oblong, mucronate: pedicels terminal, somewhat threefold, one-flowered: lobes of the calyx 6-9, bristly, glabrous, erect: tube of corolla twice as long as the calyx, lobes linear, sublanceolate, nearly equalling the tube: flowers largish, white, with a faint tinge of red.

Coromandel forests.

(12) J. MYRTOPHYLLUM. *(Zenk.)*

Ident. Zenk. plant. Ind. p. 6. t. 7.—Dec. prod. VIII. p. 304.

SPEC. CHAR. Shrub: leaves elliptic, subacute at the base, obtuse and mucronulate, membranaceous: flowers axillary and terminal, solitary or in threes: pedicels somewhat shorter than the calyx: lobes of the calyx 4-6, linear, subacute, erect, the length of the tube: tube of the corolla three times longer than the calyx, lobes 8, oblong, mucronate, somewhat shorter than the tube, white.

Neilgherries.

(13) J. CORDIFOLIUM. *(Wall.)*

Ident. Dec. prod. VIII. p. 304.—Wall. Cat. No. 2849.

SPEC. CHAR. Scandent: leaves broad ovate-cordate, triplinerv-ed, acuminate, afterwards glabrous: corymbs 3 times trichotomous, pubescent: lobes of the campanulate calyx subulate, of the corolla eight, elliptic, mucronate: flowers white.

Neilgherries.

(14) J. PERROTTETIANUM. *(Dec.)*

Ident. Dec. prod. VIII. p. 304.

SPEC. CHAR. Glabrous, shrubby: leaves coriaceous, shining above, dotted below; nerves 5-7, lower ones elliptic, obtuse at both ends, upper ones lanceolate, acuminated at both ends: pedun-

cles axillary, short, few-flowered: lobes of the calyx 7, linear-subulate, longer than the tube: berries ellipsoid; flowers subsessile,
white.

In woods near Pondicherry.

(15) J. STENOPETALUM. *(Lindl.)*

Ident. Dec. prod. VIII. p. 304.

Syn. J. trinerve, Roxb. *fl. Ind.* I. p. 91.

SPEC. CHAR. Scandent: branches terete: leaves ovate-lanceolate, acuminate, 3-nerved: petioles jointed at the middle: lobes of
the calyx 6–7, subulate; of the corolla 6–8, somewhat filiform,
longer than the tube; flowers axillary and terminal, subsessile,
solitary or 3–9 together, white.

Silhet.

(16) J. RIGIDUM. *(Zenk.)*

Ident. Zenk. plant. Ind. p. V. t. 6.—Dec. prod. VIII. p. 305.

Engrav. Wight's Icon. t. 1247.

SPEC. CHAR. Glabrous shrub, not scandent: leaves ovate or
oval, obtuse at the base, somewhat mucronate at the apex: branches
axillary and terminal, 3–6-flowered: pedicels the length of the
calyx-tube: calycine lobes 4–6, linear-subulate, erect: tube of the
corolla 3 times longer than the calyx; lobes 6, elliptic, submucronate, 3 times shorter than the tube: flowers white.

In dry stony places on the Neilgherries.

(17) J. LIGUSTRIFOLIUM. *(Wall.)*

Ident. Dec. prod. VIII. p. 305.

SPEC. CHAR. Scandent, glabrous: leaves ovate-acute: petioles
short jointed: pedicels terminal, 1-flowered: lobes of the calyx 6,
subulate, 3 times shorter than the tube of the corolla; lobes of the
corolla 8–9, linear-lanceolate, about the length of the tube: flowers
middle-sized, pure white.

Khasia Hills, flowering in February.

(18) J. ROTTLERIANUM. *(Wall.)*

Ident. Wall. Cat. No. 2865.—Dec. prod. VIII. p. 305.

Engrav. Wight's Icon. t. 1249.

SPEC. CHAR. Shrubby: whole plant hirsute except the flowers:
leaves elliptic, obtuse at the base, acute at the apex: petiole jointed
in the middle: peduncles 3, terminal, bearing fascicles of flowers
at the apex: bracts linear-lanceolate, much acuminated: calyx

pubescent: lobes 5-7. subulate: tube of the corolla 3 times longer than the calycine lobes: lobes 5-7, oblong, mucronate, 3 times shorter than the tube: flowers white.

Sivapore jungles. Travancore.

(19) J. LATIFOLIUM. *(Roxb.)*

Ident. Roxb. flor. Ind. I. p. 93.—Dec. prod. VIII. p. 308.
Syn. J. trichotomum, *Var.* latifolium, *Roth.*
Engrav. Wight's Icon. t. 703.

SPEC. CHAR. Shrubby, climbing: leaves cordate and oblong-acute, glabrous: corymbs terminal, diffuse: calycine lobes 5-7, subulate: lobes of the corolla 8-12, linear-cuspidate: berries kidney-shaped and oblong: flowers large, white, fragrant.

Common on the Western Ghauts. Circar mountains. Keamery jungles. Flowering in March and April.

(20) J. GARDNERIANUM. *(R. W'.)*

Ident. Wight's Ill. Ind. Bot. vol. II. p. 159.—Walp. Ann. III. p. 25.
Engrav. Wight l. c. t. 153.

SPEC. CHAR. Shrubby, climbing and twining, glabrous: leaves ovate, undulated, retuse, pointed: petiole articulated near the base: cymes terminal, panicled: peduncles about the length of the leaves: calyx 5-toothed, teeth acute, short: corolla about 8-lobed: lobes lanceolate, acute, the length of the tube: berry oval-oblong, dark purple: flowers white.

Coimbatore, flowering nearly all the year.

(21) J. AMPLEXICAULI. *(Wall.)*

Ident. Wall. Cat. No. 2875.—Dec. prod. VIII. p. 306.

SPEC. CHAR. Scandent, glabrous: leaves ovate, acute: peduncles axillary and terminal, 3-flowered, extreme ones corymbose: lobes of the calyx 7, subulate, long, of the corolla 7, lanceolate, acute.

Goalpara.

(22) J. SCANDENS. *(Vahl.)*

Ident. Vahl. Symb. III. p. 2.—Enum. I. p. 27.—Dec. prod. VIII. p. 306.—Roxb. flor. Ind. I. p. 83.
Syn. Nyctanthes scandens. *Retz.*—Mogorium scandens, *Lam.*

SPEC. CHAR. Climbing: leaves petioled, ovate-oblong, acuminated, subcordate at the base, glabrous: corymbs terminating the branches, glomerate, trichotomous, and with the calyxes hirsute: lobes of the calyx subulate, spreading, reflexed, of the corolla lanceolate, very acute, shorter than the tube; flowers white, fragrant.

Silhet and Chittagong, climbing over lofty trees, flowering in February and March.

(23) J. BREVILOBUM. *(Dec.)*

Ident. Dec. prod: VIII. p. 307.
Engrav. Wight's Icon. t. 1254.

SPEC. CHAR. Scandent: branches terete, pubescent: leaves ovate, very obtuse at the base, subcordate, acute or obtuse at the apex, glabrescent above, below and with the margin pubescent; flowers condensed at the top of the branchlets, subsessile; calyx pubescent, campanulate, half 2-cleft, lobes erect, ovate, obtuse: tube of the corolla five times longer than the calyx, lobes elliptic, 3-4 times shorter than the tube: flowers white, very fragrant.

Neilgherries. A variety of the above, *J. Mollissimum*, (*Wall.*) with the leaves softy hirsute beneath grows in Assam.

(24) J. COARCTATUM. *(Roxb.)*

Ident. Roxb. fl. Ind. I. p. 91.—Dec. prod. VIII. p. 308.

SPEC. CHAR. Shrubby: leaves oblong, acute, glabrous: corymbs terminal, peduncled: flowers subsessile, ternate, bracteate: calyx 5-cleft.

Chittagong Hills, flowering in April and May.

(25) J. ATTENUATUM. *(Wall.)*

Ident. Wall. Cat. No. 2864.—Dec. prod. VIII. p. 309.

SPEC. CHAR. Quite glabrous, smooth: branches subscandent, terete: leaves elliptic-oblong, attenuated at both ends, petioled, longish, somewhat twisted: racemes axillary, nearly equalling the leaf, lax-flowered: flowers somewhat opposite, remote, long-pedicelled: calyx with 5 short teeth: lobes of the corolla 5, oblong, acute, a half shorter than the tube.

Khasia Hills.

(26) J. CORDATUM. *(Wall.)*

Ident. Wall. Cat. No. 2884.—Dec. prod. VIII. p. 310.
Engrav. Lindl. Bot. Reg. 1842. t. 26.

SPEC. CHAR. Scandent, glabrous: leaves petioled, trifoliolate; leaflets petiolulate, ovate-lanceolate, long-acuminated, terminal one longer: panicles terminal, many-flowered: calyx somewhat truncated, sharply 5-toothed: lobes of the corolla oblong-linear, slightly obtuse, shorter than the tube: flowers large, white, not fragrant.

Khasia Hills.

(27) J. FLEXILE. (*Vahl.*)

Ident. Vahl. Symb. III. p. 1.—Dec. prod. VIII. p. 310.
Engrav. Wight's Icon. t. 1253.

SPEC. CHAR. Scandent, glabrous: leaves petioled, trifoliolate; leaflets petiolulate, ovate-oblong, acuminated, shining, the lateral ones a half less: petioles flexuose: racemes axillary, brachiate, 3 times longer than the leaf: calyx campanulate, minutely and sharply 5-6-toothed.

Courtallum, in dense jungles near the bottom of the falls, flowering nearly all the year.

(28) J. LANCEOLARIA. (*Roxb.*)

Ident. Roxb. fl. Ind. I. p. 97.—Dec. prod. VIII. p. 310.
SPEC. CHAR. Erect shrub: leaves trifoliolate; leaflets lanceolate; corymbs terminal.

Khasia Hills, flowering in May.

(29) J. COURTALLENSE. (*R. W.*)

Ident. Wight's Icon. vol. IV. t. 1252.

SPEC. CHAR. Fruticose, scandent, glabrous: leaves petioled, trifoliolate; leaflets broadly-ovate, rounded at the base, blunt, terminal one larger: panicles axillary, numerous towards the ends of the branches, many-flowered: calyx campanulate, 5-toothed: corolla 5-lobed: lobes obtuse; berries globose.

Courtallum, flowering in August and September.

(30) J. MALABARICUM. (*R. W.*)

Ident. Wight's Icon. vol. IV. t. 1250.

SPEC. CHAR. Scandent, everywhere except the inflorescence glabrous: leaves broad-cordate, suborbicular, cuspidately acuminate; petiole jointed in the middle: peduncles axillary and terminal, cymose, 7-9-flowered: flowers crowded, subsessile, erect: bracts subulate: calyx campanulate: lobes 5, subulate, reflexed at the apex; of the corolla ovate, cuspidate, about half the length of the tube.

Malabar Coast, near Calicut, flowering in March and April.

(31) J. OVALIFOLIUM. (*R. W.*)

Ident. Wight's Icon. vol. IV. t. 1236.

SPEC. CHAR. Scandent, villous: leaves trifoliolate; leaflets ovato-oblong, acuminate, villous on both sides, lateral pair sublanceolate, much smaller; corymbs axillary, 3-9-flowered: calyx

campanulate, slightly 5-lobed : lobes of the corolla about 7, oval or
subovate, obtuse, about one-third the length of the tube : style the
length of the tube : stigma clavate or subcapitate.

Malabar, flowering in April.

(32) J. SILHETENSE. (*Blume.*)

Ident. Walp. Ann. III. p. 22.

Syn. J. trinerve, *Wall. Dec. prod.* VIII. p. 304. *(partly.)*

SPEC. CHAR. Scandent, quite glabrous : leaves ovate-oblong or
ovate-lanceolate, long-acuminated, chartaceous, 3-nerved : peduncles axillary and terminal, somewhat 1-flowered, 2 or many-bracte-
oled, a little longer than the petiole : lobes of the calyx 6-7, subulate, of the corolla 6-8, linear-lanceolate, a little shorter than the
tube.

Silhet.

GENUS III. NYCTANTHES.

Diandria Monogynia. *Sen : Syst :*

Deriv. From *Nyx, Nyctos,* night, and *Anthos,* a flower ; the
flowers expand at night, and fall off at break of day, whence *Arbor-
tristis,* the name of the species.

GEN. CHAR. Calyx campanulate, slightly 5-toothed : corolla
salver-shaped : stigma capitate, glandular : capsule superior, obcordate, compressed, 2-celled, 2-valved : cells 1-seeded : embryo
erect, exalbuminous.

(1) N. ARBOR-TRISTIS. (*Linn.*)

Ident. Linn. Spec. pl. 8. Syst. 56.—Dec. prod. VIII. p. 814.
—Roxb. flor. Ind. I. p. 86.

Syn. Scabrita scabra, *Vahl.*—S. triflora, *Linn. Mant.*—Parill-
um arbor-tristis, *Garcin. fr.* I. p. 234.

Engrav. Bot. Reg. V. t. 399.—Rheede Mal. I. t. 21.

SPEC. CHAR. Shrub or small tree : leaves opposite, short-petioled, cordate, acuminated, entire, or coarsely serrated, scabrous :
branches tetragonal : panicles terminal, leafy, composed of small
5-flowered, terminal umbellets : involucre 4-leaved : corolla with
an orange-coloured tube and white border : segments of the limb
6-7, twisted, triangular, or obliquely lobed, very fragrant.

Supposed to be a native of Arabia, but found by Dalzell in a wild
state in the forest in Khandeish. Common in gardens all over
India. Flowering in July.

GENUS IV. SCHREBERA.

Diandria Monogynia. *Sex: Syst:*

Deriv. Named in honor of J. C. Schreber, a Botanical Author.

GEN. CHAR. Calyx tubular, 2-lipped: lips nearly equal, emarginate, occasionally with a tooth in each fissure which separates the lips: corolla salver-shaped with a cylindrical tube, border spreading 5—6—7 cleft: segments truncated, cuneate: anthers oblong, included: stigma 2-cleft, acute: capsule large, pear-shaped, woody, 2-celled, 2-valved, scabrous, opening from the top: seeds 4 in each cell, oval, compressed, with a long membranous wing.

(1) S. SWIETENIOIDES. *(Roxb.)*

Ident. Roxb. flor. Ind. I. p. 109.—Dec. prod. VIII. p. 674.
Engrav. Roxb. Cor. II. t. 101.

SPEC. CHAR. Large tree, glabrous: leaves unequally pinnate: leaflets 3—4 pair, obliquely ovate-acuminate: panicles terminal, trichotomous, minutely bracteated: flowers whitish.

Circar mountains. Tull Ghaut, near Bhewndy. Flowering in February and March. The timber is heavy and close-grained and useful for various purposes.

ORDER CVI. SALVADORACEÆ.

Small trees: leaves opposite, coriaceous, entire: flowers small, in loose panicles: calyx inferior, 4-leaved, minute: corolla membranous, monopetalous, 4-partite: stamens 4, connecting the petals into a monopetalous corolla: anthers round, 2-celled, bursting lengthwise: ovary superior, 1-celled, with a single sessile stigma: ovule solitary, erect: pericarp berried, 1-celled, indehiscent: seed solitary, erect: embryo amygdaloid, without albumen: cotyledons fleshy, plano-convex, fixed below their middle to a long axis, the radicle of which is enclosed within their bases.

GENUS I. SALVADORA.

Tetrandria Monogynia. *Sex: Syst:*

GEN. CHAR. Same as that of the Order.

(1) S. INDICA. (R. W.)

Icon. Wight's Ill. vol. II. p. 229. t. 181.

SPEC. CHAR. Tree: leaves broad ovate-oval, obtuse, glabrous: panicles terminal and axillary, diffuse: flowers longish-pedicelled: bracts somewhat persistent: berry about twice the length of the calyx, red, embraced by the withering corolla: flowers white.

Common everywhere, in low damp ground.

ORDER CVII. APOCYNACEÆ.

Calyx free, 5-parted, persistent, lobes usually furnished within with scales; æstivation contorto-imbricated; stamens 5, arising from within the tube of the corolla, alternate with its lobes; filaments distinct; anthers adhering firmly to the stigma, 2-celled, opening longitudinally; pollen granular, globose, or 3-lobed, immediately applied to the stigma; ovary free, usually embraced at the base by a fleshy nectary composed of 5 glands placed opposite the lobes of the calyx, single or double; when single, 2-or rarely 1-celled; when double, united at the apex into the single style; ovules usually numerous, amphitropous or nearly anatropous; style simple; stigma frequently enlarged at the base, expanding into a ring or campanulate membrane, contracted in the middle, and simple or 2-cleft, pointed or dilated at the apex; fruit follicular, capsular, baccate or drupaceous; seed usually pendulous, sometimes ascending, naked, or variously comose, sometimes winged, often albuminous; embryo straight; radicle usually superior, cotyledons flat, rarely convolute.—Trees, shrubs or undershrubs, rarely herbaceous, with milky juice; stems frequently twining; leaves opposite, or whorled, rarely alternate, simple, entire, rarely stipuled, but often having glands in the place of stipules; flowers usually cymose, sometimes racemose, regular, often large and handsome.

GENUS I. WILLUGHBEIA.

Pentandria Monogynia. *Sex: Syst:*

Deriv. In honor of Francis Willougbby, F. R. S., a friend and pupil of Ray.

GEN. CHAR. Climbing shrubs, often tendril-bearing, lactescent: leaves opposite, entire: cymes axillary and terminal: calyx 5-parted, or deeply 5-cleft: corolla cup-shaped, 5-cleft, tube puberulous within, lobes oblong, convolute in æstivation: stamens towards the middle of the tube of the corolla, anthers longer than the filament, ovate-acute: ovary free, ovoid, 1-celled: ovules indefinite, inserted on two parietal placentæ: style cylindric: stigma thickened, ovoid, striated: berry size and shape of an orange, globose or broadly ovoid, pulpy: seeds numerous, nestling in pulp, shape of a bean, seed-skin soft: albumen none: cotyledons fleshy, planoconvex: radicles very short.

(1) W. EDULIS. *(Roxb.)*

Ident. Roxb. flor. Ind. II. p. 57.—Dec. prod. VIII. p. 321.

Engrav. Roxb. Cor. III. t. 260.—Wall. Pl. As. Rar. III. t. 292.

SPEC. CHAR. Leaves oblong, acuminated, subacute at the base: peduncles axillary and terminal and with the pedicels shorter than the leaf, 3–5-flowered: pedicels the length of the calyx: calycine lobes ovate, ciliated: flowers pale-rose: fruit yellow, ovoid-globose.

Chittagong. Silhet. Flowering in the hot season. Fruit eaten by the natives.

GENUS II. MELODINUS.

Pentandria Monogynia. *Sex: Syst:*

Deriv. From *Melon*, an apple, and *Dineo*, to turn round. The fruit resembles an apple and the stems are twining.

GEN. CHAR. Calyx 5-cleft, eglandulose, lobes ovate, two outer ones: corolla cup-shaped, tube cylindric: throat with a corona, segments of the corona ten in one series: lobes 5, imbricately twisted in æstivation: stamens inserted below the middle of the tube: anthers subsessile, oblong, acute: ovary ovoid-conical, glabrous, 2-celled: style filiform: stigma thick-conical, laterally 10-ribbed: berry fleshy, globose, pulpy within: seeds numerous, nestling, compressed, with a ventral umbilicus: embryo straight in the axis of fleshy albumen: cotyledons oblong, subfoliaceous.

(1) W. MONOGYNUS. *(Carey.)*

Ident. Carey in Hort. Beng. p. 50.—Dec. prod. VIII. p. 329.
—Roxb. flor. Ind. II. p. 56.
Engrav. Wight's Icon. t. 394.—Bot. Mag. 25. t. 2327.—Bot. Reg. X. t. 834.

SPEC. CHAR. Climbing shrub, quite glabrous: leaves opposite, oblong-lanceolate, short-petioled, acuminated: cymes terminal, shorter than the leaf: peduncles spreading: pedicels 3–5 at the apex of peduncles: lobes of the calyx ovate, obtuse, slightly hairy at the margin: corolla glabrous outside, tube hairy within above the stamens, lobes oblong, segments of the corona lanceolate, ciliated; flowers white, fragrant; berry size of a small orange, eatable.

Silhet, flowering in March and April.

GENUS III. CARISSA.

Pentandria Monogynia. *Sex: Syst*:

Deriv. The Sanscrit name latinised.

GEN. CHAR. Ramous shrubs or small trees, lactescent: branches dichotomous, spreading: leaves opposite, entire: spines opposite, sometimes bifurcate at the forks of the branches, changed above into flower-bearing peduncles: calyx 5-parted, or deeply 5-lobed, without glands at the base, two of the lobes exterior: corolla salver-shaped, lobes twisted in æstivation, tube hairy within, throat sometimes bearded: stamens 5, anthers lanceolate, obtuse or apiculate: ovary single, 2-celled, with 2 ovules in each: style filiform, glabrous, thicker above: stigma 2-lobed, hairy, caducous: ovules few, attached to the partition, amphitropal: berry globose or ellipsoid, 2–4-seeded: seeds peltate, rough, albuminous: embryo axile, straight: radicle inferior: cotyledons ovate.

(1) C. PAUCINERVIA. *(Dec.)*

Ident. Dec. prod. VIII. p. 333.
Engrav. Wight's Spicil. II. t. 158.—Icon. t. 1290.

SPEC. CHAR. Branches subdichotomous, armed: leaves elliptic, oblong, acute at both ends, mucronate, glabrous, short petioled, few veined, oblique: peduncles terminal and axillary, much shorter than the leaves, 3–5-flowered; pedicels longer than the calyx, puberulous: calyx 5-cleft, slightly pilose, laminæ lanceolate, acuminate: flowers white with a slight dash of rose: berries oval, dark purple.

Neilgherries, flowering in the hot season.

(2) C. CARANDAS. *(Linn.)*

Ident. Linn. Mant. p. 52.—Dec. prod. VIII. p. 332.—Roxb. flor. Ind. II. p. 523.

Syn. Capparis Carandas, *Gmel.*—Echites spinosa, *Burm.*

Engrav. Lam. Ill. t. 118. fig. 1.—Roxb. Cor. I. t. 77.—Wight's Icon. t. 426.—Rumph. Amb. VII. t. 25.

Spec. Char. Shrub : branches thorny : leaves oval, short-petioled, coriaceous, glabrous, shining above : peduncles terminal, 3–5-flowered, shorter than the leaf: pedicels puberulous : calyx deeply 5-cleft, slightly hairy : segments lanceolate, acuminate : flowers pure white, inodorous : berry ellipsoid, darkish.

Peninsula. Hindostan. Lower Kemaon. Flowering in the hot season. Fruit used for preserves and pickles. This shrub makes strong fences on account of its sharp thorns.

(3) C. DIFFUSA. *(Roxb.)*

Ident. Roxb. flor. Ind. II. p. 524,—Dec. prod. VIII. p. 233.

Engrav. Wight's Icon. t. 427.

Spec. Char. Shrub : branches dichotomous, diffuse, thorny : leaves round-oval, retuse at the base, very shortly-petioled, mucronulate at the apex, coriaceous, glabrous : peduncles terminal and axillary, many times shorter than the leaf, 5–7-flowered : calyx deeply 5-cleft, lobes lanceolate-acuminate, puberulous : drupe ellipsoid : flowers white.

Ganjam. Mouths of the Hoogly. Flowering in the hot season.

(4) C. SPINARUM. *(Linn.)*

Ident. Linn. Mant. p. 559.—Dec. prod. VIII. p. 332.

Engrav. Lam. Ill. t. 118. fig. 2.

Spec. Char. Branches somewhat pubescent, thorny: leaves oval or ovate, mucronate, coriaceous, glabrous, subsessile : peduncles terminal, many times shorter than the leaf, 5–6-flowered : calyx 5-parted, segments lanceolate-acuminate, puberulous.

Coromandel Coast.

(5) C. HIRSUTA. *(Roth.)*

Ident. Roth. Nov. Sp. p. 128.—Dec. prod. VIII. p. 333.— Dals. Bomb. flor. p. 143.

Syn. C. villosa, *Roxb. fl. Ind.* II. p. 525.

Engrav. Wight's Icon. t. 437.

Spec. Char. Branches thorny, tomentose : leaves roundish or ovate, hairy on both sides : peduncles terminal and axillary, 5–7-

flowered, shorter than the leaf: berry globose, size of a pea, smooth, dark purple.

Hills eastward of Belgaum, flowering in the hot season.

(6) C. CONGESTA. (*R. W.*)

Ident. Wight's Icon. vol. IV. t. 1289.

SPEC. CHAR. Fruticose, erect: branches dichotomous, thorny: leaves broad-ovate or suborbicular, obtuse, glabrous, very smooth: peduncles terminal, about 3-flowered, congested on the points of the branches, slightly pubescent: calycine lobes ovate, acute, ciliated, much shorter than the corolla: corolla hairy within: filaments and capitate stigma hairy.

Coorg.

GENUS IV. OPHIOXYLON.

Pentandria Monogynia. *Sax: Syst:*

Derio. From *Ophis*, a serpent, and *Xylon*, wood; in allusion to its supposed healing properties.

GEN. CHAR. Calyx 5-parted, without glands: lobes linear, oblong, or lanceolate, erect: corolla salver-shaped, much longer than the calyx: tube cylindrical, narrower at the throat, hairy within: lobes five, ovate, obtuse, twisted to the right: stamens 5, inserted within the throat, incluse: anthers oblong, acute, longer than the filaments: nectary cup-shaped, entire, undulated on the margin: ovaries 2, compressed, connate at the base: ovules 2 in each, attached above the base: style 1: stigma ovoid, capitate, bituberculate at the apex, and fimbriate round the base and crown: berries connate at the base, ovoid, 1-seeded, with a more or less rugous testa: embryo nearly as long as the seed: albumen fleshy: cotyledons oval, lanceolate, or suborbicular: radicle pointing to the apex.—Lactescent, erect, or twining shrubs: leaves opposite or verticelled, oblong, acute at both ends, paler beneath, glabrous or sparingly pubescent: cymes axillary, dichotomous, shorter than the leaves, many-flowered: pedicels short: flowers white, or with the calyx, reddish at the base: berries black or red.

(1) O. SERPENTINUM. (*Willd.*)

Ident. Willd. Syst. 4. p. 079.—Dec. prod. VIII. p. 342.—Roxb. flor. Ind. I. p. 604.—Dalz. Bomb. flor. p. 143.

Engrav. Bot. Mag. 70. t. 784.—Rheede Mal. VI. t. 47.—Rumph. Amb. t. 16.—Wight's Icon. t. 849.

SPEC. CHAR. Very small shrub: leaves opposite or verticelled in threes, oblong-acute, undulated: cymes subterminal from the uppermost axils ; flowers numerous, small, white or rose-coloured, the peduncles and pedicels at length bright red ; berries ovoid, 1-seeded, oblong.

Common in the Concans, Hindostan. Peninsula. Circar mountains. Bengal. Flowering nearly all the year.

(2) O. NEILGHERRENSE. (*R. W.*)

Ident. Wight's Icon. vol. IV.—Dals. Bomb. flor. p. 144.

Engrav. Wight's Spicil. II. t. 159.—Icon. t. 1262.

SPEC. CHAR. Shrubby, erect, glabrous; leaves confined to the terminal branchlets, older branches naked, oblong-elliptic, broader towards the apex, acute at both ends ; corymbs axillary, cymose, trichotomous ; flowers white ; berries connate at the base, ovoid, dark brownish-purple when ripe.

Neilgherries, about Conoor and Kotagherry, fl wering from July to September.

(3) O. MACROCARPUM. (*R. W.*)

Ident. Wight's Spicil. II. p. 53.—Icon. vol. IV.

SPEC. CHAR. Shrubby, glabrous : leaves broad-obovate, elliptic, abruptly acuminated, acute : corymbs axillary, lax : calycine lobes linear-subulate : nut obovate, slightly compressed, tubercled.

Pulney Hills.

(4) O. BELGAUMENSE. (*R. W.*)

Ident. Wight's Spicil. II. p. 53.—Icon. vol. IV.

SPEC. CHAR. Erect glabrous shrub : leaves elliptic, oblong, obtuse or acuminate ; corymbs long-peduncled, compact, many-flowered : flowers on long pedicels ; calycine lobes dilated : tube of corolla long, slender, lobes of the limb before expansion involutely imbricated, forming a round capitulum.

Belgaum.

GENUS V. WRIGHTIA.

Pentandria Monogynia. *Sex : Syst :*

Deriv. Named after the late Dr. William Wright, a Scotch Physician and Botanist, resident in Jamaica.

GEN. CHAR. Calyx 5-parted, with 5 scales or glands at the base, of which 2 are opposite the base of the 2 interior lobes, and the fifth opposite the edge of another lobe, hence they are all nearly alternate with the lobes of the calyx : corolla 5-cleft : tube usually short : lobes twisted at the right in æstivation : throat crowned with appendages, equal or unequal, in the latter case the larger ones

opposite the lobes of the corolla: stamens 5, inserted on the middle
or throat of the tube, protruding: filaments short: anthers sagittate, adhering to the middle of the stigma, ending in a short acute
hairy point: nectary none: ovaries 2, adpressed, glabrous; style
filiform, dilated at the apex: stigma obtuse, sometimes bifid: follicles two, long, either cohering at distinct, sometimes cohering at the
apex only: seeds numerous, oblong, furnished with a tuft of hair
at the interior extremity: coat of the seed double, exterior one
somewhat striated longitudinally, soft with 1 furrow: albumen
none: radicle superior, short: cotyledons oval.—Shrubs or trees :
leaves opposite, entire: cymes terminal.

(1) W. WALLICHII. (*Dec.*)

Ident. Dec. prod. VIII. p. 405.

Engrav. Wight's Icon. t. 1296.—Spicil. II. t. 157.

SPEC. CHAR. Small tree or shrub : leaves elliptic-obovate, obtusely acuminated, covered all over with dark-brown tomentum :
calycine lobes broadly ovate-rounded : scales inside ovate-rounded,
half the length of the lobes : follicles partly connate, cylindric, scabrous with white scales, acute : flowers white.

Slopes of the Neilgherries. Warree country.

(2) W. TOMENTOSA. (*Roem. & Schult.*)

Ident. Roem. & Schult. Syst. IV. p. 414.—Dec. prod. VIII.
p. 404.

Syn. W. pubescens, *Rath.*—Nerium tomentosum, *Roxb. flor.
Ind.* II. p. 6.

Engrav. Wight's Icon. t. 445.—Rheede Mal. 9. t. 3, 4.

SPEC. CHAR. Small tree : leaves elliptic-lanceolate, or elliptic,
attenuated at the base, pubescent with dark-coloured tomentum :
corymbs dense, rigid, terminal : follicles 8-9 inches long, scabrous :
flowers yellowish-white with an orange-coloured throat.

Northern Ghauts. Circars. Concans. Flowering in the hot
season.

(3) W. TINCTORIA. (*R. Br.*)

Ident. R. Br. in Mem. Wern. Soc. I. 73.—Dec. prod. VIII.
p. 406.

Syn. Nerium tinctorium, *Roxb. fl. Ind.* II. p. 4.

Engrav. Burm. Zeyl. t. 77.—Wight's Icon. t. 444.

SPEC. CHAR. Small tree : leaves elliptic-lanceolate and ovate,
obtusely acuminated, membranaceous : panicles terminal, lax, many-
flowered : follicles very long and slender, pendulous : flowers white,
fragrant.

Coromandel. Western Coast. Flowering in April.

(4) W. COCCINEA. *(Sims.)*

Ident. Dec. prod. VIII. p. 407.
Syn. Nerium coccineum, *Roxb. for. Ind.* II. p. 2.
Engrav. Sim's Bot. Mag. t. 2697.—Wight's Icon. t. 442.

SPEC. CHAR. Small tree: leaves ovate-lanceolate, obtuse at the base, acute at the apex, acuminated: cymes 1-3-flowered: calycine lobes rounded, ciliolate: scales linear, many times less than the lobes: appendages of the corona short, obovate, 3-lobed: lobes of the corolla obovate: anthers hispid: follicles linear, spotted white: flowers green without, deep orange-red within, very fragrant.

Khasia Hills, flowering in April.

(5) W. ROTHII. *(G. Don.)*

Ident. Dec. prod. VIII. p. 406.—Don's Syst. Gard. 4. p. 86.
Syn. W. tinctoria, *Roth.*
Engrav. Wight's Icon. t. 1319.

SPEC. CHAR. Leaves oval-lanceolate, acuminate, and, with the cymes, pubescent on both sides: lobes of the calyx oblong-obtuse, pubescent, shorter than the tube of the corolla: scales lanceolate-subulate, pubescent: scales of the crown linear, cleft, scarcely pubescent, about the length of the anthers: anthers pubescent on the points: branchlets pubescent, brownish, the pubescence on the new leaves purplish, on the older ones greyish: corymbs lax, dichotomous: pedicels about an inch long: corolla everywhere pubescent: lobes oblong-obtuse, nearly half an inch long.

Nuggur Hills, near Madras.

GENUS VI. HUNTERIA.

Pentandria Monogynia. *Sex: Syst:*

Deriv. Named after Dr. William Hunter of the Bengal Medical Establishment, a Botanist and Author of the History of Pegu, &c. &c.

GEN. CHAR. Trees or shrubs with opposite or ternate, entire leaves: cymes panicle-shaped, terminal or axillary, few-flowered: calyx 5-cleft, without glands: corolla funnel-shaped, throat naked, lobes five: stamens inserted a little above the middle of the tube of the corolla: anthers ovate, acute, included, twice as long as the filament: ovaries two, ovoid: ovules 2-4 in each ovary: style filiform: stigma ovoid, short, bicuspidate at the apex: berries two, distinct, ovoid, 1 or 2-seeded by abortion: seed with copious albumen: radicle cylindric: cotyledons leafy, elliptic.

(1) H. FASCICULARIS. *(Wall.)*

Ident. Doc. prod. VIII. p. 350.—Wall. Cat. No. 1619.

SPEC. CHAR. Tree : leaves opposite or ternate, elliptic-oblong, obtusely acuminate, acute at the base, reflexed at the margin, glabrous: cymes terminal and axillary, many-flowered; trichotomous : peduncles, pedicels and ovate-acute bracts pubescent : calycine lobes ovate-acute, ciliolate : corolla three times longer than the calyx.

Pundua, Silhet.

(2) H. GRACILIS. *(Wall.)*

Ident. Doc. prod. VIII. p. 350.—Wall. Cat. No. 1013.

SPEC. CHAR. Tree; leaves ternate, oblong, obtusely acuminated, acute at the base, revolute at the margin: cymes axillary and terminal, the length of the leaf; lateral peduncles ternate along the rachis, trichotomous at the apex : bracts ovate-acuminate : calycine lobes ovate-acuminate, ciliolate ; corolla 3 times longer than the calyx.

Silhet.

(3) H. ROXBURGHIANA. *(R. W.)*

Ident. Wight's Icon. vol. IV.

Engrav. Wight I. c. t. 1294.

SPEC. CHAR. Shrubby, branches slender, glabrous : leaves long-petioled, narrow elliptic-lanceolate, slightly involute on the margin, finely veined, shining above, dull below : corymbs axillary, much shorter than the leaves, many-flowered : bracts ovate-acute : lobes of the calyx ovate-acute : tube of the corolla about three times the length of the calyx, hairy within at the insertion of the stamens, lobes ovate obtuse ; berries ovoid, tapering at both ends, 2-seeded.

Courtallum, flowering in August and September.

GENUS VII. CERBERA.

Pentandria Monogynia. *Sex. Syst.*

Deriv. So named from its poisonous qualities ; alluding to the dog Cerberus, whose bite was poisonous.

GEN. CHAR. Small trees with alternate, entire leaves : cymes terminal, 2-3-chotomous, panicle-shaped : calyx 5-partite, without glands : corolla cup-shaped : tube cylindric, 5-ribbed at the throat, ribs longitudinal, alternating with the lobes, sometimes hairy : lobes 5, ovate-acute : stamens inserted towards the middle of the tube ;

anthers linear-lanceolate, cuspidate, many times longer than the filament: ovaries 2: ovules 4, namely two superposed in either part of the ovary, inserted by the middle, erect, amphitropal: style filiform: stigma conical, 10-furrowed at the base, 2-lobed at the apex: drupe often single by abortion of the ovary, elliptic-globose: seeds 2, namely one in each imperfect cell, or solitary, (one abortive), free at the apex, ovate-acuminate: albumen none: embryo inverted: cotyledons ovate-oblong, fleshy: radicle very short, rough.

(1) C. ODALLAM. *(Gærtn.)*

Ident. Gœrtn. fr. II. p. 193.—Dec. prod. VIII. p. 353.—Roxb. fl. Ind. II. p. 692.

Engrav. Wight's Icon. t. 441.—Rheede Mal. I. t. 39.—Sim's Bot. Mag. t. 1845.

Spec. Char. Small tree: leaves approximate at the top of the branches, obversely lanceolate, acuminate, long-narrowed at the base: lateral nerves perpendicular to the centre: calycine lobes lanceolate, long-acuminate, at length revolute, nearly equalling the tube of the corolla: flowers white, fragrant: drupe ovoid-globose.

Sea shores of both Coasts, flowering nearly all the year.

GENUS VIII. TABERNÆMONTANA.

Pentandria Monogynia. *Sex: Syst.*

Deriv. In honor of J. T. Tabernæmontanus, a celebrated Physician and Botanist.

Gen. Char. Trees or shrubs: leaves opposite, entire: cymes axillary, usually twin at the apex of the branchlets: calyx 5-parted: glands linear, 4–7-verticelled.: corolla cup-shaped: tube inflated, often narrower in the middle, without appendages: throat naked, lobes obtuse: stamens inserted on the inflated part of the tube: filaments very short or none: anthers often sagittate, long acuminate, rarely linear, generally included: ovaries two, adpressed: style single, sometimes double at the base near the ovaries: stigma usually annular at the base, 2-lobed at the apex: fruits two (solitary by abortion) linear-oblong, oblong or subglobose, more or less fleshy, pulpy, divaricate: ovules numerous: seeds few or many, nestling in cellulose pulp, obovoid: albumen fleshy: cotyledons leafy, recurved at the apex: radicle superior, cylindric.

(1) T. DICHOTOMA. *(Roxb.)*

Ident. Roxb. fl. Ind. II. p. 21.—Dec. prod. VIII. p. 366.

Engrav. Wight's Icon. t. 433.—Lindl. Bot. Reg. t. 53.

Spec. Char. Shrub: leaves oblong, acute at the base, obtuse at the apex, coriaceous: petiole dilated at the base, amplectant: cymes terminal, dichotomous, many-flowered, nearly equalling the leaf: peduncles naked, long: bracts ovate, small: calycine lobes ovate, obtuse: segments of the corolla somewhat longer than the tube: follicles recurved, acutish, orange-coloured: flowers yellowish-white, slightly fragrant.

Malabar, flowering nearly all the year.

(2) T. RECURVA. (*Roxb.*)

Ident. Roxb. flor. Ind. II. p. 28.—Dec. prod. VIII. p. 371.
Syn. T. gratissima, *Lindl. Bot. Reg.* t. 1084.
Engrav. Wight's Icon. t. 470.

SPEC. CHAR. Shrub: leaves broad-lanceolate, acute at the base, obtusely acuminate at the apex: cymes twin in the forks of the branches, spreading, recurved, much shorter than the leaf, many-flowered: bracts linear-lanceolate: calycine lobes linear-lanceolate, unequal: segments of the corolla oblique, oblong, nearly equalling the tube: flowers yellowish-white, fragrant.

Chittagong, flowering in March and April.

(3) T. CHIPA. (*Roxb.*)

Ident. Roxb. fl. Ind. II. p. 24.—Dec. prod. VIII. p. 371.
Syn. T. alternifolia, *Linn.*
Engrav. Wight's Icon. t. 470.—Rheede Mal. l. t. 46.

SPEC. CHAR. A shrub, with dichotomous branches: leaves oblong-acute, undulated, glabrous: peduncles arising from the forks, few-flowered: pedicels elongated: flowers large, white, the margins of the petals crisped and curled: follicles curved, oblong-acute, 2 inches long, yellow when ripe.

On the Ghauts. Coromandel. Travancore. Flowering from April to June.

(4) T. CORONARIA. (*R. Br.*)

Ident. Dec. prod. VIII. p. 373.—R. Br. in Ait. Hort. Kew. II. p. 72.—Roxb. flor. Ind. II. p. 23.
Syn. T. divaricata, *R. Br.*—Nerium coronarium, *Ait.*—N. divaricatum, *Linn. flor. Zeyl.*
Engrav. Wight's Icon. t. 477.—Rheede Mal. II. t. 54 & 55.—Burm. Zeyl. t. 30.—Sim's Bot. Mag. t. 1865.

SPEC. CHAR. A shrub: leaves opposite, unequal, elliptic-oblong, acute at the base, obtusely acuminated: peduncles from the forks twin, erect, dichotomous, 4 to 6-flowered: flowers white, fragrant at night.

Neera Hills near Penn. Peninsula. Silhet. Bengal. Flowering nearly all the year.

GENUS IX. VINCA.

Pentandria Monogynia. *Sex : Syst :*

Derio. From *Vinculum*, a band; in allusion to the suitableness of the shoots for making bands.

GEN. CHAR. Calyx 5-partite: corolla 5-cleft or 5-lobed, tube narrow funnel-shaped or cylindric, hairy inside, throat angular: filaments short: anthers inflexed, much longer than the filament: ovaries 2: ovules numerous, amphitropal: style usually thickened at the apex, terminating with a reflexed cup-shaped membrane: stigma glandulosely viscid above the membrane, conical or cylindric, hispid at the apex, capitate, obscurely 2-lobed: follicles 2, erect, or diverging, narrow cylindric, striated: seeds numerous, truncated at both ends, darkish, granularly tubercled: albumen fleshy: embryo central: radicle cylindric: cotyledons ovate.

(1) V. PUSILLA. (*Murray.*)

Ident. Dec. prod. VIII. p. 382.—Linn. f. Suppl. p. 166.

Syn. V. parviflora, *Retz.*—*Roxb. fl. Ind.* II. p. 1.—Catharanthus pusillus, *G. Don.*

Engrav. Rheede Mal. IX. t. 33.

SPEC. CHAR. Herbaceous, annual: stem 1 foot, suberect, branched, quadrangular, glabrous: leaves lanceolate, very short-petioled, long-narrowed, scabrous at the edges: calycine lobes narrow linear-acuminate, without glands: lobes of the corolla elliptic-rotund: flowers solitary, white: follicles slender, 1¼ inch long.

Peninsula, in sandy and cultivated soils. Travancore. Common in the Deccan. Flowering in the rainy season.

GENUS X. VALLARIS.

Pentandria Monogynia. *Sex : Syst :*

Derio. From *Vallo*, to enclose; it being used for fences in Java.

GEN. CHAR. Twining shrubs: leaves opposite, entire: cymes solitary, axillary, racemose: calyx 5-partite: lobes lanceolate, without glands: corolla deeply 5-cleft, tube cylindric, without scales, lobes spreading: stamens inserted on the upper part of the tube of the corolla: filament ligulate, hairy: anthers sagittate, adhering to the middle of the stigma, gibbous on the back at the base with a fleshy tubercle: nectary cup-shaped, surrounding the ovary, 5-cleft or partite, ciliated at the apex: ovary globose, pubescent at the apex, 2-celled: style filiform, pubescent: stigma winged at the base, conical, ovate above: ovules numerous: follicle 2-celled, oblong, acute, dehiscing lengthwise: seeds compressed, rough: albumen sparing: embryo straight: radicle superior: cotyledons leafy.

(1) V. Pergularia. *(Barm.)*

Ident. Burm. Ind. p. 51.—Dec. prod. VIII. p. 599.
Syn. Pergularia glabra, *Linn.*—Emericia Pergularis, *Rœm. & Schult.*—Echites hircosa, *Roxb. flor. Ind.* II. p. 18.
Engrav. Wight's Icon. t. 429.—Rumph. Amb. 5, 20. f. 2.— Hook. Icon. L. 153.
Spec. Char. Twining: leaves broad-elliptic, acute at both ends, glabrous; axils glanduliferous: cymes pubescent: lobes of the corolla broad-ovate, acute: flowers white, of unpleasant odour.
Bengal, (rare.) Peninsula. Flowering in May and June.

(2) V. dichotoma. *(Wall.)*

Ident. Dec. prod. VIII. p. 399.—Wall. Cat. No. 1621.
Syn. Echites dichotoma, *Roxb. flor. Ind.* II. p. 10.—V. Heynei, *Spr. Dec. l. c.*—Peltanthera solanacea, *Roth.*
Engrav. Wight's Icon. t. 438.
Spec. Char. Twining: leaves elliptic, acuminated, glabrous: racemes subcorymbose, pubescent: lobes of the corolla obtuse: follicles large, oblong, six inches long, two inches thick: flowers largish, white, rotate.
Concans. Deccan. Bengal. Silhet. Flowering nearly all the year.

GENUS XI. PARSONSIA.

Pentandria Monogynia. *Sex: Syst:*

Deriv. Named after Dr. James Parsons, a Scotch Botanist.
Gen. Char. Calyx somewhat 5-partite, lobes surrounded at the base by a membranaceous scale, without glands: corolla narrow funnel-shaped, 5-cleft, throat and tube without scales, lobes oblong: stamens inserted at the bottom of the corolla: filaments slender: anthers sagittate, exserted: ovary 2-celled: style one: stigma surrounded at the base by a reflexed cup-shaped membrane, 5-cornered in the middle, apiculate and 2-lobed at the apex: follicle 2-celled.

(1) P. spiralis. *(Wall.)*

Ident. Dec. prod. VIII. p. 402.—Wall. Cat. No. 1631.
Spec. Char. Climbing: leaves opposite, broad-ovate or elliptic, very obtuse at the base, acuminated at the apex: peduncles axillary, solitary, many-flowered at the apex and with the flowers externally puberulous: calycine lobes ovate-oblong: follicles ovate-lanceolate.
Silhet.

GENUS XII. BEAUMONTIA.

Pentandria Monogynia. *Ser: Syst*

Deris. In honor of Lady Diana Beaumont of Bretton Hall.

GEN. CHAR. Calyx 5-partite, lobes leafy, with glands: corolla campanulate, narrowed at the base into a tube scarcely longer than the calyx, throat widened, apex 5-lobed, lobes ovate, subacute: stamens inserted at the summit of the tube: filaments filiform, subcompressed: anthers elliptico-sagittate, adhering to the middle of the stigma, with two narrow auricles at the base, terminated at the apex by an acute membrane: nectary from five fleshy glands attenuating with the calycine lobes; ovary 2-celled, immersed in a disk, scarcely exceeding the ovary: ovules numerous: style filiform: stigma oblong, with two tubercles at the apex: fruit oblong, woody, coriaceous, consisting of two connate follicles distinct when ripe: seeds numerous, pendulous, imbricated, ovoid-oblong; albumen fleshy: radicle superior: cotyledons oblong.

(1) B. GRANDIFLORA. (*Wall.*)

Ident. Wall. tent. flor. Nep. p. 15.
Syn. Echites grandiflora, *Roxb. II. B. p.* 20.
Engrav. Wall. l. c. t. 7.—Bot. Reg. t. 911.—Hook. Bot. Mag. t. 3213.

SPEC. CHAR. Climbing, of great extent: leaves opposite, oblong-obovate, narrowed at the base, cuspidate at the apex, puberulous on the nerves below: axils gland-bearing: cymes terminal, 8–12-flowered; peduncles short, thick; bracts large-ovate, villous; flowers white, very fragrant.

Eastern Bengal. Khasia Hills. Flowering in the hot season.

(2) B. JERDONIANA. (*R. W.*)

Ident. Dalz. Bomb. flor. p. 147.
Engrav. Wight's Icon. t. 1314, 1315.

SPEC. CHAR. Climbing: leaves obovate, abruptly acuminate: cymes terminal, many-flowered: corolla large, funnel-shaped: follicles cylindric, 9–10 inches long, 1-thick: calycine lobes narrow-lanceolate.

Coorg jungles. Warree. Canara. Flowering in June.

GENUS XIII. ALSTONIA.

Pentandria Monogynia. *Ser: Syst.*

Deriv. Named after Dr. Alston, Professor of Medicine in Edinburgh.

GEN. CHAR. Calyx 5-partite, without glands: corolla cup-shaped, usually pubescent outside: tube cylindric, without scales: stamens inserted at the middle or above the middle of the tube: filaments very short: anthers oblong-lanceolate, cordate at the base: ovaries two: ovules numerous, compressed: stigma ovoid, 2-lobed at the apex: follicles two, long: seeds oblong, comose at both ends, skin rough: albumen sparing: radicle superior: cotyledons oblong, smooth.

(1) A. SCHOLARIS. *(R. Br.)*

Ident. Dec. prod. VIII. p. 408.

Syn. Echites scholaris, *Linn.*

Engrav. Wight's Icon. t. 422.—Rheede Mal. I. t. 45.—Rumph. Amb. II. t. 82.

SPEC. CHAR. Large tree: leaves verticelled, 5-7, obovate-oblong, acute at the base, obtuse at the apex, glabrous, shining above: cymes globose, composite, many-flowered: peduncles pubescent: flowers greenish-white, subsessile, fascicled, pubescent: margin of the corolla slightly hairy: follicles very long.

Coromandel. Travancore. Hilly parts of the Concan. Bengal as far north as Mirzapore. Flowering in January and February. The flowers emit a peculiar sickening smell, which pervades the whole air about sunset. The bark is a powerful tonic, and the wood as bitter as Gentian.

(2) A. VENENATA. *(R. Br.)*

Ident. R. Brown. Mem. Wern. Soc. I. p. 75.

Syn. Blaberopus venenatus, *Dec. prod.* VIII. p. 411.—Echites venenata, *Roxb. H. B. p. 20.*

Engrav. Wight's Icon. t. 456.—Lodd. Bot. Cat. t. 180.

SPEC. CHAR. Large tree: leaves in fours, oblong-lanceolate, acuminated at both ends: follicles shorter than the leaf, attenuated at both ends: calycine lobes ovate-acute: flowers largish, pure white.

Peninsula. Bengal. Flowering nearly all the year. The juice of the tree is highly poisonous.

GENUS XIV. HOLARRHENA.

Pentandria Monogynia. *See Syst.*

Deriv. From *Olos*, entire, and *Arrhen*, male; in reference to the entire anthers.

GEN. CHAR. Shrubs or small trees: leaves opposite, entire: cymes terminal, corymbose: calyx somewhat 5-partite, lobes lanceolate, acuminate, all or the inner ones only with 1-2 glands at the base on both sides, sometimes without glands: corolla cup-shaped, tube dilated between the base and middle, throat without appendages, contracted, lobes oblong or linear: stamens inserted between the base and middle of the tube: filaments very short, slender: anthers lanceolate, apiculated: ovaries two, ovate-acute: stigma oblong, abruptly acuminated, simple: follicles slender: seeds numerous, pendulous, comose above, oblong-fusiform, striated longitudinally: albumen none: embryo straight: radicle superior, cylindric: cotyledons long-elliptic, cordate at the base.

(1) H. ANTIDYSENTERICA. (*Wall.*)

Ident. Wall. Cat. No. 1672.—Dec. prod. VIII. p. 413.
Syn. Echites antidysenterica, Roxb. H. B. p. 20.
Engrav. Wight's Icon. t. 439.

SPEC. CHAR. Shrub: branches, leaves, and pedicels glabrous: leaves elliptic, very obtuse at the base, acute or abruptly acuminated at the apex: cymes many-flowered, terminal: flowers puberulous, white: follicles 1 foot long.

Concans. Silhet. Chittagong. Flowering in April and May.

(2) H. CODAGA. (*G. Don.*)

Ident. Don. Syst. Gard. 4. p. 78.—Dec. prod. VIII. p. 414.
Syn. H. pubescens, Dec. l. c. p. 413.
Engrav. Wight's Icon. t. 1297.—Rheede Mal. I. t. 47.

SPEC. CHAR. Shrub: leaves ovate-elliptic, short-petioled, obtuse at the base, acute or acuminate at the apex, pubescent: cymes many-flowered: lobes of the corolla oblong, about the length of the tube: cells of the ovary separate: follicles 8-12 inches long, glabrous, tapering near the extremity.

Malabar, flowering in March and April.

GENUS XV. ECHALTIUM.

Pentandria Monogynia. *Lin. Syst.*

Deriv. From *Echalut*, the name in Silhet.

GEN. CHAR. Calyx 5-parted, segments acute: corolla inferior, hypocrateriform, limb 5-parted, tube crowned with 5-forked scales, alternate with the segments of the limb: stamens inserted near the bottom of the tube, included: anthers oblong, pointed, slightly sagittate at the base: ovary 2-lobed, 2-celled, with numerous ovules in each, attached to an elevation down the centre: style short: stigma capitate, bifid: follicles ovate, inflated: seeds numerous, compressed with a broad membranaceous margin: albumen thin, membranous: cotyledons round-cordate, radicle cylindrical.

(1) E. PISCIDIUM. (R. W.)

Ident. Dec. prod. VIII. p. 410.
Syn. Nerium placidium, Roxb. *fl. Ind.* II. p. 7.—Wrightia ? piscidia, *G. Don.*
Engrav. Wight's Icon. t. 472.

SPEC. CHAR. Climbing to a great extent: leaves opposite, elliptic-oblong, acute, obtuse at the base, coriaceous, glabrous: cymes terminal, trichotomous, many-flowered: flowers pale yellow, large.

Silhet. Flowering in May and June. The fibres of the bark are used as a substitute for hemp. The juice is poisonous.

GENUS XVI. STROPHANTHUS.

Pentandria Monogynia. *Sex. Syst.*

Deriv. From *Strophos*, a twisted thong, and *Anthos*, a flower; the segments of the corolla are long, narrow and twisted.

GEN. CHAR. Shrubs often sarmentose: leaves opposite, entire: cymes terminal: calyx 5-parted, lobes with glands: corolla with a funnel-shaped tube, lobes very long, linear-subulate, with 2 appendages between them: stamens inserted in the lower part of the tube: filaments linear, thickish: anthers linear-sagittate, mucronate or long-awned at the apex, awns soft, connivent, not twisted: ovaries two, subglobose: ovules numerous: stigma cylindric: follicles horizontal, thick, obtuse: seeds oblong, comose.

(1) S. WALLICHII. *(Dec.)*

Ident. Dec. prod. VIII. p. 416.
Syn. S. dichotomus, *Wall. No.* 1641.
Engrav. Bot. Reg. t. 469.
Spec. Char. Climbing: leaves elliptic-obovate, subacute at the base, shortly acuminated at the apex: cymes dichotomous, many-flowered: bracts and calycine lobes linear-lanceolate, long, spreading, reflexed at the apex: flowers greenish white.
Silhet. Coromandel. Flowering in the hot season.

(2) S. WIGHTIANUS. *(Wall.)*

Ident. Wight's Icon. vol. IV.
Engrav. Wight I. c. t. 1301.
Spec. Char. Shrubby, twining, glabrous: bark warty: leaves elliptic, acute at both ends, shortly acuminate: lobes of the calyx ovate, about one-third the length of the tube of the corolla: corolla glabrous within, appendices exserted, deeply 2-cleft, lobes filiform: arista of the stamens filiform, longer than the anthers: style somewhat ligulate, with a crisp marginal wing: follicles large, obtuse, warty all over, seed with a long apiculus.

Travancore, frequent about Quilon.

GENUS XVII. CHONEMORPHA.

Pentandria Monogynia. *Sex: Syst.*

Deriv. From *Chone*, a funnel, and *Morphe*, form; alluding to the shape of the corolla.

Gen. Char. Calyx 5-cleft, tubular, funnel-shaped, glandular at the base, lobes acuminate, erect: corolla salver-shaped, tube narrower at the base, cylindric in the middle, without appendages, lobes obovate: anthers sessile, sagittate, lobes at the base short, acute, inflexed, with a very acute point: nectary cup-shaped, thick, entire, or somewhat crenated: ovaries two, oblong: stigma oblong, winged at the base, 2-toothed at the apex: follicles long, linear-acuminate: seeds obovate-oblong, comose above: albumen small; radicle superior.

(1) C. MACROPHYLLA. *(G. Don.)*

Ident. Don. Syst. Gard. IV. p. 76.—Dec. prod. VIII. p. 430.
Syn. Echites macrophylla, *Roxb. fl. Ind.* II. p. 13.
Engrav. Wight's Icon. t. 432.—Rheede Mal. IX. t. 5 & 6.
Spec. Char. Climbing shrub: leaves opposite, entire; cymes

terminal and axillary, composite; peduncles pointed, stiff: flowers very large, white.

Silhet. Banda in the Warree country. Flowering in May and June.

GENUS XVIII. RHYNCOSPERMUM.

Pentandria Monogynia. *Ses: Syst:*

Deriv. From *Rhynchos*, a beak, and *Sperma*, seed.

GEN. CHAR. Calyx deeply 5-cleft, tube campanulate, with many truncated glands at the base inside, lobes oblong: corolla 5-cleft, tube cylindric, lobes obliquely obovate: filaments manifestly adnate with the base of the corolla, anthers hastate: nectary cup-shaped, 5-cleft: ovaries two, longer than the nectary: stigma oblong: follicles long, compressed, narrow: seeds numerous, obovate, compressed below, above narrowed into a slender neck, ending in silky coma: albumen none: embryo straight: radicle superior: cotyledons oblong.

(1) R. ELLIPTICUM. *(Dec.)*

Ident. Dec. prod. VIII. p. 431.

Syn. Echites ellipltica, *Wall.*

SPEC. CHAR. Climbing: leaves opposite, entire, elliptic, cuspidate, acute at the base, glabrous: calyx somewhat 5-parted: lobes subacute, three times shorter than the tube of the corolla; peduncles dichotomous at the apex, and with the pedicels puberulous: flowers white, rose-coloured at the base.

Silhet.

GENUS XIX. AGANOSMA.

Pentandria Monogynia. *Ses: Syst:*

Deriv. From *Aganos*, mild, and *Osme*, smell; alluding to the scent of the flowers.

GEN. CHAR. Shrubs: leaves opposite, entire: cymes terminal and axillary, many-flowered: calyx 5-partite, lobes velvety on both sides, long-lanceolate, more or less glanduliferous within: corolla 5-cleft, tube cylindric, narrower quite at the base, contracted at the apex, about the length of the calyx, lobes long-lanceolate: anthers sessile, inserted in the lower part of the tube, lanceolate-acuminate: nectary cup-shaped or cylindric, 5-lobed: ovaries two, often puberescent: stigma fusiform or cylindric at the base, broader than the style, acuminate, simple or 2-lobed above: follicles two, long: seeds linear-obovate, compressed, comose above: albumen small: embryo straight, axile: radicle superior: cotyledons oblong.

(1) A. CARYOPHYLLATA. *(G. Don.)*
Ident. Dec. prod. VIII. p. 432.
Syn. Echites caryophyllata, *Roxb. flor. Ind.* II. p. 11.—A. Roxburghii, *G. Don.* IV. p. 77.
Engrav. Rheede Mal. 9. t. 14.—Wight's Icon. t. 440.
SPEC. CHAR. Climbing: leaves elliptic, sharply acuminated or obtuse: cymes nearly equalling the leaf: bracts lanceolate, white-puberulous at the back, calycine lobes linear-lanceolate, whitish pubescent outside: flowers white, very fragrant.
Hilly parts of India near Mongbyr.

(2) A. ELEGANS. *(G. Don.)*
Ident. Dec. prod. VIII. p. 433.—Don. Syst. IV. p. 77.
Syn. Echites elegans, *Wall.*
Engrav. Wight's Icon. t. 1304.
SPEC. CHAR. Erect ramous shrub: leaves obovate-elliptic, acute, cuspidate, subacute at the base, glabrous: cymes shorter than the leaves: flowers crowded: bracts lanceolate acuminate, the length of the pedical: pedicels and flowers externally whitish-pilose: lobes of the calyx as long as the pedicels, long-lanceolate, about the length of the tube of the corolla: flowers pale-yellow.
Courtallum. Foot of the Neilgherries. Malabar.

(3) A. BLUMII. *(Dec.)*
Ident. Dec. prod. VIII. p. 433.
Syn. Echites caryophyllata, *Blume.*
Engrav. Rheede Mal. VII. t. 35.—Wight's Icon. t. 1305.
SPEC. CHAR. Leaves oval, acutish at both ends, beneath and the ramuli pubescent: corymbs terminal, spreading, lobes of the calyx as long as the tube of the corolla.
Balaghaut mountains near Madras.

(4) A. CYMOSA. *(G. Don.)*
Ident. Dec. prod. VIII. p. 433.
Syn. Echites cymosa, *Roxb. flor. Ind.* II. p. 16.—E. conferta, *Wall.*
Engrav. Wight's Icon. t. 393.
SPEC. CHAR. Shrubby, hairy: leaves elliptic, acuminated: cymes terminal, shorter than the leaves, segments of the corolla oblique-ensiform: nectary cup-shaped, 5-toothed: flowers small, white, calyx and corolla hoary outside.
Silhet.

(5) A. MARGINATA. *(G. Don.)*

Ident. Don. Syst. Gard. IV. p. 77.—Dec. prod. VIII. p. 433.
Syn. Echites marginata, Roxb. flor. Ind. II. p. 15.
Engrav. Wight's Icon. t. 425.
SPEC. CHAR. Leaves lanceolate, smooth: panicles terminal, lax, corymbose, at first subtrichotomous, afterwards dichotomous, glabrous: segments of corolla linear, falcate; nectary annular.
Silhet. Chittagong.

(6) A. ACUMINATA. *(G. Don.)*

Ident. Dec. prod. VIII. p. 434.—Don's Syst. Gard. IV. p. 77.
Syn. Echites acuminata, Roxb. flor. Ind. II. p. 15.
Engrav. Wight's Icon. t. 424.
SPEC. CHAR. Leaves from oblong to broad-lanceolate, acuminated, glabrous: panicles axillary, longer than the leaves, trichotomous, diffuse: segments of corolla linear, falcate, curled.
Silhet.

(7) A. DONIANA. *(R. W.)*

Ident. Wight's Icon. vol. IV.—Dalz. Bomb. flor. p. 146.
Engrav. Wight's Icon. t. 1306.
SPEC. CHAR. Everywhere glabrous except the inflorescence: leaves elliptic, cuspidately acuminate: corymbs terminal, compact, pilose: lobes of the calyx linear-lanceolate, pilose, longer than the externally pilose tube of the corolla: lobes of the corolla shorter than the tube, nectarial scales all united, about the length of the very hairy ovary: follicles terete, tomentose, divaricated.
Phoonda Ghaut, (Bombay.)

(8) A. CONCANENSIS. *(Dalz.)*

Ident. Dalz. Bomb. flor. p. 147.
Engrav. Hook. Icon. Pl. IX. t. 841.
SPEC. CHAR. Climbing, glabrous: leaves broad-elliptic, ovate, very shortly acuminated, cordate at the base, membranaceous: peduncles axillary, shorter than the leaf: cymes compound, dense: sepals triquetrous, acuminate: corolla tube short, divisions of the limb oblong-obtuse, spreading: stamens exserted, glands 5, large, obtuse.
South Concans.

GENUS XX. ICHNOCARPUS.

Pentandria Monogynia. *Ser: Syst:*

Deriv. From *Ichnos*, a vestige, and *Karpos*, a fruit; alluding to the slender follicles.

GEN. CHAR. Calyx 5-cleft: corolla salver-shaped: segments of the limb dimidiate, twisted, hairy; throat and tube without any scales: stamens inclosed: anthers sagittate, free from the stigma: ovaries twin; style filiform: stigma ovate-acute: hypogynous threads 5, capitate, alternating with the stamens: follicles slender.

(1) I. FRUTESCENS. *(R. Br.)*

Ident. Dec. prod. VIII. p. 435.—R. Br. in H. Kew. Vol. II. p. 69.

Syn. Apocynum frutescens, *Linn.*—Echites frutescens, *Roxb. flor. Ind.* II. p. 12.

Engrav. Wight's Icon. t. 430.—Burm. Zeyl. t. 12. f. 1.

SPEC. CHAR. Climbing shrub: leaves opposite, small, elliptic-acute at the ends, hairy beneath: cymes terminal, many-flowered, covered with reddish tomentum: flowers very small, white: follicles 6 inches long.

Common in the Warree country, climbing over trees. Southern Mahratta country. Dharwar. Bengal. Silhet. Flowering in the rainy season.

GENUS XXI. POTTSIA.

Pentandria Monogynia. *Ser: Syst:*

GEN. CHAR. Calyx 5-parted: corolla salver-shaped, tube cylindric: lobes ovate-oblong: filament inserted on the throat, decurrent to the middle of the tube and there densely pubescent: anthers exserted, hastate, bicaudate at the base, adhering to the middle of the stigma: nectary of 5 linear-lanceolate glands, longer than the calyx and ovaries and alternating with the calycine lobes: ovaries two, obovoid, pubescent without, especially at the apex: style narrow fusiform: stigma ovoid, pentagonal, acute at the apex, simple.

(1) P. OVATA. *(Dec.)*

Ident. Dec. prod. VIII. p. 442.

Syn. Parsonsia ovata, *Wall.*—Pottsia Cantoniensis, *Hook. & Arn. in Beech. Voy.* p. 198. *(Cum fig.)*

SPEC. CHAR. Shrub: leaves opposite, entire, petioled, glabrous, ovate-acuminate, somewhat cordate at the base: cymes terminal, longer than the leaf, dichotomous or trichotomous, lax: bracts minute, ovate-acuminate.

Khasia Hills.

GENUS XXII. ECDYSANTHERA.

Pentandria Monogynia. *Ser. Syst.*

Deriv. From *Ecdysis*, escaping out from, and *Anthera*, the anthers protrude from the tube of the corolla.

GEN. CHAR. Shrubs: leaves opposite, entire: peduncles axillary and twin at the tops of the branches, many times trichotomous: bracts lanceolate, opposite: calyx 5-parted: corolla subcampanulate, 5-cleft, throat and tube without scales, lobes ovate, equal-sided, spreading: stamens inserted on the middle of the tube, Included: filaments very short: anthers sagittate-oblong, scarcely exceeding the tube of the corolla: ovaries two, villous: style short: stigma conico-capitate, bicuspidate: follicles long, slender, remotely inflated: seeds comose above.

(1) E. MICRANTHA. (*Dec.*)

Ident. Dec. prod. VIII. p. 442.

SPEC. CHAR. Leaves lanceolate, acuminated, glabrous: cymes somewhat puberulous: calycine lobes ovate-acuminate, reflexed at the apex, pubescent externally, twice as short as the tube of the corolla: flowers small, rose-coloured.

Pundua, Silhet.

(2) E. BRACHIATA. (*Dec.*)

Ident. Dec. prod. VIII. p. 443.

SPEC. CHAR. Climbing: leaves lanceolate, acuminate, glabrous: cymes somewhat puberulous at the apex: calycine lobes ovate, subacute, somewhat puberulous externally: flowers small, rose-coloured.

Pundua, Silhet.

GENUS XXIII. ANODENDRON.

Pentandria Monogynia. *Ser. Syst.*

Deriv. From *Ano*, above, and *Dendron*, a tree; alluding to its high climbing habits.

GEN. CHAR. Calyx 5-parted: corolla deeply 5-cleft:' tube cylindric, slightly hairy inside above the stamens, lobes oblong, ciliolate: anthers subsessile, sagittate, inserted below the margin of the tube, terminated by a point: nectary cup-shaped, entire, waved above: ovaries two: style short: stigma ovoid-acute, surrounded at the base by a reflexed membrane, 2-lobed at the apex: follicles attenuated from an ovoid base, somewhat woody, polished: seeds obovate, compressed, attenuated above into a neck, comose, albuminous: embryo axile: cotyledons ovate, subcordate: cymes axillary and terminal, trichotomous, panicled: flowers very numerous, small.

(1) A. PANICULATUM. *(Dec.)*

Ident. Dec. prod. VIII. p. 423.
Syn. Echites paniculata, Roxb. *flor. Ind.* II. p. 17.
Engrav. Wight's Icon. t. 398.

SPEC. CHAR. Large climbing shrub, glabrous: leaves opposite, entire, obtusely cuspidate, coriaceous: cymes axillary and terminal, trichotomous, panicled: flowers very numerous, small, pale-yellow: follicles attenuated upwards from an ovoid base, somewhat woody, smooth, 4 inches long.

Very common on the Ghauts. Silhet. Flowering in March and April.

GENUS XXIV. ECHITES.

Pentandria Monogynia. *Sat: Syst:*

Deriv. From *Echis*, a viper; alluding to the smooth twining shoots.

GEN. CHAR. Calyx 5-partite, lobes with glands or scales: corolla salver-shaped, tube cylindric, sometimes funnel-shaped at the apex only, usually bispid within above the insertion of the stamens: anthers inserted where the tube of the corolla is broader, subsessile, sagittate: nectary of 5 glands alternating with the calycine lobes, free or more or less connate, sometimes 2 or 3 connate, the others distinct: ovaries two: ovules numerous: stigma capitate, ovoid, surrounded at the base with an entire umbrella-shaped reflexed membrane, simple or 2-lobed at the apex: follicles two, long, cylindric or torulose, coriaceous: seeds linear-oblong, keeled below, comose above: albumen sparing: embryo axile: cotyledons smooth: radicle superior.

(1) E. MALABARICA. *(Lam.)*

Ident. Lam. Dict. II. p. 342.—Dec. prod. VIII. p. 477.
Engrav. Rheede Mal. IX. t. 12.

SPEC. CHAR. Climbing: leaves opposite, entire, elliptic, obtuse at both ends : cymes axillary, dichotomous, few-flowered : calycine lobes lanceolate ; follicles linear, narrow : flowers small, white.

Malabar.

(2) E. PARVIFLORA. (*Roxb.*)

Ident. Roxb. flor. Ind. II. p. 20.—Dec. prod. VIII. p. 478.
Engrav. Wight's Icon. t. 423.

SPEC. CHAR. Climbing: leaves opposite, lanceolate: panicles terminal and axillary, divaricate: tube of the corolla gibbous at the base, lobes linear-falcate, nectary with an entire ring : follicles large, ovate: flowers small, white.

Northern Sircars. Silhet.

GENUS XXV. ELLERTONIA.

Pentandria Monogynia. *Ser: Syst:*

Deriv. Named in honor of J. Ellerton Stocks, of the Bombay Medical Establishment.

GEN. CHAR. Calyx 5-cleft, lobes ovate-acute, without glands: corolla hypocrateriform, 5-lobed, tube ventricose near the middle : stamens included, filaments short, anthers lanceolate, cohering round the stigma, cordate at the base, longer than the filaments ; ovaries 2, distinct, united at the apex by the style, oblong, furrowed, 2-cleft at the apex : style filiform : stigma conical, pronged into a 2-cleft apiculus : follicles terete, divaricated ; with two rows of seed : seed compressed, peltate, winged at each end : radicle superior.

(1) E. RHEEDII. (*R. W.*)

Ident. Dalz. Bomb. flor. p. 146.—Wight's Icon. vol. IV.
Engrav. Wight's Icon. t. 1295.—Rheede Mal. IX. t. 14.

SPEC. CHAR. Scandent shrub: leaves opposite or 3-4-verticelled, elliptic, acuminate : corymbs axillary or several from the ends of the branches, longish-peduncled, cymose, many-flowered.

Malabar. Quilon. Warree country.

GENUS XXVI. HELIGME.

Pentandria Monogynia. *Ser: Syst:*

Deriv. From *Helix*, a screw ; alluding to the spirally twisted filaments.

GEN. CHAR. Calyx 5-cleft : corolla rotate, tube short, ventrigous, segments oblique ; stamens exserted ; style spirally circumvolute: anthers sagittate: stigma clavate: ovary didymous, surrounded by five hypogynous scales: follicles cohering: seeds comose at the upper extremity.

(1) H. Rheedii. *(R. W.)*

Ident. Dalz. Bomb. flor. p. 146.
Engrav. Wight's Icon. t. 1303.

Spec. Char. Twining, glabrous: leaves ovate-acute, short petioled: corymbs trichotomous, many-flowered: calyx lobes ovate, obtuse, ciliated: corolla hairy within: filaments twisted into a spiral column: follicles 2-celled: seeds comose at the apex.

Banda, in the Warree country. Travancore.

ORDER CVIII. ASCLEPIADACEÆ.

Calyx 5-parted, sepals usually furnished with glands at the base within: æstivation quincuncial: corolla monopetalous, hypogynous, 5-cleft: throat naked or variously crowned with glands or appendages, below more or less extensively adnate with the tube of the stamens: æstivation subvalvate (the very edge overlapping and therefore, strictly speaking, imbricate), rarely contorted: stamens 5, inserted into the base of the corolla, alternate with its lobes: filaments cohering, forming a tube round the pistil (stylostegium or gymnostegium), rarely free: anthers erect, introrse, 2-celled or incompletely 4-celled: cells perpendicular or transverse, apex simple, truncated, acuminate, or fringed with a fine membrane: pollen at the period of the dehiscence of the anther, either cohering in masses equal to the number of the cells, or occasionally cohering in pairs, or four together, or granular: when simply equal to the cells, attached by pairs, one from each of two adjoining anthers, to the descending process of the stigmatic corpuscules: when more numerous (as in Periploceæ), adhering to the dilated apex of the corpuscules: ovaries two: ovules numerous: styles two, closely approaching each other, usually very short; stigma common to both styles, dilated, 5-cornered, the corners corpusculiferous: corpuscules either cartilaginous, bright shining brown,

oblong, sulcated down the middle, and produced below into two slender processes (in Asclepiadeæ veræ), or contracted below into a slender neck and dilated into a membranaceous expansion above (as in Periploceæ): fruit follicular, follicles 2 or 1 by abortion; placentæ attached to the suture, separating in dehiscence: seeds numerous, imbricated, pendulous, almost always comose at the hilum: albumen wanting or thin: embryo straight: cotyledons foliaceous: radicle superior: plumule inconspicuous.
—Twining or erect shrubs with milky juice, or herbaceous, or very succulent perennials with watery juice; leaves entire, opposite, rarely whorled: often furnished at the insertion with glands or hairs in lieu of stipules: inflorescence extra axillary, racemose, corymbose or more generally umbelled: flowers presenting various shades of red, yellow or white but rarely blue; sometimes fragrant, occasionally, as in nearly the whole tribe of Stapleæ, exceedingly fetid.

GENUS I. CRYPTOSTEGIA.

Pentandria Digynia. *Sex: Syst:*

Deriv. From *Kryptos*, hidden, and *Stege*, a covering; alluding to the corona being concealed within the tube of the corolla.

GEN. CHAR. Calyx 5-partite: segments long, lanceolate, undulated at the margin; corolla campanulately funnel-shaped: tube furnished within with five included narrow bipartite scales (segments subulate) covering the anthers and opposite to them; stamens included: filaments distinct, very short, inserted at the bottom of the tube; anthers included, adhering at the base to the margin of the stigma: pollen-masses solitary, attached to the spathulate (at length free appendage) of each corpuscule which is agglutinated to the lowest margin of the nearest anthers: stigma globose-conical, obscurely bipiculate at the apex: follicles much divaricated, acutely triquetrous, obtusely attenuated at the apex, incurved: seeds comose.

(1) C. GRANDIFLORA. *(R. Br.)*

Ident. Dec. prod. VIII. p. 402.—Wight's Contrib. p. 66.
Syn. Nerium grandiflorum, *Roxb. fl. Ind.* II. p. 10.
Engrav. Bot. Reg. V. t. 45.—Wight's Icon. t. 832.

SPEC. CHAR. Climbing shrub, milky; leaves opposite, shortpetioled, elliptic, obtusely acuminate, shining on both sides: corymbs trichotomous, terminal: flowers large, internally white, externally pale-rose.

Roxburgh states that the native country of this plant is the Peninsula, but Dr. Wight had never seen it except in a state of cultivation. It is in flower nearly all the year.

GENUS II. FINLAYSONIA.

Pentandria Digynia. *Ses: Syd:*

Deriv. In honor of Dr. Finlayson, Surgeon in the E. I. C.) an ardent Naturalist.

GEN. CHAR. Corolla rotate; throat crowned with 5 awned tubercles alternating with the segments: awns capillaceous, erect, hooked at the apex: filaments distinct, inserted at the throat between the tubercles: anthers agglutinated at the base to the middle of the stigma, terminated by a broad-ovate apiculum, beardless: pollen-masses 20, granular, attached by fours to the dilated appendage of each slender corpuscule: stigma large-ovate, 5-cornered at the base: follicles much divaricated, ventricosely ovate, beakedly hooked at the apex, smooth: seeds sparingly fleshy.

(1) F. OBOVATA. *(Wall.)*

Ident. Dec. prod. VIII. p. 494.—Wight's Contrib. p. 63.

Syn. Gurus obovata, *Buch.*

Engrav. Wall. Pl. As. Rar. II. t. 162.

SPEC. CHAR. Twining shrub, fleshy, milky, glabrous: leaves opposite, petioled, obovate, very obtuse: corymbs largish, many-flowered, shorter than the leaf, fructiferous ones much elongated: flowers small, yellowish brown.

Boonderbunds.

GENUS III. HEMIDESMUS.

Pentandria Digynia. *Ses: Syd:*

Deriv. From *Hemisus*, half, and *Desmos*, a tie; alluding to the filaments.

GEN. CHAR. Corolla rotate: scales 5, mutic, inserted under the sinuses: filaments connate at the base, distinct above, inserted on the tube of the corolla: anthers cohering, free from the stigma, beardless, simple at the apex: pollen-masses 20, granular, attached by fours to the kidney-shaped solitary appendage of each corpuscule: stigma mutic: follicles cylindric, much divaricated, smooth: seeds comose.

(1) H. INDICUS. (R. Br.)

Ident. Dec. prod. VIII. p. 494.—Wight's Contrib. p. 63.

Syn. Periploca Indica, *Willd.*—Asclepias pseudosarsa, *Roxb. flor. Ind.* II. p. 39.

Engrav. Rheede Mal. 9. t. 34.—Burm. Zeyl. t. 83. f. 1.—Wight's Icon. t. 594.

Spec. Char. Twining: leaves opposite, from cordate-ovate cuspidate to narrow-linear, acute, often oblong-lanceolate : cymes often subsessile, somewhat peduncled : scales of the corolla obtuse, cohering to the tube from the base to the apex: follicles slender, straight : flowers on the outside pale-green, on the inside dark blood-coloured.

Peninsula everywhere, flowering in the rainy season.

(2) H. PUBESCENS. (*W. & A.*)

Ident. Wight's Contrib. p. 63.—Dec. prod. VIII. p. 495.

Engrav. Wight's Icon. t. 1320.

Spec. Char. Twining: leaves opposite, lanceolate, and with the branchlets, peduncles and calyx pubescent : cymes sessile : scales of the corolla affixed to the tube: flowers dark purple.

Kulhutty, Neilgherries. Vendalore.

GENUS IV. BRACHYLEPIS.

Pentandria Digynia. *Sex : Syst :*

Deriv. From *Brachys*, short, and *Lepis*, a scale.

Gen. Char. Corolla subrotato, 5-partite : segments spreading : throat crowned with 5 short truncated mutic scales alternating with the segments : tube short, clothed within with a broad fleshy adnate ring : filaments very short, broad, distinct, inserted with the scales at the top of the throat : anthers agglutinated at the base to the margin of the stigma, simple and cohering at the apex, but otherwise distinct, beardless : pollen-masses 20, granular, attached to the free dilated appendage of each corpuscule : stigma mutic : follicles much divaricated, cylindric, smooth.

(1) B. NERVOSA. (*W. & A.*)

Ident. Wight's. Contrib. p. 64.—Dec. prod. VIII. p. 495.

Engrav. Wight's Icon. t. 1284.

Spec. Char. Twining shrub : branchlets pubescent : leaves opposite, oval, abruptly acuminate, younger ones somewhat pubescent, adult ones glabrous, parallel-nerved below: cymes interpetiolar, often bifid, small, tomentose, shorter than the leaf: corolla hirsute outside: flowers small, purple, surrounded with much whitish hair.

Neilgherries about Conoor and Kotagherry.

GENUS V. DECALEPIS.

Pentandria Digynia. *Sex; Syst*.

Deriv. From *Decas*, ten, and *Lepis*, a scale.

GEN. CHAR. Corolla subrotate, 5-partite: throat crowned with 5 oblong-oval mutic scales alternating with the segments: tube furnished within at the base with 5 linear-obtuse scales, the segments opposite: filaments distinct, inserted at the top of the throat with 5 superior scales and cohering with the same at the middle: anthers agglutinated at the base to the margin of the stigma, distinct above, terminated by a somewhat dilated appendage, beardless: pollen-masses granular, solitary, attached to the dilated free appendage of each corpuscule: stigma mutic.

(1) D. HAMILTONII. (*W. & A.*)

Ident. Wight's Contrib. p. 64.—Dec. prod. VIII. p. 495.
Syn. Apocynum reticulatum, *Wall. asclep. No.* 139.
Engrav. Wight's Icon. t. 1285.

SPEC. CHAR. Twining: branchlets thickened at the joints: leaves opposite, obovate-cuneate, retusely acuminate, coriaceous: cymes racemose: bracteoles numerous, small, ovate, pubescent: exterior lobes of the corolla pubescent, densely hairy within: flowers small.

Balaghaut mountains, near Madras.

GENUS VI. STREPTACAULON.

Pentandria Digynia. *Sex; Syst*.

Deriv. From *Streptos*, twisted, and *Kaulon*, the stem.

GEN. CHAR. Corolla rotate, 5-parted: throat crowned with 5 short-awned scales alternating with the segments: awns flexuously erect, filiform, straight at the apex: filaments distinct, inserted on the tube: anthers adhering at the base to the margin of the stigma, free above, simple at the apex, beardless: pollen-masses granular, solitary, attached to the dilated free appendage of each corpuscule: stigma mutic: follicles cylindric, much divaricated, smooth: seeds comose.

(1) S. KLEINII. (*W. & A.*)

Ident. Wight's Contrib. p. 63.—Dec. prod. VIII. p. 495.

Spec. Char. Twining: branches pubescent: leaves opposite, subsessile, cuneate-oblong or obovate, mucronate, cordate at the base, hispid above, softly white tomentose below: cymes diffuse, shorter than the leaf, and with the calyx and pedicels densely tomentose.

Locality unknown.

GENUS VII. SECAMONE.

Pentandria Digynia. *Ser : Syst :*

Deriv. Altered from *Squamona*, the Arabic name of *S. Ægyptica*.

Gen. Char. Corolla rotate; staminal corona 5-leaved: leaflets laterally compressed, affixed by the longitudinal margin, averted, simple: pollen-masses 20, erect, affixed by fours to the apex of each corpuscule: stigma narrowed at the apex: follicles smooth.

(1) S. EMETICA. (*R. Br.*)

Ident. Dec. prod. VIII. p. 501.—Wight's Contrib. p. 60.

Syn. Periploca emetica, *Retz.*

Engrav. Wight's Icon. 1. 1283.

Spec. Char. Twining, glabrous: leaves opposite, from elliptic to narrow-lanceolate: cymes dichotomous, shorter than the leaf, few or many-flowered: corolla glabrous: leaflets of the staminal corona, cultriform: follicles slender, attenuated at the apex: flowers small.

Peninsula. Bengal.

GENUS VIII. CONIOSTEMMA.

Pentandria Digynia. *Ser : Syst :*

Deriv. From *Gonia*, an angle, and *Stemma*, the crown; alluding to the shape of the corona.

Gen. Char. Corolla rotate, 5-partite: staminal corona, gamophyllous, tubular, 5-cornered, 5-lobed, adhering to the base of gynostegium: pollen-masses 20, erect, affixed by fours to the apex of each corpuscule: stigma beaked.

(1) O. ACUMINATUM. (R. W.)

Ident. Wight's Contrib. p. 62.—Dec. prod. VIII. p. 504.

Spec. Char. Twining shrub: bark somewhat warty: leaves opposite, oblong-elliptic, acuminated at the apex, shining above: cymes panicle-shaped, lax, many-flowered: segments of the corolla ligulate, pubescent within to the middle, glabrous above: staminal corona shorter than the gynostegium, fleshy: stigma cylindric.

Silbet.

GENUS IX. TOXOCARPUS.

Pentandria Digynia. *Sex: Syst:*

Deriv. From *Toxos*, a bow, and *Karpos*, fruit; alluding to the shape of the follicle.

Gen. Char. Corolla rotate, 5-cleft: staminal corona 5-leaved: leaflets nearly flat at the back, increased within by a small incision: pollen-masses 20, erect, affixed by fours to the apex of each corpuscule; stigma beaked, rarely apiculate, undivided: follicles smooth, divaricate: seeds comose.

(1) T. KLEINII. (*W. & A.*)

Ident. Wight's Contrib. p. 61.—Dec. prod. VIII. p. 505.

Engrav. Wight's Icon, t. 886.

Spec. Char. Twining: stems glabrous: younger branchlets pubescent: leaves elliptic, shortly and suddenly acuminated: corymbs subaxile: branches divaricate, longer than the leaf: flowers pedicelled: segments of the corolla ligulate: throat somewhat hairy: leaflets of the staminal corona bidentate-truncate, bearing at the apex an inner flat linear long-exserted incision exceeding the anthers and somewhat 3-toothed at the apex: stigma beaked, shortly 2-cleft at the apex, slightly exceeding the tube of the corolla: follicles archedly reflexed.

Common in hedges at Vellagany, near Negapatam.

(2) T. GRIFFITHSII. (*Dec.*)

Ident. Dec. prod. VIII. p. 505.

Spec. Char. Twining: branchlets slender, clothed with short deep brown pubescence: leaves oblong, acuminate, paler beneath: cymes peduncled, shorter than the petiole: segments of the corolla ligulate, hairy at the throat: leaflets of the staminal corona combined at the base, acuminated, acute inwardly with a small incurved toothlet, equalling the anthers: stigma beaked, thickened at the apex, 2-lobed: lobes roundish.

Near Madras.

(3) T. ROXBURGHII. *(W. & A.)*

Ident. Wight's Contrib. p. 61.—Dec. prod. VIII. p. 505.
Syn. Asclepias longistigma, *Roxb. flor. Ind.* II. p. 46.
Engrav. Wight's Icon. t. 475.

SPEC. CHAR. Twining: branchlets rusty-pubescent: leaves broad-oval, suddenly and shortly acuminate: corymbs short-peduncled, with divaricate branches nearly equalling the leaf: flowers subsessile: throat of the corolla hirsute: segments ligulate, glabrous: leaflets of the staminal corona ovate, acutish, with an interior short thick acute, scarcely exserted incision, equalling the anthers: stigma beaked, twisted, equalling the tube of the corolla.

Circars.

(4) T. CRASSIFOLIUS. *(R. W.)*

Ident. Wight's Contrib. p. 61.—Dec. prod. VIII. p. 506.
Syn. Secamone crassifolia, *Wall. asclep. No.* 101.

SPEC. CHAR. Branchlets sparingly pubescent; leaves short-petioled, oval, acuminate, coriaceous, glabrous: corymbs panicle-shaped, sessile, the branches elongated, and with the calyx, brownish-pubescent: flowers sessile, fascicled, small: segments of the corolla reflexed, densely white-villous: leaflets of the staminal corona with reflexed sides and forming a dorsal keel, the inner incision acuminated, long-exserted, somewhat equalling the stigma: stigma beaked.

Silhet.

(5) T. LAURIFOLIUS. *(R. W.)*

Ident. Wight's Contrib. p. 61.—Dec. prod. VIII. p. 506.
Syn. Asclepias laurifolia, *Roxb. flor. Ind.* II. p. 49.—A. micrantha, *Roxb. l. c.* p. 50.
Engrav. Wight's Icon. t. 598.

SPEC. CHAR. Twining: leaves oval, often acuminated, coriaceous: corymbs sessile, divaricately trichotomously branched; flowers very numerous, small: segments of the corolla reflexed, whitish-bearded within: leaflets of the staminal corona somewhat fleshy, short, round-ovate, acutish, the inner incision shortly exserted, somewhat equalling the gynostegium: tops of the anthers broad, somewhat crested, incurved above the obconical apiculated obtuse included stigma: follicles slender, horisontally diverging.

Chittagong. Tipperah. Cawnpore. Flowering in the rainy season.

GENUS X. RAPHISTEMMA.

Pentandria Digynia. *Sex: Syst:*

Deriv. From *Raphis*, a needle, and *Stemma*, a crown.

GEN. CHAR. Corolla campanulate, limb 5-partite; staminal corona 5-leaved: leaflets compressed, elongated: anthers terminated by a membrane: pollen-masses affixed under the apex, pendulous; stigma obtusely conical: follicles often (by abortion) solitary, subventricose: seeds comose.

(1) R. PULCHELLUM. *(Wall.)*

Ident. Wight's Contrib. 34.—Dec. prod. VIII. p. 416.
Syn. Asclepias pulchella, Roxb. flor. Ind. II. p. 54.
Engrav. Wall. Pl. As. Rar. II. t. 163.
SPEC. CHAR. Glabrous twining shrub; leaves largish, cordate; flowers corymbose, whitish.

Silhet.

GENUS XI. CYNOCTONUM.

Pentandria Digynia. *Sex: Syst:*

Deriv. From *Kuon*, a dog, and *Kteno*, to kill; alluding to its poisonous qualities.

GEN. CHAR. Perennial herbs or shrubs often twining: leaves cordate, often with diphyllous axils: peduncles extra-axillary, many-flowered: calyx 5-parted: corolla rotate, 5-partite: staminal corona tubular, plicate, simple, 5-crenate or 10-cleft at the mouth: anthers terminated by a membrane: pollen-masses club-shaped, slightly compressed, affixed to the attenuated apex, pendulous: stigma flat, 2-lobed, or attenuated into a 2-cleft papillose point: follicles slender, smooth, reflexed: seeds comose.

(1) C. CORYMBOSUM. *(Dec.)*

Ident. Dec. prod. VIII. p. 528.
Syn. Cynanchum corymbosum, *Wight's Contrib.* p. 56.

SPEC. CHAR. Twining; leaves ovate, cordate, with parallel auricles, attenuated at the apex, glaucous below, long-petioled, glanduliferous above the petiole; peduncles many-flowered: staminal corona 10-crenated at the mouth: toothlets alternate, shorter: inner keels simple: pollen-masses affixed below the apex; stigma with an 8-cleft point.

Silhet.

(1) C. PAUCIFLORUM. (*Dec.*)

Ident. Dec. prod. VIII. p. 528.
Syn. Cynanchum pauciflorum, R. Br.—*Wight's Contrib.* p. 50.
—Asclepias tunicata, *Roxb. fl. Ind.* II. p. 36.—Periploca tunicata, *Retz.*
Engrav. Wight's Icon. t. 854.

SPEC. CHAR. Twining: leaves ovate, reniform-cordate at the base, with diverging auricles: petioles with reniform leaflets stalked at the base: peduncles few-flowered, shorter than the petiole: flowers short-pedicelled: staminal corona plicate, 10-lobed at the mouth: lobes opposite the anthers lanceolate-acuminate, 2-cleft, the alternate ones very short, truncated and emarginate: pollen-masses affixed below the apex: stigma apiculate, obtuse: follicles attenuated, smooth: flowers rust-coloured.

Poonah. Peninsula. Flowering in the rainy season.

(3) C. CALLIALATUM. (*Dec.*)

Ident. Dec. prod. VIII. p. 528.
Syn. Cynanchum callialatum, *Ham.* in *Wight's Contrib.* p 56.
Engrav. Wight's Icon. t. 1279.

SPEC. CHAR. Twining: leaves ovate or oval, cordate at the base, with a narrow sinus, glaucous below, glanduliferous above at the petiole: axils diphyllous: peduncles many-flowered: flowers subumbellate: staminal corona 10-lobed at the mouth: lobes opposite the anthers, 2-cleft at the apex, alternate ones very short: stigma subapiculate, entire: follicles winged.

Colomala. Bengal.

(4) C. ALATUM. (*W. & A.*)

Ident. Dec. prod. VIII. p. 529.
Syn. Cynanchum alatum, *Wight's Contrib.* p. 57.
Engrav. Wight's Icon. t. 1280.

SPEC. CHAR. Twining: leaves of the older branches cordate-auriculate, of the younger floriferous branchlets oval, cuspidate, cordate or emarginate at the base, glaucous beneath: umbels nearly equalling the petiole: pedicels longer than the peduncle: staminal corona with a truncated mouth, 10-crenated: toothlets alternate, a little shorter: stigma apiculate, 2-cleft: follicles flattish on one side, winged at the margins.

Neilgherries. Courtallum. Flowering in the rainy season.

(5) C. ANGUSTIFOLIUM. *(Dec.)*

Ident. Dec. prod. VIII. p. 526.
Syn. Cynanchum angustifolium, *Wight's Contrib.* p. 57.
SPEC. CHAR. Twining: leaves linear-lanceolate, cuspidate, more or less cordate at the base; petioles puberulous, glanduliferous above; peduncles many-flowered; staminal corona 10-lobed at the mouth; lobes opposite the anthers, equalling the corolla, broadly linear-oval, alternate ones very short, inconspicuous, all truncated and emarginate; stigma apiculate, subemarginate.

Coromandel Coast.

GENUS XII. HOLOSTEMMA.

Pentandria Digynia. *Ser: Syst:*

Deriv. From *Holos*, entire, and *Stemma*, a crown.

GEN. CHAR. Corolla subrotate, 5-cleft; staminal corona inserted at the bottom of the gynostegium, simple, annular, obsoletely 5-lobed; anthers terminated by a membrane; pollen-masses pendulous, compressed, affixed to the attenuated apex; stigma mutic; follicles ventricose, smooth.

(1) H. RHEEDII. *(Spr.)*

Ident. Wight's Contrib. p. 55.—Dec. prod. VIII. p. 532.—Spr. Syst. I. p. 851.
Syn. H. Ada Kodien, *Roem. & Schult.*—Asclepias annularia, *Roxb. flor. Ind.* II. p. 37.—Sarcostemma annulare, *Roth.*
Engrav. Wight's Icon. t. 597.
SPEC. CHAR. Glabrous, twining: leaves opposite, broad-ovate, cordate; umbels interpetiolar, short-peduncled; flowers showy, darkish red, green and white-mixed.

Malabar. Mysore. Travancore. N. Circars. Foot of the Himelyah. Flowering in the rainy season.

GENUS XIII. CALOTROPIS.

Pentandria Monogynia. *Ser: Syst:*

Deriv. From *Kalos*, beautiful, and *Tropis*, a keel; alluding to the keel of the flower.

GEN. CHAR. Erect milky shrubs: leaves opposite; umbels interpetiolar; corolla subcampanulate, tube angled, angles saccate within, limb 5-parted; staminal corona 5-leaved, leaflets keel-shaped, vertically adnate to the gynostegium, recurved at the base: anthers terminated by a membrane: pollen-masses compressed, affixed to the attenuated apex, pendulous: stigma mutic: follicles ventricose, smooth; seeds comose.

(1) C. GIGANTEA. (*R. Br.*)

Ident. R. Br. in Hort. Kew. II. p. 78.—Wight's Contrib. p. 53.—Dec. prod. VIII. p. 535.

Syn. Asclepias gigantea, *Willd.*—*Roxb. flor. Ind.* II. p. 30.

Engrav. Rheede Mal. II. t. 31.—Rumph. Amb. 7. t. 14. f. 1.—Wight's Ill. t. 155.—Bot. Reg. XVII. t. 58.

SPEC. CHAR. Leaves oblong-ovate, downy beneath: segments of the corolla spreading, withering, reflexed, revolute at the margin: leaflets of the staminal corona shorter than the gynostegium, base obtuse, acuminately recurved, apex incurved, somewhat 3-toothed: flowers pale-purple.

Common in southern part of the Peninsula, especially in sandy and barren soils, flowering at all seasons. This is the Madar plant, from which is obtained Mudarine, which possesses the property of coagulating by heat, and becoming again fluid by exposure to cold.

(2) C. PROCERA. (*R. Br.*)

Ident. R. Br. l. c.—Dec. prod. VIII. p. 535.

Syn. C. Wallichii, *Wight's Contrib.* p. 53.—C. Hamiltonii, *Wight.* l. c.—C. heterophylla, *Wall.*—*Wight.* l. c.—C. prosera, *Willd.*

Engrav. Lindl. Bot. Reg. t. 1792.—Wight's Icon. t. 1278.

SPEC. CHAR. Leaves ovate or oval, cordate at the base: segments of the corolla spreading, revolute at the margin: leaflets of the staminal corona equalling the gynostegium: umbels peduncled: follicles obovoid, downy: flowers pale purple.

Deccan and Guzerat. Flowering in March and April.

(3) C. HERBACEA. (*Wall.*)

Ident. Wight's Contrib. p. 54.—Dec. prod. VIII. p. 536.

Syn. Asclepias herbacea, *Roxb. flor. Ind.* II. p. 50.

Engrav. Wight's Icon. t. 492.

SPEC. CHAR. Segments of the corolla somewhat erect: leaflets of the staminal corona equalling the gynostegium, base obtuse, recurved: apex incumbent on the margin of the stigma, 2-lobed, lobes thick, somewhat diverging: flowers largish, pale purple.

Interior of Bengal, flowering in the cold season.

GENUS XIV. PENTATROPIS.

Pentandria Digynia. *Sar ; Syst .*

Deriv. From *Pente*, five, and *Tropis*, a keel.

GEN. CHAR. Corolla rotate, 5-cleft: staminal corona 5-leaved; leaflets opposite the anthers, vertically adnate to the gynostegium, averted, free at the apex: pollen-masses ventricose, pendulous, affixed below the apex: stigma mutic: follicles smooth: seeds comose.

(1) P. MICROPHYLLA. *(W. & A.)*

Ident. Wight's Contrib. p. 52.—Dec. prod. VIII. p.
Syn. Asclepias microphylla, Roxb. flor. Ind. II. p. 35.
Engrav. Wight's Icon. t. 352.

SPEC. CHAR. Twining undershrub: leaves somewhat fleshy, flat: umbels interpetiolar, few-flowered, submassile; follicles obsoletely 3-cornered.

Coromandel Coast. Bengal.

GENUS XV. SARCOSTEMMA.

Pentandria Digynia. *Sar; Syst ,*

Deriv. From *Sarx*, flesh, and *Stemma*, a crown.

GEN. CHAR. Twining or decumbent shrubs: stems leafless, jointed, or with opposite distant leaves: umbels lateral or terminal: corolla rotate: staminal corona double, outer one cup-shaped or annular, crenated, inner one 5-leaved, exceeding the outer one, leaflets fleshy: anthers terminated by a membrane: pollen-masses affixed by the apex, pendulous: stigma apiculate or mutic: follicles slender, smooth: seeds comose.

(1) S. BREVISTIGMA. *(W. & A.)*

Ident. Wight's Contrib. p. 59.—Dec. prod. VIII. p. 538.
Syn. S. viminale, *Wall.*—Asclepias acida, Roxb. fl. Ind. II. p. 31.
Engrav. Wight's Icon. t. 595.

SPEC. CHAR. Twining, leafless: umbels terminal or terminating the short lateral branchlets: calyx and pedicels glabrous: divisions of the corolla ovate: exterior staminal corona with 10 folds, interior ones gibbous on the back, equalling the gynostegium: stigma mutic: flowers white.

Common in the Deccan in stony places, flowering in June.

(2) S. INTERMEDIUM. *(Dec.)*

Ident. Dec. prod. VIII. p. 538.
Syn. S. viminale. *R. Br. Wight's Contrib.* p. 59.
Engrav. Wight's Icon. t. 1281.
Spec. Char. Twining, leafless: branches round: peduncles terminal or axillary: umbels many-flowered, subglobose: segments of the corolla oblong-lanceolate, waved: follicles linear or oblong, obtuse: flowers white.

Deccan. South Coromandel. Flowering in the rainy season.

(3) S. BRUNONIANUM. *(W. & A.)*

Ident. Wight's Contrib. p. 59.—Dec. prod. VIII. p. 538.
Engrav. Wight's Icon. t. 1282.
Spec. Char. Twining: umbels lateral, sessile: pedicels and calyx canescent: segments of the corolla ovate-lanceolate: exterior staminal corona subplicate, 10-crenated, the alternate toothlets somewhat obsolete: stigma apiculate, subentire.

Coimbatore, flowering in the hot season.

GENUS XVI. OXYSTELMA.

Pentandria Digynia. *Rex: Syst.*

Deriv. From *Oxys*, sharp, and *Stelma*, a crown; the leaflets of the corona are acute.

Gen. Char. Corolla subrotate, spreading with a short tube: gynostegium exserted: staminal corona 5-leaved, leaflets acute, compressed, undivided: anthers terminated by a membrane: pollen-masses compressed, affixed to the attenuated apex, pendulous: stigma muticous: follicles smooth: seeds comose.

(1) O. ESCULENTUM. *(R. Br.)*

Ident. Wight's Contrib. p. 54.—Dec. prod, VIII. p.
Syn. Periploca esculenta, *Linn.*—Asclepias rosea, *Roxb. fl. Ind.* II. p. 40.
Engrav. Roxb. Cor. I. t. 11.—Pluk. t. 350. fig. 6.
Spec. Char. Twining undershrub: leaves opposite, linear-lanceolate: corolla ciliated at the margin: racemes interpetiolar: follicles oblong, acuminated.

All over India.

(2) O. WALLICHII. *(R. W.)*

Ident. Wight's Contrib. p. 54.—Dec. prod. VIII. p.
Syn. O. esculentum, *Wall. asclep.* p. 95.

SPEC. CHAR. Twining: leaves opposite, leaves narrow linear-lanceolate; corolla ciliated at the margin: follicles short-oval, obtuse.

Poodoocottah, in Tanjore.

GENUS XVII. DÆMIA.

Pentandria Digynia. *Sex: Syst.*

Deriv. The Arabic name latinised.

GEN. CHAR. Corolla subrotate, with a short tube: staminal crown double, outer short, 10-parted, segments alternate, dwarf, inner 5-leaved, leaflets free at the base, subulate above: anthers terminated by a membrane: pollen-masses compressed, affixed at the apex, pendulous: stigma mutic: follicles ramentaceous; seeds comose.

(1) D. EXTENSA. *(R. Br.)*

Ident. Wight's Contrib. p. 59.—Dec. prod. VIII. p. 544.
Syn. Cynanchum extensum, *Ait. Hort. Kew.*—C. cordifolium; *Retz.*—Asclepias echinata, *Roxb. flor. Ind.* II. p. 44.
Engrav. Wight's Icon. t. 596. and Bot. Rep. t. 562.

SPEC. CHAR. Fruticose, twining; leaves opposite, somewhat round-cordate, acute, auricled at the base, pubescent, glaucous below: peduncles and pedicels elongated, filiform: corolla ciliated at the margin: flowers pale green, internally tinged with purple.

Peninsula. Bengal. North Concan. Guzerat. Flowering nearly all the year.

GENUS XVIII. TYLOPHORA.

Pentandria Digynia. *Sex: Syst.*

Deriv. From *Tylos,* a swelling, and *Phoreo,* to bear; alluding to the ventricose pollen-masses.

GEN. CHAR. Herbs or shrubs, twining: umbels interpetiolar, solitary or arranged alternately together with the long flexuose peduncle: corolla rotate, 5-parted: staminal corona 5-leaved, leaflets simple, fleshy: anthers terminated by a membrane: pollen-masses transverse or sub-ascendent, minute, ventricose: stigma mutic: follicles smooth, attenuated at the apex, compressed, sometimes slightly angled; seeds comose.

(1) T. CARNOSA. (*Wall.*)

Ident. Wight's Contrib. p. 49.—Dec. prod. VIII. p. 607.
Engrav. Wight's Icon. t. 351.

SPEC. CHAR. Twining; leaves fleshy, ovate or subcordate, mucronate, shining, pale beneath: peduncles flexuose bearing at the flexures many filiform pedicels: leaflets of the staminal crown suborbicular; pollen-masses ascending; stigma convex: follicles often solitary by abortion: flowers small, purplish inside.

In sandy localities in the Tanjore district.

(2) T. TENUISSIMA. (*W. & A.*)

Ident. Wight's Contrib. p. 49.—Dec. prod. VIII. p. 607.
Syn. Asclepias tenuissima, Roxb. flor. Ind. II. p. 41.
Engrav. Wight's Icon. t. 588.

SPEC. CHAR. Twining: leaves oblong-lanceolate, subcordate at the base, cuspidate, veinless, subrevolute at the margin: leaflets of the staminal crown ovate-oblong: pollen-masses ascending: stigma convex: flowers small, dull purple.

Peninsula. Bengal. Flowering in the rainy season. Probably a mere variety of the last species.

(3) T. PAUCIFLORA. (*W. & A.*)

Ident. Wight's Contrib. p. 49.—Dec. prod. VIII. p. 607.
Engrav. Wight's Icon. t. 1274.

SPEC. CHAR. Twining; leaves largish petioled, broad from the base, ovate or subcordate, gradually attenuated, somewhat undulated at the margin; peduncles shorter than the leaf, flexuose, bearing 2-3 filiform pedicels at the flexures: leaflets of the staminal crown broad-elliptic, obtuse; pollen-masses ascending: stigma convex.

Courtallum. Malabar. Northern Provinces of Bengal. Flowering in August and September.

(4) T. MOLLISSIMA. (*Wall.*)

Ident. Wight's Contrib. p. 49.—Dec. prod. VIII. p. 607.
Engrav. Wight's Icon. t. 1275.

SPEC. CHAR. Twining; whole plant pubescent: leaves oval or cordate-ovate, acuminate, mucronate, peduncles 2-3 times longer than the leaf, flexuose, bearing a sessile umbel at the fissures: pedicels filiform: segments of the corolla ovate, obtuse, undulated: leaflets of the staminal crown truncated or rounded at the apex: pollen-masses transverse: stigma obtuse: follicles pubescent.

Neilgherries. Dindigul mountains. Flowering nearly all the year.

(5) T. FASCICULATA. *(Ham.)*

Ident. Wight's Contrib. p. 50.—Dec. prod. VIII. p. 606.

Engrav. Wight's Icon. t. 848.

SPEC. CHAR. Root fascicled, stems woody, erect, fascicles approximate: leaves ovate, somewhat fleshy, decreasing towards the tops of the branches; peduncles erect, flexuose bearing 2-3 few-flowered fascicles at the flexures: leaflets of the staminal crown oblong-ovate: pollen-masses transverse: stigma apiculate: flowers largish: segments of the corolla ligulate.

Neilgherries.

(6) T. IPHISIA. *(Dec.)*

Ident. Dec. prod. VIII. p. 610.

Syn. Iphisia multiflora, *Wight's Contrib.* p. 53.

Engrav. Wight's Icon. t. 1276.

SPEC. CHAR. Twining: leaves ovate, or ovate-lanceolate or subcordate at the base, acute at the apex, succulent: petioles with glands at the origin of the limb: peduncles about the length of the smaller leaves, subflexuose, usually with short secondary peduncles, bearing two or three flowers: pedicels short, stout: pollen-masses globose: stigma mutic: follicles swollen at the base: flowers small, dark dull purple.

Common in clumps of jungle about Ootacamund; flowering in August and September.

(7) T. ASTHMATICA. *(W. & A.)*

Ident. Wight's Contrib. p. 51.—Dec. prod. VIII. p. 611.

Syn. Cynanchum vomitorium, *Lam.*—C. viridiflorum, *Sims.*—Asclepias vomitoria, *Koen.*—A. asthmatica, *Roxb. fl. Ind.* II. p. 33.—Cynanchum Ipecacuanha, *Willd.*

Engrav. Bot. Mag. t. 1929.—Wight's Icon. t. 1277.

SPEC. CHAR. Twining, glabrous or pubescent: leaves ovate or roundish, abruptly acuminate, often cordate at the base, glabrous above: petioles without glands: peduncles shorter than the leaves with 2-3 sessile, few-flowered umbels towards the extremity: leaflets of the staminal crown fleshy, depressed, embracing the base of the gynostegium and prolonged at the apex into a tooth equalling the gynostegium; pollen-masses transverse, small, globose; stigma obtuse: follicles divaricating, attenuated, glabrous: flowers largish, long-pedicelled, externally pale greenish, with a faint tinge of purple, internally light purple.

A very abundant and widely diffused plant to be met with in nearly all situations and in flower at all seasons. The roots partake in an eminent degree of the properties of Ipecacuanha.

(8) T. LONGIFOLIA. (R. W.)

Ident. Wight's Contrib. p. 50.—Dec. prod. VIII. p. 608.

SPEC. CHAR. Twining: leaves oblong-lanceolate, acute, subcordate at the base: panicles shorter than the leaf, many-flowered: branches flexuose, bearing a small umbel at the flexures: leaflets of the staminal crown nearly a half shorter than the gynostegium, broad-elliptic, obtuse: pollen-masses transverse: stigma convex.

Silhet.

(9) T. EXILIS. (*Colebr.*)

Ident. Wight's Contrib. p. 50.—Colebr. in Linn. Soc. trans. p. 12.—Dec. prod. VIII. p. 608.

Syn. Pergularia exilis, *Spr.*

SPEC. CHAR. Twining: leaves ovate-lanceolate, acuminate: panicles 2-3 times longer than the leaf, many-flowered: branches flexuose, bearing umbellate pedicels at the flexures: segments of the corolla acutish: leaflets of the staminal crown broad-elliptic, very obtuse, shorter than the gynostegium: pollen-masses transverse: stigma apiculate.

Silhet.

(10) T. CAPPARIDIFOLIA. (*W. & A.*)

Ident. Wight's Contrib. p. 51.—Dec. prod. VIII. p. 511.

Syn. Asclepias tenuis, *Wall.*

SPEC. CHAR. Twining: leaves elliptic-oblong, mucronately acuminate, coriaceous: racemes umbelliform, subsessile: flowers few and small: segments of the corolla obtuse: leaflets of the staminal crown rounded, fleshy, increased at the apex within by a short, sharp toothlet: pollen-masses subascendent: stigma obtuse.

Mysore. (?)

GENUS XIX. COSMOSTIGMA.

Pentandria Digynia. *Sex: Syst:*

Deriv. From *Kosmos*, beautiful, and *Stigma*, from the beautiful appearance of that organ.

GEN. CHAR. Corolla rotate, 5-parted: staminal corona 5-leaved, leaflets compressed, 2-cleft: anthers ventricose, terminated by a membrane: pollen-masses erect, affixed at the base to the corpuscule of the stigma by the assistance of a long pendulous appendage: stigma mutic, crowned by a repand flexuose narrow wing: follicles large, linear-oblong, obtuse, smooth: seeds comose.

(1) C. RACEMOSUM. (*R. W.*)

Ident. Wight's Contrib. p. 42.—Dec. prod. VIII. p. 613.

Syn. Asclepias racemosa, *Roxb. flor. Ind.* II. p. 32.

Engrav. Wight's Icon. t. 591.—Deless. Ic. Sel. V. t. 64.

SPEC. CHAR. Shrubby, climbing: leaves broadly ovate, or roundly acuminate, obtuse or cordate at the base: peduncles interpetiolar, corymbosely racemose at the apex: flowers small, yellow, marked with rusty dots.

Silhet. Chittagong. North Concans, common in hedges. Flowering in the rainy season.

(2) C. ACUMINATUM. (*R. W.*)

Ident. Wight's Icon. vol. IV. t. 1270.

SPEC. CHAR. Twining: leaves broad-ovate or cordate at the base, acuminate, sparingly hairy on both sides: peduncles rigid, hairy: pedicels short, coriaceous, stout: corolla marked with purple spots.

Balaghaut hills, near Madras.

GENUS XX. MARSDENIA.

Pentandria Digynia. *Ser. Syst.*

Deriv. In honor of William Marsden, late Secretary to the Admiralty, Author of a History of Sumatra.

GEN. CHAR. Shrubs often twining: leaves opposite, membranaceous, broadish: cymes or thyrses interpetiolar: corolla urceolate, 5-cleft, occasionally subrotate: staminal corona 5-leaved, leaflets compressed, simple within: anthers terminated by a membrane: pollen-masses erect, affixed by the base: stigma mutic or beaked: follicles smooth: seeds comose.

(1) M. BRUNONIANA. (*W. & A.*)

Ident. Wight's Contrib. p. 40.—Dec. prod. VIII. p. 614.

Engrav. Wight's Icon. t. 356.

SPEC. CHAR. Twining, glabrous: leaves broadly cordate-ovate, acuminate: peduncles shorter than the petiole: flowers cymose, largish; segments of the corolla obtuse: leaflets of the staminal crown equalling the gynostegium: stigma obtusely apiculated.

Columala.

(2) M. TINCTORIA. *(R. Br.)*

Ident. Wight's Contrib. p. 40.—Dec. prod. VIII. p. 615.

Syn. Pergularia tinctoria, *Spreng.*—Asclepias tinctoria, *Roxb. fl. Ind.* II. p. 43.

Engrav. Wight's Icon. t. 589.—Rheede Mal. IX. t. 8.

SPEC. CHAR. Twining: leaves ovate or oblong, acuminated, cordate at the base furnished above with a small gland towards the base: thyrses lateral: flowers small, yellow, throats bearded: leaflets of the staminal crown nearly equalling the gynostegium, subulate: stigma mutic.

Silhet. Malabar. Goalpara. Flowering nearly all the year.

(3) M. TENACISSIMA. *(W. & A.)*

Ident. Wight's Contrib. p. 41.—Dec. prod. VIII. p. 616.

Syn. Asclepias tenacissima, *Roxb. flor. Ind.* II. p. 31.—Gymnema tenacissima, *Spreng.*

Engrav. Wight's Icon. t. 590.—Roxb. Cor. III. t. 240.

SPEC. CHAR. Twining: leaves cordate, acuminate, tomentose on both sides: cymes large: segments of the corolla broad, obtuse: leaflets of the staminal crown broad, truncated at the apex, subentire or two-forked: stigma obtusely apiculate: flowers small, greenish-yellow.

Rajmahal Hills. Chittagong. Mysore. Flowering in April.

GENUS XXI. DISCHIDIA.

Pentandria Digynia. *Ser. Syst.*

Nerio. From *Dis*, double, and *Schizo*, to split; alluding to the leaflets of the staminal crown.

GEN. CHAR. Corolla urceolate, 5-cleft: staminal corona 5-leaved, leaflets 2-cleft, lobes spreading, recurved at the apex: anthers terminated by a membrane: pollen-masses erect, affixed by the base: stigma mutic; follicles smooth; seeds comose.

(1) D. Benghalensis. *(Colebr.)*

Ident. Wight's Contrib. p. 43.—Dec. prod. VIII. p. 631.—Culebr. in Linn. Soc. trans. XII. p. 357.

Syn. D. lanceolata, *Wall.* asclep. *No.* 62.

Engrav. Hook. Bot. Mag. t. 2016.

Spec. Char. Parasitical on trees, milky: stem rooting at the lower joints: leaves opposite, fleshy, lanceolate, glabrous: cymes somewhat capituliform, short-peduncled: flowers small, greenish-yellowish-white.

Silhet. South Concan.

GENUS XXII. PERGULARIA.

Pentandria Digynia. *Sex: Syst:*

Deriv. From *Pergula*, trellis-work; twining plants fit for arbours.

Gen. Char. Twining plants: leaves opposite: cymes interpetiolar: flowers yellowish, fragrant: corolla hypocrateriform, tube urceolate: staminal corona 5-leaved, leaflets compressed, undivided at the apex, increased by an incision inside: anthers terminated by a membrane: pollen-masses erect: stigma mutic: follicles smooth, ventricose: seeds comose.

(1) P. pallida. *(W. & A.)*

Ident. Wight's Contrib. p. 42.—Dec. prod. VIII. p. 619.

Syn. P. fimbata, *Wall.*—Asclepias pallida, *Roxb. flor. Ind.* II. p. 48.

Engrav. Wight's Icon. t. 585.

Spec. Char. Branches softly pubescent: leaves cordate acuminate: cymes subsessile, many-flowered: segments of the corolla subulate, tube glabrous within: leaflets of the staminal crown dilated, somewhat 3-lobed, internal ligulæ flat: flowers smallish, pale yellow, inodorous.

Bengal. Monghyr. Flowering in the rainy season.

(2) P. odoratissima. *(Sm.)*

Ident. Wight's Contrib. p. 43.—Dec. prod. VIII. p. 618.

Syn. Cynanchum odoratissimum, *Lour. Coch.* 1. p. 206.—Asclepias odoratissima, *Roxb. flor. Ind.* II. p. 47.

Engrav. Smith. Ic. pict. t. 16.—Andr. Bot. Rep. t. 183.—Wight's Icon. t. 414.—Rumph. Amb. VII. t. 26. fig. 1.

Spec. Char. Branches softly pubescent: leaves cordate acuminate, pubescent at the veins: cymes short-peduncled: segments of the corolla short, obtuse: tube twice the length of the gynostegium: external leaflets of the staminal crown obtuse, increased with an internal concave arched ligula: flowers whitish outside, inside greenish yellow orange, very fragrant.

Patna. Peninsula. ? Flowering nearly all the year.

(3) P. COROMANDELIANA. (*Dec.*)

Ident. Dec. prod. VIII. p. 619.

SPEC. CHAR. Branches softly puberulous: leaves cordate acuminate: cymes subsessile, many-flowered: calycine leaflets oblong-lanceolate: segments of the corolla linear-oblong or ligulate: tube glabrous within: leaflets of the staminal corona dilated, somewhat 3-lobed: internal ligulæ dilated and continuous with the lower point.

Coromandel.

GENUS XXIII. GYMNEMA.

Pentandria Digynia. *Sm: Syst:*

Deriv. From *Gymnos*, naked, and *Nema*, a thread; in reference to the stamens.

GEN. CHAR. Shrubs or undershrubs often twining: leaves opposite, flat: umbels interpetiolar, cymiform: corolla sub-urceolate, 5-cleft, throat sometimes crowned with five small scales or toothlets inserted on sinuses.

(1) G. SYLVESTRE. (*R. Br.*)

Ident. R. Br. in Wern. Soc. Mem. I. p. 33.—Wight's Contrib. p. 44.—Dec. prod. VIII. p. 621.

Syn. G. parviflorum, *Wall. tent. flor. Nep. p.* 50.—Periploca sylvestris, *Willd.*—Asclepias geminata, *Roxb. flor. Ind.* II. p. 45.

Engrav. Wight's Icon. t. 349.

SPEC. CHAR. Climbing: all, except the upper side of the leaves, softly pubescent: leaves ovate, or ovate-lanceolate, attenuated at both ends, or obscurely cordate at the base: peduncles as long as the petioles: umbels twin, many-flowered: follicles slender, attenuated: flowers crowded, small, yellow.

South Mahratta country. Coromandel. Silhet. Assam. Flowering nearly all the year.

(2) G. HIRSUTUM. (*W. & A.*)

Ident. Wight's Icon. vol. IV.
Engrav. Wight's Icon. t. 1272.

SPEC. CHAR. Twining: leaves ovate or subcordate, hirsute above, tomentose beneath: umbels short-peduncled, many-flowered: tube of the corolla furnished with leafy scales: filaments with 2 black fleshy glands at the base: stigma depressed, scarcely exceeding the anthers.

Southern Peninsula, in subalpine jungles.

(3) G. DECAISNEIANUM. (*R. W.*)

Ident, Wight's Icon. vol. IV.
Syn. G. hirsutum, Dec. prod. VIII. p. 622.—*Wight's Contrib.* p. 44.
Engrav. Wight's Icon. t. 1271.

SPEC. CHAR. Young branches and under surface of the leaves shortly-tomentose, upper surface, calyx and petioles hirsute: leaves ovate, subacuminate: peduncles axillary, about the length of the petioles: umbels compact, many-flowered: throat of the corolla furnished with fleshy prominences: filaments without glands at the base: stigma conical, prolonged beyond the anthers.

Neilgherries, about Cunoor and Kaity, flowering nearly all the year.

(4) G. ACUMINATUM. (*Wall.*)

Ident. Wall. tent. flor. Nep. p. 50.—Wight's Contrib. p. 45.—Dec. prod. VIII. p. 622.
Syn. Asclepias acuminata, Roxb. flor. Ind. II. p. 53.

SPEC. CHAR. Bark corky: younger branchlets subtomentose: leaves oval-oblong, acuminated, tomentose beneath: peduncles shorter than the petiole: umbels subcapitate, few-flowered: scales of the throat of the corolla exserted, a half shorter than the segments: follicles conical from a thick base, short, woody-coriaceous: seeds large, margined with a membranaceous wing: flowers small, white.

Chittagong, Silhet. Flowering from May to July.

(5) G. LATIFOLIUM. (*Wall.*)

Ident. Wight's Contrib. p. 45.—Dec. prod. VIII. p. 623.

SPEC. CHAR. Younger branchlets densely pubescent: leaves broad-oval, acuminate, sparingly hairy on both sides, parallel-veined: umbels shorter than the petiole, cymiform, many or few-flowered: scales of the corolla included: follicles slender, warty.

Silhet.

GENUS XXIV. BIDARIA.

Pentandria Digynia. *Sex : Syst.*

GEN. CHAR. Corolla campanulate, 5-cleft ; throat destitute of scales, tube furnished within with twin hairy lines, the sinuses opposite : anthers terminated by an ovate membrane : pollen-masses oblong-clavate : stigma conoid, fleshy, apex entire or obscurely emarginate.

(1) B. ELEGANS. *(Dec.)*

Ident. Dec. prod. VIII. p. 623.
Syn. Gymnema elegans, *Wight's Contrib. p.* 46.
Engrav. Wight's Icon. t. 830.

SPEC. CHAR. Branches glabrous, twining: leaves opposite, cordate, ovate or oval, acuminated, acute, somewhat undulated at the margin, quite glabrous, glanduliferous above the petiole : petioles slender, furrowed, puberulous : umbels short-peduncled, afterwards spirally elongated : stigma obtuse, exceeding the stamens : follicles often solitary : flowers small.

Dindigul hills. Salem. Columala.

(2) B. INODORA. *(Dec.)*

Ident. Dec. prod. VIII. p. 624.
Syn. Cynanchum inodorum, *Lour. Coch. p.* 166.

SPEC. CHAR. Twining : branches fistular, glabrous : leaves opposite, roundish or lanceolate-ovate, acuminated, glanduliferous above the petiole, glabrous on both sides : petioles slender, furrowed above : umbels short-peduncled : pedicels longer : flowers largish, puberulous : tube hairy within : stigma conical, emarginate at the apex, exceeding the anthers : follicles twin, oblong.

In sandy places near Pondicherry.

GENUS XXV. GONGRONEMA.

Pentandria Digynia. *Sex : Syst.*

Deriv. From *Gongros,* an eel, and *Nema,* a thread ; in allusion to the stamens.

GEN. CHAR. Twining shrubs : leaves opposite, glanduliferous above the petiole : cymes solitary, many-flowered in brachiate branchlets : calyx 5-parted : corolla rotate, 5-cleft, throat and tube naked : staminal crown none : gynostegium furnished at the bottom of the base with small fleshy glands : anthers terminated by a membrane : pollen-masses erect, affixed by the base, ovoid : follicles smooth : seeds comose.

(1) G. SAGITTATUM. *(Dec.)*

Ident. Dec. prod. VIII. p. 624.

Syn. Gymnema sagittatum, *Wall. tent. flor. Nep.* p. 50.— Wight's Contrib. p. 46.

SPEC. CHAR. Leaves oval or oblong-ovate, acuminate, acute, subcordate at the base, glabrous on both sides, white glaucous beneath and with the veins of the same colour: peduncles thin, shorter than the leaf or nearly equal: cymes composite: flowers small, often caducous: younger follicles thin, glabrous.

Silhet.

(2) G. ? ATTENUATUM. *(Dec.)*

Ident. Dec. prod. VIII. p. 625.

Syn. Gymnema attenuatum, *Wall. tent. flor. Nep.* p. 50.

SPEC. CHAR. Leaves lanceolate, long attenuated: corymbs slender, flexuose, slightly longer than the leaf: pollen-masses divaricate, ovate, nearly horizontal.

Silhet.

GENUS XXVI. SARCOLOBUS.

Pentandria Digynia. *Sex: Syst :*

Deriv. From *Sarx*, flesh, and *Lobos*, a pod; the seed in vesicels are fleshy.

GEN. CHAR. Twining shrubs: leaves opposite, firm: umbels or corymbs interpetiolar: corolla rotate, 5-parted: gynostegium sub-globose: staminal crown wanting: anthers terminated by a membrane: pollen-masses erect, attached by the base: stigma mutic: follicles ventricose, fleshy or coriaceous, sub-solitary by abortion: seeds margined, scabrous.

(1) S. GLOBOSUS. *(Wall.)*

Ident. Wight's Contrib. p. 47.—Dec. prod. VIII. p. 625.— Wall. in As. Res. 12. p. 577.

Engrav. Wall. l. c. t. 4.—Wight's Icon. t. 1273.

SPEC. CHAR. Leaves ovate-oblong: cymes bifid: corolla with a very short tube, villous within: follicles large, fleshy, globose, retuse on both sides, muricated.

Banks of the Hooghly.

(2) S. carinatis. (*Wall.*)

Ident. Wight's Contrib. p. 47.—Dec. prod. VIII. p. 625.— Wall. in As. Res. 12. p. 578. t. 5.

Spec. Char. Leaves oval or oblong, somewhat fleshy: peduncles few-flowered: corolla glabrous within with a very short tube: follicles oblong, acute, smooth, keeled beneath.

Banks of the Hooghly.

GENUS XXVII. ORTHANTHERA.

Pentandria Digynia. *Sex : Syst :*

Deriv. From *Orthos*, straight, and *Anthera*, the anther.

Gen. Char. Corolla urceolate, 5-cleft, throat naked, tube subventricose: staminal corona none: anthers simple at the apex, erect, simple, acute: pollen-masses erect, affixed by the base, apex narrowed, pellucid: stigma apiculated.

(1) O. viminea. (*R. W.*)

Ident. Wight's Contrib. p. 48.—Dec. prod. VIII. p. 626.
Syn. Apocynum vimineum, *Wall.*

Spec. Char. Leafless shrub: umbels short-peduncled, few-flowered: calyx 5-parted, segments subulate at the apex: corolla villous externally, glabrous within, twice as long as the gynostegium: filaments furnished at the base with a small gland or fleshy leaflet: anthers free, incumbent on the apiculate stigma.

Doab, Hindustan.

GENUS XXVIII. PENTASACME.

Pentandria Digynia. *Sex : Syst :*

Deriv. From *Pente*, five, and *Achme*, a point; alluding to the pointed segments of the corolla.

Gen. Char. Corolla subrotate, deeply 5-cleft, throat crowned with 5 small scales opposite the segments: staminal crown none: anthers free, simple or acuminate at the apex: pollen-masses pendulous, attached above the middle, apex narrowed, pellucid: stigma mutic or apiculate.—Slender erect somewhat branched herbs: leaves opposite, flat, membranaceous: umbels subsessile, few-flowered: segments of the corolla 2-3 times longer than the tube, narrow linear.

(1) P. CAUDATUM. *(Wall.)*

Ident. Wight's Contrib. p. 60.—Dec. prod. VIII. p. 627.—Wall. Asclep. No. 5.

SPEC. CHAR. Leaves lanceolate, long-attenuated: segments of the corolla subulate, scales of the throat largish, 4-cleft: anthers obtuse: stigma mutic.

Silhet mountains.

(2) P. WALLICHII. *(R. W.)*

Ident. Wight's Contrib. p. 60.—Dec. prod. VIII. p. 627.—Wall. Asclep. No. 74.

SPEC. CHAR. Leaves lanceolate, acuminate: umbels subsessile, few-flowered: segments of the corolla linear, obtuse: scales of the throat small, gland-shaped: anthers acuminate: stigma subrostrate.

Silhet.

GENUS XXIX. LEPTADENIA.

Pentandria Digynia. *Sex. Syst.*

Deriv. From *Leptos*, slender, and *Aden*, a gland.

GEN. CHAR. Corolla subrotate: tube short, scales of the throat five, alternating with the segments: staminal corona none: anthers free, simple at the apex: pollen-masses erect, attached by the base, apex narrowed, pellucid: stigma mutic, or rarely apiculate: follicles smooth: seeds comose.—Twining shrubs, often ashy-puberulous: leaves flat: umbels interpetiolar: corpuscules of the stigma minute.

(1) L. RETICULATA. *(W. & A.)*

Ident. Wight's Contrib. p. 47.—Dec. prod. VIII. p. 628.

Syn. Cynanchum reticulatum, *Retz.*—Asclepias suberosa, *Roxb. fl. Ind.* II. p. 38.—A. volubilis, *Wall.*—Secamone canescens, *Smith in Rees' Cycl.*

Engrav. Wight's Icon. t. 350.

SPEC. CHAR. Bark corky: younger branches ash-coloured, pubescent: leaves ovate or lanceolate, acute, sometimes with short white pubescence: umbels lateral, many-flowered, as long as the petiole: segments of the corolla revolute on the margin, bearded within: follicles cylindrical-oblong, obtuse, often solitary by abortion: flowers greenish-yellow.

Common near the sea in various parts of the Peninsula. Negapatam. Northern India.

(2) L. IMBERBIS. *(R. W.)*

Ident. Wight's Contrib. p. 48.—Dec. prod. VIII. p. 628.

SPEC. CHAR. Leaves broad-ovate or cordate, acuminate; umbels lateral, many-flowered, nearly equalling the petiole: segments of the corolla revolute at the margin, beardless within, throat with simple scales; stigma apiculate.

Meerut. Hedges at Adjunta.

(3) L. SPARTIUM. *(R. W.)*

Ident. Wight's Contrib. p. 48.—Dec. prod. VIII. p. 629.

Syn. Cymnema spartium, *Wall.*

SPEC. CHAR. Suffruticose: stems thin, much branched; leaves narrow-linear, attenuated at the apex: umbels subsessile: segments of the corolla beardless within, revolute at the margin, throat with broad exserted scales: stigma mutic.

Banks of the Jumna.

(4) L. APPENDICULATA. *(Dec.)*

Ident. Dec. prod. VIII. p. 628.

SPEC. CHAR. Younger branches hoary: leaves ovate-lanceolate or more rarely oblong-lanceolate, acutish: umbels many-flowered: peduncles equally or much exceeding the petiole: pedicels slender, longer than the flowers: segments of the corolla bearded within, increased in the middle by a hooked appendage: scales depressed, short.

Pondicherry.

(5) L. JACQUEMONTIANA. *(Decaisne.)*

Ident. Dec. prod. VIII. p. 628.—Dalz. Bomb. flor. p. 152.

SPEC. CHAR. Erect, much branched shrub: branches twiggy, slender: leaves narrow-linear (on the younger branches only): umbels few-flowered, very shortly peduncled: segments of the corolla keeled within at the middle, glabrous: scales fleshy, depressed, somewhat 2-lobed: flowers subsessile, very small, yellow.

In dry jungles near Agra. Sea-shore south of Gogo.

GENUS XXX. HETEROSTEMMA.

Pentandria Digynia. *Bor; Syst:*

Deriv. From *Heteros*, variable, and *Stemma*, the crown.

GEN. CHAR. Corolla rotate, 5-parted: staminal crown 5-leaved: leaflets dilated, increased by a process within: anthers terminated by a membrane incumbent on the stigma: pollen-masses erect, obsoletely 5-sided, sometimes pellucid at the margin: follicles smooth: seeds comose.—Twining shrubs: leaves opposite, furnished with a minute gland above towards the base: umbels small, interpetiolar, short-peduncled.

(1) H. TANJORENSIS. *(W. & A.)*

Ident. Wight's Contrib. p. 42.—Dec. prod. VIII. p. 530.
Engrav. Wight's Icon. t. 348.
SPEC. CHAR. Leaves broad ovate or oblong, shortly acuminated, obtuse or cordate at the base: peduncles shorter than the petiole, few-flowered: leaflets of the staminal crown spreading, broad, truncated, interior process tongue-shaped; follicles divaricate, thin, glabrous, hooked at the apex.

Sandy places in the Tanjore district.

(2) H. WALLICHII. *(R. W.)*

Ident. Wight's Contrib. p. 42.—Dec. prod. VIII. p. 630.— Dals. Bomb. flor. p. 152.
SPEC. CHAR. Branches with two opposite lines of hairs: leaves ovate-acuminate: peduncles very short, few-flowered: follicles smooth, purple, blunt pointed: flowers fuscous within.

Sheapore. Near Malwan. Flowering in September.

(3) H. URCEOLATUM. *(Dalz.)*

Ident. Dalz. Bomb. flor. p. 152.—Hook. Journ. Bot. IV. p. 295.
SPEC. CHAR. Stem purple, puberulous: leaves broadly ovate, acute, cordate at the base; umbels very short-peduncled, few-flowered: corolla deeply urceolate, reddish-purple.

Hills near Belgaum, flowering in July.

GENUS XXXI. PTEROSTELMA.

Pentandria Digynia. *Sex. Syst.*

Deriv. From *Pteron*, a wing, and *Stelma*, a crown; referring to the leafy crown-leaves.

GEN. CHAR. Corolla rotate, 5-cleft: staminal corona 5-leaved; leaflets membranaceous, sides reflexed, inner angle prolonged with an erect subulate tooth: anthers terminated by a membrane: pollen-masses erect, approximated, affixed by the base to the back of the corpuscule: stigma apiculate.

(1) P. ACUMINATA. (R. W.)

Ident. Wight's Contrib. p. 39.—Dec. prod. VIII. p. 633.

SPEC. CHAR. Twining or decumbent? leaves opposite, oblong, acuminate, fleshy: segments of the corolla linear-lanceolate, tube hairy: leaflets of the crown broad, sides narrowly reflexed and convolvent with the margins: flowers largish.

Silhet.

GENUS XXXII. HOYA.

Pentandria Monogynia. *Sm; Syn:*

Deriv. In honor of Thomas Hoy, F. L. S., late Gardener to the Duke of Northumberland at Sion House.

GEN. CHAR. Corolla rotate, 5-cleft: staminal corona 5-leaved, leaflets depressed, spreading, fleshy, inner angle produced into a tooth incumbent on the anther: anthers terminated by a membrane: pollen-masses affixed by the base, conniveat, compressed: stigma mutic or subapiculate: follicles smooth: seeds comose.—Shrubs or undershrubs, twining, climbing, or decumbent: leaves opposite, fleshy or membranaceous: umbels lateral, many-flowered.

(1) H. VELUTINA. (R. W.)

Ident. Wight's Contrib. p. 35.—Dec. prod. VIII. p. 635.

SPEC. CHAR. Twining: leaves oval, shortly acuminated, revolute at the margin, glabrous above, velvety below: corolla pubescent externally, glabrous within, segments broad, revolute at the margin: leaflets of the staminal crown suborbicular, depressed, shining above, inner angle obtuse, incumbent on the umbilicate stigma.

Silhet.

(2) H. PENDULA. (W. & A.)

Ident. Wight's Contrib. p. 36.—Dec. prod. VIII. p. 635.

Syn. Asclepias pendula, *Roxb. flor. Ind.* II. p. 36.

Engrav. Wight's Icon. t. 474.—Rheede Mal. IX. t. 13.

SPEC. CHAR. Twining: leaves fleshy, glabrous, from oblong-oval acute to broad-ovate acuminate, revolute at the margin: peduncles a little exceeding the petiole, pendulous, many-flowered: corolla pubescent within: leaflets of the staminal crown oboval, very obtuse, depressed, inner angle short, apex truncated: stigma apiculate: flowers middle sized, white, fragrant.

Circars. Malabar. South Concans. Neilgherries. Flowering in the rainy season.

(3) H. PALLIDA. (*Lindl.*)

Ident. Lindl. Bot. Reg. vol. XI.—Dec. prod. VIII. p. 636.
—Dals. Bomb. flor. p. 152.

Engrav. Lindl. l. c. t. 951.

SPEC. CHAR. Parasitic, climbing: leaves ovate-lanceolate, acuminate, fleshy: umbels compact, hemispherical: flowers whitish, fragrant.

Common on trees on the Western Coast. Very like *H. carnosa*, but much paler, and the segments of the corolla more acute.

(3) H. RETUSA. (*Dals.*)

Ident. Dals. Bomb. flor. p. 153.—Hook. Journ. Bot. IV. p. 294.

SPEC. CHAR. Parasitic, pendulous: branches long, filiform: leaves short-petioled, linear, 3-sided, fleshy, glabrous, retuse at the apex: flowers on a very short axillary peduncle, solitary or twin, fascicled, long-pedicelled, white, shining.

Dandelly jungles (Bombay Presidency), flowering in the rainy season.

(5) H. VIRIDIFLORA. (*R. Br.*)

Ident. R. Br. in Wern. Soc. Mem. I, p. 26.—Wight's Contrib. p. 39.—Dec. prod. VIII. p. 639.

Syn. Asclepias volubilis, *Linn. suppl.*—*Roxb. fl. Ind.* II. p. 36.—Apocynum tiliæfolium, *Lam.*

Engrav. Wight's Icon. t. 580.—Rheede Mal. IX. t. 15.—Hook. Bot. Misc. II. suppl. t. 1.

SPEC. CHAR. Twining: leaves ovate or cordate, acuminate, membranaceous, glabrous on both sides: corolla glabrous, segments ovate, acutish: leaflets of the staminal crown flat above, obovol, very obtuse, inner angle short, obtuse: follicles divaricate, thick, obtuse, rusty farinaceous: flowers smallish, green, inodorous.

All over India, flowering from March till May.

(6) H. HOOKERIANA. (*R. W.*)

Ident. Wight's Contrib. p. 37.—Dec. prod. VIII. p. 630.

SPEC. CHAR. Twining: leaves fleshy, oblong-lanceolate, 3-nerved: peduncles short, many flowered: corolla glabrous, segments obtuse: leaflets of the staminal crown ovate, acute, inner angle short: stigma obtuse.

Chittagong. Silhet.

(7) H. LANCEOLATA. *(Wall.)*

Ident. Wight's Contrib. p. 38.—Dec. prod. VIII. p. 637.— Don. prod. flor. Nep. p. 130.

SPEC. CHAR. Twining: stems thin, leafy: leaves fleshy, lanceolate, acuminated; peduncles shorter than the leaf, few-flowered: corolla pubescent within, segments obtuse: leaflets of the staminal crown convex above, recurved at the margin, the inner angle elongated, obtuse, erect, exceeding the apiculate stigma.

Khasia Hills.

(8) H. PARASITICA. *(Wall.)*

Ident. Wight's Contrib. p. 37.—Dec. prod. VIII. p. 637.
Syn. Asclepias parasitica, Roxb. flor. Ind. II. p. 42.
Engrav. Wight's Icon. t. 587.

SPEC. CHAR. Parasitic, climbing, rooting: leaves fleshy, glabrous, shining, 3-nerved at the base, oblong-lanceolate, attenuated; peduncles shorter than the leaf, many-flowered: corolla glabrous, deeply 5-cleft: leaflets of the staminal crown ovate-acute, inner angle incumbent on the stigma: flowers white, fragrant.

Woods on the Banks of the Ganges.

(9) H. OVALIFOLIA. *(W. & A.)*

Ident. Wight's Contrib. p. 37.—Dec. prod. VIII. p. 638.
Engrav. Wight's Icon. t. 847.

SPEC. CHAR. Twining, rooting: leaves fleshy, oval, acuminated at both ends: peduncles shorter than the leaf, many-flowered: corolla puberulous within, segments ovate, acute: leaflets of the staminal crown oval, obtuse, inner angle short: stigma mutic.

Neilgherries,

(10) H. FUSCA. *(Wall.)*

Ident. Wall. Pl. As. Rar. II. p. 78.—Dec. prod. VIII. p. 639.
Engrav. Wall. l. c. t. 175.

SPEC. CHAR. Twining: leaves coriaceous, scarcely fleshy, linear-oblong, acuminate, transversely veined; peduncles short, many-flowered: corolla glabrous with obtuse segments: leaflets of the staminal crown ovate, obtuse, inner angle attenuated, erect, exceeding the apiculated stigma; flowers small, brownish.

Silhet.

(11) H. parviflora. *(R. W.)*
Ident. Wight's Contrib. p. 37.—Dec. prod. VIII. p. 637.
Engrav. Wight's Icon. t. 1269.

Spec. Char. Climbing: leaves approximated, fleshy, glabrous, narrow-lanceolate, blunt-pointed: flowers few, generally in pairs from a short thick peduncle: pedicels shorter than the leaves: corolla glabrous: leaflets of the crown ovate, pointed, the apex resting on the stigma.

Coortallum, flowering in September.

GENUS XXXIII. CEROPEGIA.

Pentandria Digynia. *Ser: Syst.*

Deriv. From *Keros*, wax, and *Pege*, a fountain; literally a fountain of wax.

Gen. Char. Calyx 5-parted: corolla more or less ventricose from the base, funnel-shaped: segments of the limb narrow, cohering at the apex: gynostegium included: staminal crown gamophyllous, 5-10-15-lobed, in a single or double series: lobes placed before the stamens, ligulate: anthers simple at the apex: pollen-masses erect, affixed by the base, margins simple: stigma mutic: follicles cylindraceous, smooth: seeds comose.—Herbs or undershrubs, often twining, sometimes milky.

(1) C. lucida. *(Wall.)*
Ident. Wight's Contrib. p. 50.—Dec. prod. VIII. p. 641.
Engrav. Wall. Pl. As. Rar. II. t. 139.

Spec. Char. Twining: root fibrous: leaves from broad ovate to oblong-lanceolate, acuminate: peduncles many-flowered: calycine segments subulate, recurved at the apex: corolla clavate, scarcely ventricose at the base, segments of the limb attenuated, equalling the tube: lateral lobes of the leaflets of the staminal crown united above the middle, hairy at the margin, the primary ones recurved at the apex: flowers green, purple-dotted.

Khasia Hills, flowering in the cold season.

(2) C. hirsuta. *(W. & A.)*
Ident. Wight's Contrib. p. 30.—Dec. prod. VIII. p. 641.

Spec. Char. Suffruticose, hirsute, twining: leaves from cordate-ovato obtusely acuminate to narrow-lanceolate, acute: peduncles few-flowered: calycine segments filiform: corolla clavate, ventricose at the base, segments of the limb broader upwards, shorter than the tube: lateral lobes of the leaflets of the staminal crown united at the middle, the primary ones hooked at the apex: follicles straight, glabrous.

Neilgherries.

(3) C. JACQUEMONTIANA. (*Dec.*)

Ident. Dec. prod. VIII. p. 641.

SPEC. CHAR. Hirsute, herbaceous, twining: leaves ovate or ovate-lanceolate, subcordate or round, pubescent on both sides, younger ones subtomentose; peduncles hispid, shorter than the leaf, many-flowered; sepals linear-lanceolate, erect; corolla clavate, ventricose at the base, greenish, livid-violet below, spotted above, segments of the limb broader upwards, nearly equalling the tube; lateral lobes of the leaflets of the staminal crown united at the middle, ciliated, primary ones elongated, uncinately reflexed at the apex: follicles straight, quite glabrous.

Karlee in the Deccan.

(4) C. SPHENANTHERA. (*W. & A.*)

Ident. Wight's Contrib. p. 31.—Dec. prod. VIII. p. 643.

SPEC. CHAR. Twining: leaves distant, lanceolate, attenuated; peduncles few-flowered: calyx minute, segments filiform; corolla ventricose at the base, tube clavate, limb hemispherical, segments broader upwards, long-ciliated; outer lobes of the staminal crown short, acute, inner ones slightly exceeding the gynostegium; follicles slender, glabrous.

Neilgherries.

(5) C. ACUMINATA. (*Roxb.*)

Ident. Wight's Contrib. p. 32.—Dec. prod. VIII. p. 643.— Roxb. flor. Ind. II. p. 20.

Engrav. Roxb. Cor. I. t. 8.

SPEC. CHAR. Twining, slightly fleshy: leaves linear-lanceolate, attenuated at the apex: peduncles few-flowered: corolla ventricose at the base, tube sub-clavate, segments of the limb much shorter than the tube: outer lobes of the staminal crown minute, inner ones subulate.

Samulcottah.

(6) C. CANDELABRUM. (*Linn.*)

Ident. Wight's Contrib. p. 33.—Dec. prod. VIII. p. 643.

Engrav. Rheede Mal. 9. t. 16.

SPEC. CHAR. Twining: leaves ovate-lanceolate, round or obtusely cordate at the base, mucronate at the apex: peduncles spreading, slender, many-flowered, shorter than the leaf: pedicels erect: calyx very short, sepals subulate, tube of the corolla cylin-

dric above, ventricose below above the gynostegium, streaked with
purple, segments of the limb oblong, short, connate at the apex,
ciliated: gynostegium stalked: outer leaflets of the staminal crown
adnate, quite glabrous, inner ones ligulate, more or less cohering at
the apex.

Sandy places near Mangalore.

(7) C. JUNCEA. *(Roxb.)*

Ident. Wight's Contrib. p. 50.—Dec. prod. VIII. p. 641.

Engrav. Wight's Icon. t. 1260.—Roxb. Cor. I. t. 10.

SPEC. CHAR. Glabrous, somewhat fleshy, twining, milky: leaves
small, sessile, lanceolate, acute: peduncles few-flowered: calycine
segments subulate: corolla club-shaped, curved, ventricose at the
base, segments of the limb broader upwards nearly equalling the
tube, hairy: lateral lobes of the leaflets of the staminal crown united at the middle, primary ones hooked at the apex: flowers large,
greenish-yellow, variegated with purple.

Circars. Hedges near Samulcottah. Negapatam. Flowering
in the cold season.

(8) C. CILIATA. *(R. W.)*

Ident. Wight's Icon. vol. IV. & t. 1262.

SPEC. CHAR. Suffruticose, twining: root tuberous: leaves
ovate-lanceolate, coarsely pubescent on both sides, hairy on the
veins beneath, ciliate on the margin: peduncles about half the
length of the leaves, bispid, umbels 6–10-flowered: calycine segments subulate, shorter than the ventricose base of the corolla:
lobes of the corolla shorter than the tube: outer lobes of the staminal crown emarginate, ciliate, inner ones clavate, recurved at the
points.

Rocks at Katie falls, Neilgherries, flowering in June and July.

(9) C. DECAISNEANA. *(R. W.)*

Ident. Wight's Icon. vol. IV. & t. 1259.

SPEC. CHAR. Twining, glabrous: leaves lanceolate, acuminate
at both ends, acute, bispid above from short scattered rigid hairs,
glabrous beneath; umbels pendulous, 6-flowered, pedicels divaricated, longer than the peduncles: calycine lobes bristly: corolla
clavate, largely ventricose at the base, lobes cohering at the points:
secondary lobes of the staminal crown about half the length of the
primary, erect, slightly cleft at the apex, tipped with purple: flowers large, ascending, mottled with purple spots: follicles long and
very slender.

Sisparah Ghaut, Neilgherries. Flowering in March and August.

(10) C. PUSILLA. *(W. & A.)*

Ident. Wight's Icon. vol. IV. & t. 1261.

SPEC. CHAR. Herbaceous, glabrous, erect, 2–6 inches high: root tuberous: leaves linear-lanceolate, succulent: flowers axillary, solitary, erect: corolla ventricose at the base, tube cylindrical, longer than the lobes of the limb: exterior lobes of the staminal crown ciliate, shorter, the inner ones longer than the gynostegium: follicles erect, attenuated at the point.

Neilgherries in pasture ground. Banks of the Pycarrah river.

(11) C. INTERMEDIA. *(R. W.)*

Ident. Wight's Icon. vol. IV. & t. 1263.

SPEC. CHAR. Fruticose, twining: leaves ovate-lanceolate, acute, glabious on both sides: peduncles shorter than the leaves, several flowered: sepals subulate, about the length of the ventricose part of the corolla: limb of the corolla shorter than the tube, lobes sub-spathulate, ciliate, united at the point, forming a globose head: exterior lobes of the crown obsolete, inner ones long, spathulate, hairy towards the base.

Seeramallie Hills near Dindigul, flowering in October.

(12) C. MUNROSII. *(R. W.)*

Ident. Wight's Icon. vol. IV. & t. 1264.

SPEC. CHAR. Fruticose, slender, twining: leaves short-petioled, narrow-lanceolate, acute: flowers large, solitary, short-peduncled: corolla ventricose at the base, tube short, contracted in the middle, limb long, deeply cleft into five slender lobes, ciliated with glanduliferous hairs: outer lobes of the crown inconspicuous, inner ones ligulate, twice the length of the column.

Neilgherries and Coorg jungles.

(13) C. ELEGANS. *(Wall.)*

Ident. Wight's Contrib. p. 31.—Dec. prod. VIII. p. 642.

Engrav. Bot. Mag. t. 3015.—Bot. Reg. 20. t. 1700.—Wight's Icon. t. 1265.

SPEC. CHAR. Twining, glabrous: leaves ovate-oblong or oblong-lanceolate, attenuated or shortly acuminate, somewhat succulent, ciliolate: peduncles equalling the petioles, few-flowered: tube of the corolla ventricose, curved at the base, purplish speckled: lobes subdeltoid, acuminate, cohering at the apex, often ciliate: exterior lobes of the staminal crown ligulate, approximated, interior ones longer, inflexed, more or less united at the points: follicles very long, slender, glabrous, sub-torulose.

Dindigul. Neilgherries. Flowering in the cold season.

(14) C. Mysorensis. (*B. W.*)

Ident. Wight's Icon. vol. III. & t. 846.

Spec. Char. Suffruticose, twining, glabrous: leaves broad cordate-ovate, acuminated; peduncles about the length of the petioles, 4–8-flowered; segments of the calyx acute, much shorter than the dilated base of the corolla; tube of the corolla short, suddenly expanding into a large 5-cleft limb, segments short, broad-ovate, adhering at the point, glabrous on the margins; lobes of the crown ligulate, lateral ones about equalling the primary; follicles long, slender, irregularly curved; flowers pale straw-coloured.

Mysore, flowering in December.

(15) C. spiralis. (*R. W.*)

Ident. Wight's Icon. vol. IV. & t. 1267.

Spec. Char. Suffruticose, erect, glabrous: root tuberous; leaves long, narrow-lanceolate, acute; flowers large, solitary, short-peduncled; tube of the corolla ventricose at the base, lobes long, subulate, spirally twisted, ciliate at the base; exterior lobes of the crown shorter than the column, interior dilated at the base, ligulate and free above.

Balaghaut Hills near Madras, flowering in July and August.

(16) C. bulbosa. (*Roxb.*)

Ident. Wight's Contrib. p. 32.—Roxb. flor. Ind. l. p. 28.— Dec. prod. VIII. p. 643.

Engrav. Roxb. Cor. I. t. 7.—Wight's Icon. t. 843.—Hook. Bot. Misc. suppl. II. t. 2.

Spec. Char. Twining, glabrous, rather fleshy: root tuberous; leaves from suborbicular to lanceolate-acuminated; peduncles many-flowered; calycine segments much shorter than the ventricose base of the corolla; tube of the corolla subclavate, segments enlarging upwards, much shorter than the tube, ciliated; middle lobes of the leaflets of the corona subulate, incurved at the top, acuminated, lying on the primary ones.

Point Calymere. Samulcottah. Allahabad.

(17) C. tuberosa. (*Roxb.*)

Ident. Wight's Contrib. p. 32.—Dec. prod. VIII. p. 644.

Syn. C. mucronata, *Roth.*—C. candelabrum, *Roxb. flor. Ind.* II. p. 27.

Engrav. Roxb. Cor. I. t. 9.—Wight's Icon. t. 353.

Spec. Char. Herbaceous, glabrous, twining: leaves from nearly orbicular to oval or ovate, cuspidate, sometimes lanceolate-acuminated; peduncles usually twin, few or many-flowered, longer or shorter than the leaves : calyx small with subulate segments : corolla ventricose at the base, the tube widened upwards, segments narrow, nearly linear, villous : gynostegium stalked : middle lobes of the leaflets of the crown ligulate, lateral ones short, cohering with the primary one : follicles slender, round.

Coromandel. Negapatam. Concans.

(18) O. VINCÆFOLIA. (*Dalz.*)

Ident. Dalz. Bomb. flor. p. 153.
Engrav. Bot. Mag. t. 3740.

SPEC. CHAR. Twining, pubescent : leaves subcordate or broadly ovate-acuminate, shortly-petioled ; peduncles with spreading hairs, few-flowered : tube of corolla ventricose at the base, white spotted ; divisions oblong, erect, connivent, ciliated, dark purple.

Near Bombay.

(19) C. LUSHII. (*Grah.*)

Ident. Dalz. Bomb. flor. p. 154.
Engrav. Bot. Mag. t. 3300.

SPEC. CHAR. Twining, glabrous : leaves linear-acuminate, fleshy, channelled, glaucous : base of the corolla tube globose, inflated, greenish : divisions linear-ciliated, cohering at the apex, violet-coloured within.

Kaseraye Jungles, Bombay Presidency.

(20) C. ANGUSTIFOLIA. (*Dalz.*)

Ident. Dalz. Bomb. flor. p. 154.—Hook. Journ. Bot. II. p. 159.

SPEC. CHAR. Herbaceous, erect, pubescent, 5 to 6 inches high : root tuberous : stem round : leaves narrow, linear-lanceolate, acute, hairy on the margins of the upper side, glabrous and pale beneath : flowers outside of the axils, solitary, ascending : corolla slightly ventricose at the base : tube cylindrical, segments of the limb narrow-linear, spathulate, as long as the tube : flowers purple, with a green base.

Malwan District, flowering in July.

(21) C. OPHIOCEPHALA. (*Dalz.*)

Ident. Dalz. Bomb. flor. p. 154.—Hook. Journ. l. c.

SPEC. CHAR. All hispid, twining : leaves broad-lanceolate, rounded or cordate at the base, acuminated at the apex, hispid on both sides : peduncles outside of the axils, longer than the petiole, hispid, 3 to 4-flowered : sepals linear-subulate, spreading : corolla tube ascending, ventricose at the base, dark-purple, glabrous : divisions of the limb one-third shorter, oblong-obtuse, attenuated towards the apex, purple, yellow and green: follicles linear, smooth, 4 to 5 inches long, spotted with purple.

On Caranjah Hill, Bombay Presidency.

(22) C. OCULATA. *(Hook.)*

Ident. Dals. Bomb. flor. p. 154.
Engrav. Bot. Mag. t. 4093.

SPEC. CHAR. Stem herbaceous, twining, glabrous: leaves cordate-ovate, acuminate, rather hairy, cilliated, with glands at the base: peduncles with spreading hairs, 4 to 6-flowered: tube of the corolla much inflated at the base, globose, broader than the limb: segments of the limb oblong, erect, connivent, ciliated, yellow below, with black spots, deep-green above: lobes of the outer staminal crown attenuated, emarginate, of the inner, narrow-linear, straight, entire.

Bombay. Annamullay Hills.

(23) C. ATTENUATA. *(Hook.)*

Ident. Dals. Bomb. flor. p. 154.
Engrav. Hook. Ic. pl. IX t. 867.

SPEC. CHAR. Erect: leaves linear, long and slenderly attenuated: younger ones slightly pilose: peduncle axillary, solitary, 1-flowered: calycine lobes subulate, ciliated, spreading: corolla tube long, inflated at the base: lobes of the limb as long as the tube, slender filiform.

South Concan near Vingorla.

(24) C. MACULATA. *(R. H. B.)*

Ident. Beddome in Madr. Journ. No. I. 1864.

SPEC. CHAR. Root fibrous: twining stems terete, glabrous, maculate: leaves ovate-acuminate, maculate and minutely punctated, glabrous, furnished with a minute gland on the lamina just above the insertion of the petiole: petioles very minutely pilose, channelled: peduncles a little shorter than the petioles, glabrous: pedicels umbelled, 7-10, as long or longer than the peduncles, sepals subulate: corolla dull greenish purple, segments ciliated, exterior lobes of the staminal crown of the same length as the inner ones, alternate with them and bifid to the base: follicles terete, slender.

Annamullays in moist woods.

(25) C. ENSIFOLIA. *(R. H. B.)*

Ident. Beddome l. c.

SPEC. CHAR. Root tuberous: stem twining, glabrous: leaves very narrow, linear, tapering at the apex, mucronate, above a few adpressed hairs, below pale, glabrous, very short petioled : peduncles axillary from half as long to nearly as long as the leaves, pubescent, furnished with several ovate pointed bracts at the apex, umbelliferous, bearing several flowers on simple pedicels and a second peduncle which is again umbelliferous or sometimes much elongated and paniculate, pedicels pubescent; calycine lobes glabrous, subulate, half the length of the ventricose base of the glabrous corolla: segments of the corolla as long or longer than the tube, exterior lobes of the staminal crown short, emarginate, ciliate, alternate with the inner long ligulate lobes: follicles long, slender, terete; flowers greenish-white.

Annamullays, rocky places at 2,500 to 3,500 feet.

(26) C. FIMBRIIFERA. (*B. H. B.*)

Ident. Beddome l. c.

SPEC. CHAR. Root tuberous: stem erect, minutely pubescent; leaves subsessile, narrow ensiform, tapering to the apex, minutely ciliated, above pubescent, beneath glabrous, except on the midrib and minutely frosted: peduncles axillary, short, nearly glabrous, about 4-flowered, flowers opening in succession: pedicels about as long as peduncles, furnished at the base with a few subulate bracts: calycine segments subulate, acute: corolla tube narrowed upwards (but not ventricose at the base), segments about the length of the tube with tufts of numerous long purple gland tipped hairs between the segments: outer lobes of the staminal crown short, sharply bifid, and ciliated, inner lobes long, ligulate and adnate to the centre of the outer ones: follicles terete: flower tube greenish, outside deep purple striated, inside, segments pale greenish purple, fringe deep purple.

Annamullay hills, at 3,000 feet, rare.

(27) C. GRACILIS. (*R. H. B.*)

Ident. Beddome l. c.

SPEC. CHAR. Root fibrous: stems twining, glabrous: leaves short petioled, ovate-elliptic, acuminate, minutely ciliate, above furnished with a few distant short hairs, minutely pellucid dotted, beneath shining, glabrous, except on the midrib: peduncles longer than the petioles, 2–5-flowered: flowers very large, tube short ventricose at the base, segments longer than the tube, very narrow at the middle, broader upwards: outer lobes of the staminal crown short, deeply bifid, ciliate, with long fine hairs, inner lobes alternate with them, distant, long, ligulate.

Annamullays in moist woods at 4,000 feet, rare.

GENUS XXXIV. ERIOPETALUM.

Pentandria Digynia. *Sw; Syst.*

Deriv. From *Erion*, wool, and *Petalon*, the petal.

GEN. CHAR. Corolla subcampanulate, 5-cleft: segments narrow-linear, sinus broad: staminal corona gamophyllous, 15-lobed, 5 interior lobes incumbent on the anthers, the rest erect, approximate, adhering to the inner row: anthers simple at the apex: pollen-masses attached near the base, erect, incumbent on the stigma: follicles slender, glabrous.

(1) E. LAEVIGATUM. (R. W.)

Ident. Wight's Contrib. p. 35.—Dec. prod. VIII. p. 846.

Syn. Gomphocarpus laevigata, *Ham.*—Microstemma, *Wall. ex Delp. No. 23.*

SPEC. CHAR. Herbaceous, erect: leaves small, scale-shaped, adpressed: segments of the corolla linear, undulated at the margin, almost glabrous, nearly twice as long as the tube: staminal corona equalling the gynostegium: umbels lateral or terminal, sessile: flowers small.

Jungles of Cossala, Goruckpore.

GENUS XXXV. CARALLUMA.

Pentandria Digynia. *Ser: Syst:*

Deriv. The Indian name Latinised.

GEN. CHAR. Corolla rotate, deeply 5-cleft: gynostegium exserted: staminal crown 10-lobed in a simple row, five lobes opposite the stamens, fleshy, incumbent on the anthers, alternate ones ligulate, bipartite at the apex: anthers simple at the apex: pollen-masses erect, four-cornered: stigma mutic: follicles slender, smooth: seeds comose.—Erect, fleshy, leafless herbs: stems tetragonal, toothed at the angles: flowers towards the tops of the branches: peduncles solitary, 1-flowered, rising from the axils of the teeth of the branches.

(1) C. ASCENDENS. (R. Br.)

Ident. Wight's Contrib. p. 33.—Dec. prod. VIII. p. 647.

Syn. Stapelia ascendens, *Roxb.*

Engrav. Roxb. Cor. I. t. 30.

SPEC. CHAR. Flowers often nodding: segments of the corolla acuminate, glabrous.

Peninsula.

(2) C. ATTENUATA. (*R. W.*)

Ident. Wight's Icon. vol. IV.
Engrav. Wight's Icon. t. 1268.

SPEC. CHAR. Erect: stems tetragonal at the base, subterete towards the apex, sparingly branched: flowers confined to the ends of the branches, drooping: lobes of the corolla lanceolate, fimbriate on the margin.

Dry plains at the foot of the Neilgherries, flowering in March and April.

(3) C. FIMBRIATA. (*Wall.*)

Ident. Wight's Contrib. p. 34.—Dec. prod. VIII, p. 647.
Engrav. Wall. Pl. As. Rar. I. t. 9.

SPEC. CHAR. Flowers nodding or ascending, short-pedicelled; sepals short, subulate: segments of the corolla linear-oblong, cuspidate, fimbriated: flowers yellowish above, transversely purple-streaked, underneath yellow.

Peninsula. Deccan. Flowering in June.

GENUS XXXVI. BOUCEROSIA.

Pentandria Digynia. *Sax: Syst.*

Deriv. From *Bous*, an ox, and *Keras*, a horn.

GEN. CHAR. Corolla subcampanulate, 5-cleft: segments broad triangular, sinus acute: gynostegium scarcely exserted: staminal corona gamophyllous, 15-lobed in two rows, five inner lobes opposite the stamens incumbent on the anthers, the rest outer, erect or somewhat incurved at the apex adhering to the back of the inner ones: anthers simple at the apex: pollen-masses erect, 4-cornered: stigma mutic: follicles smooth, terete, attenuated at the apex: seeds comose.—Fleshy, leafless, erect, 4-sided plants, with dentate angles: flowers numerous, terminal, umbelled.

(1) B. UMBELLATA. (*W. & A.*)

Ident. Wight's Contrib. p. 34.—Dec. prod. VIII. p. 648.
Syn. Stapelia umbellata, *Roxb.*—Caralluma umbellata, *Spr.*—*Wall.*
Engrav. Roxb. Cor. III. t. 241.—Wight's Icon. t. 495.

SPEC. CHAR. Segments of the corolla glabrous: flowers long-pedicelled, externally whitish, with dark purple confluent spots, internally yellowish with dark purple circles.

Southern parts of the Peninsula, flowering in May.

(2) B. HUTCHINIA. *(Dec.)*

Ident. Dec. prod. VIII. p. 649.
Syn. Hutchinia Indica, *Wight's Contrib.* p. 34.
Spec. Char. Branches diffuse: teeth reflexed: flowers terminal, subumbellate: lobes of the corolla wrinkled, purplish-spotted, scattered with purple hairs inside.

Southern Peninsula, probably on mountain tracts.

(3) B. CAMPANULATA. *(R. W.)*

Ident. Wight's Icon. vol. IV. & t. 1287.
Spec. Char. Angles of the stem somewhat dilated: tube of corolla conical, glabrous: gynostegium short, not exserted beyond the tube.

Station unknown.

(4) B. LAXIANTHA. *(R. W.)*

Ident. Wight's Icon. vol. IV. & t. 1286.
Spec. Char. Much branched: flowers umbelled, long-pedicelled: corolla rotate, 4-lobed, externally glabrous, densely pubescent within, lobes at first ciliated with longish, jointed, caducous hairs.

Nuggur hills, near Madras.

(5) B. DIFFUSA. *(R. W.)*

Ident. Wight's Icon. vol. VI. & t. 1599.
Spec. Char. Branched, diffuse, procumbent: floriferous branchlets ascending, angles subacute, with minute teeth: umbels terminal, simple, many-flowered: calyx small, 5-parted, lobes subulate: corolla tubular, delicately transversely wrinkled, fimbriated on the edge: flowers subsessile, dark purplish brown, variegated within with fine whitish lines.

Rocky hills near Coimbatore, flowering in April and May.

GENUS XXXVII. CRYPTOLEPIS.

Pentandria Digynia. *Sex. Syst.*

Deriv. From *Krypto*, to hide, and *Lepis*, a scale; referring to the scales in the tube of the corolla.

Gen. Char. Corolla funnel-shaped: tube enclosing five obtuse, wedge-shaped scales, alternating with the segments of the limb:

throat naked: stamens inclosed, inserted in the bottom of the tube: anthers sagittate: ovaries two: style wanting: stigma dilated, 5-cornered, ending in a conical point: hypogynous scales five: follicles lanceolate, horizontal.—Twining shrubs: leaves opposite: corymbs interpetiolar, almost sessile, very short.

(1) C. BUCHANANI. *(Roem. & Schult.)*

Ident. R. & S. Syst. IV. p. 409.—Dals. Bomb. flor. p. 148.

Syn. C. reticulata, *Wall.*—Nerium reticulatum, *Roxb. flor. Ind.* II. p. 9.

Engrav. Wight's Icon. t. 494.—Rheede Mal. IX. t. 11.

SPEC. CHAR. Milky shrub, climbing, smooth: leaves short-petioled, broad-elliptic, with a short subulate point, bright green above, whitish and glaucous below, transversely veined: corymbs axillary, short-peduncled: flowers subsessile, pale yellow.

Coromandel and Western Coast. Bengal. Flowering in the rainy season.

(2) C. GRANDIFLORA. *(R. W.)*

Ident. Wight's Icon. vol. III. & t. 831.

SPEC. CHAR. Leaves from oval to obovate, spathulate: cymes axillary, diffuse, longer than the leaves: corolla funnel-shaped, throat furnished with five inflexed capitate processes: anthers acuminate, 5, hypogynous, emarginate: scales alternate with the stamens: follicles divaricated.

Balaghaut Mountains near Naggary.

(3) C. ELEGANS. *(Wall.)*

Ident. Wall. Cat. No. 1639.—Don. L c.

SPEC. CHAR. Leaves oblong, mucronate, glaucous beneath, somewhat cordate at the base: peduncles terminal, panicled: flowers pure white, very fragrant.

Silhet. Assam. Flowering in February and March.

GENUS XXXVIII. FREREA.

Pentandria Digynia. *Lin. Syst.*

Deriv. Named in honor of Sir Bartle Frere, Governor of Bombay, a great promoter of scientific research.

GEN. CHAR. Calyx 5-parted: corolla glabrous, 5-lobed, lobes broadly triangular, margin fimbriated, sinus small, acute between the lobes: staminal corona gamophyllous, 10-lobed in a single row, namely five lobes opposite the segments of the corolla, broad, very short, sinuately truncated, five ligulate, alternate with them, incumbent on the anthers, truncated at the apex: anthers simple at the margin: pollen-masses erect, attached above the base, furnished on the inner margin with a golden pellucid line: stigma mutic.

(1) F. INDICA. (*Dals.*)

Ident. Dalzell in Journ. Proc. Linn. Soc. VIII. p. 10.

SPEC. CHAR. Herbaceous, perennial, low, cæspitose, branched: stems terete, whitish, smooth: leaves short-petioled, oblong, fleshy: flowers extra-axillary, solitary, very shortly peduncled, purplish-red, with a small pale-yellow spot in the middle of each lobe.

Concan at 3000 feet.

ORDER CIX. LOGANIACEÆ.

Calyx free, 5, rarely 4-lobed: corolla regular or rarely irregular, hypogynous, 5, rarely 4-lobed, or many lobed: æstivation valvate, twisted, or imbricated: stamens inserted on the tube of the corolla, sometimes 5, alternate with the lobes, or rarely 1, or 10, or 12, then opposite the lobes; or lastly 3 alternate and 2 opposite the lobes of the corolla; anthers 2-celled, dehiscing lengthwise; pollen vittato-three-lobed: nectary none: ovary free, 2, rarely 3-celled or 1-celled: ovules amphitropous or rarely anatropous, style simple: stigma simple or 2-lobed: fruit sometimes capsular, the margin curved inward, and bearing the placentas, sometimes drupaceo-baccate, placentas in the capsules often at length free: seed usually peltate, rarely erect, sometimes winged: albumen fleshy or cartilaginous: embryo straight with the radicle next the hilum: cotyledons 2, foliaceous.—Shrubs or small trees, rarely herbs: leaves opposite, entire, penninerved, petioled: stipules between, or within the petioles, often united into a sheath, sometimes wanting: flowers racemose or corymbose, rarely solitary, terminal or axillary.

GENUS I. MITREOLA.

Pentandria Monogynia. *Sar; Syd;*

Deriv. From the Latin, signifying a small mitre in reference to the shape of the capsule.

GEN. CHAR. Calyx 5-partite, lobes lanceolate; corolla tubular, short, half-5-cleft, deciduous, tube somewhat ventricose, throat hairy: stamens inserted at or a little below the middle of the tube, included: filaments slender, anthers ovato-acute: ovary 2-celled: ovules numerous; style short; stigma capitellate, hairy: capsule deeply 2-cleft, mitre or moon-shaped, membranaceous: seeds ovate or oblong-compressed, not angled; albumen fleshy: embryo axile, linear: radicle inferior.

(1) M. OLDENLANDOIDES. (*Wall.*)

Ident. Wall. Cat. No. 4350.—Dalz. Bomb. flor. p. 155.—Dec. prod. IX. p. 9.

Syn. M. paniculata, *Wall.*

Engrav. Wight's Icon. t. 1600.—Hook. Ic. pl. t. 827.

SPEC. CHAR. Herbaceous, erect: stem somewhat quadrangular, glabrous; branches roughish, hairy; leaves opposite, entire, ovate-oblong, acuminate, narrowing at the base, margin and veins roughish with hairs, capsule lunate with the lobes inflexed, rough on the inner angle: seed elongated, compressed: bracts and lobes of the calyx lanceolate, margins and back slightly pilose: flowers subsessile, secund, small, white.

Concans. Mysore. Bombay. Sukanaghur.

GENUS II. MITRASACME.

Tetrandria Monogynia. *Sar; Syd;*

Deriv. From *Mitra*, a mitre, and *Acme*, a point; alluding to the form of the capsule.

GEN. CHAR. Calyx 4-cleft, rarely bifid, without glands; corolla 4-lobed, usually campanulate, lobes inflexed at the margin with valvular æstivation, afterwards spreading: stamens usually inserted at the base of the corolla and included, rarely in the throat and then exserted: filaments slender; anthers ovate-acute, cordate at the base: ovary 2-celled; ovules numerous: style 2-cleft at the base: stigma capitellate, somewhat 2-lobed; capsule sub-globose, 2-celled: seeds numerous, ovate-globose, wrinkled; albumen fleshy: embryo straight: radicle inferior.

(1) M. CRYSTALLINA. *(Griffith.)*

Ident. Griffith Notulæ IV, p. 87.

Syn. M. Indica, *Wight's Icon. vol.* IV, *t.* 1601.—M. pusilla, Dals. *Bomb. flor.* p. 155.—*Hook. Journ. Bot.* II. p. 136.

Spec. Char. Erect, branched, glabrous: branches somewhat flexuose, compressed, 2-edged: leaves opposite, entire, sessile, ovate-lanceolate, acute: peduncles longer than the leaves, 1-flowered: corolla longer than the calyx, pilose inside: seeds peltate, scrobiculate: flowers white.

Jamlash. Arcot. Coimbatore. Malwan. Flowering in August and September.

(2) M. NUDICAULIS. *(Retsm.)*

Ident. Retsw. in Dec. prod. IX. p. 12.—Benth. Logan. in Linn. Journ. I. p. 92.

Spec. Char. Stem slightly hairy at the base: leaves somewhat radical, spreading: scape glabrous with few small erect leaves: style undivided at the base while flowering, afterwards 2-cleft.

Khasia. Assam.

(3) M. CAPILLARIS. *(Wall.)*

Ident. Dec. prod. IX. p. 11.

Syn. M. trinervis, *Spanoghe in Linn.* XV. p. 335.—M. Malaccensis, R. *Br.*—Limnophila campanuloides, *Benth. in Wall. Cat. No.* 3908.

Engrav. Wight's Icon. t. 1601.

Spec. Char. Stem slightly hairy at the base, leafy: leaves lanceolate: peduncles umbelliferous, nearly naked, glabrous: style undivided at the base while flowering, afterwards 2-cleft.

Canara.

GENUS III. STRYCHNOS.

Pentandria Monogynia. *Ser. Syst.*

Deriv. The Greek name of the *Solanum.*

Gen. Char. Calyx 5-lobed: corolla tubular, hypocrateriform, or infundibuliform with an abbreviated tube, throat naked or bearded, limb 5-partite: stamens inserted in the throat, filaments very short, anthers slightly exserted: ovary 2-celled: style filiform : stigma capitate, undivided or obscurely somewhat 2-lobed ; ovules numerous : berry with a rind, 1-celled, many-seeded, or 1-seeded by abortion : seeds nestling in pulp, discoidly compressed : embryo somewhat bilamellate at the base of cartilaginous albumen, excentric, straight, short : cotyledons sessile, leafy : radicle terete, vague.—Trees or shrubs, usually climbing ; leaves opposite, short-petioled, quite entire, 3–5-nerved at the base, petioles connate at the base : another leaf, often abortive, and putting forth from the axil a simple tendril-shaped branchlet : arils in some species bearing a straight thorn : corymbs axillary or terminal : flowers white, or whitish green, usually fragrant.

(1) S. WALLICHIANA. *(Steud.)*

Ident. Dec. prod. IX. p. 13.

Syn. S. lucida, *Wall.*

SPEC. CHAR. Climbing, unarmed, glabrous: tendrils thickened above : leaves ovate, acuminate, shining, somewhat 3-nerved at the base : corymbs axillary, trichotomous, 3 times shorter than the leaf : pedicels somewhat velvety : calyx acutely 5-lobed : throat of the corolla naked.

Pundua, Silhet.

(2) S. AXILLARIS. *(Colebr.)*

Ident. Colebr. trans. Linn. Soc. 12. p. 356. & t. 15.—Dec. prod. IX. p. 13.—Wall. in Roxb. flor. Ind. II. p. 206.

SPEC. CHAR. Stem climbing : leaves ovate, acuminate, triplinerved : branchlets, petioles and middle nerve of the leaf somewhat velvety-pubescent at the base : tendrils axillary : corymbs axillary, scarcely longer than the petiole : berry 1-seeded.

Khasia mountains. Southern Ghauts ? Flowering in the hot season.

(3) S. COLUBRINA. *(Linn.)*

Ident. Linn. Sp. 271.—Dec. prod. IX. p. 14.—Wall. in Roxb. flor. Ind. II. p. 264.

Syn. S. bicirrhosa, *Lind.*—Dec. prod. IX. p. 16.

Engrav. Wight's Icon. t. 434.—Rheede Mal. VIII. t. 24.— Rumph. Amb. II. t. 48.

SPEC. CHAR. Scandent, unarmed : tendrils simple, lateral, solitary : leaves elliptic, or oblong, obtusely acuminated, triplinerved, glabrous : corymbs terminal, few-flowered ; berries globose, 2–12-seeded : flowers small, greenish-yellow.

South Concan. Coromandel. Khasia mountains. Flowering in April.

(4) S. NUX VOMICA. *(Linn.)*

Ident. Linn. Sp. 271.—Dec. prod. IX. p. 15.—Roxb. flog. Ind. II. p. 261.

Engrav. Roxb. Cor. I. t. 4.—Rheede Mal. I. t. 37.—Rumph. Amb. I. t. 25.

Spec. Char. Arboreous, without thorns or tendrils: leaves ovate, petioled, 3–5-nerved, quite glabrous; corymbs terminal: calyx shortly 5-toothed: corolla glabrous within: berry globose, many-seeded, beak hard, brownish-red: flowers greenish-white: seeds light-grey, silky.

Malabar. Coromandel. Hilly parts of the Concans. Flowering in February. The seeds furnish the poisonous principle known as Strychnine.

(5) S. POTATORUM. *(Linn.)*

Ident. Linn. f. Suppl. 148.—Dec. prod. IX. p. 15.—Roxb. Flor. Ind. II. p. 263.

Engrav. Roxb. Cor. I. t. 5.

Spec. Char. Arboreous, without thorns or tendrils: leaves very shortly petioled, elliptic, acute, glabrous, membranaceous, 5 and almost penninerved: corymbs axillary, opposite, shorter than the leaf: corolla hirsute within: berry 1-seeded: flowers greenish-yellow, fragrant.

Coromandel. Concans. Western Ghauts. Flowering in April.

GENUS IV. GARDNERIA.

Pentandria Monogynia. *Sex: Syst.*

Deriv. In honor of Hon. E. Gardner, Resident in Nepaul, an ardent Botanist.

Gen. Char. Calyx minute, persistent, cup-shaped, 4–5-toothed: corolla rotate, 4–5-parted, throat naked, lobes ovate, coriaceous and thick, somewhat thickened at the margins and apex: stamens 4–5, alternate with the lobes, inserted at the throat, filaments scarcely any, anthers erect, free or somewhat concrete at the base and free at the tops: ovary free, ovoid: style filiform, shorter than the anthers: stigma obscurely 2-lobed: berry globose, slightly depressed, 2-celled: ovules solitary in each cell: seeds convex at the back, flat in front: albumen horny: embryo erect: cotyledons lanceolate.—Glabrous shrubs: branches 4-sided, afterwards terete: leaves opposite, quite entire, petioled: glands ciliary, axillary: flowers springing from the axils, yellow or white: berries red.

(1) G. ovata. *(Wall.)*

Ident. Wall. in Roxb. flor. Ind. I. p. 400.—Dec. prod. IX. p. 20.

Syn. G. Wallichii, *R. W. in Wall. Pl. As. Rar.*

Engrav. Wall. Pl. As. Rar. III. t. 281.—Wight's Icon. t. 1313.

Spec. Char. Extensively climbing: leaves oval, subacuminate; panicles axillary, oppositely branched, many-flowered, flowers tetramerous and tetrandrous; anthers concrete into a tube: flowers yellow.

Khasia hills. Common on the Neilgherries. Flowering in March and April.

GENUS V. FAGRÆA.

Pentandria Monogynia. *Ser: Syst:*

Deriv. Named by Thunberg after his friend J. T. Fagræus, M.D.

Gen. Char. Calyx bibracteate at the base, 5-parted, lobes imbricated, obtuse: corolla funnel-shaped, tube somewhat amplified above, lobes oblique: stamens inserted in the middle of the tube, filaments subulate, subexserted, anthers 2-celled, subincumbent: ovary 2-celled: style filiform: stigma peltately depressed: berry with a rind, oval, 2-celled: seeds numerous, immersed in pulp, small, crustaceous: albumen horny or fleshy?—Shrubs or trees: leaves opposite, oval, petioled, entire, coriaceous: stipules interpetiolar: flowers white, arranged in a terminal corymb or trichotomous raceme.

(1) F. obovata. *(Wall.)*

Ident. Wall. in Roxb. flor. Ind. II. p. 33.—Dec. prod. IX. p. 29.

Spec. Char. Arboreous: leaves ovate or obovate, rounded at the apex, apiculate: peduncles terminal, 3-flowered, subcorymbose: berry brown, shining.

Silhet, flowering in the rainy season.

(2) F. Coromandeliana. *(R. W.)*

Ident. Wight's Icon. vol. IV. & t. 1316.

Spec. Char. Arboreous: leaves succulent, spathulate-oblong: peduncles ternate, 3-flowered: corolla subcampanulate, lobes revolute: berry elliptic, tapering at both ends, pointed with the persistent base of the style: seeds subglobose, rough.

Courtallum. Coonoor. Flowering in the rainy season.

(3) F. MALABARICA. (*R. W.*)

Icon. Wight's Icon. vol. IV. & t. 1317.
Engrav. Rheede Mal. IV. t. 38.

SPEC. CHAR. Arboreous: leaves obovate-cuneate, subapiculate, longish-petioled: peduncles axillary and terminal, about 3 together, elongated, 3-flowered: corolla funnel-shaped, tube slender at the base, limb dilated, lobes spreading: stamens and style about the length of the corolla.

Malabar. Travancore.

(4) F. KHASIANA. (*Benth.*)

Icon. Benth. in Journ. Proc. Linn. Soc. 1. p. 99.

SPEC. CHAR. Arboreous: leaves longish-petioled, oblong-elliptic, acuminate, narrowed at the base, thick, almost veinless: cymes loosely trichotomous, few-flowered: segments of the calyx membranaceous at the margin: tube of the corolla dilated almost from the base: ovary 1-celled above: berry ovoid-oblong: seeds half immersed in fleshy pulp, numerous.

Khasia hills.

ORDER CX. GENTIANACEÆ.

Calyx free, persistent, 4-5-lobed, rarely 6-12: sepals cohering, valvate in æstivation: corolla monopetalous, hypogynous, persistent, regular, or bilabiate, the lobes alternate with the segments of the calyx, contorted to the left in æstivation: stamens inserted on the tube of the corolla, alternate with its lobes: filaments free: anthers 2-celled, erect or incumbent, occasionally at length twisted: ovary single, 1-celled, composed of two carpels with introflexed margins, hence half-2-celled: ovules numerous, anatropous, attached to the margins of the valves: style simple: stigmas two or one: capsule one or imperfectly 2-or 4-celled, septicidal, placentas parietal: seeds usually numerous: albumen fleshy: embryo axile, minute, cylindrical, straight: radicle thickened, directed to the hilum: cotyledons fleshy, short.—Herbaceous or frutescent plants, usually glabrous, bitter, not milky: leaves opposite, rarely alternate, simple, 3–5-ribbed, entire or 3-lobed: petioles often confluent at the base: flowers terminal, or axillary, regular or rarely irregular.

GENUS I. EXACUM.

Tetrandria Monogynia. *Sex : Syst :*

Deriv. From *Ex*, out of, and *Ago*, to drive; it is said to possess the property of expelling poison.

GEN. CHAR. Calyx 4–5-parted, segments keeled or winged at the back: corolla rotate, withering, tube afterwards globose, limb 4–5-parted: stamens 4–5, inserted on the throat of the corolla, suberect: anthers dehiscing at the apex by the poriform opening of a cleft: ovary 2-celled with introflexed valves, ovules attached to either side of the central suture: style distinct, declinate, deciduous, stigma undivided, capitulate or slightly transversely furrowed: capsule 2-celled, 2-valved, septicidal, placentae central, sometimes separating, sometimes jointed into one, afterwards free from the valves: seeds very minute, immersed in the placenta.—Herbs usually annual, straight, quite glabrous: cymes terminal.

(1) E. TETRAGONUM. *(Roxb.)*

Ident. Roxb. flor. Ind. I. p. 398.—Dec. prod. IX. p. 44.
Syn. E. Hamiltonii, *Don.*
Engrav. Wall. Ic. III. t. 274.

SPEC. CHAR. Stem 4-cornered, somewhat simple: leaves stem-clasping, ovate-oblong, acuminate, 5-nerved, smooth at the margin: segments of the calyx ovate-acute: lobes of the corolla elliptic, twice as long as the tube: flowers large, blue with gold coloured anthers.

Bengal, flowering in the rainy season.

(2) E. TENUE. *(Wall.)*

Ident. Roxb. flor. Ind. (Ed. Wall.) I. p. 414.—Dec. prod. IX. p. 44.

SPEC. CHAR. Stem terete: leaves very short, stem-clasping, attenuated at the base, linear-lanceolate, long-acuminated, 3-nerved: segments of the 4-parted winged calyx ovate, acute: lobes of the corolla obovate-elliptic, apiculate: flowers blue.

Silhet, flowering in February and March.

(3) E. SULCATUS. *(Roxb.)*

Ident. Roxb. flor. Ind. I. p. 416.—Dec. prod. IX. p. 45.

SPEC. CHAR. Stem four-cornered: leaves sessile, lanceolate, acuminated, 3-nerved: segments of the winged calyx ovate, acute: lobes of the corolla obovate, apiculate, blue.

Bengal.

(4) E. GRANDIFLORUM. *(Wall.)*

Ident. Wall. Cat. No. 4358.—Dec. prod. IX. p. 47.

SPEC. CHAR. Stem four-cornered: leaves ovate-lanceolate and lanceolate, acute, 3-5-nerved: peduncles axillary and terminal, trichotomous or 3-flowered, arranged in a corymb: segments of the corolla ovate-oblong, acute.

Courtallum.

(5) E. WIGHTIANUM. *(Arn.)*

Ident. Wight's Icon. vol. III. & t. 840.

SPEC. CHAR. Stems very racemose and with the branches broadly winged: leaves oblong-lanceolate, acuminated, subsessile: corymbs leafy: corolla five-cleft, segments oval or acuminated: fructiferous pedicels recurved: capsule globosely ellipsoidal.

Locality not specified.

(6) E. PEROTTETII. *(Griseb.)*

Ident. Dec. prod. IX. p. 45.

Engrav. Wight's Icon. t. 1322.

SPEC. CHAR. Stem straight, 4-angled, simplish: leaves sessile, oblong-lanceolate, acuminate, 5-nerved with smooth margins: calyx deeply 4-cleft, segments subulate with semi-lanceolate wings: corolla rose-coloured or blue, lobes obovate-elliptic, cuspidate, 4 times longer than the tube.

Neilgherries. Coonoor.

(7) E. PEDUNCULATUM. *(Linn.)*

Ident. Dec. prod. IX. p. 46.—Linn. Sp. I. p. 103.

Syn. E. pedunculare, *Wight.*—E. carinatum, *Roxb.*—Sebæa carinata, *Spreng.*

Engrav. Wight's Icon. t. 336.

SPEC. CHAR. Stem erect, ramous 4-sided: leaves lanceolate: corymbs nearly naked (not leafy): corolla 4-cleft: segments oval, capsule globose, flowers smallish, blue or yellow?

Noorungabad. Pondicherry in rice-fields.

(8) E. BICOLOR. (Roxb.)

Ident. Roxb. flor. Ind. I. p. 397.—Dec. prod. IX. p. 45.
Engrav. Wight's Icon. t. 1321.

SPEC. CHAR. Stem 4-angled: leaves sessile, ovate, subacute, 5-nerved with smooth margins; calyx deeply 4-cleft, segments subulate with ovate-lanceolate wings; corolla white, tipped with blue; lobes elliptic-oblong, cuspidate, three times longer than the tube, which is a little shorter than the calyx; flowers white and blue.

Neilgherries, below Kolaghery, rare. Nedawuttim. Flowering during the autumnal months.

(9) E. COURTALLENSE. (*Arn.*)

Ident. Dec. prod. IX. p. 47.
Engrav. Wight's Icon. t. 1323.

SPEC. CHAR. Stem dichotomously branched, branches with 4 very narrow wings: leaves oblong-lanceolate, acuminate; inflorescence leafy, segments of the corolla obovate, obtuse, fructiferous pedicels straight: capsule oblong-ovate, narrowing towards the apex: flowers deep blue.

Courtallum. Cuttack. Flowering in the cold season.

(10) E. PUMILUM. (*Griseb.*)

Ident. Dec. prod. VIII. p. 46.
Engrav. Wight's Icon. t. 1324. f. 3.

SPEC. CHAR. Stem 4-sided: leaves sessile, oblong-lanceolate, bluntish, 3-nerved, the last shorter, one-nerved: calyx 4-parted, segments subulate, wingless: corolla small, purplish, lobes roundish-ovate, obtuse: style elongated.

Bombay, flowering in the rainy season.

(11) E. PETIOLARE. (*Griseb.*)

Ident. Dec. prod. IX. p. 46.
Engrav. Wight's Icon. t. 1324. f. 2.

SPEC. CHAR. Stem simple, 4-sided: leaves long-petioled, broad ovate, obtuse, 5-nerved: calyx 4-parted, segments acute, with truncated, semi-ovate, transversely-veined wings at the base; lobes of the corolla elliptic, acute: flowers pedicelled: flowers pale bluish-purple.

Belgaum. Island of Caranjab. Flowering in the rainy season.

GENUS II. ERYTHRÆA.

Pentandria Monogynia. *Sex: Syst:*

Deriv. From *Erythros*, red; alluding to the colour of the flowers.

GEN. CHAR. Calyx 5-4-parted, segments flattish, wingless: corolla funnel-shaped, naked, twisted and withering above the capsule, tube cylindric, limb 5-4-parted: stamens 5-4, inserted above the tube: anthers erect, spirally twisted, exserted: ovary 1-celled or semibilocular with the valves a little introflexed, ovules inserted at the suture: style distinct, deciduous, stigma bilamellate or undivided, capitulate: capsule 2-valved, septicidal, 1-half 2-celled, placentæ very cellular, sutural: seeds immersed, subglobose, smooth, minute.

(1) E. ROXBURGHII. *(Don.)*

Ident. Don. Syst. Gard. 4. p. 203.—Dec. prod. IX. p. 59.
Syn. Chironia centauroides, Roxb. flor. Ind. II. p. 584.
Engrav. Wight's Icon. t. 1325.
SPEC. CHAR. Herbaceous: stem erect: lowermost leaves roundate, obovate-oblong, obtuse: cymes 1-2-dichotomous, spreading; flowers lateral, ebracteate, pink, starlike.

Peninsula. Common in cultivated fields after the rains. Bengal. Flowering in January and February.

GENUS III. PLADERA.

Tetrandria Monogynia. *Sex: Syst:*

Deriv. From *Pladeros*, abounding in juice.

GEN. CHAR. Calyx 4-parted, segments wingless: corolla funnel-shaped, naked, withering, tube campanulate-ventricose with a somewhat continuous 4-parted limb: stamens inserted on the throat, three upper ones abortive, destitute of anthers, lower one fertile: anther erect, oblong, included: ovary 1-celled, ovules inserted at the suture: style distinct, deciduous, stigma undivided, capitulate: capsule 2-valved, septicidal, 1-celled, placentæ sutural: seeds immersed.

(1) P. PUSILLA. *(Roxb.)*

Ident. Roxb. flor. Ind. I. p. 419.—Dec. prod. IX. p. 63.

Syn. Hopea dichotoma, *Vahl.*—Canscora pusilla, *Roem. & Schult.*

SPEC. CHAR. Low, herbaceous plant, dichotomous from the base: leaves short, ovate or oblong: cymes many times dichotomous, outermost branchlets 3-flowered: calyx and corolla nearly equal.

Moist places in the Peninsula.

GENUS IV. CANSCORA.

Tetrandria Monogynia. *Sex: Syst:*

Deriv. The Malabar name Latinised.

GEN. CHAR. Calyx tubular, 4-toothed: corolla 2-lipped, naked, at last usually deciduous, upper lip deeply 2-lobed, triandrous at the base, lower one emarginate monandrous, lower stamen longer: anthers erect, three upper ones linear, destitute of filaments, supreme one intermediate with the lobes, lateral ones placed near and below the upper lip of the corolla, lower one subrotund, less than the filament: ovary 1-celled, ovules inserted at the suture: style distinct, deciduous, stigma bilamellate or bi-globose or undivided, capitulate or having two legs: capsule 2-valved, septicidal, subunilocular, placentæ spongy, sutural: seeds immersed in the placentæ, minute.—Annual herbs: stem 4-winged, branched: panicle dichotomous, very rarely reduced to a spike: flowers rose-coloured or white, thin: calyx cylindric, equalling the tube of the corolla.

(1) C. DIFFUSA. *(R. Br.)*

Ident. R. Br. prod. p. 451.—Dec. prod. IX. p. 64.

Syn. Gentiana diffusa, *Vahl.*—Exacum diffusum, *Willd.*—Pladera virgata, *Roxb.*—C. foliosa, *G. Don.*—C. tenella, *Wall.*—C. diffusa, *Do.*—C. Lawii, *Wight.*

Engrav. Wight's Icon. t. 1327.

SPEC. CHAR. Stem obtusely winged, very much branched: leaves ovate-acute: centre flowers pedicelled, sometimes wanting: calyx wingless; corolla pink or rosy.

Rocky parts of the Concans. Coromandel. Flowering nearly all the year.

(2) C. SESSILIFLORA. (*Roem. & Schult.*)

Ident. Dec. prod. IX. p. 64.—Roem. & Schult. Mant. p. 230.
Syn. Gentiana heteroclita, *Linn. Mant.*—Exacum heteroclitum, *Willd.*—Pladera sessiliflora, *Roxb. fl. Ind.* I. p. 416.—Centaurium Malabaricum, *Borkh. Gent.* p. 27.
SPEC. CHAR. Stem broadly 4-winged: leaves ovate, obtuse: flowers central, sessile: calyx wingless; corolla pale, rose-coloured.
Moist places on the Coromandel Coast and Bengal.

(3) C. ALATA. (*Wall.*)

Ident. Wall. Cat. No. 4363.—Dec. prod. IX. p. 65.
SPEC. CHAR. Stem 4-winged above, below simple, tetragonous: leaves ovate-oblong, acutish, floral ones orbiculate, sometimes kidney-shaped: flowers central, sessile, usually deficient: wing of the calyx half-ovate.
Madras.

(4) C. PERFOLIATA. (*Lam.*)

Ident. Dec. prod. IX. p. 65.—Lam. Enc. I. p. 601.
Syn. Pladera perfoliata, *Roxb.*
Engrav. Rheede Mal. X. t. 52.—Wight's Icon. t. 1327. f. 2.
SPEC. CHAR. Stem 4-winged, ramous from the base: leaves oblong-lanceolate, acute, floral ones roundish; central flowers pedicelled: calyx wingless.
Mysore and Malabar.

(5) C. GRANDIFLORA. (*R. W.*)

Ident. Wight's Icon. vol. IV. & t. 1326.
SPEC. CHAR. Stems alone furnished with 4 narrow wings, diffusely ramous: leaves lanceolate, acute, 3-nerved, floral ones orbiculate, perfoliate: flowers ternate, subsessile: calyx broadly winged.
Coorg and Western provinces of Mysore, flowering in May and June.

(6) C. TENELLA. (*Wall.*)

Ident. Wight's Icon. vol. IV. & t. 1327. f. 3.
SPEC. CHAR. Stems obsoletely winged, diffuse, and very ramous: lower leaves broad-ovate, acute, those of the floriferous branchlets linear-lanceolate or minute, subulate: flowers long and

slenderly pedicelled ; calyx wingless, dentate, teeth acute, about one-third the length of the lobes of the corolla : style scarcely the length of the tube, stigma inclusive.

Malabar and Mysore.

(7) C. DECURRENS. (*Dalz.*)

Ident. Dalz. in Hook. Journ. Bot. II. p. 156.—Bomb. for. p. 137.

SPEC. CHAR. Stem erect, broadly 4-winged : branches opposite and alternate : leaves decurrent, lower ones oblong, attenuated towards the base, upper ovate or lanceolate, acute : calyx without wings : corolla small, pale rose-coloured or white.

In rice fields, Southern Concan, flowering in October and November.

(8) C. PAUCIFLORA. (*Dalz.*)

Ident. Dalz. ut supra.

SPEC. CHAR. Stem erect, 4-winged, scarcely branched ; leaves very small, lower ovate-obtuse, upper oblong-acute, all sessile, 3-nerved, rough on the margin alone : panicle lax, few-flowered : flowers long-pedicelled, solitary : pedicels 4-winged, thickened upwards.

Malwan, in grassy places, flowering in September.

GENUS V. SLEVOGTIA.

Pentandria Monogynia. *Bar ; Syst ;*

Deriv. In honor of J. H. Slevogt, a Botanist and Author.

GEN. CHAR. Calyx 5-cleft, ebracteate, lobes wingless : corolla funnel-shaped, naked, withering and twisted above the capsule, tube cylindric from the bottom enlarged into a campanulate throat, limb 5-parted : stamens five, inserted, included, sheaths attached above to the short tube of the corolla, and increased between the filament by five short teeth : anthers erect, produced by a connectivum into a small point : ovary 1-celled : style distinct, deciduous, stigma undivided, capitulate : capsule 2-valved, septicidal, 1-celled, placentae sutural : seeds minute, subglobose.

(1) S. ORIENTALIS. (*Dec.*)

Ident. Dec. prod. VIII. p. 65.

Syn. Gentiana verticillaris, *Retz.*—G. verticillata, *Linn.*—Exacum hyssopifolium, *Willd.*—Hippion hyssopifolium, *Spreng.*—

Adouema hyssopifolium, *G. Don.*—Cicendia hyssopifolia, *W. & A. in Hook. Comp. Mag.*

Engrav. Wight's Icon. t. 600.

SPEC. CHAR. Stem smooth: leaves opposite, lanceolate, subsessile, attenuated at the base with the petiole, obtuse at the apex: lobes of the calyx ovate, acute, erect, shorter than the capsule: calyx with bracts: flowers small, white, sessile in the opposite axils.

Common in Goaerat. Concans. Coromandel. Bengal. Flowering in the rainy season.

GENUS VI. GENTIANA.

Pentandria Digynia. *Lin. Syst.*

Deriv. From Gentius, king of Illyria, who first experienced the virtues of the plant.

GEN. CHAR. Calyx 4-5-parted, or cleft, valvate in æstivation: corolla marcescent (withering on the stalk), funnel-shaped, or salver-shaped, naked or furnished with a crown: limb 4-5-parted, or, counting the folds, spuriously 10-cleft: stamens 4-5, inserted on the tube of the corolla: anthers incumbent, or erect; sometimes united into a tube, opening externally: ovary sometimes bound with a spurious interrupted disk, 1-celled, ovules near the sutures: stigmas 2, terminal, revolute or, if contiguous, funnel-shaped: style none, or with the stigma, persistent: capsule 2-valved, septicidal, 1-celled: placentas membranaceous, inserted along the edge of the sutures: seed immersed in the placentas.—Herbaceous perennials, erect, or procumbent, with racemose-like cymes, or terminal flowers.

(1) G. PEDICELLATA. (*Wall.*)

Ident. Dec. prod. IX. p. 107.
Syn. G. abscondita, *Zenker.*
Engrav. Wight's Icon. t. 1328.—Spicil. II, t. 164.

SPEC. CHAR. Stem loosely ramous, glabrous: leaves elliptic-lanceolate, the broader ones aristate at the apex, smooth on the margins, the lowest ones rosulate: flowers pedicelled: calyx campanulate, 5-cleft, lobes ovate, cuspidate, recurved at the apex, shorter than the clavate tube of the corolla: corolla blue, the tube furnished with 5 projecting, triangular, acutely mucronate lobes: plicæ emarginate: capsule obovate, rounded at the apex.

Neilgherries. Pulneys. Common in pastures. Flowering all the year.

GENUS VII. CRAWFURDIA.

Pentandria Monogynia. *Sex : Syst :*

Deriv. In honor of John Crawford, Governor of Singapore, author of a history of the Indian Archipelago.

GEN. CHAR. Calyx 5-cleft or 5-toothed, teeth distant : corolla withering, clavate, destitute of pits or glands, naked within, limb 5-parted, increased by exserted folds : stamens inserted at the bottom of the tube of the corolla : anthers erect, included : ovary surrounded at the base by a 5-lobed hypogynous disc, 1-celled : style distinct, persistent, stigmas twin, terminal, oblong, revolute : capsule stalked, 2-valved, septicidal, valves very shortly involute, subunilocular : seed immersed in the placenta, testa winged.

(1) C. FASCICULATA. *(Wall.)*

Ident. Dec. prod. IX. p. 120.
Syn. C. affinis, *Wall.*—Gentiana volubilis, *Don. Nep.* p. 126.
SPEC. CHAR. Twining : leaves oblong-lanceolate, alternately acuminated : flowers very shortly pedicelled, subaggregate in the axils : lobes of the 5-cleft calyx bristly equalling the tube, of the corolla ovate, acute, folds short, broad, round, eroded at the apex : capsule obovate, three times longer than the stalk : flowers blue.

Silhet.

GENUS VIII. OPHELIA.

Pentandria Digynia. *Sex : Syst :*

Deriv. From the Greek *Ophelcia*, service ; the plants being useful in Medicine.

GEN. CHAR. Calyx 5-4-parted, segments connected at the lowest base : corolla withering, rotate, 5-4-parted, destitute of folds and continuous crown, furnished above the base with glanduliferous pits which are sometimes naked and sometimes covered with a small scale, often fimbriated : stamens 5, 4, inserted on the throat, filaments sometimes dilated at the base and monadelphous, sometimes equal at the base and free : anthers incumbent, nodding, usually greenish : ovary 1-celled, ovules many, inserted on the sutures : stigmas twin, terminal, short, often revolute, style none or short : capsule 2-valved, septicidal, 1-celled : seeds immersed in the placenta, very numerous, small, usually wingless. —Herbs, annual or rarely perennial, straight, branched, panicled, with nearly equal internodes, opposite leaves, and terminal umbelliform cymes.

(1) O. CORYMBOSA. (*Griseb.*)

Ident. Griseb. Gent. p. 317.—Dec. prod. IX. p. 125.
Engrav. Wight's Icon. t. 1329.

SPEC. CHAR. Stem 4-sided, ascending, branches divaricate: leaves spathulate-elliptic, roughish, 3-nerved, the lower ones largest, the stem ones short, sessile: cymes sub-fastigiate, few-flowered, pedicels spreading, segments of the calyx linear-acuminate, half the length of the corolla: corolla 4-parted, blue, segments obovate-elliptic, mucronate: foveæ minute, orbicular, solitary, covered with a scale, fimbriate at the apex, and themselves bound with short fimbriæ: filaments linear.

Neilgherries.

(2) O. MINOR. (*Griseb.*)

Ident. Dec. prod. IX. p. 126.
Engrav. Wight's Icon. t. 1332.

SPEC. CHAR. Stems subterete, erect, filiform, glabrous, sparingly ramous: branches erect, 1-3-flowered: leaves short cordate-ovate, or ovate, glabrous, obscurely 3-nerved, cauline ones sessile: cymes terminal, lax, 3-5-flowered, the axillary pedicels shorter: segments of the calyx lanceolate-oblong, acute, about half the length of the corolla: corolla 4-parted, blue, segments elliptic-oblong, acute, subersct: foveæ orbicular, paired, distant, most minute: margins naked: filaments linear, shorter than the corolla.

Neilgberries, in wet marshy ground.

(3) O. ELEGANS. (*R. W.*)

Ident. Wight's Icon. vol. IV. & t. 1331.

SPEC. CHAR. Erect, ramous above, obsoletely 4-sided: leaves sessile, narrow-lanceolate, tapering to a slender point, 3-nerved: lateral nerves close to the margin: branches ascending, slender, bearing, at each joint lateral few-flowered cymes, forming together a large many-flowered panicle: calyx-lobes narrow lanceolate, acute, about two-thirds the length of the corolla: lobes of the corolla obovate cuspidate: foveæ bound with longish coarse hairs: flowers pale blue.

Palneys, flowering in August and September.

(4) O. GRISLBACHIANA. (*R. W.*)

Ident. Wight's Icon. vol. IV. & t. 1330.

SPEC. CHAR. Erect, simple below, ramous above, fastigiate: leaves opposite or ternate, lanceolate, acute or sometimes narrow-linear, 3-nerved: corymbs many-flowered, compact: calyx lobes tubulate-pointed, nearly as long as the corolla; corolla 4-cleft, divisions lanceolate acute; foveæ covered with a scale and bound with long fimbriæ round the margin.

Puluey mountains, among long grass, flowering in September and October.

(5) O. MULTIFLORA. (*Dalz.*)

Ident. Dalz. in Hook. Journ. Bot. II. p. 135.—Bomb. flor. p. 156.

SPEC. CHAR. Stem quadrangular, 4-winged, ascending, densely leafy; leaves round-ovate, stem-clasping, 3-nerved, macronulate, glabrous, decussate: cymes many-flowered: calyx divisions lanceolate-acuminate: corolla white, 4-divided: segments ovate-elliptic, their rounded pits surrounded by long fringes: filaments united at the very base.

Mahableshwar.

(6) O. PAUCIFLORA. (*Dalz.*)

Ident. Dalz. l. c. III. p. 111.—Bomb. flor. p. 156.

SPEC. CHAR. Stem erect, 4-winged, glabrous, branched towards the top: leaves sessile, lanceolate-acuminate, 3-nerved: cymes few-flowered: calyx segments subulate, as long as the corolla: corolla white, 4-divided, the segments obovate-elliptic, their pits large, round, covered with a fringed scale, and surrounded by a short fringe.

The Ghauts, flowering in September.

GENUS IX. HALENIA.

Pentandria Digynia. *Sav: Syst:*

GEN. CHAR. Calyx 4-5-partite, segments connected at the base; corolla withering, shortly campanulate, 4-5-cleft, lobes erect, equalling the tube, destitute of folds or fringes, glanduliferous pits solitary, produced into spurs: stamens 4-5, inserted on the throat, filaments equal at the base: anthers minute, incumbent: ovary 1-celled, ovules numerous, inserted on the sutures: stigmas twin, terminal, often connate and confluent with the ovary: capsule 2-valved, septicidal, 1-celled: seeds numerous, immersed in the placenta.

(1) H. PERROTTETTII. *(Dec.)*

Ident. Dec. prod. IX. p. 129.
Engrav. Wight's Icon. t. 1334.—Spicil. II. t. 160.

SPEC. CHAR. Stem erect, ramous; leaves ovate-lanceolate, acute, 5-nerved, subsessile: pedicels axillary and terminal, unequal, filiform: segments of the calyx lanceolate, acute: spurs thickish, half the length of the corolla, corniculato-obtuse, spreading and ascending at the point: corolla pale blue, lobes ovate, mucronate: stigmas small, distinct at the apex.

Neilgherries. Pulney mountains. Common in long grass.

GENUS X. LIMNANTHEMUM.

Pentandria Monogynia. *Ser: Syst:*

Deriv. From *Limne*, a marsh, and *Anthemos*, flowering; alluding to the habitat of the species.

GEN. CHAR. Calyx 5-partite, segments connected at the base into a tube: corolla deciduous, rotate, somewhat membranaceous, 5-parted, segments variously fringed, furnished with glands occasionally on the petals: stamens inserted on the tube, filaments equal at the base; anthers erect: ovary surrounded by five hypogynous glands, 1-celled, ovules inserted on the suture: style persistent with the 2-lobed stigma: capsule 1-celled, without valves, afterwards to be opened by maceration, placentæ sutural: seeds two or numerous with a smooth or muricated testa.—Perennial herbs, floating, with axillary or petiolar inflorescence: leaves long-petioled, floating, peltate or cordate, orbiculate, entire: umbels unequal, sessile, emersed.

(1) L. KLEINIANUM. *(Griseb.)*

Ident. Grisb. Gent. I. p. 344.—Dec. prod. IX. p. 139.

Syn. Villarsia Indica, *Wall. Cat. No.* 4352.—V. macrophylla, *Roem. & Schult.* ?

Engrav. Wight. in Hook. Bot. Misc. III. p. 96. Suppl. t. 30.

SPEC. CHAR. Leaves cordate-orbiculate, smooth or roughish above, glanduliferous below, afterwards rough with depressions, 3-nerved, nerves prominent beneath: segments of the calyx ovate-lanceolate: segments of the white-yellow corolla irregularly fringed at the margin and within, without glands: style abbreviated, thick: stigma 2-lobed: capsule many-seeded: seeds shining, smooth, bluntly keeled.

Silhet.

(2) L. Wightianum. *(Grisb.)*

Ident. Grisb. Gent. p. 244.—Dec. prod. IX. p. 139.

Spec. Char. Leaves cordate-orbiculate, smoothish or roughish above, nerves prominent beneath; segments of the calyx ovate; corolla white? segments irregularly fringed at the margin and disk, without glands: style long, thin: stigma 3-2-lobed: capsule many-seeded: seeds muricated, not keeled.

Madras.

(3) L. cristatum. *(Grisb.)*

Ident. Grisb. Gent. p. 342.

Syn. Menyanthes cristata, *Roxb.*—Villarsia cristata, *Spreng.*— M. Indica, *Borg.*

Engrav. Roxb. Cor. II. t. 105.

Spec. Char. Aquatic: leaves cordate-orbicular, roughish above, glandular beneath: calyx segments ovate-lanceolate: segments of the corolla waved, ciliated, with a longitudinal crest within: flowers white, cymose, inserted on the petiole: capsule 1-2-seeded: seeds muricated.

Tanks in the Concan.

(4) L. aurantiacum. *(Dalz.)*

Ident. Dalz. in Hook. Journ. Bot. II. p.136.—Bomb. flor. p. 158.

Spec. Char. Umbels axillary: leaves small, orbicular, deeply cordate, shining above, glandular dotted and purple beneath: corolla orange-coloured: segments of the limb wedge-shaped, broadly and deeply emarginate, fringed on the margin, bearded at the base: seeds lenticular, muricated: capsule ovate, obtuse, 12-seeded.

Near Malwan, flowering in September.

(5) L. Indicum. *(Grisb.)*

Ident. Grisb. Gent. p. 343.—Dalz. Bomb. flor. p. 158.

Syn. Menyanthes Indica, *Linn.*—Villarsia Indica, *Vent.*

Spec. Char. Leaves cordate orbicular, membranaceous, roughish: calyx segments ovate: segments of the corolla fringed on the margin, destitute of a crest within: flowers white, arising from the petiole: capsule many-seeded: seeds muricated.

Tanks in the Concans and Deccan.

(6) L. PARVIFOLIUM. *(Grisrb.)*

Ident. Dalz. Bomb. flor. p. 158.—Dec. prod. IX. p. 141.
Syn. Villarsia parvifolia, *Wall.*
SPEC. CHAR. Very minute: leaves cordate-orbicular, small, membranaceous: petioles bearing the flowers immediately below the leaf: capsule many-seeded; seeds minute, rough.
Malwan. Surat. Common in tanks but difficult to find on account of its small size.

ORDER CXI. BIGNONIACEÆ.

Calyx lobed or entire, sometimes spathaceous: corolla monopetalous, hypogynous, deciduous, irregular, 4-5-lobed or sub-bilabiate, lobes imbricating in æstivation: stamens usually 4, fertile, didynamous, with a sterile filament, sometimes all fertile; anthers 2-celled, cells parallel and contiguous, or separate and diverging, opening longitudinally: disk glandulose, tumid, embracing the base of the ovary: ovary 2, rarely 1-celled, ovules several or numerous, attached to the lateral placentæ, usually united in the axis by a short process which, with the thickened placentæ, afterwards becomes the spongy partition: style filiform, stigma bilamellate or bifid: lamellæ anticous and posticous: capsule 2-valved, 2-celled, often long, compressed, sometimes spuriously 4-celled, the septum either parallel to the valves, or contrary to them, finally separating and bearing the seeds: seeds transverse, compressed, winged, exalbuminous: embryo straight next the hilum, cotyledons flat, foliaceous or fleshy.—Trees or shrubs: stems erect, scandent, or twining: leaves opposite, sometimes simple, usually compound, the petiole sometimes produced into a tendril: stipules none, but sometimes replaced by accessory leaflets: inflorescence usually panicled or racemose.

GENUS I. BIGNONIA.

Didynamia Angiospermia. *Sts: Syst:*

Deriv. In memory of Abbé Bignon, Librarian to Louis XIV.

GEN. CHAR. Calyx 5-toothed at the margin, rarely entire or 5-parted or 2-3-lobed : corolla 2-lipped or nearly equal, 5-cleft : stamens four, fertile, didynamous, fifth sterile : anthers with glabrous cells, very often distinct : stigma bilamellate : capsule with the valves scarcely convex, or flat, partition flat, parallel to the valves : seeds in a single row at each side of the partition, winged on both sides, wing pellucid : stems frutescent, or trees, erect, sometimes fruticous-scandent : leaves almost always opposite, petioled, but very various.

(1) B. XYLOCARPA.

Ident. Dec. prod. IX. p. 170.—Roxb. flor. Ind. III. p. 108.
—Dalz. Bomb. flor. p. 159.

Engrav. Wight's Icon. t. 1335-6.

SPEC. CHAR. Arboreous, glabrous : leaves bi-tri-pinnate: petiole sharply-angular, leaflets petioled, ovate or oblong-acuminated, membranaceous, retivalately veined : panicle corymbose: branches dichotomous: calyx unequally 5-toothed : corolla campanulate, shortly tubular: lobes rounded : capsule round-linear, incurved, woody, tubercled : flowers white, fragrant.

Tull Ghaut. Klaudeish. Neilgherries. Flowering in August.

GENUS II. CALOSANTHES.

Didynamia Angiospermia. *Sts: Syst:*

Deriv. From *Kalos*, beautiful, and *Anthos*, a flower.

GEN. CHAR. Calyx coriaceous, tubular, truncated, afterwards cut as the fruit increases : corolla with a short tube, campanulate throat, and 5-lobed somewhat 2-lipped limb : stamens five, fertile, scarcely exserted, of which two are longer : anthers with cells pendulous from the connectivum : stigma bilamellate : capsule pod-like, very long, compressed, 2-valved, partition parallel to the valves : seed surrounded with a semicircular, membranaceous wing.

(1) C. Indica. *(Blume.)*

Ident. Dec. prod. IX. p. 177.—Dals. Bomb. flor. p. 161.
Syn. Bignonia Indica, *Linn.*—*Roxb. flor. Ind.* II. p. 110.—B. pentandra, *Lour.*—Spathodea Indica, *Pers.*
Engrav. Rheede Mal. L t. 43.—Wight's Icon. t. 1337-8.

Spec. Char. Tree: leaves opposite, pinnate, leaflets on the branches of the petiole 2 to 3 pair, petioletted, subcordate, ovate-acuminated: panicle terminal, erect: flowers fleshy, of a dark, lurid appearance, fœtid: pod 2 feet long, 3 inches broad, straight and flat.

Coromandel. Malabar. Concan. Flowering in the rainy season.

GENUS III. MILLINGTONIA.

Didynamia Angiospermia. *Sex: Syst.*

Deriv. In honor of Thomas Millington, an English Botanist and Author.

Gen. Char. Calyx campanulate, very shortly and equally 5-lobed: tube of the corolla slender, terete, elongated, limb 5-partite, two upper lobes connected at the middle: stamens four, exserted, none sterile: anthers affixed by the base, lobes parallel, 1-spurred at the base: style filiform, fistular within: stigma 2-lipped: capsule pod-like, elongated, attenuated at both ends, smooth, 2-celled with a parallel partition, valves flat: seeds broadly winged.

(1) M. hortensis. *(Linn.)*

Ident. Linn. f. Suppl. p. 291.—Dec. prod. IX. p. 182.
Syn. Bignonia azedarachta, *Koen.*—B. suberosa, *Roxb. flor. Ind.* III. p. 111.
Engrav. Roxb. Cor. III. t. 214.

Spec. Char. Tree: leaves opposite, bipinnate with an odd one, leaflets entire: panicle large, oppositely branched, many-flowered: flowers white, fragrant.

Said to be a native of Ajmere, flowering in the cold season.

GENUS IV. SPATHODEA.

Didynamia Angiospermia. *Ser: Spe.*

Derio. From *Spathe*, a spathe; alluding to the calyx.

GEN. CHAR. Calyx spathaceous, younger one closed, afterwards longitudinally cleft, then toothed or entire: corolla somewhat funnel-shaped, limb 5-cleft, slightly unequal: stamens four, with the fifth sterile: anthers with the cells separate: stigma bilamellate: capsule pod-shaped, 2-celled, loculicidally dehiscing: seeds corky, membranaceously winged, attached to the partition, not immersed in pits.—Trees or shrubs often climbing; leaves opposite, rarely alternate, simple, conjugate, digitate or unequally pinnate: fruit almost unknown.

(1) S. RHEEDII. (*Wall.*)

Ident. Dec. prod. IX. p. 206.

Syn. S. longiflora, *Vent.*—Bignonia longiflora, *Willd.*

Engrav. Wight's Icon. t. 1339.—Rheede Mal. VI. t. 29.

SPEC. CHAR. Arboreous, glabrous: leaves unequally pinnate, 3-paired: leaflets oval-lanceolate, acuminate, petiolulate, entire: racemes terminal, short, about 3-flowered: corolla with a long slender tube, capsule siliquiform, subcylindrical, erect or more or less curved: corolla white, 5-6-inches, capsule about 8 inches long: septum thickened in the middle, hence the capsule is somewhat 4-celled: wing of the seed thickish, opaque, truncated.

Malabar, near Tellicherry.

(2) S. FALCATA. (*Wall.*)

Ident. Dec. prod. IX. p. 206.—Dalz. Bomb. flor. p. 160.

Syn. Bignonia falcata, *Roem. Mss.*—B. spathacea, *Linn. f. suppl.*—*Roxb. flor. Ind.* III. p. 103.—S. Rheedii, *Spreng.*—S. longiflora, *Pers.*

Engrav. Roxb. Cor. II. t. 144.

SPEC. CHAR. Small tree: leaves unequally pinnate, 2 to 3 pair, oval, rounded entire, slightly hairy: racemes terminal, few-flowered: calyx cylindrical, oblique: flowers white, about 1 inch long, fragrant: capsule linear, 1 foot oblong, various twisted.

Khandalla Ghaut. S. Mahratta country. Coromandel Coast.

(3) S. CRISPA. (*Wall.*)

Ident. Dec. prod. IX. p. 206.—Dalz. Bomb. flor. p. 160.
Syn. Bignonia crispa, *Roxb.*—B. atrovirens, *Roth.*—S. atrovirens, *Spreng.*

SPEC. CHAR. Arboreous: leaves unequally pinnate: leaflets 1 to 3 pair; branchlets and racemes pubescent and velvety: leaflets oval-oblong, acuminated at both ends, quite entire: raceme terminal, few-flowered: corolla tube slender, elongated: lobes much curled and crispid: flowers pure white, fragrant: capsule pod-like, elongated, obtusely-acuminated, pendulous.

Duddi on the Gutporba river. Mysore. Flowering in May and June.

(4) S. ARCUATA. (*R. W.*)

Ident. Wight's Icon. vol. IV. & t. 1340.

SPEC. CHAR. Arboreous: leaves unequally pinnate, 4–5 pairs, leaflets from ovate subacute to orbicular, unequal at the base, entire, softly pubescent when young, afterwards glabrous: racemes terminal, elongated, many-flowered: calyx cylindrical, oblique, pubescent externally: tube of the corolla slender, limb funnel-shaped, 5-lobed, fimbriated on the margin; capsule acute, compressed, 8-12 inches long by about 1 broad.

Coimbatore, flowering in September.

GENUS V. HETEROPHRAGMA.

Didynamia Angiospermia. *Bar. Syst.*

Derio. From *Heteros*, different, and *Phragma*, a division; alluding to the partitions of the capsule.

GEN. CHAR. Calyx campanulate, 3-lobed: corolla with a broad tube, limb spreading, lobes five, equal, obtuse, somewhat undulating: stamens four, fertile, fifth sterile: cells of the anthers glabrous, divaricate: a small gland surrounding the base of the ovary; style filiform: stigmas two, subulate: capsule rigid, oblong, acuminated, 2-valved, 4-celled, namely, the partition thick, cruciate, the longer lobes stretching to the commissure, the shorter ones to the middle of the valves: seeds attached to the shorter lobes of the partitions, broadly winged.

(1) H. ROXBURGHII. (*Dec.*)

Ident. Dec. prod. IX. p. 210.—Dalz. Bomb. flor. p. 160.
Syn. Bignonia quadrilocularis, *Roxb.*—Spathodea Roxburghii, *Spreng.*

Engrav. Roxb. Cor. II. t. 145.

Spec. Char. Large tree: branches round; leaves opposite or tern, glabrous, simply pinnated, 4-5 pair, leaflets ovate, acute, serrated; panicle terminal, tomentose, and velvety: flowers whitish, with a pink margin: pod thick, linear, about a foot long and 2 inches broad, 4-celled.

Circar mountains. S. Mahratta country. Travancore. Flowering in the hot season.

GENUS VI. STEREOSPERMUM.

Didynamia Angiospermia. *Sex: Syst:*

Deriv. From *Stereos*, hard, and *Sperma*, seed.

Gen. Char. Calyx coriaceous, cup-shaped, cylindric, subtruncate, obtusely 5-toothed: corolla with a straight campanulate tube, limb 2-lipped, roundly 5-lobed: four stamens fertile, the fifth sterile, small: anthers 2-lobed, naked: disk fleshy, 5-lobed: ovary cylindric; stigma bilamellate: capsule obtusely 4-cornered or cylindric, elongated, membranaceous, partition contrary to the valves, very cellular and corky, thick: seeds bony, immersed in the flesh of the partition, or if 2-celled, laterally very thinly winged.

(1) S. CHELOSOIDES. *(Dec.)*

Ident. Dec. prod. IX. p. 110.

Syn. Bignonia chelonoides, *Linn. f. suppl.*—*Roxb. flor. Ind.* III. p. 106.

Engrav. Rheede Mal. 6. t. 26.—Wight's Icon. t. 1341.

Spec. Char. Arboreous, glabrous: branches terete: leaves unequally pinnate, 4-paired: leaflets elliptic, cuspidato-acuminate: panicles terminal, loose, the extreme branchlets 3-flowered: calyx coriaceous, 2-3-lobed or toothed: corolla campanulate-bilabiate, ciliate: capsule very long, roundish, glabrous, with a spongy septum: flowers fragrant, yellow: capsules a foot or more in length.

Ghauts. Coromandel. Silhet. Assam. Flowering in May and June.

(2) S. SUAVEOLENS. *(Dec.)*

Ident. Dec. prod. IX. p. 311.

Syn. Bignonia suaveolens, *Roxb.*—Tecoma suaveolens, *G. Don.*

Engrav. Wight's Icon. t. 1342.

Spec. Char. Arboreous: leaves unequally pinnate, 2-4 pairs; leaflets oval, acuminate, entire: panicles terminal, loose, subbrachiate: calyx 5-toothed: corolla hairy or woolly, capsule siliquiform, cylindrical: septum corky, cylindrical: flowers dull purplish, very fragrant: leaves vary from broad ovate, shortly and abruptly acuminate, to oval lanceolate long acuminate, entire, or subserrate, pubescent or glabrous: panicles pilosely viscid or glabrous.

Deccan. Bengal. Western Coast. Flowering in March and April.

GENUS VII. TECOMA.

Didynamia Angiospermia. *Ser: Syst:*

Deriv. From *Tecomaxochitl*, the Mexican name of the species.

GEN. CHAR. Calyx campanulate, 5-toothed: corolla with a short tube, throat dilated, limb 5-lobed, somewhat 2-lipped or equal: four stamens with the rudiment of a fifth: anthers 2-celled, cells diverging: stigma bilamellate: capsule 2-celled, 2-valved, partition contrary to the valves: seeds imbricated, winged, transverse.

(1) T. UNDULATA. *(Don.)*

Ident. Don. Gen. Syst. 4. p. 223.—Dec. prod. IX. p. 222.— Dals. Bomb. flor. p. 161.

Syn. Bignonia undulata, Roxb. flor. Ind. III. p. 101.

Engrav. Smith's Exot. Bot. I. t. 19.

SPEC. CHAR. Tree: leaves opposite, petioled, simple, linear-lanceolate, obtuse, waved, entire: racemes terminating the lateral branchlets, few-flowered: capsule pod-shaped, linear-compressed, smooth: flowers orange-yellow.

Western Khandeish. Allahabad.

GENUS VIII. PAJANELIA.

Didynamia Angiospermia. *Ser: Syst:*

Deriv. The Malabar name Latinised.

GEN. CHAR. Calyx coriaceous, oblong, 5-cornered, cleft into 5 acute teeth: corolla coriaceous, tube short, broad, throat widely campanulate, gaping, lobes five, roundish: four stamens fertile, fifth sterile, scarcely shorter than the rest: cells of the anthers divaricate, somewhat reflexed; stigma 2-lobed, clavate: capsule flat, lanceolate, appendiculate on both sides with broad wings, partition contrary to the valves: seeds roundish, flat, winged, except at the base, membrane very thin.

(1) P. MULTIJUGA. (Dec.)

Ident. Dec. prod. IX. p. 227.
Syn. Bignonia multijuga, *Wall. Pl. As. Rar.* L. p. 81.—B. Pajanelia, *Ham. in trans. Linn. Soc.* 13. p. 316.—B. Indica, *Lour. Cor*.—B. longifolia, *Willd.*—Spathodea Indica, *Pers.*
Engrav. Wall. l. c. t. 95 & 96.
Spec. Char. Large tree: leaves opposite, unequally pinnate, 10-12 pair of pinnae, petiole somewhat 4-cornered, leaflets quite entire, half-cordate, acuminate: thyrse panicled, erect, many-flowered: calyx somewhat pulveraceous externally, obscurely 2-lipped: corolla lurid purple, white within, beset with hairs outside.
Silhet.

(2) P. RHEEDII. (R. W.)

Ident. Wight's Icon. vol. IV.
Engrav. Wall. l. c. t. 1343-4.—Rheede Mal. I. t. 44.
Spec. Char. Arboreous: leaves unequally pinnate, leaflets unequal sided, acuminate: calyx campanulate, 5-lobed, lobes emarginate at the points: corolla campanulate, sub-bilabiate, 5-lobed, lobes dilated and crisp on the outer margin, furnished on the edges with a line of dense woolly tomentum externally, tube glabrous, limb pubescent: longer stamens connivent: capsule 12-15 inches long, winged, cuspidate: seed orbicular, compressed, winged.

Malabar, not unfrequent in the jungles between Coimbatore and Paulghaut, flowering in July and August.

ORDER CXII. PEDALIACEÆ.

Calyx equally 5-lobed: corolla monopetalous, hypogynous, irregular, throat ventricose, limb bilabiate, the limb subvalvate in æstivation: disk hypogynous, fleshy, or sometimes glandular: stamens included within the tube, didynamous, with the rudiment of a fifth: anthers adnate, 2-celled: connective articulated with the filament, slightly prolonged beyond the cells, glandular at the point: ovary seated on a glandular disk, formed of 2 capellary leaves, anterior and posterior as regards the axis, at first 1-celled, afterwards divided into 2, 4 or 6 spurious cells: style 1, simple: stigma bilamellate: fruit capsular or drupaceous, dehiscent or indehiscent, few or many-seeded: seed (in Sesamum attached to an easily separable 4-sided central placenta) winged or wingless, exalbuminous: embryo straight: cotyledons plano-convex, longer than the radicle.—Herbaceous plants, often with soft texture and heavy smell, covered with glandular hairs or quaternary vesicles; leaves opposite or alternate, undivided or lobed, without stipules: flowers axillary, solitary or clustered, usually large, furnished in many cases with conspicuous bracts, sometimes with glands on the pedicels.

GENUS 1. SESAMUM.

Didynamia Angiospermia. *Bra. Syst.*

Deriv. From *Sempsen*, the Egyptian name of one of the species.

GEN. CHAR. Calyx 5-parted, persistent, upper lobe smaller; tube of the corolla enlarged upwards, limb folded, somewhat 2-lipped, upper lip emarginate, lower one half 3-cleft, intermediate segment elongated; stamens four with the rudiment of a fifth; anthers oval-oblong: stigma narrowly bilamellate: capsule oblong, obtusely four-cornered, 4-furrowed, acuminated with the base of the style, 2-valved, 2-celled, valves recurved inwards as if 4-celled: seeds numerous, in one series, thick, obovoid, wingless.—Herbaceous annuals with opposite, or alternate, undivided or lower 3-lobed or trisect, entire or toothed leaves: flowers solitary in the axils: seeds oily.

(1) S. INDICUM. (*Dec.*)

Ident. Dec. prod. IX. p. 250.

Syn. S. orientale, *Linn.* Roxb. fl. Ind. III. p. 100.—S. luteum, *Retz.*

Engrav. Bot. Mag. t. 1688.—Rheede Mal. IX. t. 54, 55.—Rumph. Amb. 5. t. 76. fig. 1.—Burm. Zeyl. t. 38. fig. 1.

SPEC. CHAR. Stem erect, pubescent: leaves ovate-oblong or lanceolate, lower ones often 3-lobed or cut: capsule mucronate with the persistent base of the style, velvety-pubescent: flowers white, suffused with rose.

All over India, flowering in the hot season. The seeds contain a fixed oil, known as Gingely. It has never been found in a wild state.

(2) S. LACINIATUM. *(Klein.)*

Ident. Klein. in Willd. Sp. III. p. 359.—Dec. prod. IX. p. 250.
Engrav. Wight's Icon. t. 1345.

SPEC. CHAR. Stem prostrate, hispid: all the leaves tripartite, laciniated.

Balaghaut mountains in the Carnatic.

(3) S. PROSTRATUM. *(Retz.)*

Ident. Wight's Icon. vol. IV. & t. 1346.

SPEC. CHAR. Leaves orbiculate, crenated, hispid above, white-tomentose beneath: flowers solitary, peduncled: stem diffuse.

Sand hills along the Sea Coast. Near Madras, towards the Adyar.

GENUS II. PEDALIUM.

Didynamia Angiospermia. *Syst: Syst:*

Deriv. From *Pedalion*, a rudder; in reference to the dilated angles of the fruit.

GEN. CHAR. Calyx 5-parted, persistent, upper lobe shorter: corolla tubular, 5-cleft, lobes rounded, lowest one larger: stamens four, included: anthers approximated crosswise: style filiform: stigma 2-cleft: fruit nut-like, indehiscent, ovately pyramidal, somewhat 4-sided, increased at the base by four conical simple spines, irregularly 3-celled, 2 upper cells 2-seeded, third sterile: seeds pendulous, ovate-oblong.

(1) P. MUREX. *(Linn.)*

Ident. Linn. Sp. 892.—Dec. prod. IX. p. 256.—Roxb. flor. Ind. III. p. 114.
Engrav. Wight's Icon. t. 1615.—Rheede Mal. X. t. 72.—Burm. flor. Ind. t. 45. fig. 2.

SPEC. CHAR. Herbaceous annual, succulent, exhaling the odour of musk: leaves opposite, petioled, oval, dentately cut, obtuse, somewhat cuneate at the base: pedicels axillary, 1-flowered, shorter than the petiole, furnished with a gland at the base on both sides: flowers yellow.

Malabar. Cape Comorin. Kattywar. Deccan. Sea shores of the Peninsula. Flowering nearly all the year. The fresh leaves, if stirred in water, render it mucilaginous.

ORDER CXIII. CYRTANDRACEÆ.

Calyx 5-cleft or 5-parted, usually equal: corolla hypogynous, tubular at the base, more or less enlarged above; limb 5-lobed, unequal or somewhat 2-lipped: lobes rounded, imbricated in æstivation: stamens 4–5, adnate to the corolla, often only 2, rarely 4, didynamous, posterior one usually barren or wanting: anthers 2-celled; cells parallel or usually separating at the base, sometimes confluent into a single cell: ovary free, surrounded by a glandular ring: style simple; stigma bilamellate, 2-lobed or concave and undivided; fruit usually capsular, rarely baccate, 2-valved, 2-celled, falsely 4-celled, the partitions rising from the middle valves, 2-lobed, the lobes revolute and seed-bearing, the dehiscence is therefore loculicidal: seeds numerous, minute, pendulous, ovate or cylindric, naked, comose, or rarely winged: albumen none: embryo minute, terete, inverted: cotyledons oblong, shorter than the terete radicle.—Herbaceous perennials, very rarely annuals, sometimes frutescent: leaves usually opposite, entire or toothed: flowers umbellate, cymose, racemose or solitary: corolla purple, white or very rarely yellow.

GENUS I. BABACTES.

Didynamia Angiospermia. *Sax: Syst.*

Deriv. Alluding to a name of Bacchus, who is said to have travelled in India.

GEN. CHAR. Calyx ebracteate, campanulate, 5-parted: tube of the corolla obliquely gibbous; limb unequally 5-parted: stamens 2 fertile, 2 sterile, thick, villous: anthers 2-lobed, not furnished with a bristle or spur: capsule oblong, thickish: seeds flat, membranaceously winged at the margin.

(1) B. OBLONGIFOLIA. (*Dec.*)

Ident. Dec. prod. IX. p. 260.

Syn. Incarvillea oblongifolia. *Roxb. fl. Ind.* II. p. 113.—? Æchynanthus oblongifolius, *G. Don.*

Spec. Char. Tomentose shrub: branches sometimes 4-angled: leaves opposite or in threes, long-petioled, ovate-oblong, acuminate, serrulated: cymes axillary, dichotomous: flowers largish, whitish-red, mixed with yellow.

Common in the moist valleys of Chittagong.

GENUS II. ÆSCHYNANTHUS.

Didynamia Angiospermia. *See: Syst.*

Deriv. From *Aischuno*, to be ashamed, and *Anthos*, a flower.

Gen. Char. Calyx ventricosely tubular, 5-lobed at the apex, or 5-cleft or 5-parted: lobes equal; corolla tubular, incurved; limb oblique, somewhat unequally 5-cleft, sub-bilabiate; four anther-bearing stamens, often exserted, with the rudiment of a fifth: cells of the anthers parallel: ovary surrounded at the base by a cup-shaped ring: style filiform; stigma entire, depressed, hollow: capsule silicle-shaped, elongated, acuminated: valves 2, straight: placentæ 2-cleft, bilamellate, revolute at the margin as if 4-celled: seeds numerous, minute, oblong, pendulous, with long, few or solitary bristles at both ends.—Pseudo-parasitic under-shrubs, scandent, often rooting: stems terete, geniculate, glabrous: leaves opposite, petioled, fleshy, quite entire, very often glabrous: pedicels terminal or axillary, 1-2, rarely many-flowered: corolla red.

(1) Æ. GRANDIFLORA. (*G. Don.*)

Ident. Don. Syst. IV. p. 656.—Dec. prod. IX. p. 261.

Syn. Incarvillea parasitica, *Roxb.*—Trichosporum grandiflorum, *Don. prod. flor. Nep. p. 125.*—Æ. parasitica, *Wall.*

Engrav. Bot. Reg. 1861. t. 49.—Roxb. Cor. III. t. 291.

Spec. Char. Pseudo-parasitic on trees; leaves entire or repandly toothed, long-lanceolate, acuminated at both ends, nerveless: pedicels 1-flowered, deflexed, collected into a many-flowered terminal umbel: calycine lobes linear-oblong: corolla ventricose at the base: lobes rounded, nearly equal; stamens exserted: flowers orange-red.

Silhet mountains.

(2) Æ. BRACTEATA. (*Wall.*)

Ident. Dec. prod. IX. p. 261.

Spec. Char. Leaves elliptic, subacute at the base, acuminated at the apex, feather-veined; corymb terminal, somewhat panicled, many-flowered: bracts oval, acuminate, longer than the pedicel: calyx 5-partite: lobes lanceolate, erect: stamens included: seeds terminated at both ends by a simple hair: flowers red.

Pundua.

(3) Æ. ACUMINATA. *(Wall.)*

Ident. Dec. prod. IX. p. 263.

SPEC. CHAR. Leaves ovate-subobovate, acute at the base, cuspidately acuminate at the apex, glabrous, somewhat veinless: pedicels twin at the apex of the branch: calyx 5-partite: seeds with a single hair at both ends.

Silhet mountains.

(4) Æ. PERROTTETII. *(Alph. Dec)*

Ident. Dec. prod. IX. p. 361.—Dalz. Bomb. flor. p.

Syn. Æ. Zeylanica, *Gardner.*

Engrav. Wight's Icon. t. 1347.

SPEC. CHAR. Leaves lanceolate, rather obtuse at the base, obtusely acuminated at the apex, glabrous: lateral nerves few, oblique: umbels 3–5-flowered: pedicels twice the length of the calyx: flowers red spotted, two inches long: capsule three inches long.

Courtallum, Travancore Mountains. Parwar Ghaut, flowering in October.

GENUS III. DIDYMOCARPUS.

Didynamia Angiospermia. *Sys. Syst.*

Deriv. From *Didymos*, twin, and *Karpos*, a fruit; in allusion to the twin capsules.

GEN. CHAR. Calyx 5-cleft or 5-parted: corolla funnel-shaped: limb 5-lobed, somewhat irregular, rarely 2-lipped: stamens 4, of which 2 are anther-bearing: anthers kidney-shaped: ovary elongated: style short: stigma orbiculate, undivided: capsule silicle-shaped, 2-valved: valves bent inwards, falsely 4-celled: seeds naked, smooth, pendulous.—Undershrubs or herbs, with or without stems: leaves radical or cauline, alternate or usually opposite, unequal: peduncles axillary, branched or dichotomously cymose: flowers violet or white.

(1) D. MISSIONIS. *(Wall.)*

Ident. Br. in Horsf. Pl. Jav. p. 119.—Dec. prod. IX p. 266.

SPEC. CHAR. Stem very short: leaves cordate-ovate: peduncles axillary, scapiform: calyx persistent: lobes acute: capsule dehiscing at both sides simultaneously.

Peninsula.

(2) D. Punduana. *(Wall.)*

Ident. Br. in Horsf. Pl. Jav. p. 118.—Dec. prod. IX. p. 267.

Spec. Char. Stemless: leaves ternate, oblong, entire: peduncles and pedicels scattered with glandular pubescence: fertile stamens 2.

Silhet mountains.

(3) D. Rottleriana. *(Wall.)*

Ident. Dec. prod. IX. p. 268.

Syn. Gratiola montana, *Rottl in Litt.*—Rottlera incana, *Vahl.*—Henckelia incana, *Spreng.*

Engrav. Wight's Icon. t. 1348.

Spec. Char. Stemless, incanous leaves spathulato-obovate, crenate, densely clothed with white tomentium: scapes erect, hairy, subcorymbose, drooping at the apex: flowers smallish, corolla subinfundibuliform, 5-lobed.

Shevagherry hills, near Courtallum, flowering in August.

(4) D. lyrata. *(R. W.)*

Ident. Wight's Icon. vol. IV. & t. 1350.

Spec. Char. Stemless: leaves large, lyrate, finely crenate, pubescently pilose on both sides, especially on the veins: scapes erect, shorter than the leaves, dichotomous: branches racemose, hairy: calyx 5-parted, lobes lanceolate acute, pilose: corolla tubular, curved, somewhat ventricose beneath, equally 5-lobed: capsule terete or slightly compressed, 8-10 inches long, hairy: splitting along one side only.

Courtallum, flowering in August and September.

(5) D. ovalifolia. *(R. W.)*

Ident. Wight's Icon. vol. IV. & t. 1351.

Spec. Char. Leaves petioled, oval, obtuse at both ends, crenato-serrate, penninerved, slightly pilose on both sides, more densely so on the veins beneath: scapes about the length of the leaves, umbellately 3-6-flowered, villous: calyx deciduous, 5-parted, lobes lanceolate, pilose: corolla tubular, ventricose beneath, contracted at the throat: stigmas scarcely dilated: capsule long, slender, pubescent, dehiscing on one side.

Courtallum, flowering in August and September.

(6) D. TOMENTOSA. *(R. W.)*

Ident. Wight's Icon. vol. IV. t. 1349.

SPEC. CHAR. Leaves obovate-spathulate, densely crenate, dull whitish tomentose above, densely rusty-tomentose or woolly beneath; scapes erect, dichotomously cymose, many-flowered, hairy above; calyx 5-cleft, lobes linear obtuse, clothed with glandular hairs; corolla subcampanulate, 5-cleft, lobes suborbicular; fertile stamens shorter than the tube; ovary about the length of the calyx; stigma dilated; capsule cylindrical, about 1½ inch long, pointed, hairy, splitting along one side only; flowers bluish purple.

Neilgherries. Hills near Coimbatore. Flowering in February.

(7) D. GRIFFITHII. *(R. W.)*

Ident. Wight's Illust. vol. II. p. 182. & t. 159.

SPEC. CHAR. Herbaceous, erect, 4-sided, furrowed on the sides, angles roundish, blunt, pilose above; leaves ovate, acuminate, crenate-serrated, longish-petioled, pilose on both sides; peduncles axillary, longer than the petioles, cymosely 5-7-flowered; flowers longish pedicelled; calyx deeply 5-parted, segments lanceolate; corolla tubular, five-lobed; stamens didynamous, the fifth rudimentary one sometimes wanting, stigma 2-lobed; cymes furnished with a pair of lanceolate bracts at each division; tube of the corolla hairy without.

Khasia.

(8) D. CRISTATA. *(Dalz.)*

Ident. Dalz. Bomb. flor. p. 134.—Hook. Journ. Bot. III. p. 225.

SPEC. CHAR. Stem herbaceous, 8 to 9 inches high, simple, erect, round, fleshy; leaves large, opposite, petioled, broadly cordate, ovate-obtuse, slightly hairy on both sides. Inflorescence in the opposite axils and connate with the petioles, crested, hairy, shorter than the leaf, consisting of numerous pedicels rising upwards, and united below into a short thick peduncle; corolla white, half an inch long; capsule long, slender, curved, pubescent; seeds 5-angled, oblong.

On rocks near Parwar Ghaut, flowering in September and October.

GENUS IV. RHYNCHOGLOSSUM.

Didynamia Angiospermia. *Sex ; Syst :*

Deriv. From *Rhynchos*, a beak, and *Glossa*, a tongue; the lower lip of the flower is in the shape of a tongue-like beak.

GEN. CHAR. Calyx tubular, 5-cleft, lobes valvate in æstivation: corolla tubular, personate, shortly 2-lipped, upper lip short, 2-lobed, lower lengthened out half 3-lobed, lateral lobes very short: stamens included, 2 lower ones with kidney-shaped anthers, 2 upper ones (with a small rudiment of a third) sterile; vaginula incomplete, surrounding the base of the ovary: stigma capitate, scarcely divided: capsule topped with the persistent filiform style, ovate, 2-valved, enclosed by the calyx: placentæ two, parietal, adnate, split into two lamellæ: seeds numerous, small, elliptic-oblong.—Annual, glabrous, or slightly puberulous herbs: stem succulent: leaves alternate, petioled, ovate, occasionally deeply cut at the base, acuminated at the apex: racemes terminal, secund, simple: pedicels solitary, 1-bracteate: flowers deflexed, blue.

(1) R. OBLIQUM. *(Dec.)*

Ident Dec. prod. IX. p. 274.

Syn. Loxotis obliqua, *R. Br.*—Wulfenia obliqua, *Don. flor. Nep.*

Engrav. Wall. tent. flor. Nep. t. 35.

SPEC. CHAR. Stem slightly glabrous, or sparingly puberulous: adult leaves quite glabrous: raceme longer than the leaves, lower lip of the corolla 3-lobed.

Silhet.

GENUS V. KLUGIA.

Didynamia Angiospermia. *Sex ; Syst :*

Deriv. Named after Fr. Klug, M. D.

GEN. CHAR. Calyx loosely tubular, unequal at the base, occasionally gibbous above, 5-winged, 5-cornered, 5-cleft, lobes valvate in æstivation, wings or folds of the tube alternating with the lobes: corolla personate, tube subcylindric, throat closed; upper lip shortened, 2-lobed, lower lengthened, undivided, or half 3-lobed: stamens inserted on the tube, included, all fertile: anthers 2-celled, kidney-shaped, cohering into a little crown: ovary surrounded by a complete annular disk, 1-celled: placentæ

2, parietal, 2-lobed, many ovules on both sides : stigma depresso-capitate, simple : capsule ovate, inclosed by the calyx, valves 2 : seeds numerous, elliptic-oblong, furrowed, transversely wrinkled.

(1) K. NOTONIANA. *(Dec.)*

Ident. Dec. prod. IX. p. 276.
Syn. Wulfenia Notoniana, *Wall. tent. flor. Nep.*—Glossanthus Malabaricus, *Klein in Beat. seroj. Ind.* p. 57.—G. Notoniana, *Br. in Horsf. Pl. Jav.* p. 121.
Engrav. Wight's Icon. t. 1353.
SPEC. CHAR. Herbaceous, annual : stem fleshy, occasionally marked with a densely villous line : leaves alternate, oblong-ovate, dimidiate-cordate at the base : calyx coarsely and obtusely spurred at the base : flowers racemose, subsecund, blue.

In marshy places on the Neilgherries.

(2) K. SCABRA. *(Dalz.)*

Ident. Dalz. Bomb. flor. p. 134.—Hook. Journ. Bot. III. p. 140.
SPEC. CHAR. Stem terete, herbaceous, scabrous : leaves obliquely ovate, acute, entire, penninerved, upperside of the leaf and nerves beneath scabrous : flowers disposed in a long terminal raceme, alternate, drooping, of a deep blue colour : pedicels shorter than the filiform bract, lower lip of corolla subentire, with a triangular acute apex : leaves 4 inches long.

Warree country, flowering in the rainy season.

GENUS VI. ISANTHERA.

Polygamia Diœcia. *Sm : Syst.*

Deriv. From *Isos*, equal, and *Anthera*, an anther ; in reference to the stamens being five and equal.
GEN. CHAR. Flowers polygamous. HERMAPHRODITE : Calyx 5-cleft; corolla rotate, 5-cleft, shorter than the calyx : anthers 1-celled, dehiscing by a longitudinal or vertical cleft, embracing a semicircular connectivum : ovary 1-celled : placentæ prominent, bilamellate, ovuliferous at the margin, falsely 2-celled : stigma truncate : capsule almost 2-celled : seeds small, 4. FEMALE : Corolla none : rudiments of stamens tubercle-shaped : ovary as in the hermaphrodite.

(1) I. PERMOLLIS. *(Nees.)*

Ident. Nees in Trans. Linn. Soc. 17. p. 82.—Dec. prod. IX. p. 280.

Engrav. Wight's Icon. t. 1355.

SPEC. CHAR. Suffruticose, erect, simple, glabrous below, woolly tomentose above; leaves congested towards the apex, short-petioled, alternate, obovate-cuneiform, acute or shortly acuminate, minutely serrated, penninerved, pubescent above, tomentose beneath: tomentum in the dried specimen rusty or tawny coloured: peduncles axillary, about the length of the petioles, slender, drooping: cymes many-flowered.

Courtallum, in moist shady jungles. Western slopes of the Shevaghurry hills. Flowering in August.

GENUS VII. JERDONIA.

Didynamia Angiospermia. *Sex: Syst.*

Deriv. Named after Dr. Jerdon, Madras Medical Service, an eminent Naturalist.

GEN. CHAR. Calyx 5-parted, lobes narrow lanceolate: corolla subinfundibuliform, 4-lobed, the posterior one larger, emarginate: stamens 4, all fertile: filaments dilated, anterior pair broader, furnished with a broad descending tooth: anthers 2-celled, and, cohering at the apex, form a disk-like crown over the stigma: cells divaricating: ovary embraced at the base by a cup-shaped disk, 1-celled, with 4 parietal placentæ, 2 at each side: ovules attached to the slender filiform podosperm: style short: stigma dilated, peltate, concealed under the cohering anthers.

(1) J. INDICA. *(R. W.)*

Ident. Wight's Icon. vol. IV. & t. 1352.

SPEC. CHAR. A small herbaceous stemless plant: leaves petioled, oval, obtuse at both ends or slightly cordate at the base, the younger ones pubescent all over, the veins and margins only of the older ones clothed with long reddish hairs: scapes erect, filiform, longer than the leaves: pedicels short, subumbellate, surrounded with subulate pilose bracts: calyx winged with moniliform hairs, lobes narrow-lanceolate or subulate: corolla infundibuliform, limb somewhat bilabiate: filaments incurved at the apex, dilated below, the anterior pair pubescent.

Western slopes of the Neilgherries, flowering in March and April.

GENUS VIII. EPITHEMA.

Didynamia Angiospermia. *Linn. Syst.*

Deriv. A Greek word, signifying a lid or cover; alluding to the circumscissile capsule.

GEN. CHAR. Calyx tubular, 5-cleft: corolla funnel-shaped; limb somewhat 2-lipped, 5-lobed: stamens 2, included, cleft at the base, connivent (*Blume*); upper ones 2, fertile, lower ones 2, barren (*Benth.*): style 1; stigma capitate: capsule enclosed by the calyx, globose, cut circularly round, 1-celled, many-seeded: seed-bearing receptacles 2, free, clavate, diverging.

(1) E. ZEYLANICA. (*Gardn.*)

Ident. Wight's Icon. vol. IV.—Dalz. Bomb. flor. p. 135.
Engrav. Wight l. c. t. 1354.

SPEC. CHAR. Pilosely hispid all over: leaves opposite, or solitary by abortion, petioled, broad ovate cordate, doubly serratodentate, the upper ones opposite, sessile: peduncles terminal, 1-3, elongated, spicate at the apex: spikes dense, secund, circinate, bracteate at the base: bracts cordate, cucullate, obtuse, dentate.

Neilgherries on rocks. Southern Ghaut. Courtallum. Flowering in the rainy season.

ORDER CXIV. CONVOLVULACEÆ.

Calyx 5-sepaled, sepals persistent, equal or unequal, arranged in a single, double, or triple series, often enlarging with the fruit: corolla monopetalous, hypogynous, regular, with the limb 5-plaited or 5-lobed, twisted in æstivation; stamens 5, alternate with the lobes of the corolla, filaments often unequal, dilated at the base, anthers long, adnate, sagittate, 2-celled: pollen granular, spherical or annular: nectary annular, embracing the base of the ovary of most species: ovary usually simple, 2-4-celled, rarely partially, or altogether, 1-celled, occasionally double or quadruple, each cell with one or two erect ovules: style one, entire, more or less deeply bifid, rarely double: stigma acute, flattened or globose, terminating each branch of the style, hence 2-lobed or a single style: fruit capsular, variously dehiscing or dry-baccate, indehiscent, 1-4-celled: cells 1-2-seeded: dehiscence of the capsules valvate: seed subtriangular, rounded on the back, glabrous or villous, testa mostly hard: albumen mucilaginous: cotyledons foliaceous, corrugated, radicle incurved, inferior.—Herbs, undershrubs, shrubs or trees: stems straight, procumbent or twining, parasitical and leafless (in *Cuscuta*): leaves alternate, simple, entire or lobed, sessile or petioled: flowers one or several on axillary peduncles forming cymes, racemes, umbels or corymbs or often capitula: pedicels often bibracteate; bracts sometimes enwrapping the flowers: roots simple, or tuberous: pubescence often shining and beautiful.

GENUS I. RIVEA.

Pentandria Monogynia. *Sex: Syst:*

Deriv. Dedicated by Choisy to Auguste de la Rive, a Physiologist of Geneva.

GEN. CHAR. Sepals 5; corolla showy, tubular or funnel-shaped: style one: stigma capitate or lamelliform, 2-lobed: ovary 4-celled, 4-ovuled: capsule baccate.—Twining suffruticose plants.

(1) R. TILIÆFOLIA. (*Choisy*.)

Ident. Choisy. Conv. or. p. 25.—Dec. prod. IX. p. 325.

Syn. Convolvulus tiliæfolius, *Desv. Enc.*—C. gangeticus, *Roxb.*—Ipomœa tiliæfolia, *Roem. & Schult.*—I. gangetica, *Sweet.* —I. melanosticta, *Don.*

Engrav. Wight's Icon. t. 1358.

SPEC. CHAR. Twining, greyish pubescent; leaves roundish-cordate, sometimes obtuse, sometimes acuminate, pubescent beneath, petioled: peduncles short, 1-3-flowered: sepals roundish obtuse, afterwards enlarging: corolla inflato-cylindrical: fruit coriaceous, enclosed within the enlarged calyx: flowers pale rose, with a dark purple eye.

Coimbatore and elsewhere, in low moist soil, flowering during the autumnal rains.

(2) R. ORNATA. *(Choisy.)*

Ident. Choisy. Conv. or. p. 27. &. t. 3.—Dec. prod. IX. p. 326.

Syn. Convolvulus candicans, *Roth.*—Lettsomia ornata, *Roxb.*—Argyreia ornata, *Sweet.*

Engrav. Wight's Icon. t. 1356.

Spec. Char. Stems climbing: leaves petioled, orbiculato-cordate or reniform, glabrous above, whitish tomentose beneath: peduncles elongated, spicato-panicled or umbellate: sepals ovate-lanceolate, bluntish, coriaceous, externally villous: corolla slender, tubular: berry smooth: flowers pure white with a pale greenish eye.

Balaghaut mountains, Madras. Monghyr. Cawnpore. High ghauts, west of Joaneer. Flowering in the rainy season.

(3) R. BONANOX. *(Choisy.)*

Ident. Choisy. Conv. or. p. 27.—Dec. prod. IX. p. 326.

Syn. Lettsomia bonanox, *Roxb.*—Argyreia bonanox, *Sweet.*—R. fragrans, *Nimmo.*

Spec. Char. Twining: leaves rounded-cordate, emarginate, sometimes hairy: peduncles shorter than the petiole, commonly 3-flowered: sepals ovate-cordate, obtuse: corolla large, pure white, expanding at sun set, very fragrant.

Concans. Guzerat. Bengal. Flowering in the rainy season. It has the scent of cloves.

(4) R. HYPOCRATERIFORMIS. *(Choisy.)*

Ident. Choisy. Conv. or. p. 26.—Dec. prod. IX. p. 326.

Syn. Convolvulus hypocrateriformis, *Lam. Enc.*—C. candicans, *Roem. & Schult.*—Lettsomia uniflora, *Roxb.*—Argyreia uniflora, *Sweet.*

Spec. Char. Stems twining, pubescent: leaves rounded, obtuse, cordate, covered on the underside with white hairs: peduncles 1-flowered, sometimes axillary, solitary: sometimes disposed like a spike at the apex of the branchlets: sepals ovate, obtuse, unequal, hairy outside: corolla with a very narrow tube: flowers pure white.

Bombay and the Concans. Tanjore. Flowering in the cold season.

GENUS II. ARGYREIA.

Pentandria Monogynia. *Bot. Syst.*

Deriv. So named from the Greek word signifying *Silvery*; in reference to the appearance of the leaves.

GEN. CHAR. Sepals 5: corolla campanulate: style one: stigma capitate, 2-lobed: ovary 2-celled, 4-seeded: fruit baccate, surrounded by the sepals, often rubescent within and hardened.— Herbs or undershrubs, usually silvery, silky or tomentose.

(1) A. BRACTEATA. (*Choisy*.)

Ident. Choisy. Conv. or. p. 30.—Dec. prod. IX. p. 328.
Engrav. Hook. Bot. Comp. t. 3.

SPEC. CHAR. Covered with silky hairs: leaves oblong-rotund, acute at the apex, glabrous above, silky rufescent below: peduncles cymosely many-flowered, much shorter than the leaf; bracts lanceolate surrounding the cyme, somewhat villous externally: sepals ovate-lanceolate or elliptic, villous externally at the middle, membranaceous at the margin: corolla hirsute externally, greenish-white, with dark purple eye.

Common about Madras.

(2) A. SPECIOSA. (*Sweet*.)

Ident. Sweet. Hort. Sub. p. 289.—Dec. prod. IX. p. 328.
Syn. Convolvulus nervosus, *Burm.*—C. speciosus, *Linn.*—Ipomœa speciosa, *Pers.*—Lettsomia nervosa, *Roxb.*
Engrav. Burm. Ind. t. 20. fig. 1.—Rheede Mal. XI. t. 61.— Wight's Icon. t. 851.

SPEC. CHAR. Large climber: stem tomentose: leaves very large, cordate-acute, smooth or nearly so above, covered with white silky hairs beneath: peduncles as long as the petioles: flowers somewhat umbellate or capitate: sepals ovate, very obtuse, tomentose: corolla 2 inches long, somewhat cylindric, pale rose-coloured: fruit berried, 4-celled: cells 1-seeded.

Coromandel. Malabar. Deccan. Hindostan. Flowering nearly all the year. It is called the Elephant creeper.

(3) A. LESCHENAULTII. (*Choisy*.)

Ident. Dec. prod. IX. p. 320.—Choisy. Conv. or. p. 31.

SPEC. CHAR. Ash-coloured, silky: leaves ovate-elliptic, stringently hirsute above and green, ashy-tomentose below: peduncles cymose, many-flowered: bracts ovate or linear hirsute externally, intermixed with the flowers; sepals ovate, scariose on the margin; corolla hirsute externally.

Mysore. Neilgherries.

(4) A. POMACEA. *(Choisy.)*

Ident. Dec. prod. IX. p. 329.—Choisy. p. 31.
Syn. Lettsomia pomacea, *Roxb. fl. Ind.* II. p. 83.—Ipomæa Zeylanica, *Gærtn.*
Engrav. Wight's Icon. t. 858.

SPEC. CHAR. Ashy-tomentose: leaves ovate-elliptic, obtuse, ashy-velvety on both sides especially below, sometimes undulate on the margin: peduncles cymose, many-flowered: bracts linear-lanceolate, sometimes sub-cuneate, adpressed to each flower: sepals ovate-lanceolate, obtuse, slightly villous, adpressed: flowers large, rose-coloured.

Mysore. Diadigul at 3,000 feet. Flowering from May till November.

(5) A. NELLYGHERRYA. *(Choisy.)*

Ident. Dec. prod. IX. p. 329.—Choisy. p. 32.

SPEC. CHAR. Hirsutely rufescent: leaves cordate-orbiculate or oblong-acuminate, sinuated, strigosely hirsute on both sides, scabrous and greener above: peduncles umbellately many-flowered: bracts ovate or linear, externally villous, intermixed with the flowers: sepals ovate-elliptic, obtuse, membranaceous at the margin, hirsute externally.

Neilgherries.

(6) A. POPULIFOLIA. *(Choisy.)*

Ident. Choisy. Conv. or. p. 33.—Dec. prod. IX. p. 329.

SPEC. CHAR. Glabrous: leaves cordate-orbiculate, shortly acuminated: peduncles often exceeding the leaves, cymosely dichotomous: flowers lax: bracts linear-lanceolate, numerous, glabrous: sepals small, ovate-rotund, very obtuse, villous: corolla showy, tube narrowed at the base.

Goalparah.

(7) A. SPLENDENS. *(Sweet.)*

Ident. Sweet. Hort. Sub. (2d. ed.) I. p. 289.—Dec. prod. IX. p. 329.
Syn. Lettsomia splendens, *Roxb. flor. Ind.* II. p. 75.—Ipomæa splendens, *Sims.*
Engrav. Bot. Mag. t. 2628.

Spec. Char. Stem slightly glabrous or hoary: leaves ovate-oblong, or ovate-elliptic, entire, or pandurate sinuate, sometimes somewhat 3-lobed, smooth above, silvery-silky below: peduncles exceeding the petioles, spiked or corymbose, many-flowered: bracts none: sepals ovate, obtuse, hoary tomentose: flowers large, pale rose with a white eye.

Chittagong, flowering nearly all the year.

(8) A. ARGENTEA. *(Choisy.)*

Ident. Choisy. Conv. or. p. 330.—Dec. prod. IX. p. 330.
Syn. Lettsomia argentea, Roxb. flor. Ind. II. p. 79.

Spec. Char. Stem pubescent: leaves cordate, rounded, shortly acuminated at the apex, glabrous above, or rarely very shortly hairy, silvery-silky below: peduncles equalling the petioles, rigid, loose, umbelliferous at the apex: bracts lanceolate or linear, externally white-silky, intermixed with the flowers: sepals lanceolate, outer ones larger, revolute at the margin, villous externally: corolla glabrous, large, deep rose coloured.

Chittagong. Silhet. Flowering in the rainy season.

(9) A. ROXBURGHII. *(Choisy.)*

Ident. Dec. prod. IX. p. 330.—Choisy. Conv. or. p. 37.
Syn. Ipomœa Roxburghii, *Sweet.*—I. multiflora, *Roxb. flor. Ind.* II. p. 89.

Spec. Char. Similar to the preceding, but with the leaves ashy-villous on both sides, especially below: flowers large, rose-coloured with a bright red eye.

Bengal, in woods and hedges, flowering in the cold season.

(10) A. PILOSA. *(W. & A.)*

Ident. W. & A. pug. pl. Ind. or. p. 38.—Dec. prod. IX. p. 330.

Spec. Char. Stem, petioles and peduncles strigosely hirsute with spreading hairs: leaves ovate-round, obtuse at the base, even subcordate, covered on both sides, especially below, with few adpressed strigose hairs: peduncles exceeding the petioles, not equalling the leaves, many-flowered at the apex, flowers approximated: bracts linear or lanceolate, hirsute, intermixed with the flowers: sepals very acute: corolla strigose externally.

Southern mountains of the Peninsula.

(11) A. HIRSUTA. *(W. & A.)*

Ident. W. & A. pug. pl. Ind. or. p. 38.—Dec. prod. IX. p. 330.

Engrav. Wight's Icon. t. 891.

SPEC. CHAR. Stem petioles and peduncles as in the preceding: leaves ovate-acuminate, obtuse at the base, ashy-villous on both sides, paler above, somewhat silky below: peduncles loosely many-flowered: bracts linear or lanceolate, strigosely hirsute, mixed with the flowers: sepals ovate, obtuse, darkish, hirsute with few spreading hairs.

Southern mountains of the Peninsula.

(12) A. CYMOSA. *(Sweet.)*

Ident. Sweet. Hort. Sub. (2d. ed.) p. 289.—Dec. prod. IX. p. 333.

Syn. Lettsomia cymosa, *Roxb. fler. Ind.* II. p. 82.

SPEC. CHAR. Pruinosely pubescent; leaves ovate-rotund or cordately reniform, obtuse, very shortly mucronulate, glabrous or pruinosely pubescent on both sides: peduncles leafy at the top, cymosely many-flowered: bracts ovate-rotund, obtuse, plicately recurved, pubescent: sepals obtuse, outer ones similar to the bracts, inner ones ovate-linear: corolla tubular, funnel-shaped, large, rose-coloured.

Mountains of Malabar. Wynaad. Flowering in the cold season.

(13) A. FULGENS. *(Choisy.)*

Ident. Choisy. Conv. or. p. 33.—Dec. prod. IX. p. 329.

SPEC. CHAR. Tomentose or villous: leaves lanceolate long acuminate, glabrous, silvery tomentose beneath: peduncles shorter than the petioles, brachiately and loosely many-flowered: bracts narrow lanceolate or wanting: sepals villous, ovate, very obtuse, the exterior ones the smallest, clothed with white villi.

Courtallum and Quilon, flowering in August and September.

(14) A. CUNEATA. *(Ker.)*

Ident. Dec. prod. IX. p. 330.

Syn. Convolvulus cuneatus, *Willd.*—Lettsomia cuneata, *Roxb.*—Ipomœa atrosanguinea, *Sims.*—Rivea cuneata, *Wight.* in *Hook. Journ.* III. p. 199.

Engrav. Bot. Reg. t. 661.—Bot. Mag. 47. t, 2170.—Wight's Icon. t. 890.

SPEC. CHAR. An erect growing shrub, glabrous: leaves obovate-cuneate, emarginate, glabrous above, hairy beneath, scarcely petioled; peduncles shorter than the leaf, 3 to 6-flowered: bracts minute, linear; sepals ovate-obtuse; corolla tubular, an inch long, deep purple.

Common in the Mawul districts and Deccan. Mysore. Flowering in the cold season.

(15) A. AGGREGATA. *(Choisy.)*

Ident. Choisy. Conv. or. p. 45.—Dec. prod. IX. p. 333.
Syn. Lettsomia aggregata, *Roxb.*—Ipomœa imbricata, *Roth.*—Conv. imbricata, *Spreng.*
Engrav. Wight's Icon. t. 1359.

SPEC. CHAR. Hoary, tomentose; leaves ovate-cordate, smooth above, hoary beneath, very obtuse: peduncles longer than the petioles: flowers numerous, capitate: bracts ovate-orbiculate, hoary, obtuse: sepals ovate-obtuse; corolla small, pale rose-coloured.

Southern Mahratta country. Mysore, Coromandel. Orissa. Flowering in the cold season.

(16) A. MALABARICA. *(Choisy.)*

Ident. Choisy. Conv. or. p. 38.—Dec. prod. IX. p. 331.
Syn. Conv. Malabaricus, *Sims.*—Ip. Malaharica, *Roem. & Schult.*
Engrav. Rheede Mal. XI. t. 51.

SPEC. CHAR. Stem pubescent: leaves cordate, rounded, acute, glabrous or slightly hairy: peduncles as long as or longer than the petiole, many-flowered at the apex: sepals lanceolate-acute, hoary, the margins revolute: corolla white or cream-coloured, the bottom deep purple.

Western Ghauts. Mysore.

(17) A. SETOSA. *(Choisy.)*

Ident. Choisy. Conv. or. p. 43.—Dec. prod. IX. p. 332.
Syn. Ip. strigosa, *Roth.*—Lettsomia setosa, *Roxb.*—Conv. strigosus, *Spreng.*
Engrav. Wight's Icon. t. 1360.

SPEC. CHAR. Covered with close-pressed hairs; leaves cordate-ovate, or rounded-acuminate, smooth above, strigose beneath: peduncles longer than the petioles, rigid: flowers numerous, corymbose, rose-coloured: bracts reniform, orbiculate-obtuse: sepals of the same shape: corolla tubular, rather hairy, whitish.

Near Viziadroog. Surat, Malabar. Flowering in the cold season.

(18) A. SERICEA. (*Dalz.*)

Ident. Dalz. Bomb. flor. p. 169.

SPEC. CHAR. Twining, tomentose; leaves ample, broad, cordate-acuminate, hispid on the upper surface, white and silky, with adpressed pubescence beneath: petiole 2 inches long, peduncles axillary, simple, bearing a head of 6 to 8 flowers, enveloped in large foliaceous linear-oblong bracts: calyx and outer surface of corolla with long white hairs: flowers large, pink: ovary 4-celled: berry small, orange-coloured.

South Concan. High hills west of Joonum. Flowering in September.

(19) A. ELLIPTICA. (*Choisy.*)

Ident. Choisy. Conv. or. p. 35.—Dec. prod. IX. p. 330.

Syn. Conv. cilipticus, *Spreng.*—C. laurifolius, *Roxb.*—Ip. elliptica, *Rath.*—Ip. laurifolia, *Sweet.*

SPEC. CHAR. Climbing: leaves ovate or obovate-elliptic, villous: peduncles very long, bearing at the apex a corymbose panicle of flowers: sepals very obtuse, hairy outside, corolla an inch long, rose-coloured: fruit a berry, size of a large pea, orange-coloured, 2-celled.

Common on the Ghauts.

GENUS III. QUAMOCLIT.

Pentandria Monogynia. *Ser: Syst:*

Deriv. From *Kyamos*, a kidney-bean, and *Klitos*, dwarf; the species resemble the kidney-bean in their climbing stems.

GEN. CHAR. Sepals five, often mucronate: corolla tubular cylindric: stamens exserted: style one: stigma capitate, 2-lobed: ovary 4-celled, cells 1-seeded.—Twining herbs.

(1) Q. PHŒNICEUM. (*Choisy.*)

Ident. Choisy. Conv or. p. 51.—Dec. prod. IX. p. 336.

Syn. Ipomæa phænicea, *Wall. in Roxb. flor. Ind.* II. p. 92.—Convolvulus phæniceus, *Spreng.*

SPEC. CHAR. Leaves cordately subreniform, angularly toothed at the margin and sometimes 3-lobed, lobes acute, dentato-sinuate at the margin, glabrous on both sides: peduncles many-flowered, loosely spiked: sepals equal, awned: flowers large, bright scarlet.

Coromandel. Bengal. Common in most parts of the country. Flowering in the cold season.

(2) Q. VULGARIS. *(Choisy.)*

Ident. Choisy. Conv. or. p. 52.—Dec. prod. IX. p. 336.

Syn. Ipomœa quamoclit, *Linn.*—*Roxb. fl. Ind.* I. p. 503.—Convolvulus pennatus, *Desv. Enc.*—C. quamoclit, *Spreng.*

Engrav. Bot. Mag. 7. t. 244.—Rheede Mal. XI. t. 60.

Spec. Char. Leaves pinnatifid, segments linear parallel, acute: peduncles 1-flowered: sepals ovate-lanceolate: flowers scarlet.

Common everywhere as far Northwards as Dheyra Dhoon; flowering in the rainy season, but probably not a native of India.

GENUS IV. BATATAS.

Pentandria Monogynia. *Sex. Syst.*

Deriv. Either a Malayan, or according to others, a Mexican name.

Gen. Char. Sepals five: corolla campanulate: stamens included: ovary 4-celled, or by abortion, 3-2-called.—Herbs or undershrubs.

(1) B. PANICULATA. *(Choisy.)*

Ident. Choisy. Conv. or. p. 54.—Dec. prod. IX. p. 339.

Syn. Convolvulus paniculatus, *Linn.*—*Roxb. fl. Ind.* I. p. 478.—Ipomœa paniculata, *R. Br.*—I. gossipifolia, *Willd.*—I. eriosperma, *Beauv.*—I. quinqueloba, *Willd.*—I. insignis, *Andr.*—C. gossipifolius, *Spreng.*—C. insignis, *Spreng.*—C. roseus, *H. B. & Kth.*

Engrav. Bot. Reg. I. t. 62.—Bot. Repos. t. 686.—Bot. Reg. I. t. 75.—Bot. Mag. t. 1790.

Spec. Char. Stem twining, glabrous, thick: leaves large, palmate, 5-7-cleft, glabrous, petioled, lobes ovate-lanceolate, seldom acuminate: peduncles much exceeding the petioles, many-flowered, dichotomous: sepals ovate-rounded, concave, very obtuse, equal: corolla purple, narrowed at the base, showy: seeds long, hairy.

Coromandel. Western Coast. Silhet. Assam. Flowering in the rainy season.

(2) C. PENTAPHYLLA. *(Choisy.)*

Ident. Dec. prod. IX. p. 339.—Dalz. Bomb. flor. p. 167.

Syn. Convolvulus pentaphyllus, *Linn.*—C. monitus, *Wight's Ill.*—C. hirsutus, *Roxb. flor. Ind.* I. p. 470.—Ipomœa pentaphylla, *Jacq. Ic. Rar.* t. 319.—I. pilosa, *Cav. Ic.* 4. t. 323.—C. nemorosus, *Roem. & Schult.*

Engrav. Wight's Icon. t. 834.

SPEC. CHAR. Twining, hirsute: leaves digitate, leaflets 5, elliptic-lanceolate, entire: peduncles as long as the petioles, hirsute, 1-3-flowered: sepals very hairy, lanceolate, acute: corolla a little longer than the calyx, white: capsule and seeds smooth.

Peninsula. Hindostan. Flowering in the rainy season.

GENUS V. PHARBITIS.

Pentandria Monogynia. *Sex: Syst:*

Derio. Meaning not explained.

GEN. CHAR. Calyx 5-sepaled; corolla campanulate or campanulately funnel-shaped: style one, stigma capitately granulate: ovary 3, rarely 4-celled, cells 2-seeded.—Twining herbs.

(1) P. NIL. (*Choisy.*)

Ident. Dec. prod. IX. p. 343.

Syn. Convolvulus Nil, *Linn.*—Ipomæa Nil, *Roth.*—I. cærulea, *Koen.*—*Roxb. fl. Ind.* 1. p. 401.

Engrav. Bot. Reg. 4. t. 276.

SPEC. CHAR. Stem retrorsely hairy: leaves cordate, 3-lobed, middle lobe dilated at the base, lateral ones shorter, acute: petioles long: peduncles 2-3-flowered, longer than the common petiole, divaricate at the apex: sepals ovate-lanceolate, bispid at the base: corolla large, pale blue.

Common in most parts of India, flowering in the rainy season.

(2) P. LACINIATA. (*Dals.*)

Ident. Dals. in Hook. Journ. Bot. III. p. 178.—Bomb. flor, p. 167.

SPEC. CHAR. Stem filiform, creeping or twining, angular-twisted: leaves short-petioled, 7-lobed, lobes narrow-linear, between serrated and pinnatifid, teeth unequal, mucronate: peduncles axillary, solitary, angular-clavate, 1-3-flowered, shorter than the leaf: sepals oblong, mucronate, thick, fleshy, 3-ribbed, wrinkled: capsule 3-celled, cells 2-seeded; seeds silky: corolla white, purple inside, tube long, slender.

Malwan, flowering in August. The flowers open only at sunset.

GENUS VI. CALONYCTION.

Pentandria Monogynia. *Sar ; Syst ;*

Deriv. From *Kalos,* beautiful, and *Nyx,* night; the flowers do not open till sunset.

GEN. CHAR. Sepals five: corolla funnel-shaped: stamens exserted: style one, stigma capitate, 2-lobed: ovary 2-celled or somewhat 4-celled by the radiment of another dissepiment, 4-ovuled : pedicels fleshy.—Twining herbs.

(1) C. SPECIOSUM. (*Choisy.*)

Ident. Choisy. Conv. or. p. 59.—Dec. prod. IX. p. 345.

Syn. Ip. bonanox, *Linn.*—I. grandiflora, *Roxb.*—I. longiflora, *Willd.*—I. Roxburghii, *Steudel.*—Conv. Roxburghii, *Don.*—C. muricatus, *Linn.*

Engrav. Bot. Mag. t. 752.—Wight's Icon. t. 1861.

SPEC. CHAR. Stem sometimes prickly, climbing to a great height; leaves large, quite smooth, cordate, pointed; peduncles very long, 1-3-flowered; flowers very large, pure white, opening at sunset.

Everywhere in India. A very variable plant, flowering in the rainy season.

(2) C. ASPERUM. (*Choisy.*)

Ident. Dec. prod. IX. p. 346.—Choisy. Conv. or. p. 60.

SPEC. CHAR. Stem tortuous, thick, rough with short retroflexed somewhat thorny tubercles; leaves cordate-acuminate, glabrous: peduncles thick, 1-flowered, short; outer sepals shorter, shortly acuminated.

Silhet.

GENUS VII. LEPISTEMON.

Pentandria Monogynia. *Sar ; Syst ;*

Deriv. From *Lepis,* a scale, and *Stemon,* a stamen.

GEN. CHAR. Sepals five, equal; corolla tubular, inflated at the base: stamens furnished at the base with five scales forked above the ovary: stigma capitate, 2-lobed; ovary 2-celled, cells 1-2-ovuled.—Twining herbs.

(1) L. WALLICHII. (*Choisy.*)

Ident. Choisy. Conv. or. p. 61.—Dec. prod. IX. p. 348.

Syn. Vallaris controversa, *Spreng.*—C. binectariferus, *Roxb. fl. Ind.* II. p. 47.

SPEC. CHAR. Stem hairy; leaves cordate-acuminate, younger ones sometimes almost 3-lobed, large, hairy, petioled; peduncles very short, umbellately many-flowered; sepals oblong-lanceolate, acute, hirsute externally; corolla dilated at the base; scales ovate-rotund, glabrous.

Silhet.

GENUS VIII. IPOMÆA.

Pentandria Monogynia. *See : Syst:*

Deriv. From *Ips*, bindweed, and *Homoios*, similar; alluding to the twining habit of the plants.

GEN. CHAR. Sepals five; corolla campanulate; stamens included; stigma capitate, often 2-lobed; ovary 2-celled, cells 2-seeded; capsule 2-celled.—Herbs, undershrubs or even trees.

(1) I. RUMICIFOLIA. (*Choisy.*)

Ident. Choisy. Conv. or. p. 65.—Dec. prod. IX. p. 351.
Engrav. Visiani III. Ic. t. 1. fig. 2.
SPEC. CHAR. Stem herbaceous, diffuse, hairy, hairs dark-glandular; leaves cordate-reniform, obtuse at the apex, glabrous, entire at the margin, long-petioled; peduncles axillary, solitary, two or three, 1-2-flowered, not equalling the petioles; sepals linear lanceolate, hairy, very acute; corolla glabrous, scarcely exceeding the calyx; seeds glabrous, or shortly tomentose.

Travancore.

(2) I. ANCEPS. (*Roem. & Schult.*)

Ident. Roem. & Schult. 4. p. 231.—Dec. prod. IX. p. 360.
Syn. Convolvulus anceps, *Linn. Mant.* 13.
SPEC. CHAR. Stem 4-winged; leaves cordate-oblong, subsagittate or ovate, mucronulate at the apex, often glabrous; peduncles 2-3-flowered, bracteate at the apex.

Bengal.

(3) I. TUBEROSA. (*Linn.*)

Ident. Linn. Sp. 227.—Dec. prod. IX. p. 362.
Syn. Convolvulus tuberosus, *Spreng.*—C. Major, *Sloane.*
Engrav. Bot. Reg. t. 768.—Sloane Jam. I. t. 96. f. 2.
SPEC. CHAR. Stem thick, glabrous; leaves palmately 7-parted, lobes elliptic-lanceolate, entire, acute, petioles long; peduncles exceeding the petiole, rigid, smooth, bifurcately many-flowered;

pedicels thickened: sepals ovate, very obtuse, glabrous, adpressed: capsule globose, large; seeds black: flowers large, bright yellow, fragrant.

Madras. Bombay. Flowering in the cold season.

(4) I. WIGHTII. *(Choisy.)*

Ident. Choisy. Conv. or. p. 88.—Dec. prod. IX. p. 364.
Syn. Convolvulus Wightii, *Wall. Pl. As. Rar.* II. p. 55. t. 171.

SPEC. CHAR. Stem retrorsely hairy: leaves cordate-acuminate, acute or mucronulate at the apex, lower ones oblong, sinuately toothed at the margin, younger ones 3-lobed, all downy above, white-tomentose beneath: petiole hirsute: peduncles exceeding the petioles, 2–5-flowered: bracts linear-aristate, acute, hirsute and as if involving the capitulum: sepals oblong-linear, awned, very acute, hirsute: corolla campanulate, rose-coloured: capsule pubescent.

Neilgherries.

(5) I. CAPITELLATA. *(Choisy.)*

Ident. Choisy. Conv. or. p. 75.—Dec. prod. IX. p. 365.
Syn. I. tamnifolia, *Burm. Ind.* p. 50.

SPEC. CHAR. Stem hairy: leaves cordate-acuminate, hairy; peduncles equalling the petioles: bracts unequal, obtuse, ovate-linear: sepals linear-lanceolate, very acute: corolla violet.

Monghyr.

(6) I. RUBENS. *(Choisy.)*

Ident. Choisy. Conv. or. p. 81.—Dec. prod. IX. p. 371.

SPEC. CHAR. Stem twisted, pubescent: leaves cordate-acuminate above, often glabrous below, ashy-pubescent, long-petioled: peduncles stiff, much exceeding the petioles, umbellately many-flowered: sepals ovate acuminate or lanceolate-mucronate, equal, acute or obtuse, outer ones ashy-silky: corolla purple.

Silhet, Goalpara.

(7) I. CYMOSA. *(Roem. & Schult.)*

Ident. Roem. & Schult. 4. p. 241.—Dec. prod. IX. p. 371.

Syn. Convolvulus cymosus, *Dess.*—C. bifidus, *Vahl.*—C. blandus, *Roxb. flor. Ind.* I. p. 470.—Ipomœa bifida, *Roth.*—I. Heynei, *Roem. & Schult.*—I. blanda, *Sweet.*—C. pentagonus, *(var. pilosus.) Wall. in Roxb. Ind. flor.* I. p. 485.

SPEC. CHAR. Leaves ovate-cordate, acute at the apex, darkish above, short-petioled: peduncles axillary, 2-cleft, many-flowered, short, bracteolate: sepals coriaceous, obtuse, or rarely acutish, yellowish-dark, glabrous, outer ones often shorter: corolla with the tube narrowed at the base; capsule glabrous, conical, seeds rufous woolly with short hairs: flowers large, pure white.

Circar mountains, flowering in February and March.

(8) I. RACEMOSA. (*Roth.*)

Ident. Roth. Nov. Sp. p. 115. (not Poir.)—Dec. prod. IX. p. 371.

Syn. I. staphylina, *Roem. & Schult.*—Convolvulus Malabaricus, *Roxb. flor. Ind.* I. p. 469.—C. racemosus, *Roth.* (not *Spreng.*) C. Kleinii, *Spreng.*

SPEC. CHAR. Suffruticose, glabrous: bark often roughly wrinkled: leaves ovate-oblong, subcordate, acute, glabrous, reticulated beneath: flowers racemosely panicled, very numerous: sepals equal, very obtuse, ovate-rotund: corolla tubular-cylindric, long, rose-coloured with a dark-purple eye: capsule conical, glabrous, seeds long, woolly.

Coromandel, flowering in the cold season.

(9) I. STIPULACEA. (*Sweet.*)

Ident. Sweet. Hort. (2d. ed.) p. 289.—Dec. prod. IX. p. 370.

Syn. Convolvulus stipulaceus, *Roxb. flor. Ind.* II. p. 71.

SPEC. CHAR. Stem smooth: leaves cordately-sagittate, subacuminate, auricles often repand, glabrous: petioles long, 2-stipuled at the base: peduncles equalling the petioles, 2-6-flowered: pedicels clavate: sepals ovate, concave, glabrous: corolla white: capsule glabrous: seeds covered with olivaceous wool.

Chittagong. The Concans. Flowering in March and April.

(10) I. GEMELLA. (*Roth.*)

Ident. Roth. Nov. Sp. p. 110. ?—Dec. prod. IX. p. 380.

Syn. Convolvulus gemellus, *Berm.*

Engrav. Burm. Ind. 46. t. 21. fig. 1.

SPEC. CHAR. Ashy-pubescent, hairs adpressed: leaves cordate-acuminate, entire, or somewhat 3-lobed, ashy-villous beneath, with an obtuse mucronulate point: auricles sometimes sinuately crenated: petioles long, villous: peduncles much exceeding the petioles, loose, sometimes 2-flowered, sometimes dichotomously many-flowered, with one flower in each fork: sepals coriaceous, obtuse, somewhat torn at the margin, ovate, glabrous or externally villous; corolla striated, twice exceeding the calyx, 5-lobed at the apex.

Tranquebar.

(11) I. FASTIGIATA. (*Sweet.*)

Ident. Sweet. Hort. p. 288.—Dec. prod. IX. p. 380.

Syn. Convolvulus fastigiatus, *Roxb. flor. Ind.* II. p. 48.—C. platanifolius, *Vahl.*—C. emarquebensis, *Spreng.*—Ipomœa platanifolia, *Roem. & Schult.*

SPEC. CHAR. Glabrous, elongated: leaves cordate-acuminated and mucronulate, entire, sinuate, pandurate or 3-lobed, glabrous: petiole long: peduncles exceeding the petiole, 3-12-flowered, cymose, glabrous or somewhat villous at the base: sepals lanceolate, mucronately awned, glabrous, outer ones shorter: capsule smooth: seeds glabrous: flowers large, purple.—Varies in the stem being angular, peduncles few or many-flowered, sepals shorter or longer awned, and leaves pubescent on the veins.

Concans. Bengal. Flowering in February and March.

(12) I. DASYSPERMA. (*Jacq.*)

Ident. Jacq. Ecl. 1. p. 132. t. 89.—Dec. prod. XI. p. 360.

Syn. I. tuberculata, *Ker.* (*not Roem. & Schult.*)—Convolvulus dasyspermus, *Spreng.*—C. pedatus, *Roxb. flor. Ind.* II. p. 63.

Engrav. Ker. Bot. Reg. I. t. 86.

SPEC. CHAR. Herbaceous, smooth: leaves 3-parted, segments 3-cleft, longer ones pinnately 5-lobed, lower ones bifid, all glabrous, linear-lanceolate, mucronulate, petioles stipulaceous at the base: peduncles shorter than the leaf, 1-3-flowered: sepals ovate, obtuse, as if cordately spurred at the base before flowering, outer ones shorter: capsule glabrous: seeds with long orange silky hairs: flowers deep cream-coloured, tinged with yellow and with a lilac-purple eye.

Peninsula, flowering in the cold season.

(13) I. TUBERCULATA. (*Roem. & Schult.*)

Ident. Roem. & Schult. 4. p. 208.—Dec. prod. IX. p. 386.

Syn. Convolvulus tuberosus, *Desv.*—C. digitatus, *Roxb.*—I. stipulata, *Jacq. II. Schœnb.*—Batatas Loureirii, *Don.*

Engrav. Bot. Reg. t. 86.—Jacq. l. c. II. t. 199.

SPEC. CHAR. Stem glabrous, smooth or warty, muricated: leaves quinate; lobes lanceolate, entire at the margin, mucronate, outer ones often 2-cleft; petioles stipulaceous: peduncles often 1-flowered: sepals obtuse, unequal, ovate-oblong, membranaceous at the margin: capsule smooth; seeds glabrous: flowers very large, white lilac-purple.

Peninsula, flowering nearly all the year.

(14) I. REPTANS. *(Poir.)*

Ident. Poir. Enc. Suppl. 3. p. 460.—Dec. prod. IX. p. 349.

Syn. Conv. reptans, *Linn.*—C. repens, *Willd.*—*Roxb.*—C. Adansonii, *Desv. Eng.*—I. repens, *Roth.*

Engrav. Rumph. Amb. 5. t. 155. fig. 1.—Rheede Mal. XI. t. 52.

SPEC. CHAR. Stems creeping and rooting, fistulous, smooth: leaves sagittate, lanceolate: petioles glabrous: peduncles 1 to 3-flowered, nearly as long as the petioles: sepals ovate, glabrous: corolla tubulose, campanulate, of a pretty rose-colour.

Common in Guzerat. Coromandel. Conran. Flowering nearly all the year.

(15) I. PESCAPRE. *(Sweet.)*

Ident. Dec. prod. IX. p. 349.

Syn. Conv. pescapræ, *Linn.*—C. maritimus, *Desv.*—C. bilobabatus, *Roxb.*—C. carnosus, *Spreng.*—I. rotundifolia, *Don.*—I. maritima, *R. Br.*

Engrav. Rumph. Amb. V. t. 159. f. 1.—Rheede Mal. XI. t. 57.—Bot. Mag. t. 319.

SPEC. CHAR. Stems creeping to a great length: leaves subrotund, bilobed, parallel-veined, rather fleshy: peduncles 1 to many-flowered, a little longer than the petiole: sepals ovate-lanceolate: corolla rosy or purple.

Peninsula. Soonderbuns. Common on sandy beaches. Flowering nearly all the year.

(16) I. RUGOSA. *(Choisy.)*

Ident. Choisy. Conv. or. p. 64.—Dec. prod. IX. p. 350.

Syn. Conv. rugosus, *Rottl.*—C. flagelliformis, *Roxb.*—Ip. repens, *Lam.* (*not Roth.*)

Engrav. Wight's Icon. t. 887.—Rheede Mal. XI. t. 58.

SPEC. CHAR. Stems creeping: leaves cordately reniform, glabrous, obtuse, mucronulate: peduncles usually shorter than the leaves: sepals ovate, outer ones shortest and rugosely plicate.

Frequent in moist soil as about the banks of water-courses and under the bunds of tanks. Flowers usually pink, sometimes pure white.

(17) I. reniformis. (*Choisy.*)

Ident. Choisy. Conv. or. p. 64.—Dec. prod. IX. p. 351.

Syn. Conv. reniformis, *Roxb.*—C. gangeticus, *Linn.*—Evolvulus emarginatus, *Burm.*—E. gangeticus, *Linn.*

Engrav. Burm. flor. Ind. t. 30. f. 1.

Spec. Char. Stem creeping and rooting: leaves kidney-shaped, waved and dentate on the margin, obtuse: petioles hairy: peduncles very short, 1 to 2-flowered: corolla small, yellow.

Concan and Deccan, common in places where water has lodged, flowering in the cold weather.

(18) I. tridentata. (*Roth.*)

Ident. Roth. Cat. II. p. 19.—Dec. prod. IX. p. 353.

Syn. Conv. tridentatus, *Linn.*—Evolvulus tridentatus, *Linn.*

Engrav. Burm. Ind. t. 16. f. 3.

Spec. Char. Herbaceous, annual; stem filiform, angular: leaves sessile, oblong-linear, truncate at the apex, often 3-toothed, auricled and toothed at the base, scarcely an inch long, smooth: peduncles 1-flowered, longer than the leaf: sepals ovate, awned: corolla pale-yellow.

Near Hussein and Ghorebunder. Peninsula. Flowering in the rainy season.

(19) I. filicaulis. (*Blume.*)

Ident. Dec. prod. IX. p. 353.

Syn. Conv. filicaulis, *Vahl.*—C. hastatus, *Desr.*—C. simplex, *Pers.*—C. medium, *Lour.* (*not Linn.*)—C. denticulatus, *Spreng.*—Ip. denticulata, *R. Br.* (*not Choisy.*)—I. angustifolia, *Jacq.* (*not Choisy.*)—I. bidentata, *Don.*

Engrav. Rheede Mal. XI. t. 55.—Bot. Reg. t. 117.—Jacq. Ic. Rar. t. 317.

Spec. Char. Stem elongated, rarely twining, filiform, angular: leaves linear or linear-lanceolate, shortly petioled, hastate and denticulate at the base, quite smooth: peduncles longer than the leaf, 1 to 2-flowered: pedicels clavate: sepals ovate-acuminate: corolla small, pale-yellow, with a crimson eye.

Common in the Concans and Deccan. Travancore. Flowering nearly all the year.

(20) I. CAMPANULATA. (*Linn.*)

Ident. Dec. prod. IX. p. 359.
Syn. Conv. campanulatus, *Spreng.*
Engrav. Rheede Mal. XI. t. 50.—Wight's Icon. t. 1375.

SPEC. CHAR. Stem straight, glabrous, ramous: leaves cordate-acute, large, glabrous, reticulated beneath with reddish veins, long petioled: peduncles many-flowered, spicately racemose, as long as the petioles: pedicels afterwards thickening, black: sepals about half an inch long, ovate-orbicular, equal, glabrous: seeds silky, flowers white, tinged with rose, purplish near the bottom of the tube.

Eastern sides of the Neilgherries.

(21) I. TURPETHUM. (*R. Br.*)

Ident. R. Br. prod. p. 485.—Dec. prod. IX. p. 360.
Syn. Conv. Turpethum, *Linn.*
Engrav. Bot. Reg. t. 279.—Bot. Mag. t. 2093.—Wight's Ill. supp. t. 38.

SPEC. CHAR. Stems angular: leaves cordate, sometimes entire, sometimes sinuate-angled or crenated, pubescent and velvety on both sides: peduncles thick, 1 to 4-flowered: bracts ovate-lanceolate, velvety, deciduous: exterior sepals large: flowers white, with a tinge of crimson colour.

Guzerat, very common. Deccan. Peninsula. Flowering nearly all the year.

(22) I. VITIFOLIA. (*Sweet.*)

Ident. Dec. prod. IX. p. 361.
Syn. Conv. vitifolius, *Linn. Mant.* p. 203.
Engrav. Burm. Ind. t. 18. f. 1. and t. 19. f. 2.

SPEC. CHAR. Stem round, hairy or pubescent: leaves cordate, palmately pinnatifid: lobes unequal, irregularly crenate and dentate: peduncles many-flowered: flowers large, handsome, yellow.

Hilly parts of the Concan. Travancore. Silhet. Assam. Flowering in January and February.

(23) I. PESTIGRIDIS. (*Linn.*)

Ident. Dec. prod. IX. p. 363.
Syn. Conv. pestigridis, *Spreng.*
Engrav. Wight's Icon. t. 836.—Rheede Mal. XI. t. 50.

SPEC. CHAR. Stems round, hairy; leaves palmately 5 to 7-lobed: lobes ovate-acute, silky and hairy: peduncles many-flowered, as long as the leaf: heads of flowers surrounded by 6 to 8 ovate-linear hairy bracts: corolla white, hairy.

Common in hedges in the Peninsula, flowering nearly all the year.

(24) I. PILOSA. (*Sweet.*)

Ident. Dec. prnd. IX. p. 363.
Syn. Conv. pilosus, *Roxb.* I. p. 473.—*Ed. Cer.* II. p. 55.

SPEC. CHAR. Stems hairy, herbaceous: leaves broadly-cordate, entire or slightly 3-lobed, the middle lobe acuminate, 2 to 6 inches long, long-petioled: peduncles longer than the petioles: flowers many, cymose: sepals linear, hairy: corolla tubular, white or rose-coloured: capsules glabrous: seeds villous.

Mysore. Bengal. Flowering in the cold season.

(25) I. PILEATA. (*Roxb.*)

Ident. Roxb. flor. Ind. II. p. 94.—Dec. prod. IX. p. 365.
Syn. Conv. pileatus, *Spreng.*
Engrav. Wight's Icon. t. 1363.

SPEC. CHAR. Stem slender, villous: leaves cordate-acuminate, petioled, glabrous: peduncles shorter than the petiole: flowers 3 to 8, sessile, in a boat-shaped perfoliate involucre: bracts obovate, hirsute: corolla tubular campanulate: capsule glabrous; flowers rose-coloured.

Jungles in the Southern Concan. Flowering in the cold season.

(26) I. SESSILIFLORA. (*Roth.*)

Ident. Roth. Nov. Sp. p. 117.—Dec. prod. IX. p. 366.
Syn. Ip. sphærocephalus, *Sweet.*—Conv. sessilflorus, *Spreng.*— C. hispidus, *Vahl.*—C. sphærocephalus, *Roxb.*—Ip. hispida, *Roem. & Schult.*

Engrav. Wight's Icon. t. 169.

SPEC. CHAR. Stem herbaceous, covered with hairs pointing downwards: leaves cordate, ovate-lanceolate or sagittate: flowers axillary, 1 to 12, subsessile or very shortly-pedicelled, rose-coloured: sepals acuminate, subulate, hairy: corolla scarcely longer than the calyx: capsule hairy: seeds glabrous.

Severndroog. Common in most parts of the country. Flowering in the cold season.

(27) I. OBSCURA. (*Ker.*)

Idem. Dec. prod. IX. p. 370.

Syn. Convolvulus obscurus, *Linn.*—C. gemellus, *Vahl.*—Ipomaeniflolia, *Burm.*

Engrav. Bot. Reg. t. 239.

SPEC. CHAR. Stem herbaceous, elongated: leaves cordate-acuminate, glabrous or puberulous, acute, reticulated beneath, long-petioled: peduncles longer than the petioles, 1 to 3-flowered: pedicels thick, articulated: sepals oblong-ovate, glabrous or puberulous: flowers yellow; base of the tube purple.

Common about Bombay. Peninsula. Flowering nearly all the year.

(28) I. SEPIARIA. (*Koen.*)

Idem. Koen. in Roxb. flor. Ind. II. p. 90.—Dec. prod. IX. p. 370.

Syn. Conv. maximus, *Vahl.*—C. marginatus, *Lam.*

SPEC. CHAR. Twining: leaves cordate-oblong: peduncles many-flowered: heads of flowers dense: sepals oblong, ovate, acute or obtuse: corolla pinkish, tubular, funnel-shaped.

Common in hedges.

(29) I. COPTICA. (*Roth.*)

Idem. Roth. Nov. Sp. p. 110.—Dec. prod. IX. p. 384.

Syn. Conv. Copticus, *Linn. & Roxb.*—C. stipulatus, *Lam.*

SPEC. CHAR. Herbaceous, procumbent: leaves palmate or pedate, lower lobes shorter, bifid, all serrated, glabrous: petioles compressed: peduncles longer than the petiole, 1 to 2-flowered: sepals wrinkled, muricated, ovate-oblong, glabrous, mucronulate: corolla white, tubular, shortly 5-lobed: capsule glabrous.

Khandalla, creeping amongst the grass. Coromandel. Flowering in the cold season.

(30) I. BRACTEATA. (*R. W.*)

Idem. Wight's Icon. vol. IV. & t. 1374.

SPEC. CHAR. Herbaceous, twining, everywhere clothed with long pubescence: leaves long petioled, round cordate, mucronate: peduncles about the length of the petioles, cymosely 3-flowered: flowers sessile, small, the lateral ones each furnished with 3 ovate-cordate obtuse foliaceous bracts: sepals about the length of the corolla: corolla subcampanulate, tube glabrous, limb somewhat pubescent on the angles: stamens enclosed.

Quilon.

(31) I. WIGHTII. *(Choisy.)*

Ident. Choisy. Conv. or. p. 88.—Dec. prod. IX. p. 364.
Syn. Conv. Wightii, *Wall.*
Engrav. Wight's Icon. t. 1364.—Wall. Pl. As. Rar. t. 171.

SPEC. CHAR. Stem terete, elongated, retrorsely pilose: leaves cordato-acuminate, acute and mucronulate at the apex, the inferior ones oblong, with the margin sinuately dentate, the younger ones 3-lobed, all lanuginose above, whitish tomentose beneath, 2-3 inches long: petioles long, hairy: peduncles longer than the petioles, 2-3-flowered: bracts linear, aristate, acute, hairy, and as if embracing a capitulum: sepals oblong linear, aristate, acute, hairy, 5 lines long: corolla campanulate, rose-coloured, about an inch long: capsule pubescent: seed glabrous.

Neilgherries. Mysore. Flowering in the cold season.

(32) I. SALICIFOLIA. *(Roxb.)*

Ident. Roxb. flor. Ind. II. p. 88.—Dec. prod. IX. p. 367.
Syn. Ip. Buchananii, *Choisy.*

SPEC. CHAR. Glabrous; leaves linear-lanceolate, acuminate, short-petioled, very long: peduncles 1-3-flowered: sepals ovate, glabrous: corolla white; tube cylindric.

Rungpore, in Bengal.

(33) I. RHYNCORRHIZA. *(Dalz.)*

Ident. Dalz. Bomb. flor. p. 167.—Hook. Journ. Bot. III. p. 179.

SPEC. CHAR. Root an ovoid compressed beaked tuber: stem filiform, climbing, glabrous: leaves long-petioled, palmately divided into 7 lobes; lobes unequally pinnatifid, acuminate: peduncles axillary, solitary, filiform, 1 to 2-flowered, longer than the leaf; flowers middle-sized, yellow.

Near Tulkut Ghaut, flowering in August and September.

(34) I. CHRYSEIDES. *(Ker.)*

Ident. Dec. prod. IX. p. 382.
Syn. Conv. chryseides, *Spreng.*—C. dentatus, *Vahl.*—Roxb.—Ip. dentata, *Willd.*
Engrav. Wight's Icon. t. 157.—Bot. Reg. IV. t. 270.

SPEC. CHAR. Stem twisted: leaves oblong-cordate, subhastate, entire or often angled, even 3-lobed, acuminate, glabrous, auricles toothed and serrated: petioles long, muricately warted at the base: peduncles stiff, exceeding the petioles, 2-7-flowered, dichotomous, one flower in each division: sepals coriaceous, ovate-retuse, mucronulate: corolla small, yellow: capsule wrinkled, 4-cornered.

Peninsula. Assam. Flowering in the cold season.

GENUS IX. CONVOLVULUS.

Pentandria Monogynia. *Sur. Syst.*

Deriv. From *Convolvo*, to twine around; alluding to the habit of the plants.

GEN. CHAR. Sepals five: corolla campanulate; style one: stigmas two, linear-cylindric, often revolute: ovary 2-celled, 4-ovuled; capsule 2-celled.—Herbs or undershrubs.

(1) C. PLURICAULIS. *(Choisy.)*

Ident. Choisy. Conv. or. p. 95.—Dec. prod. IX. p. 403.

SPEC. CHAR. Stem suffruticose, very villous from the root, hairs spreading: leaves linear, diluted at the apex, sessile, very closely packed at the base of the stem, obtuse, villous: flowers long spiked, 2-3, subsessile in the axil of the leaves: sepals linear-acuminate, acute, hirsute: corolla twice exceeding the calyx, glabrous.

Mountains of North India. Buxar. Hurdwar. Bhagulpore.

(2) C. ROTTLERIANUS. *(Choisy.)*

Ident. Choisy. Conv. or. p. 95.—Dec. prod. IX. p. 403.

SPEC. CHAR. Stem simple, at first very leafy, rusty-coloured, shortened, afterwards elongated, scarcely rusty, ramously diverging: leaves linear, acute, subsessile, rusty-hirsute on both sides: peduncles 2-flowered, at first short, afterwards elongated into branchlets: pedicels bracteate: sepals ovate-acuminate, very acute, outer ones a little longer, hirsute: corolla tubular, scarcely exceeding the calyx, small, starlike, pink.

Near Madura, Southern Peninsula. Kattywar. Deccan. Flowering in October.

(3) C. ARVENSIS. *(Linn.)*

Ident. Linn. Sp. 218.—Dec. prod. IX. p. 406.
Syn. C. Chinensis, *Ker.*—C. Malcolmi, *Roxb.*
Engrav. Engl. Bot. V. t. 312.—Bot. Reg. IV. t. 322.
Spec. Char. Stem slender, prostrate or twining, striated, angled: leaves narrow-sagittate, subauricled: sepals ovate-obtuse: capsule smooth: peduncles 1-2-flowered, with 2 small bracts: flowers rose-coloured, fragrant.

Common in the black soil of Guzerat and the Deccan.

(4) C. PARVIFLORUS. *(Vahl.)*

Ident. Vahl. Symb. III. p. 20.—Dec. prod. IX. p. 413.—Roxb. flor. Ind. I. p. 471.
Syn. Ip. paniculata, *Burm.*—Ip. parviflora, *Pers.*
Engrav. Burm. Ind. XXI. t. 3.
Spec. Char. Stems twining, pubescent: leaves cordate-ovate, acute, glabrous, petioled: peduncles longer than the petioles: flowers umbelled, numerous, small, pure-white.

Island of Caranjah. Peninsula. Flowering in October.

(5) C. RUFESCENS. *(Choisy.)*

Ident. Choisy. Conv. or. p. 97.—Dec. prod. IX. p. 408.
Engrav. Wight's Icon. t. 1305.
Spec. Char. Stem red-rusty-coloured: leaves hastato-cordate, acute at the apex, mucronulate, sinuate at the margin: auricles crenately lobed: peduncles short, 1-3-flowered: bracts small: sepals ovate-acuminate, ciliated at the margin, acute, outer ones pubescent: capsule glabrous.

Neilgherries.

(6) C. MICROPHYLLUS. *(Seib.)*

Ident. Dec. prod. IX. p. 402.—Dalz. Bomb. flor. p. 164.
Engrav. Wight's Icon. t. 1367.
Spec. Char. Stems prostrate, elongated, hirsute: leaves lanceolate, attenuated into a very short petiole: flowers axillary, sometimes solitary, sometimes 2-3 on the rudiment of a branch, rotate, white or pale pink: capsule globose, smooth: seeds smooth.

Common in Guzerat.

GENUS X. ANISEIA.

Pentandria Monogynia. *Soz: Syst:*

Deriv. From *Anisos*, unequal; in reference to the sepals.

GEN. CHAR. Sepals five, disposed in two or three rows, namely, two outer ones larger inserted lower and decurrent into the peduncle, the third intermediate, and two inner ones less and inserted higher up: corolla campanulate: style simple: stigma 2-lobed, capitate or often flattened: ovary 2-celled, 4-ovuled: capsule 2-celled.—Herbs or undershrubs.

(1) A. CALYCINA. (*Choisy.*)

Ident. Choisy. Conv. or. p. 100.—Dec. prod. IX. p. 429.

Syn. Conv. calycinus, *Roxb.*—C. Hardwickii, *Spreng.*

Engrav. Wight's Icon. t. 833.

SPEC. CHAR. Twining, hairy: leaves oblong, cordate, acuminate, glabrous, petioled: peduncles shorter than the petioles, 1–3-flowered; exterior sepals sagittate: corolla tubular, pure white: capsule pointed: seeds silky.

Surat and Broach. Cawnpore. Flowering in the cold season.

(2) A. BARLERIOIDES. (*Choisy.*)

Ident. Choisy. Conv. or: p. 102.—Dec. prod. IX. p. 432.

SPEC. CHAR. Stem long, pubescent: leaves oblong, attenuated at the apex, often obtuse, entire at the base or scarcely cordate, very shortly petioled, villous on both sides: peduncles 1-flowered, exceeding the petioles, bibracteate at the apex about the flower: outer sepals oblong-lanceolate, acute, villously pubescent, alnate to the peduncle at the base, the rest narrower: corolla exceeding the calyx, limb gradually attenuated into the base.

Gorackpore. Sukanagur.

(3) A. UNIFLORA. (*Choisy.*)

Ident. Choisy. Conv. or. p. 101.—Dec. prod. IX. p. 431.

Syn. Conv. uniflorus, *Lam.*—C. emarginatus, *Vahl.*—C. Rheedii, *Well. in Roxb. flor. Ind.* II. *p.* 70.—Ip. uniflora, *Roem. & Schult.*

Engrav. Wight's Icon. 1. 850.—Ill. I. t. 8.—Burm. Ind. t. 21. f. 2.

SPEC. CHAR. Stem prostrate: leaves oblong-linear, very shortly petioled, mucronate, glabrous: corolla white, hairy on the outside: capsule silky within.

South Concan.

GENUS XI. HEWITTIA.

Pentandria Monogynia. *Sax, Syst.*

GEN. CHAR. Sepals five, unequal: corolla campanulate: style one: stigma 2-lobed, lobes ovate, flat: capsule 1-celled, 4-seeded.

(1) H. BICOLOR. *(W. & A.)*

Ident. Wight in Madr. Journ. 1837.

Syn. Shuteria bicolor, *Choisy. Conv. or.* p. 104.—*Dec. prod.* IX. p. 435.—Convolvulus bicolor, *Vahl. Roxb. flor. Ind.* I. p. 475.—Ipomœa bicolor, *Sweet.*—Calystegia Keriana, *Sweet.*—C. subintuatus, *Linn. suppl.*—C. involucratus, *Bot. Reg.* (not *Spreng.* or *Willd.*)—Palmia bicolor, *Endl. Dalz.*—C. bracteatus, *Vahl.*—I. bracteata, *Roem. & Schult.*

Engrav. Bot. Mag. 48. t. 2205.—Bot. Reg. 4. t. 318.

SPEC. CHAR. Stem twining, hairy: leaves ovate, cordate, entire, or with waved angles; peduncles frequently 1-flowered, longer than the leaves: bracts on the peduncle ovate-lanceolate, leafy, acute, pubescent: corolla yellow and purple: capsule hairy, 1-celled.

Concans. Bengal. Flowering nearly all the year.

GENUS XII. SKINNERIA.

Pentandria Monogynia. *Sex, Syst.*

GEN. CHAR. Sepals five: corolla small, as if urceolate: style one: stigma capitate, 2-lobed: ovary 1-celled, 4-ovuled: capsule 1-celled.

(1) S. CÆSPITOSA. *(Choisy.)*

Ident. Choisy. Conv. or. p. 105.—Dec. prod. IX. p. 435.

Syn. Convolvulus cæspitosus, *Roxb. fl. Ind.* II. p. 70.—Hewittia cæspitosa, *Stead.*

SPEC. CHAR. Stem herbaceous, diffuse: branchlets with loose spreading often hirsute hairs: leaves linear-lanceolate, entire, short-petioled, glabrous: peduncles simple or loosely brachiate, many-flowered: sepals ovate, obtuse, glabrous, outer ones smaller: corolla veined, small, very pale yellow: capsule glabrous.

Rungpore. Dinagepore. Assam. Flowering in the cold season.

GENUS XIII. PORANA.

Pentandria Monogynia. *Lin: Syst:*

Deriv. From *Poreno*, to journey; the branches extend to a great distance.

GEN. CHAR. Sepals five, middle-sized, and after flowering wonderfully and unequally increased; corolla campaculate or tubular, funnel-shaped: style one, entire or half 2-cleft: stigmas capitate: ovary 1-celled, 2–4-seeded: capsule 1-celled, often 1-seeded.—Twining herbs.

(1) P. RACEMOSA. (*Roxb.*)

Ident. Roxb. flor. Ind. II. p. 41.—Dec. prod. IX. p. 436.

Syn. P. dichotoma, *Don. flor. Nep.* p. 99.—Dinetus racemosus, *Sweet. Brit. fl. Gard.* t. 127.

Engrav. Wight's Icon. t. 1376.

SPEC. CHAR. Glabrous or rarely rough strigose: leaves cordate-acuminate, glabrous or pubescent, long-petioled: panicles racemose, leafy, loose-flowered: younger sepals very acute, afterwards ovate, scariose: corolla funnel-shaped, exceeding the calyx: flowers small, white.

Eastern slopes of the Neilgherries. Bengal. Silhet. Flowering in February and March.

(2) P. PANICULATA. (*Roxb.*)

Ident. Dec. prod. IX. p. 436.—Roxb. flor. Ind. I. p. 464.

Syn. Dinetus paniculatus, *Sweet. Brit. fl. Gard.*

Engrav. Roxb. Cor. t. 235.

SPEC. CHAR. Suffruticose, hoary-tomentose: leaves cordate-acuminate, glabrous above, hoary below, short-petioled; panicles large, much branched, leafy: younger sepals tomentose externally, linear subulate, equal: corolla tubular campanulate, exceeding the calyx: flowers small, white.

Silhet. Rajmahal. Hurdwar. Flowering in the cold season.

GENUS XIV. BREWERIA.

Pentandria Monogynia. *Lin: Syst:*

Deriv. Named by Dillenius, in memory of his friend Samuel Brewer.

GEN. CHAR. Sepals 5, equal or nearly so: corolla campanulate: style one, semibifid: stigmas two, thin, capitate: ovary 2-celled, 4-ovuled: capsule 2-celled.—Herbs or undershrubs.

(1) B. Roxburghii. (*Choisy.*)

Ident. Choisy. Conv. or. p. 111.—Dec. prod. IX. p. 438.
Syn. Convolvulus semidigynus, *Roxb. fl. Ind.* II. p. 47.
Engrav. Wight's Icon. t. 1370.

Spec. Char. Ramous: branchlets ferrugineo-villous; leaves ovate cordate, sub-acuminate, ferrugineous, long petioled: peduncles about the length of the petioles, 3 or many-flowered: sepals ovate-acuminate or ovate-rotund, subequal: corolla rufescent, narrow at the base: flowers pure white.

Travancore. Courtallum. Vingorla. Silhet. Flowering in the cold season.

(2) R. evolvuloides. (*Choisy.*)

Ident. Choisy. Conv. or. p. 112.—Dec. prod. IX. p. 439.
Syn. Seddera evolvuloides, *R. W.*
Engrav. Wight's Icon. t. 1369.

Spec. Char. Stems suffruticose, ramous: leaves ovate-lanceolate, sessile, glabrous, acute: flowers axillary, solitary, short-peduncled: sepals ovate, equal, often recurved, corolla very small.

Sea coast, near Tuticoreen, and many other places.

GENUS XV. CRESSA.

Pentandria Digynia. *Sex: Syst.*

Deriv. From *Cressa*, a native of Crete; the plant is plentiful there.

Gen. Char. Sepals five: corolla funnel-shaped, 5-cleft: stamens exserted: styles two: stigmas capitate: ovary 2-celled, 4-ovuled: capsule 2-celled, 1-4-seeded.

(1) C. Cretica. (*Linn.*)

Ident. Linn. Sp. p. 325.—Dec. prod. IX. p. 440.—Retz. obs. 4. p. 24.—Roxb. flor. Ind. II. p. 72.—Dals. Bomb. flor. p. 162.

Spec. Char. Shrubby, diffuse: leaves ovate, sessile, very small, acute, numerous, hoary or ashy-pubescent: flowers small, white or pink, subsessile in the upper axils, forming a many-flowered capitulum.

Common in cultivated fields about Bombay, flowering in the cold season.

GENUS XVI. EVOLVULUS.

Pentandria Digynia. *Ser: Syst:*

Deriv. From *Evolvo*, to roll out; not twining, opposite to *Convolvulus*.

GEN. CHAR. Sepals five: corolla campanulate or funnel-shaped: styles 2-cleft: ovary 2-celled, 4-ovuled: capsule 2-celled.— Herbs or small undershrubs, not twining.

(1) E. ALSINOIDES. (*Linn.*)

Ident. Linn. Sp. 392.—Dec. prod. IX. p. 447.—Roxb. flor. Ind. II. p. 105.
Syn. E. hirsutus, *Lam.*
Engrav. Burm. Zeyl. 9. f. 1. t. 6.—Rheede Mal. XI. t. 64.
SPEC. CHAR. Herbaceous, cæspitose, procumbent, covered with adpressed hairs: leaves ovate-oblong, subsessile, hirsute beneath: peduncles 1-flowered, as long as or longer than the leaf: flowers very small, deep blue.

Common in the Peninsula in grassy places, flowering in the cold season.

(2) E. PILOSUS. (*Rarb.*)

Ident. Roxb. flor. Ind. II. p.
SPEC. CHAR. Stems scarcely any, young parts clothed with soft hairs: leaves alternate, remote, sessile, linear-lanceolate, hairy: peduncles axillary, very short, hairy, 3-flowered: flowers nearly sessile on a common peduncle, small, white.

Hindostan, flowering in the cold season.

GENUS XVII. CUSCUTA.

Pentandria Digynia. *Ser: Syst:*

Deriv. From its Arabic name *Kechout*.

GEN. CHAR. Calyx 5, rarely 4-cleft: corolla globose-urceolate, or tubular, limb 5, rarely 4-cleft: stamens 5, rarely 4, adnate to the tube of the corolla, often furnished with epipetalous scales within at the base: ovary free, 2-celled, 4-ovuled: styles rarely combined into one: stigmas acute, clavate or capitate: fruit often capsular with a membranaceous pericarp: embryo spiral, filiform, more or less convolute round fleshy albumen.—Twining parasitic herbs, germinating in the earth, then, after the radicle dies, clinging to herbs or shrubs and nourished by their assistance; stems yellowish or reddish: leaves none or small scales in their place: flowers variously aggregate.

(1) C. REFLEXA. (*Roxb.*)

Ident. Roxb. flor. Ind. I. p. 446.—Dec. prod. IX. p. 454.
Syn. C. verrucosa, *Sweet. Brit. fl. Gard. t. 6.*
Engrav. Roxb. Cor. II. t. 104.—Hook. Exot. flor. t. 150.

SPEC. CHAR. Stem funicular: flowers loosely racemose, each flower pedicelled: calyx 5-sepaled; sepals acutish, ovate-oblong; corolla tubular: lobes minute, acute, externally reflexed: anthers subsessile at the throat of the corolla: scales inserted at the base of the corolla, fimbriated: styles short; capsule baccate; flowers small, white.

Coromandel. Concans. Guzerat. Silhet. Mahableshwur. Flowering in February and March.

(2) C. CHINENSIS. (*Linn.*)

Ident. Lam. Enc. II. p. 229.—Dec. prod. IX. p. 457.
Syn. C. sulcata, *Roxb. flor. Ind. I. p.* 447.
Engrav. Wight's Icon. t. 1373.

SPEC. CHAR. Stem filiform, very slender: fascicles of flowers lateral, sometimes glomerate, sometimes loosely panicled, few-flowered, each flower sessile or subsessile, minute: calyx 5-lobed; lobes ovate-oblong, obtuse: corolla campanulate, 5-lobed: scales minute, inserted at the throat, penicillate: flowers small, white.

Silhet. Malabar. Bengal. Flowering in March.

(3) C. CAPITATA. (*Roxb.*)

Ident. Roxb. flor. Ind. I. p. 448.
Syn. C. Major, *Dec. prod. IX. p. 452.*—C. vulgaris, *Pers.*

SPEC. CHAR. Stem capillaceous: heads of flowers sessile, dense, approximate; calyx 5-parted: lobes slightly obtuse; corolla urceolate, exceeding the calyx, withering in the top of the capsule: stamens slightly exserted: scales at the base of the filaments; styles diverging arch-like from the base: flowers small, white.

Bengal, flowering in February and March.

(4) O. STALINA. (*Roth.*)

Ident. Roth. Nov. Sp. p. 100.—Dec. prod. IX. p. 160.
Engrav. Wight's Icon. t. 1372.

Spec. Char. Stems filiform: corolla 4–5-cleft: stamens attached to the throat of the corolla with overlying scales fimbriated on the margin: peduncles about 3-flowered: flowers pedicelled: corolla hyaline, longer than the calyx: segments lanceolate.

Palamcottah.

GENUS XVIII. BRYCIBE.

Pentandria Monogynia. *Lam.*, *Syst.*,

Deris. Said to be the native name Latinised.

Gen. Char. Calyx 5-partite: lobes hollow, nearly equal: corolla campanulate: lobes 5, deeply 2-cleft, villous externally at the base: stamens alternate with the lobes of the corolla: filaments very short: anthers pyramidal, very acute, 2-celled: ovary free, ovate, 1-celled, 3-5-ovuled: style none: stigma 5-lobed: berry 1-seeded.—Scandent shrubs: leaves alternate, quite entire, glabrous, without stipules: racemes terminal or axillary: flowers yellow.

(1) B. PANICULATA. (*Roxb.*)

Ident. Roxb. flor. Ind. I. p. 585.—Dec. prod. IX. p. 464.
Syn. Erimatalla Rheedii, *Roem. & Schult.*
Engrav. Rheede Mal. VII. t. 39.—Roxb. Cor. II. t. 159.

Spec. Char. Branchlets, petioles, panicles and calyces velvety-pubescent: leaves elliptic, acutish at the base, abruptly and long-acuminate at the apex, glabrous: racemes panicled, longer than the leaf, terminal and axillary: flowers somewhat fascicled along the stalk, short-pedicelled, straw-coloured, very fragrant.

Concann. Circars. Monghyr. Silhet. Flowering in April and May.

(2) B. LAEVIGATA. (*Wall.*)

Ident. Wall. Cat. No. 1339.—Dec. prod. IX. p. 464.

Spec. Char. Branchlets glabrous: leaves oblong-lanceolate, acuminate: panicles almost supra-axillary: corolla rufous-silky outside.

Silhet.

(3) E. CORIACEA. (*Wall.*)

Ident. Wall. Cat. No. 1337.—Dec. prod. IX. p. 445.

SPEC. CHAR. Branchlets and peduncles somewhat pubescent; leaves coriaceous, elliptic, attenuated at both ends; peduncles axillary, scarcely longer than the petiole, closely panicled; corolla 4 times exceeding the calyx, rusty-silky outside.

Chittagong.

(4) E. WIGHTIANA. (*J. Grah.*)

Ident Dalz. Bomb. flor. p. 170.

SPEC. CHAR. Climbing; leaves coriaceous, shining, oblong-acuminate, attenuate at the base; racemes about the length of the leaves; corolla almost rotate, white; divisions cuneate, 2-lobed, their lower part clothed outside with rusty tomentum, very fragrant; segments of the calyx rounded, clothed with rusty hairs; stigma large, convolute.

Poonda ghaut, flowering in November.

ORDER CXV. BORAGINACEÆ.

Calyx free, often persistent, sometimes enlarging with the fruit, 5, (very rarely 4)-cleft or parted; sepals valvate in æstivation; corolla hypogynous, deciduous, monopetalous, usually 5-lobed, tube more or less elongated, limb spreading or erect, sometimes slightly unequal; æstivation various, usually imbricate with one lobe often exterior; stamens as many as the lobes of the corolla; anthers erect or incumbent, 2-celled; cells parallel, opening lengthwise; ovary consisting of 2 (anterior and posterior) more or less distinct, 2-celled carpels; cells concrete or separate, 1-ovuled; torus depressed or elongated, bearing the carpels, inserted either by the base or by the back; style between the separate carpels or springing from the apex of the concrete ones, simple, bifid, or twice bifid at the apex; fruit various, from 4 to 1-seeded; seed without, or with thin fleshy albumen; embryo straight, inverse, rarely curved; cotyledons foliaceous, entire, flat, or plicate.—Herbs, undershrubs, shrubs or trees with the surface of the leaves, calyx and branchlets usually beset with bristles, and at length with whitish scales, the indurated bases of the hairs: branches terete or irregularly angled: roots, especially of the Boragineæ, often tinged brownish-red with a peculiar resin-like colouring matter, soluble in water, spirits or oil: leaves alternate, simple, exstipulate, usually rough or variously bristly: racemes or spikes (rarely corymbs) variously disposed, often secund and circinate before evolution.

TRIBE I. CORDIEÆ.

Ovary 4-celled, undivided; style terminal, dichotomous, divided at the top into four branchlets, rarely none: fruit fleshy-drupaceous, indehiscent, cells 4, many, often abortive: cotyledons thick, somewhat fleshy, folded lengthwise: albumen none.—Trees or shrubs: flowers arranged in a raceme, spike, corymb or panicle: pulp of the fruit mucilaginous.

GENUS I. CORDIA.

Pentandria Monogynia. *Lin. Syst.*

Deriv. In honor of E. Cordus, a German Botanist of the 16th century.

Gen. Char. Calyx tubular, obovate or campanulate, 4-5-toothed, rarely 3 or 6-8-toothed: corolla funnel or cup-shaped, limb 4-5-partite, rarely 6-12-lobed; stamens as many as the lobes, inserted on the tube: style twice 2-cleft, often exserted: drupe ovate or globose, pulpy, often surrounded by the persistent calyx, ovary 4-celled, or 1-8-celled by abortion, often flowering, 1-8-celled, cells 1-seeded.—Trees or shrubs: leaves alternate, or very rarely sub-opposite, petioled, various in shape, quite entire or toothed; flowers sometimes polygamous or monoicous by abortion; corolla nearly always white.

(1) C. SUBCORDATA. (*Lam.*)

Ident. Lam. Ill. No. 1899.—Dec. prod. IX. p. 477.

Syn. C. orientalis, *R. Br.*—C. campanulata, *Roxb. Ind. fior.* II. p. 338.—C. Rumphii, *Blume.*—C. hexandra et orientalis, *Rœm. & Schult.*

Engrav. Rumph. Amb. II. t. 75.

SPEC. CHAR. Tree, glabrous: leaves long-petioled, ovate or subcordate, quite entire, somewhat undulated, bearded below at the axils of the lateral nerves: panicles lateral, racemose: calyx cylindric, juiceless, coriaceous, broadly 3–5-toothed: corolla funnel-shaped, tube longer than the calyx, mouth spreading, limb round, 6–7-lobed: drupe ovate, submucronate: flowers white, suffused with red.

Peninsula.

(2) C. OCTANDRA. *(Dec.)*

Ident. Dec. prod. IX. p. 477.

Syn. C. serrata, *Roxb. flor. Ind.* I. p. 591.—*Ed. Car.* III. p. 333.

Engrav. Wight's Icon. t. 169.

SPEC. CHAR. Tree: young petioles and peduncles sparingly villous: leaves long-petioled, ovate (or subcordate,) remotely serrated, acuminated, smoothish, 3-nerved at the base, here and there sparingly pilose at the nerves: panicles lateral and peduncled from the forks of the branches: calyx oval-cylindric, 3–5-toothed, hairy within: corolla equalling the tube of the calyx, lobes eight, oblong, recurved: stamens eight, inserted below the middle of the tube, bearded: flowers white.

Travancore, flowering in August.

(3) C. MONOICA. *(Roxb.)*

Ident. Roxb. flor. Ind. I. p. 592.—Ed. Car. II. p. 334.—Dec. prod. IX. p. 479.

Engrav. Roxb. Cor. I. t. 58.

SPEC. CHAR. Small tree: leaves ovate, remotely serrated, scabrous above, 3-nerved at the base: panicles terminal and lateral, small, bifid: flowers sessile, monoicous, males many: drupe ovate acuminated with a mucro: calyx cup-shaped: nut 4-celled with a perforated apex: flowers white.

Circars, flowering in the rainy season.

(4) C. POLYGAMA. *(Roxb.)*

Ident. Roxb. flor. Ind. I. p. 594.—Ed. Car. II. p. 337.—Dec. prod. IX. p. 480.

SPEC. CHAR. Tree: leaves petioled, somewhat round-cordate, almost quite entire, scabrous, younger ones villous: panicle terminal, pubescent, flowers polygamous: males often triandrous and hermaphrodite ones pentandrous: calyx villous externally: drupe ovate: flowers small, white.

Mountains of Coromandel, flowering in May.

(5) C. ORANDIS. *(Roxb.)*

Ident. Roxb. flor. Ind. I. p. 593.—Ed. Car. II. p. 335.—Dec. prod. IX. p. 481.

SPEC. CHAR. Tree ; leaves long-petioled, ovate, somewhat obtuse at the base, scarcely acute at the apex, quite entire, glabrous: panicle terminal, somewhat cernuous, rusty-velvety : flowers subsessile, 4-cleft : calyx oblong ; lobes of the corolla linear-oblong ; flowers small, white, fragrant ; drupe yellowish, smooth, somewhat depressed at the top, pulpa glutinous, nut 4-horned.

Chittagong, flowering in October and November.

(6) C. LESCHENAULTII. *(Dec.)*

Ident. Dec. prod. IX. p. 482.

SPEC. CHAR. Tree: quite glabrous, smooth : leaf-buds hirsute ; leaves long-petioled, elliptic, attenuated at both ends, quite entire or obtusely repand at the apex : panicles terminal, lax : flowers subsessile, 4-cleft : calyx subcampanulate, obtusely toothed ; stigmas much exserted, undulately dilated : drupe somewhat tetragonal.

Interior of the Peninsula.

(7) C. ACUMINATA. *(Wall.)*

Ident. Wall. in Roxb. flor. Ind. II. p. 339.—Dec. prod. IX. p. 499.

Syn. C. reflexa, *Roem. & Schult.* 4. p. 800.

SPEC. CHAR. Tree ; smooth, glabrous : leaves long-petioled, oblong-ovate, shortly acuminate, quite entire, coriaceous : corymbs terminal, short, dichotomous, somewhat pubescent : calyx smooth, deeply cleft ; lobes of the corolla lanceolate, ciliated : style far exceeding the stamens and throat, 2-cleft at the apex.

Silhet.

(8) C. LATIFOLIA. *(Roxb.)*

Ident. Roxb. flor. Ind. I. p. 588.—Dec. prod. IX. p. 478.

SPEC. CHAR. Tree : branchlets angled, petioles smooth : leaves petioled, ovate-rotund, sometimes slightly cordate, slightly repand, entire, smooth above, paler beneath, and with the nerves a little hairy : panicles terminal and lateral, rather shorter than the leaf : calyx irregularly toothed, hirsute within : corolla lobes 5, linear-oblong ; flowers white : drupe yellow when ripe, full of a glutinous pulp.

Goomerat. Hindostan. Silhet. Flowering in the cold season.

(9) C. MYXA. *(Linn.)*

Ident. Linn. Sp. p. 273.—Dec. prod. IX. p. 479.
Syn. C. officinalis and C. Africana, *Lam.*—C. domestica, *Roth.*
Engrav. Wight's Icon. t. 1378.—Rheede Mal. IV. t. 37.—Lam. Ill. t. 96. fig. 3.

SPEC. CHAR. Tree: branches round and smooth: leaves petioled, ovate, repand-dentate on young trees, smooth above, roughish beneath: panicles terminal, rarely lateral: flowers white.

Peninsula. Hindostan. Flowering in March and April.

(10) C. WALLICHII. *(G. Don.)*

Ident. Dec. prod. IX. p. 479.—Don's Gen. Syst. 4. p. 079.
Syn. C. tomentosa, *Wall. in Roxb. flor. Ind.*

SPEC. CHAR. Arboreous: leaves broad-ovate, rounded, more or less cordate, 3-nerved, acute, smooth above, densely tomentous and white beneath: corymbs lateral and terminal, dichotomous: calyx campanulate, velvety, irregularly toothed.

Ahmednugger and Western Ghauts.

(11) C. ROTHII. *(Roem. & Schult.)*

Ident. Dec. prod. IX. p. 480.—Roem. & Schult. Syst. 4. p. 798.
Syn. C. reticulata, *Roxb.*—C. angustifolia, *Roxb.*
Engrav. Wight's Icon. t. 1379.

SPEC. CHAR. Small tree: leaves subopposite, lanceolate obtuse, mucronate, entire, rough: panicles terminal and lateral, rather shorter than the leaf, many-flowered: flowers small, white: fruit size of a pea, orange-yellow.

Mysore. Western Coast. Flowering in May and June.

(12) C. FULVOSA. *(R. W.)*

Ident. Wight's Icon. vol. IV. &. t. 1380.

SPEC. CHAR. Tree: branches glabrous, terete, young shoots and leaves, petioles, peduncles and calyx before expansion, clothed with short fulvous pubescence: leaves petioled, ovate, obtuse at both ends, sometimes suborbicular, entire or repandly toothed, smooth, dull or somewhat hoary, being thinly clothed with very short pale fulvous pubescence: corymbs terminal and axillary: flowers congested on the branchlets: flower-buds villous on the apex, obovate: calyx subcampanulate, irregularly 4-6-toothed: corolla deeply 4-6-cleft, lobes obovate, emarginate, glabrous, tube hairy.

Peninsula.

(13) C. Perrottetii. (*Dec.*)

Ident. Dec. prod. IX. p. 482.
Engrav. Wight's Icon. t. 1381.

Spec. Char. Tree: branches terete, the younger ones, petioles and peduncles velvety-scabrous, rufescent: leaves petioled, elliptic, obtuse at both ends, entire, scabrous above, somewhat velvety beneath: the young ones tomentose: panicles terminal, subrazemose, shorter than the leaves: calyx oblong, obtuse, irregularly toothed, tomentose on both sides: tube of the corolla equal to the calyx, 4-cleft, lobes oblong, reflexed: stigmas long, exserted.

Bellary, flowering in September and October.

TRIDE II. EHRETIEÆ.

Ovary undivided or rarely lobed, many-celled, style often 2-lobed at the apex, rarely simple: fruit baccate or somewhat juiceless, indehiscent: seeds with thin fleshy or no albumen: cotyledons flat.—Shrubs, very rarely herbs.

GENUS II. EHRETIA.

Pentandria Monogynia. *Sex: Syst:*

Deriv. In honor of D. G. Ehret, a celebrated German Botanical draughtsman.

Gen. Char. Calyx 5-lobed: corolla cup-shaped or subrotate, lobes imbricated: filaments subulate, anthers ovate, 2-celled: style filiform, usually bifid or bi-partite: stigmas capitellate: ovary surrounded at the base by a hypogynous gland: berry fleshy or juiceless, sometimes with two stones, stones 2-celled, or with 4-stones, and stones 1-celled: seeds pendulous in the cells (or erect in few species): albumen sparing: embryo axile.—Shrubs or small trees: leaves alternate, sometimes fascicled, entire, rarely serrated: flowers usually corymbose, white.

(1) E. serrata. (*Roxb.*)

Ident. Roxb. flor. Ind. I. p. 597.—Dec. prod. IX. p. 503.
Engrav. Lindl. Bot. Reg. t. 1097.

Spec. Char. Tree: leaves oblong-lanceolate, acuminate, serrated: panicles terminal and axillary, compound: flowers fascicled, sessile: calycine lobes obtuse: corolla subrotate: stamens exserted: flowers small, greenish-white, fragrant.

Bengal. Hindostan, Chittagong. Silhet. Flowering in the hot season.

(2) E. ASPERA. (*Roxb.*)

Ident. Roxb. flor. Ind. I. p. 598.—Dec. prod. IX. p. 507.
Syn. E. tomentosa, *Roth.*—E. Heynii, *Roem. & Schult.*—Beurreria aspera, *Don. Gen. Syst.*
Engrav. Roxb. Cor. I. t. 55.
SPEC. CHAR. Shrubby: branchlets, petioles, leaves, peduncles and calyx rough-villous: leaves ovate, quite entire, short-petioled: corymb terminal, dichotomous: corolla campanulate, lobes patent, reflexed: stamens exserted: flowers very small, white.

Coromandel, flowering in the rainy season.

(3) E. BUXIFOLIA. (*Roxb.*)

Ident. Roxb. flor. Ind. I. p. 599.—Dec. prod. IX. p. 509.
Syn. Cordia retusa, *Vahl.*
Engrav. Roxb. Cor. I. t. 57.
SPEC. CHAR. Fruticose: branchlets at the top and younger leaves bristly on both sides: leaves fascicled from villous tubercles, obovate-cuneate, subpetioled, coriaceous, 3-toothed at the apex, adult ones glabrous, white-dotted above: peduncles axillary, 1-3-flowered, shorter than the leaf, and with the calyx covered with downy bristles: corolla subcampanulate, lobes oval: flowers small, white.

Coromandel forests, flowering in the cold season.

(4) E. LÆVIS. (*Roxb.*)

Ident. Roxb. flor. Ind. I. p. 527.—Dec. prod. IX. p. 505.
Syn. E. punctata, *Roth.*—Beurreria lævis, *G. Don.*
Engrav. Wight's Icon. t. 1382.—Roxb. Cor. I. t. 55.
SPEC. CHAR. Tree: leaves short-petioled, ovate or oval, quite entire, pubescent below at the axils of the nerves, otherwise glabrous: corymbs lateral or subaxillary, dichotomous, many-spiked: pedicels slightly hairy: calyx deeply 5-cleft: corolla subrotate: stamens exserted: flowers small, white.

Peninsula. Hindostan. Near Malwan. Flowering from October to February.

(5) E. OVALIFOLIA. (*R. W.*)

Ident. Wight's Icon. vol. IV. & t. 1383.
SPEC. CHAR. Small tree: leaves short-petioled, from oval-obtuse at both ends to somewhat obovate, or ending in a short blunt

stamen, smooth on both sides : corymbs terminal or axillary, dichotomously branched, circinate: flowers secund, short-pedicelled: calyx deeply 5-cleft, slightly hairy: corolla rotate ; limb reflexed: stamens exserted: drupe red when ripe, size of a small pea.

Coimbatore, flowering from August till October.

(6) E. WIGHTIANA. *(Wall.)*

Ident. Wall. Cat. No. 7013.—G. Don. Gen. Syst. IV. p. 388.
Engrav. Wight's Icon. t. 1384.

SPEC. CHAR. Shrubby, glabrous: branchlets slender, smooth : leaves elliptic-lanceolate, subacuminate above, tapering below into a longish slender petiole, quite entire, smooth on both sides: corymbs terminal, compact, dichotomous : branches revolute : flowers secund, subsessile : calyx 5-lobed, much shorter than the tube of the sub-hypocrateriform corolla ; stamens exserted : style equalling the tube: drupes, when immature, about the size of a large pepper corn.

Courtallum, flowering in August and September.

(7) E. CUNEATA. *(R. W.)*

Ident. Wight's Icon. vol. IV. & t. 1385.

SPEC. CHAR. Shrubby, branches virgate, terete, glabrous, nigrescent, smooth : leaves obovate-cuneate, retuse, subsessile, glabrous and smooth on both sides, quite entire, coriaceous : flowers solitary, axillary, on the ends of short leafy branches, subsessile : calyx 5-parted, lobes ovate-lanceolate, equalling the tube of the corolla, glabrous : corolla 5-cleft, lobes ovate obtuse : stamens attached near the bottom of the tube, scarcely exserted : style filiform : stigma capitate : fruit dry, 4-seeded.

Banks of the Cavery river near Errode, flowering in February.

GENUS III. RHABDIA.

Pentandria Monogynia. *Ser : Syst :*

Deriv. From *Rhabdos*, a twig; alluding to the habit of the species.

GEN. CHAR. Calyx 5, very rarely 6-parted : corolla campanulate, tube short, limb 5-cleft ; stamens included, rising from the base of the corolla : anthers oblong, fixed at the base, erect, 2-celled : ovary ovate-oblong : style simple : stigma 2-lobed : berry 4, rarely 6-stoned, stones 1-seeded : seeds albuminous.

(1) R. VIMINEA. *(Dalz.)*

Ident. Dalz. Bomb. flor. p. 170.
Syn. Ehretia cuneata, *R. W.*
Engrav. Wight's Icon. t. 1385.

SPEC. CHAR. Shrub with twiggy branches: leaves obovate-cuneate, small, smooth: flowers axillary, few, corymbiform, small, pink: berries size of a small pea, orange-red when ripe.

Common in the beds of the Concan rivers. Banks of the Cavery near Errode. Flowering in February.

GENUS IV. TOURNEFORTIA.

Syngenesia Monogamia. *Sex: Syst:*

Deriv. In honor of J. P. de Tournefort, a distinguished Botanical author.

GEN. CHAR. Calyx 5, rarely 4-parted: corolla cup-shaped, throat naked: stamens 5, rarely 4, included: style short: stigma undivided or 2-lobed, peltate, subconical: fruit of two carpels, carpels either undivided, pyreniform, 2-seeded, 2-3-celled, or 2-parted and in that case tetradynamous: seeds solitary in the cells: radicle superior.—Erect or scandent shrubs, rarely arborescent or herbaceous: leaves almost always alternate, petioled, quite entire, very rarely either partly opposite, sessile, or sinuate: spikes secund, ebracteate, often cymose: flowers white or yellowish.

(1) T. HEYNEANA. *(Wall.)*

Ident. Wall. List. No. 910.—Dec. prod. IX. p. 516.

SPEC. CHAR. Climbing: branchlets puberulous, angularly compressed at the apex: leaves petioled, elliptic or lanceolate, acuminate, quite entire, sparingly puberulous above: peduncles terminal, dichotomous: flowers sessile, approximated: calyx slightly hispid, 5-parted, acuminate: tube of the corolla puberulous, five-cornered, limb short, ovate.

Pundua, Silhet.

(2) T. ROYLEANA. *(Dec.)*

Ident. Dec. prod. IX. p. 527.
Syn. Messerschmidia hispida, *Benth. in Royle. Ill. p. 306.*

SPEC. CHAR. Stem hispid: leaves oblong-lanceolate, strigosely hispid: spikes elongated, paniculately branched: tube of the corolla twice as long as the calyx, lobes long, subulate-acuminate: berry somewhat dry with four 1-seeded drupes.
Delhi.

(3) T. VIRIDIFLORA. *(Wall.)*

Ident. Wall. in Roxb. flor. Ind. II. p. 5.—Dec. prod. IX. p. 516.

Syn. Lithospermum viridiflorum, *Roxb.*—Heliotropium viridiflorum, *Lehm.*

SPEC. CHAR. Fruticose, erect, branches terete, hairy: leaves petioled, ovate-lanceolate, wrinkled, quite entire, sparingly hairy below: peduncles bifid, diverging: spikes corymbose: lobes of the 5-parted calyx almost linear, a little hispid: corolla densely hairy outside: berry smooth, oval, deep purple, with two nuts.
Chittagong.

(4) T. RETICOSA. (*R. W.*)

Ident. Wight's Icon. vol. IV. & t. 1386.

SPEC. CHAR. Shrubby, climbing: branches terete, and with the under surface of the leaves sparingly covered with short adpressed pubescence: leaves short petioled, ovate-lanceolate acuminate, acute, round at the base, dark-green above, pale beneath and marked with a delicate net-work of brownish-purple veins: peduncles leaf-opposed, dichotomous: branches divaricating, spikes corymbose, circinate: calyx 5-parted, lobes ovate, hispid: corolla 4 or 5 times longer than the calyx, hairy, obtusely 5-lobed.

Coorg. Western slope of the Neilgherries, below Nedawuttum. Flowering in April.

(5) T. ZEYLANICA. *(R. W.)*

Ident. Wight's Icon. vol. IV. & t. 892.

SPEC. CHAR. Suffruticose, erect, ramous, hispid: leaves oblong-lanceolate, piloso-hispid on both sides: spikes elongated, geminate circinate: tube of the corolla 5-cleft, about twice the length of the calyx: lobes subulate-acuminate, toothed in the sinuses: pericarp dry, consisting of four 1-seeded nuts.

Bellary. Frequent in cultivated land about Coimbatore, flowering during the autumnal rains and throughout the cool season.

(6) T. SUBULATA. *(Hochst.)*

Ident. Dec. prod. IX. p. 528.—Dalz. Bomb. flor. p. 171.

SPEC. CHAR. Stem suffruticose, erect, sparingly-branched, clothed with stiff bristly hairs: leaves linear-lanceolate, pilose on both sides: racemes axillary, very long and slender; flowers secund: corolla tubular, segments spreading, acuminated: anthers 3-toothed at the apex: flowers yellowish.

Near Gogo, flowering in November.

TRIBE III. HELIOTROPEÆ.

Ovary many-celled, bearing a simple, terminal style; fruit somewhat juiceless, drupaceous, dividing: seeds exalbuminous.

GENUS V. HELIOTROPIUM.

Triandria Monogynia. *Ser. Syst.*

Deriv. From *Helios*, the sun, and *Trope*, twining; the flowers are said to turn towards the sun.

GEN. CHAR. Calyx 5-parted or very rarely 5-toothed, persistent; corolla cup-shaped, throat pervious, sometimes bearded, segments of the limb furnished with a simple fold or very rarely with teeth: style short: stigma sub-conical: nuts 1-celled, cohering by the base, afterwards separable; seeds exalbuminous.—Herbs or undershrubs, variously villous, rarely quite glabrous: leaves entire, or toothed, alternate or rarely opposite; spikes unilateral; corolla white or purplish.

(1) H. ROXBURGHII. *(Spreng.)*

Ident. Spreng. Cur. post. p. 54.—Dec. prod. IX. p. 549.

Syn. H. paniculatum, Roxb. *fl. Ind. (Ed. Car.)* II. p. 2.

SPEC. CHAR. Erect, branched, hairy: leaves petioled, ovate-oblong: spikes terminal, panicled, secund: tube of the corolla long, gibbous.

Chittagong.

(2) H. POLYSTACHYUM. *(Poir.)*

Ident. Poir. Suppl. III. p. 23.—Dec. prod. IX. p. 549.

SPEC. CHAR. Branches villously hairy: leaves long-petioled, alternate, ovate-lanceolate, tomentose: spikes many, terminal: calyx hispid.

Pondicherry.

335

(3) H. Coromandeliana. *(Lehm.)*

Ident. Lehm. Asp. p. 40.—Dec. prod. IX. p. 541.
Engrav. Wight's Icon. t. 1388.

Spec. Char. Stems herbaceous, erect or diffuse and with the leaves adpressed, villous: leaves obovate, oblong, entire, mucronate: spikes ternate, conjugate or solitary, ebracteate; lobes of the calyx somewhat unequal: corolla longer than the calyx: style scarcely any: nuts subglobose, hispid at the apex: flowers small, white.

Peninsula. Plains of North India. Flowering nearly all the year.

(4) H. linifolium. *(Lehm.)*

Ident. Lehm. Asp. p. 35.—Dec. prod. IX. p. 547.
Syn. Cynoglossum Zeylanicum, *Wight. Herb.*
Engrav. Wight's Icon. t. 1391.

Spec. Char. Suffruticose, erect, sparingly ramous, glabrous, 4-sided towards the apex: leaves linear, acutish, entire, revolute on the margin, sparingly adpressed, strigose on both sides: racemes sub-spicate, solitary, short, bracteolate: calyx very short, slightly hairy: lobes of the corolla acutish, tube ventricose, pilose: nuts glabrous or sometimes roughish.

Locality not given.

(5) H. scabrum. *(Retz.)*

Ident. Retz. Obs. II. p. 8.
Engrav. Wight's Icon. t. 1389.

Spec. Char. Procumbent, diffuse, strigose: leaves alternate, somewhat oblique, entire towards the ends of the branches, subopposite: flowers small, congested on the ends of the branches, concealed among the leaves: sepals somewhat unequal, hairy: corolla scarcely exceeding the calyx, subventricose: anthers apiculate: stigma dilated, shortly apiculate: nuts 4, roundish, glabrous.

Coimbatore, flowering during the rainy weather.

(6) H. Rottleri. *(Lehm.)*

Ident. Lehm. Asp. p. 66.—Dec. prod. IX. p. 549.
Engrav. Wight's Icon. t. 1392.

Spec. Char. Shrubby: stems short, dividing into many horizontal prostrate branches: branchlets and leaves strigose: leaves subsessile, ovate lanceolate, acutish: spikes lateral, 1 to 2 inches long: flowers subsessile, secund, the under side of the rachis bearing the bract: bracts ovate, strigose: calyx lobes ovate-acute, a little shorter than the corolla tube; corolla hairy externally, throat closed with hairs: fruit strigose, globose.

Bombay Presidency.

(7) H. SUPINUM. *(Linn.)*
Var. MALABARICUM. *(Benth.)*

Ident. Linn. Sp. p. 87.—Dec. prod. IX. p. 533.—Royle. Ill. 306.

Syn. H. Malabaricum, *Retz.*—Piptoclaina Malabarica, *G. Don.*

Engrav. Wight's Icon. t. 1387.—Burm. flor. Ind. t. 10. fig. 1.

Spec. Char. Herbaceous: stems ascending: leaves oval, obtuse, hoary and tomentose on both sides: spikes mostly solitary: calyx 5-toothed, short, very hirsute: corolla scarcely longer than the calyx, white with a yellow tube.

Malabar. Rice fields in the Deccan. Flowering in the cold season.

(8) H. MARIFOLIUM. *(Retz.)*

Ident. Retz. Obs. II. p. 8.—Dec. prod. IX. p. 547.

Engrav. Wight's Icon. t. 1390.

Spec. Char. Suffruticose: stems procumbent, diffuse, much branched: leaves and calyx strigose: leaves lanceolate, acute, revolute on the margin: racemes subspicate, solitary, alternate, the uppermost ones twin: flowers minute, white, with a yellow eye: nuts densely hispid.

Vingorla.

(9) H. LAXIFLORUM. *(Roth.)*

Ident. Roth. Nov. Sp. p. 102.—Dec. prod. IX. p. 548.

Spec. Char. Erect, branched, strigose with minute adpressed hairs: leaves linear-lanceolate, entire: racemes simple, subspicate, solitary or twin, elongated: flowers rather lax: lobes of the calyx strigose, as long as the tube of the corolla: nuts globose, strigose, not glabrous.

Worlee hill, Island of Bombay. Deccan.

GENUS VI. HELIOPHYTUM.

Pentandria Monogynia. *Sex. Syst.*

Deriv. From *Helios*, the sun, and *Phyton*, a plant.

Gen. Char. Calyx 5-partite or 5-cleft, persistent: corolla cup-shaped, throat contracted, 5-rayed, lobes of the limb obtuse, often undulate; anthers included; style very short; stigma capitate or conical; nuts two, separable, 2-seeded, 2-celled, (sometimes furnished with 1-2 empty cells.)

(1) H. INDICUM. *(Dec.)*

Ideal. Dec. prod. IX. p. 557.
Syn. Heliotropium Indicum, *Linn.*—H. cordifolium, *Mœnch.*—Tiaridium Indicum, *Lehm.*—H. anisophyllum, *Beauv.*—T. aniso-phyllum, *Don. Syst.*
Engrav. Sim's Bot. Mag. t. 1837.—Pluk. Alm. t. 245. f. 3.
Spec. Char. Stem herbaceous, erect, branched, hairy: leaves opposite and alternate, petioled, cordate-ovate, or oval, decurrent at the base into the petiole, subserrated, wrinkled: spikes terminal, solitary, simple; fruit glabrous, 2-cleft, mitre-shaped, segments divaricate, 4-celled, 2-seeded: flowers small, lilac-bluish.

A common weed, usually found among rubbish in the Peninsula, flowering in the rainy season.

GENUS VII. COLDENIA.

Tetrandria Trigynia. *Ser. Syst.*

Deriv. Named in honor of C. Colden, a North American Botanist.

Gen. Char. Calyx 5, (rarely 4)-parted: corolla funnel-shaped, throat enlarged, naked, lobes 5, (rarely 4)-parted, subrotund, flat, spreading: stamens as many as the lobes of the corolla: style 2-cleft: nuts 4, 1-celled, somewhat 3-cornered, externally convex, acuminated, coherent at the base into a pyramidal fruit: seeds with fleshy albumen: embryo inverse: cotyledons conduplicate.

(1) C. PROCUMBENS. *(Linn.)*

Ideal. Linn. Sp. p. 182.—Dec. prod. IX. p. 558.
Engrav. Lam. Ill. t. 69.—Pluk. Alm. t. 64. f. 6.
Spec. Char. Stems procumbent, hirsute: leaves short-petioled, obovate, unequally produced at the base above the petiole, plicate, coarsely toothed, with adpressed villous hairs above, beneath hir-sute: flowers axillary, solitary, sessile, small, white: nuts wrinkled, rough.

Common in rice fields in most parts of the country, flowering in the cold season.

TRIBE IV. BORRAGE.E.

Ovary consisting of two carpels, sometimes undivided and 2-celled, more frequently parted into two 1-celled nuts: style rising between the segments of the ovary, simple or 2-cleft at the apex: fruit dry, or very rarely somewhat fleshy, 2 or often 4-parting, segments in the former case 2, in the latter 1-seeded: seeds exalbuminous.—Shrubs or often herbs: root often red externally: leaves often bristly and afterwards scaly: racemes or spikes usually secund, and the younger ones circinate.

GENUS VIII. BOTHRIOSPERMUM.

Pentandria Monogynia. *Ser: Syst:*

Deriv. From *Bothrion*, an ulcer, and *Sperma*, seed; alluding to the appearance of that part of the plant.

Gen. Char. Calyx somewhat 5-parted, lobes ovate-lanceolate, acute: corolla cup-shaped, tube not exceeding the calyx, throat very obtusely emarginate with five scales, lobes of the limb somewhat rounded: anthers included under the middle of the tube: nectary small, annular, or a disk prominent around the ovary: ovary 4-partite: style very short: stigma capitoliate: nuts four, or by abortion two, distinct, longer than the persistent style, kidney-shaped: seed obovoid: radicle superior.—Annual or biennial herbs, strigose or hispid: flowers short pedicelled, arranged in a leafy raceme, extra-axillary, white or blue.

(1) B. TENELLUM. *(Fisch.)*

Ident. Dec. prod. X. p. 116.

Syn. Anchusa tenella, *Roem. & Schult.*—Cynoglossum diffusum, *Roxb. β. Ind. Ed. Car.* II. p. 7.—D. prostratum, *Don. β. Nep. prod.* p. 100.

Spec. Char. Stems much branched with adpressed strigæ: leaves oval-oblong, covered with adpressed hairs, lower ones petioled: fructiferous calyces longer than the pedicel, nodding: flowers small, white, often slightly reddish at the throat.

Bengal. Hindostan.

(2) B. ? MARIFOLIUM. *(Dec.)*

Ident. Dec. prod. X. p. 116.

Syn. Cynoglossum marifolium, *Roxb. flor. Ind. Ed. Car.* II. p. 8.

SPEC. CHAR. Stems branched, hairy, procumbent: leaves bifarious, subsessile, oblong or elliptic, hairy beneath: flowers pale azure, axillary or interfoliaceous: seeds rough, globose.

Eastern Bengal.

GENUS IX.   CYNOGLOSSUM.

Pentandria Monogynia.   *Lin: Syst:*

Deriv. From *Kuon*, a dog, and *Glossa*, a tongue; the leaves resemble a dog's tongue.

GEN. CHAR. Calyx 5-parted: corolla funnel-shaped, throat closed by obtuse scales, lobes very obtuse: stamens included: stigma entire or emarginate: nuts four, imperforated at the base, affixed to the base of the style, subrotund, convex or depressed, not marginate, almost everywhere echinated by hooked processes.—Herbs, very rarely undershrubs: leaves alternate, entire: racemes often spiked: pedicels 1-flowered: corolla blue, purple or white.

(1) C. HEYNII. (*G. Don.*)

Ident. Don's Gen. Syst. 4. p. 354.—Dec. prod. X. p. 150.

SPEC. CHAR. Strigosely hispid: leaves ovate-lanceolate, acuminated; racemes ebracteate, slender, terminal ones twin, axillary, solitary: nuts with hooked prickles on the margin, tubercled in the centre.

Dindigul.

(2) C. MICRANTHUM. (*Desf.*)

Ident. Desf. Cat. h. par. 1804. p. 220.—Dec. prod. X. p. 149.

Syn. C. canescens, *Willd.*—C. hirsutum, *Jacq.*—C. racemosum, *Roxb.*

SPEC. CHAR. Stem branched, hispid with spreading hairs: younger branches and racemes with hoary adpressed pubescence: leaves lanceolate-acute, denticulate, rough with hairs above, softly hairy beneath: racemes without bracts: calyx lobes ovate-obtuse, nearly as long as the corolla: flowers blue, with a white eye.

Bengal. Concan. Khandalla. Flowering in August and September.

(3) C. CŒLESTINUM. (*Lindl.*)

Ident. Dec. prod. X. p. 151.—Dalz. Bomb. flor. p. 173.

Engrav. Lindl. Bot. Reg. 1839. t. 36.—Wight's Icon. t. 1394.

SPEC. CHAR. Pubescent: stem erect, branched: radical leaves cordate-ovate, subacute: stem ones ovate-acute, cuneate at the base: racemes without bracts, often twice bifid: calyx lobes ovate, rather obtuse: corolla tube the length of the calyx, limb spreading: flowers pale blue: nuts ovate-compressed, covered with hooked bristles round the margin.

Near Vingorla. Mahableshwur.

(4) C. GLOCHIDIATUM. (*Wall.*)

Ident. Wall. Cat. No. 922.—Dec. prod. X. p. 150.

Syn. C. vesiculosum, *Wall.*

SPEC. CHAR. Stem erect, branched, hispid: hairs of the stem and the leaves spreading, tuberculate at their base, branches rather angular: leaves lanceolate-acuminate, attenuated at the base: racemes terminal, solitary or twin, without bracts: calyx lobes ovate, oblong-obtuse, silky and strigose: fruit with a hairy ring surmounted by long teeth, which have 5 to 6 recurved hooks at their apex.

Parr Ghaut, flowering in November.

(5) C. FURCATUM. (*R. W.*)

Ident. Wight's Icon. vol. IV. &. t. 1395.

SPEC. CHAR. Stems ramous, adpressed, pubescent or tomentose, the hairs of the lower part reflexed: leaves glaucescent, adpressed, pubescent: radical ones petioled, oval-lanceolate, acute at both ends: cauline ones sessile, the upper ones half-stem-clasping, ovate-cordate: racemes paired, slender, ebracteate, secund, hairy: flowers purple, scales of the throat two-lobed.

Neilgherries, very common, rising from 1 to 3 feet high, and flowering nearly all seasons.

GENUS X. TRICHODESMA.

Pentandria Monogynia. *Ser. Syst.*

Deriv. From *Thrix*, hair, and *Desma*, a bond; the anthers are bound to each other by hairs.

GEN. CHAR. Calyx 5-parted or deeply 5-cleft, often accrescent, lobes attenuated from a broad often auricled base: corolla scarcely longer than the calyx, tube broad cylindric, throat without scales, lobes broad from the base, often acuminate-subulate: stamens inserted on the tube: anthers many times greater than the filament, connivent into a cone, more or less villous at the back, usually long-subulate at the apex: ovary 4-celled, ovate-acute or depressed, somewhat 4-lobed: style filiform; stigma somewhat simple: nuts 4-1, closed at the base, aduate to a contral quadrangular column: seed obovoid, hanging from a very short funiculus, exalbuminous.—Herbs, sometimes suffrutescent at the base, erect, branched: leaves alternate or opposite, sessile, entire: pedicels lateral, hispid, arranged in racemes.

(1) T. INDICUM. *(R. Br.)*

Ident. R. Br. prod. 496.—Dec. prod. X. p. 172.
Syn. Borago Indica, *Linn.*
Engrav. Pluk. Alm. t. 76. f. 3.
SPEC. CHAR. Diffuse or erect, stem shortly villous: leaves usually opposite, narrow-lanceolate, half-stem-clasping, sessile: pedicels opposite-flowered or lateral 1-flowered: calyx villous, acutely auricled at the base, lobes increased by a subulate point: limb of the corolla spreading, reflexed: flowers pale blue.

Peninsula.

(2) T. AMPLEXICAULE. *(Roth.)*

Ident. Roth. Nov. Sp. p. 101.—Dec. prod. X. p. 172.
SPEC. CHAR. Erect, hispid with scattered hairs: lower leaves opposite, sessile, linear-oblong, upper ones alternate, cordate, stem-clasping, broadly ovate, acuminate: pedicels lateral and opposite the leaves, 1-flowered: calycine lobes shortly and obtusely auriculate: limb of the corolla with scattered hairs inside, lobes rounded, mucronulate: flowers pale blue.

Bombay.

(3) T. ZEYLANICUM. *(R. Br.)*

Ident. R. Br. prod. p. 496.—Dec. prod. X. p. 172.
Syn. Borago Zeylanica, *Linn.*—Leiocarya Klotschyana, *Horhst.*
Engrav. Jacq. Ic. rar. II. t. 314.—Burm. flor. Ind. t. 14. f. 2.
SPEC. CHAR. Stem erect, sparingly covered with bristles: leaves opposite, subsessile, oblong-lanceolate, attenuated at the base, bristly: pedicels hispid, lateral, long, 1-flowered, disposed in a raceme: calycine lobes ovate-lanceolate, villous: flowers pale blue.

Deccan. Bengal. Western Coast. Flowering in the cold season.

(4) T. SPINULOSUM. *(Voigkt.)*

Ident. Voight. Hort. Calc. p. 444.
Syn. Borago spinulosa, *Roxb. fl. Ind.* I. p. 459.—Dec. prod. X. p. 35.
SPEC. CHAR. Diffuse, dichotomous, armed with depressed bristles: leaves alternate, sessile, oblong, scabrous: pedicels 1-flowered, leaf-opposed.

Bengal, flowering in the cold season.

ORDER CXVI. HYDROLEACEÆ.

Calyx usually 5-parted, lobes equal, often spathulately dilated at the apex, persistent: corolla monopetalous, usually campanulate, 5-lobed: stamens inserted on the corolla, alternate with its lobes, incluse or exserted: ovary 2-celled, placentæ axile, covered with numerous minute amphitropal ovules: styles two: stigmata thickened or capitate: carpels 2-celled, 2-valved, dehiscence marginal, bearing on each side a spongy placenta: seeds sessile on the placentæ, very numerous: albumen fleshy, embryo straight.—Herbaceous annuals: stems straight or diffuse, glabrous, pubescent or hispid, rarely exceeding two feet in length: leaves simple, alternate, entire or dentate, often petioled: flowers corymbose or spicate, sometimes scorpioid.

GENUS I. HYDROLEA.

Pentandria Digynia. *Sur. Syst.*

Deriv. From *Hydor*, water, and *Elais*, oil; alluding to the situation and nature of the plant.

Gen. Char. Sepals five, persistent: corolla rotato-campanulate: stamens inserted on the tube: stigmas depresso-capitate: capsule globose or ovate: placentæ terete, fungose.

(1) H. Zeylanica. (*Vahl.*)

Ident. Vahl. Symb. II. p. 46.—Dec. prod. X. p. 180.
Syn. Nama Zeylanica, *Linn.*
Engrav. Wight's Icon. t. 601.—Hook. Comp. Bot. Mag. II. t. 26.—Rheede Mal. X. t. 28.

Spec. Char. Stem herbaceous, glabrous: leaves lanceolate, attenuated at both ends, glabrous, acute: panicles axillary, branched, leafy: sepals lanceolato-linear, acute, bristly pubescent at the base: flowers blue: capsule glabrous.

Borders of tanks and other wet places in most parts of India, flowering in the cold season.

ORDER CXVII. SOLANACEÆ.

Calyx 5-parted, seldom 4-parted, persistent, inferior: corolla monopetalous, hypogynous, deciduous, limb 5-cleft, seldom 4-cleft, regular or rarely somewhat unequal, the æstivation plaited or imbricate or even valvate: stamens inserted on the corolla, as many as the segments of the limb, with which they are alternate: anthers bursting longitudinally, rarely by pores at the apex, ovary 2-celled, composed of a pair of carpels right and left of the axis, rarely 4-5, or many-celled, with polyspermous placentæ: style continuous: stigma simple: ovules numerous, amphitropal: pericarp with 2, or 4, or many cells, either a capsule with double dissepiment parallel with valves, or a berry with the placentæ adhering to the dissepiment: seeds numerous: embryo straight or curved, often out of the centre, lying in fleshy albumen: radicle near the hilum.—Herbaceous plants and shrubs: leaves alternate, undivided or lobed, sometimes collateral, the floral ones sometimes double, and placed near each other: inflorescence variable, often out of the axil: the pedicels without bracts.

GENUS I. SOLANUM.

Pentandria Monogynia. *Sex. Syst.*

Deriv. A name used by Pliny, but the derivation is uncertain.

GEN. CHAR. Calyx 5, [rarely 4-6-10]-parted, cleft, toothed or crenated, even entire, regular or rarely somewhat irregular: corolla rotate, cup-shaped or pan-shaped, tube short, limb folded, 5, [rarely 4 or 6]-cleft, parted or angular: stamens five, rarely 4 or 6, adnate to the throat of the corolla, usually exserted: filaments very short, equal or rarely unequal: anthers free, dehiscent by twin pores at the apex, connivent, very rarely connate, equal or sometimes unequal, cells lateral, adnate to an inconspicuous connectivum: ovary 2, [rarely 3-4]-celled, cells many-ovuled: style simple: stigma obtuse: berry 2, [rarely 3-4]-celled: seeds many, subreniform, compressed: embryo spiral: albumen fleshy.—Annual or perennial herbs, undershrubs, shrubs or trees, unarmed or prickly, or rarely thorny, glabrous or hairy;

the underground stems sometimes with tuberiferous creeping roots, trunk or stem branched, erect, straight or flexuose, scandent or twining, branches more or less spreading, sometimes spinescent: leaves alternate, solitary or twin or in threes, rarely sessile, more often petioled, entire or variously divided: inflorescence terminal or axillary: cymes simple or dichotomous, resembling a raceme, corymb, umbel or panicle: flowers hermaphrodite, rarely polygamous, often barren with an abortive pistil, sometimes solitary.

(1) S. INCERTUM. *(Dun.)*

Ident. Dun. Sol. p. 155.—Dec. prod. XIII. p. 1. p. 57.
Syn. S. nigrum, *Forsk.*
Engrav. Rheede Mal. X. t. 73.

SPEC. CHAR. Stems glabrous, rooting or cirrhate: peduncles abortive: leaves ovate, angulately toothed, glabrous: flowers subumbellate, small, short.

Sandy places in Malabar. Bengal. Silhet. Flowering in the cold season.

(2) S. ROXBURGHII. *(Dec.)*

Ident. Dec. prod. XIII. p. 1. p. 57.
Syn. S. rubrum, *Roxb. flor. Ind.* II. p. 246.
Engrav. Wight's Icon. t. 344.—Rheede Mal. X. t. 73.

SPEC. CHAR. Stem branched, diffuse, angled: angles denticulate: leaves ovate-oblong, attenuated at both ends, ribs denticulate, repandly-toothed: common peduncle slender, fructiferous pedicels diverging, shorter than the peduncle: berries red: flowers small, white.

Common weed in all parts of the country, flowering all the year.

(3) S. SPIRALE. *(Roxb.)*

Ident. Roxb. flor. Ind. II. p. 247.—Dec. prod. XIII. p. 1. p. 146.

SPEC. CHAR. Branches dichotomous, fistular, glabrous, obscurely green, shining: leaves twin, one twice less, oblong-lanceolate, long cuneate into the petiole, acuminated, entire or somewhat repand, quite glabrous: racemes somewhat leaf-opposed, quite glabrous, spirally twisted before flowering: pedicels filiform, long, cymose, sometimes distant, sometimes subumbellate: flowers small, white.

Silhet. Assam. Flowering several times during the year.

(4) S. BIGEMINATUM. (*Nees.*)

Ident. Nees. ab. Esenb. in trans. Linn. Soc. 17. p. 42.—Dec. prod. XIII. p. 1. p. 175.—Walp. Rep. III. p. 62.

Spec. Char. Stems suffruticose: lower leaves solitary, upper ones twin, slightly bristly above, elliptic-oblong, acuminated at both ends, one less of the same shape: flowers by the sides of the leaves somewhat twin: calyx quite entire, smooth: peduncles of the fruit erect.

Travancore.

(5) S. GOUAKAI. (*Dec.*)

Ident. Dec. prod. XIII. p. 1. p. 177.

Spec. Char. Stem herbaceous, somewhat dichotomous, glabrous at the base, puberulous at the apex, subangled: leaves solitary or twin, petioled, lanceolate, acute, entire or few-toothed, teeth acute, scarcely hairy: flowers pedicelled, solitary or subumbellate: umbels 2–3-flowered, axillary: berry black.

Neilgherries.

(6) S. SUBTRUNCATUM. (*Wall.*)

Ident. Wall. Cat. No. 2620.—Dec. prod. XIII. p. 1. p. 180.

Spec. Char. Branches twiggy, terete, glabrous, puberulent at the apex: leaves ovate-lanceolate, acuminate, rather hairy above, whitish-rufescent beneath, glabrous: peduncles short, solitary or twin: calyx subtruncate: berry globose.

Silhet.

(7) S. MACRODON. (*Wall.*)

Ident. Wall. Cat. No. 2021.—Dec. prod. XIII. p. 1. p. 180.

Spec. Char. Branches greenish, elongated: leaves twin, petioled, oblong-lanceolate, acute at both ends, somewhat undulated, green above, glabrous beneath and pale whitish: peduncles solitary, twin, axillary, 1-flowered, slightly hairy: calyx crateriform-campanulate, 10-toothed, teeth filiform, very long.

Silhet.

(8) S. BARBISETUM. (*Nees.*)

Ident. Nees. ab. Esenb. trans. Linn. Soc. 17. p. 52.—Dec. prod. XIII. p. 1. p. 230.

Spec. Char. Stem herbaceous, prickly, prickles straight: leaves twin, elliptic, sinuate, very hirsute on both sides and prickly, segments angular: racemes lateral, simple, many-flowered, secund, prickly: berries glabrous, covered by the aciculate-bristled calyx.

Silhet.

(9) S. LASIOCARPUM. (*Dun.*)

Ident. Dun. Sol. 222. p. 173.—Dec. prod. XIII. p. 1. p. 251.
Engrav. Rheede Mal. II. t. 35.

Spec. Char. Stem fruticose, prickly: leaves subcordate, sinuately angled, tomentosely woolly on both sides and prickly: peduncles and calyx either prickly or unarmed.

Malabar.

(10) S. XANTHOCARPUM. (*Schrad. & Wendl.*)

Ident. Schrad. & Wendl. Sert. Hanov. I. 8. t. 2.—Dec. prod. XIII. p. 1. p. 302.

Spec. Char. Stem herbaceous, prickly; lower branches procumbent, divaricated, diffuse: leaves pinnatifid-sinuate, prickly on both sides, stellately pubescent, segments sinuate, acute: racemes few-flowered.

Var. a. SCHRADERI. Stem prickles fewer: leaves and flowers larger: segments of the leaves acuter: berry larger, yellow when ripe: flower bluish-purple.

Syn. S. diffusum, *Roxb. flor. Ind.* II. p. 250.

Coromandel. Bombay. Deccan. Flowering nearly all the year.

Var. b. JACQUINI. Stem decurrent: prickles strong, more numerous: leaves glabrous, less acute: berry smaller: flowers largish, bright blue.

Syn. S. Jacquini, *Willd.*—*Roxb. flor. Ind.* I. p. 251.—S. Virginianum, *Jacq. Ic. Rar.* I. 332.

Bengal. Flowering nearly all the year.

(11) S. UNDATUM. (*Lam.*)

Ident. Lam. III. No. 2367.—Dec. prod. XIII. p. 1. p. 359.
Engrav. Rheede Mal. II. t. 37.

Spec. Char. Stem somewhat herbaceous, dark-purple, prickly: leaves petioled, broad-ovate, sinuate-repand, lobes very obtuse, waved, with stellate hairs on both sides, green above, greenish-gray beneath, unarmed, or prickly on both sides, petioles long, nerves and veins brownish-purple, unarmed or prickly: peduncles usually 1-flowered, solitary.

Malabar.

(12) S. TROXOCUM. *(Poir.)*

Ident. Poiret, Dict. 4. p. 308.—Dec. prod. XIII. p. 1. p. 361.
Spec. Char. Stem suffruticose: leaves ovate-oblong, sinuate, prickly on both sides: flowers solitary: berry globose.

Var. a. Divaricatum. Branches divaricated, with stellate hairs above, prickly: leaves petioled, ovate-lanceolate or ovate-oblong, subrepand, greyish green above, somewhat hoary beneath.
Pondicherry.

(13) S. VERBASCIFOLIUM. *(Linn.)*

Ident. Linn. Spec. I. p. 263.—Dec. prod. XIII. p. 1. p. 114.
Syn. S. pubescens, *Roxb.*—S. erianthum, *Don. prod. flor. Nep.*—S. bicolor, *Willd.*
Engrav. Wight's Icon. t. 1398.
Spec. Char. Shrubby: leaves ovate-oblong, acuminate, entire, tomentose, surfaces discoloured: axils leafless; corymbs subterminal, dichotomous, peduncled: calyx half 5-cleft: flowers pale yellow or dirty white.
Neilgherries. Dharwar. Flowering nearly all the year.

(14) S. PUBESCENS. *(Willd.)*

Ident. Willd. phyt. I. p. 5.—Sp, p. 1026.—Dec. prod. XIII. p. 1. p. 127.
Engrav. Wight's Icon. t. 1402.
Spec. Char. Shrubby, unarmed, clothed all over with short, somewhat rigid tomentose pubescence: leaves ovate, acute, entire or subrepand: racemes corymbose, lateral: the lower anther larger: flowers blue, berries red.
Coimbatore. Peninsula.

(15) S. FEROX. *(Linn.)*

Ident. Dec. prod. XIII. p. 1. p. 255.—Linn. Spec. I. p. 267.
Engrav. Wight's Icon. t. 1309-1100.
Spec. Char. Perennial, herbaceous, woody at the base: leaves paired, cordate, sinuately angled, woolly tomentose and prickly on both sides: peduncles intrafoliaceous and like the short pedicels calyx and berries, hairy.

Courtallum, flowering in August and September. Neilgherries, always in flower. Coromandel. S. Concan. Bengal.

(16) S. GIGANTEUM. *(Jacq.)*

Ident. Jacq. Coll. Bot. 4. p. 125.—Dec. prod. XIII. p. 1. p. 258.

Syn. S. niveum, *Vahl.*—S. farinosum, *Wall. in Roxb.*

Engrav. Wight's Icon. t. 603.—Jacq. Icon. Rar. II. t. 238.—Bot. Mag. t. 1921.

SPEC. CHAR. Prickly shrub: prickles tomentose at the base: leaves large, oblong-lanceolate, entire, covered on the underside with a mealy tomentum; cymes lateral, dichotomous, many-flowered: berries round, red, size of a pea: flowers purplish-violet.

Common on the Ghauts.

(17) S. TORVUM. *(Swartz.)*

Var. INERME.

Ident. Dec. prod. XIII. p. 1. p. 261.—Swartz. prod. 47. Sor. Ind. Occ. I. p. 456.—Dals. Bomb. flor. p. 175.

Syn. S. multiflorum, *Roth.*

Engrav. Wight's Icon. t. 345.

SPEC. CHAR. Leaves slightly repand, ovate or cordate-acute, tomentose: peduncles lateral, corymbose, many-flowered: berry size of an apple.

In the southern Mahratta country, but probably has escaped from cultivated fields. (Dalzell.)

(18) S. TRILOBATUM. *(Linn.)*

Ident. Linn. Spec. I. p. 270. (partly.)—Dec. prod. XIII. p. 1. p. 287.—Roxb. flor. Ind. II. p. 253.

Syn. S. acetosæfolium, *Lam. Ill.*

Engrav. Wight's Icon. t. 854.—Burm. flor. Ind. t. 23. fig. 2.

SPEC. CHAR. Stems climbing, prickly: leaves 3-lobed, obtuse, glabrous: petioles and peduncles prickly: racemes terminal and lateral, somewhat umbellate: corolla deeply 5-divided, purple: fruit scarlet, size of a large pea.

Hedges in Guzerat. Coromandel, Deccan. Soonderbuns. Flowering nearly all the year.

(19) S. Indicum. *(Linn.)*

Ident. Nees. ab. Eseub. in Linn. trans. 17. p. 55.—Dec. prod. XIII. p. 1. p. 309.

Syn. S. violaceum, *Jacq.*—S. canescens, *Blume.*—S. pinnatifidum, *Roth.*—S. Heynei, *Roem. & Schult.*—S. agrestis, *Roth.*

Engrav. Wight's Icon. t. 346.—Burm. Zeyl. t. 102.—Rheede Mal. II. t. 36.

Spec. Char. Prickly shrub: prickles on the stem compressed, recurved: leaves solitary or twin, ovate, sinuate, lobed or pinnatifid, unequal at the base, tomentose, prickly, of two colours: racemes subcorymbose, placed between the leaves: calyx prickly: berries globose, size of a cherry, yellow when ripe: flowers blue.

All over India, flowering all the year.

(20) S. Wightii. *(Nees.)*

Ident. Dec. prod. XIII. p. 1. p. 334.—Nees. ab. Eseub. Linn. 17. p. 51.

Engrav. Wight's Icon. t. 841.

Spec. Char. Herbaceous, armed with aciculate prickles, and clothed with fascicled hairs: leaves cordate, ovate or elliptic, repando-sinuate: fructiferous peduncles elongated, reflexed: the 3 inferior anthers larger: fruit covered by the persistent calyx.

Neilgherries, rare.

(21) S. denticulatum. *(Blume.)*

Ident. Dec. prod. XIII. p. 1. p. 181.—Blume bijdr. p. 697.

Engrav. Wight's Icon. t. 1397.

Spec. Char. Stem suffruticose: inferior leaves solitary, upper ones paired, smoothish: one of them larger, acuminate at both ends, the other smaller, somewhat obovate: flowers fasciculately-aggregated, lateral: calyx minutely 10-denticulate, furrowed.

Neilgherries, not unfrequent in clumps of jungle in moist soil near springs and streams.

(22) S. Nessianum. *(Wall.)*

Ident. Dec. prod. XIII. p. 1. p. 174.—Dals. Bomb. flor. p. 175.

Spec. Char. Stem suffruticose: branches quadrangular, roughish towards the apex: lower leaves solitary, upper twin, densely and minutely scabrous above, oblong-lanceolate, acuminated at both ends, one smaller than the other: flowers lateral, fascicled; calyx quite entire; berry size of a pea, smooth and red, 2-celled.

Phoonda Ghaut.

GENUS II. PHYSALIS.

Pentandria Monogynia. *Ber. Syst.*

Deriv. From *Physa*, a bladder; alluding to the calyx.

GEN. CHAR. Calyx 5-cleft or 5-toothed, increasing after flowering, inflated like a bladder: corolla rotato-campanulate, plicate, limb 5-sinuate with as many acute angles: stamens included; filaments free, filiform: anthers erect, a half shorter than the filaments or nearly equalling them, connivent, dehiscing lengthwise: ovary 2-celled: placentas subglobose, adnate to the dissepiment, many-ovuled; style simple: stigma capitate: berry covered by a connivent inflated calyx, globose, 2-celled, placentas thick: seeds many, reniform compressed: embryo fleshy.—Annual or perennial herbs: leaves alternate or twin, entire or lobed: flowers solitary, extra-axillary.

(1) P. HERMANNI. (*Dec.*)

Ideat. Dec. prod. XIII. p. 1. p. 444.

Syn. P. pseudo-angulata, *Blume.*

SPEC. CHAR. Branched: stems slender: leaves petioled, ovate, acute, repand, somewhat entire, or sinuately toothed, glabrous: flowers small, drooping: peduncles capillaceous, hairy: calyx hirsute: corolla funnel-shaped, campanulate, marked with five spots at the base: fructiferous calyx angular, bladdery, closed, 5-toothed at the apex, glabrous: flowers yellowish.

Roadsides and waste places in Malabar.

(2) P. MINIMA. (*Linn.*)

Ident. Linn. Spec. I. 203.—Dec. prod. XIII. p. 1. p. 445.

Syn. P. parviflora, *Lag.*—P. Lagascæ, *Roem. & Schult.*

Engrav. Rheede Mal. X. t. 140.

SPEC. CHAR. Herbaceous, loosely villous with diffuse branches: leaves subcordate or ovate, acuminate, repandly toothed or almost entire, hairy: corolla without spots: anthers yellow: calyx of the fruit ovate, angled, here and there hairy: segments while flowering, triangular acuminate, shorter than the tube: flowers small, pale yellow.

Peninsula. Hindostan. Flowering nearly all the year.

GENUS III. WITHANIA.

Pentandria Monogynia. *Sat: Spl:*

GEN. CHAR. Calyx campanulate, 5-cleft or 5-parted: segments subulate, inflated after flowering, more or less covering the berry, afterwards withering: corolla campanulato, funnel-shaped or somewhat rotate, longer than the calyx, 5-cleft: stamens inserted on the tube of the corolla, included: filaments subulate, often dilated at the base, naked or scaly: anthers yellow, oblong, dehiscing lengthwise: ovary ovato-globose, glabrous, surrounded at the base by a thin glandular annular disk, 2-celled, many-ovuled: style simple, straight, subulate: stigma capitate: berry globose, 2-celled, covered more or less by a withering inflated calyx: seeds many, somewhat reniform.

(1) W. SOMNIFERA. *(Dec.)*

Ident. Dec. prod. XIII. p. 1. p. 453.
Syn. Physalis somnifera, *Nees.*
Engrav. Wight's Icon. t. 853.
SPEC. CHAR. Undershrub, more or less tomentose: hairs stellate, soft, hoary, caducous: stems erect: branches more or less flexuose: leaves entire, ovate, obovate or oblong, obtuse, attenuated into the petiole, twin: flowers short-pedicelled, pressed together: fruit-bearing pedicels more or less drooping.

Var. flexuosa. Branches more flexuose, less tomentose: flowers smaller, pale greenish.—Physalis flexuosa, *Linn. Rheede. Mal.* IX. t. 35.

Coromandel. Concans. Bengal. Flowering nearly all the year.

(2) W.? ARBORESCENS. *(Dec.)*

Ident. Dec. prod. XIII. p. 1. p. 455.
Syn. Physalis arborescens, *Lam.*
SPEC. CHAR.. Shrub: branches straight or slightly twisted: bark grey: leaves solitary or twin, repand or obsoletely angled, a dirty green above, slightly tomentose and ash-coloured below: flowers pendulous, solitary or 3–4–5–6 together, short-pedicelled.

Malabar.

352

GENUS IV. DATURA.

Pentandria Monogynia. *Sm: Syd.*

Deriv. Altered from the Arabic name *Tatorah.*

GEN. CHAR. Calyx tubular, often angled, 5-cleft at the apex or sometimes longitudinally cleft, cut round above the poltate persistent base, the cut portion deciduous: corolla funnel-shaped, limb spreading, enlarged, plicate, 5-toothed: stamens inserted on the tube of the corolla, included or slightly exserted, anthers dehiscing lengthwise; ovary incompletely 4-celled, one dissepiment deliquescent above the middle, the other complete placentiferous, placentae many-ovuled; style simple, stigma bilamellate: capsules ovate or subglobose, muricated or prickly, half-four-celled, incompletely 4-valved at the divisions: seeds many, reniform, testa sometimes crustaceous and hard, or corky, very thick.—Annual or perennial herbs, fetid and poisonous, sometimes suffrutescent or arborescent: leaves petioled, oblong or ovate, often angularly toothed; flowers axillary, solitary, often large, white, violet or carmine.

(1) D. ALBA. *(Nees. ab. Esenb.)*

Ident. Nees. ab. Es. in trans. Linn. 17. p. 73.—Dec. prod. XIII. p. 1. 541.

Syn. Datura Metel, *Rorb.*

Engrav. Wight's Icon. t. 852.—Rumph. Amb. V. t. 87. f. 2.—Rheede Mal. II. t. 28.

SPEC. CHAR. Shrub: leaves ovate-acuminate, repand-dentate, unequal at the base, smooth: stems herbaceous: capsules nodding, covered with prickles: flowers very large, white.

A common and well-known plant, of which there are several varieties. It flowers the whole year.

(2) D. NILHUMMATU. *(Dun.)*

Ident. Dunal in Dec. prod. XIII. p. 1. p. 542.—Dals. Bomb. flor. p.

Syn. D. hummatu, *Rorsh. in Linn.*

SPEC. CHAR. Shrub: leaves smooth; calyx with five angles: corolla often double and treble: flower violet-coloured outside, white within: fruit tubercled.

Almost as common as the preceding.

(3) D. fastuosa. *(Linn.)*

Ident. Linn. Spec. p. 256.—Dec. prod. XIII. p. 1. p. 542.—Roxb. flor. Ind. I. p. 561.

Syn. Stramonium fastuosum, *Manch.*

Esgrae. Wight's Icon. L. 1396.—Rheede Mal. II. t. 29, 30.

Spec. Char. Shrub: leaves ovate acuminate, repand-dentate, unequal at the base, and like the stems puberulous: fruit drooping, tubercled: flowers very large, outside dark purple, inside whitish.

Common about Coimbatore and every part of the country, flowering all the year.

GENUS V. HYOSCYAMUS.

Pentandria Monogynia. *Sex: Syst:*

Deriv. From *Hyos*, a hog, and *Kyamos*, a bean; the fruit is eaten by swine.

Gen. Char. Calyx urceolate, 5-toothed: corolla funnel-shaped, limb plicate, 5-lobed, lobes unequal: stamens inserted at the bottom of the tube of the corolla, included or exserted, declinate, anthers dehiscing lengthwise: ovary 2-celled, many-ovuled: style simple: stigma capitate: capsules hidden by the persistent calyx, narrowed from a ventricose base, membranaceous, 2-celled, cut round at the top, lid 1-2-celled: seeds many, kidney-shaped or orbicular.

(1) H. niger. *(Linn.)*

Ident. Linn. Spec. 257.—Dec. prod. XIII. p. 1. p. 546.

Spec. Char. Stem viscous, branched: leaves oblong, sinuately toothed or sinuate-pinnatifid, viscously pubescent, lower ones petioled, the rest half-stem-clasping, subdecurrent: flowers subsessile, erect, arranged in simple, unilateral, recurved, leafy terminal spikes: corollas reticulately veined.

Var. Agrestis, (Nees.)—Stem simple, few-flowered: root annual: flowers largish, minutely reticulated with purple veins on a pale rose-coloured and yellowish ground, marked with a dark purple throat.—*H. agrestis*, Kit.—Sweet. fl. Gard. I. t. 27.—*H. niger*, Wall.—Roxb. flor. Ind. Ed. Car. II. p. 237.—Bot. Mag. 50. t. 2394.

Futteyghur, Delhi. Sasseram. Flowering in February and March.

ORDER CXVIII. SCROPHULARIACEÆ.

Flowers hermaphrodite, usually irregular: calyx free, persistent, 5–4 merous: corolla monopetalous, hypogynous, pentamerous, or by union of the upper lobes, tetramerous, rarely 6–7 merous, or by the union of the lobes, 2-lipped: æstivation bilabiate or irregularly imbricate, the upper lobes either exterior or within, very rarely, in some of the diandrous or didynamous genera, plicate: stamens inserted on the corolla, alternate with its lobes, the upper one very often, and sometimes also the two anticous or posticous, sterile, or deficient, the remaining ones usually in equal pairs: anthers 2-celled, either confluent or dimidiately 1-celled: cells dehiscing longitudinally: ovary free, 2-celled: ovules numerous in each cell, inserted near the axis of the partition, anatropous or amphitropous: style simple or shortly bifid, the stigmatic portion slender or thickened, entire, or 2-lobed: fruit capsular with various dehiscences or rarely baccate: placentæ 4, either separating during dehiscence or variously united between themselves, or with the margins of the valves, or with the central column: seeds albuminous with the embryo straight, or rarely curved, sometimes indefinite with the radicle directed to the hilum, sometimes few or definite with the hilum more or less lateral, and the radicle directed to the apex of the fruit.—Herbs, undershrubs or shrubs: lower leaves opposite or whorled, the upper ones alternate, sometimes all opposite or all alternate, venation and clothing very variable: stipules usually none: flowers axillary or racemose, rarely spiked: peduncles opposite or alternate, sometimes simple, 1-flowered, sometimes many-flowered, in dichotomous cymes: bracts 2, opposite at the ramifications of the cymes, solitary under the pedicels, no bracteoles on the pedicels, or 1–2 alternate, or 2 opposite under the calyx: two opposite on a 1-flowered peduncle indicates a cyme reduced to a single flower.

GENUS I. VERBASCUM.

Pentandria Monogynia. *Ses. Syst.*

Deriv. Supposed to be from *Barbascum*, bearded ; in allusion to the filaments.

GEN. CHAR. Calyx deeply 5-cleft or parted, rarely 5-toothed : corolla flat-rotate, rarely concave, segments slightly unequal ; filaments 3, posterior ones or all woolly-bearded, rarely naked : style compressed and dilated at the apex, thickish : capsule globose, ovoid or oblong, dehiscent.

(1) V. VIRGATUM. (*With.*)

Ident. With. arrang. p. 250.—Dec. prod. X. p. 229.

Syn. V. blattarioides, *Lam.*—V. viscidulum, *Pers.*—V. glabrum, *Willd.*—Celsia viscosa, *Nees. in Linn. trans.*

Engrav. Wight's Icon. t. 1404-5.

SPEC. CHAR. Stem somewhat viscous, hispidulous or glabrous at the base: leaves alternate, oblong, glabrous or glandulosely hispid beneath, Inferior ones petioled, dentate or sinuate-pinnatifid, the upper ones sessile or cordately stem-clasping ; pedicels 2-3 together, rarely solitary, shorter or about the length of the calyx : filaments clothed with violet-coloured woolly hairs : flowers yellow, nearly sessile.

Common on the Neilgherries, flowering in the cold season.

GENUS II. CELSIA.

Didynamia Angiospermia. *Ser. Syst.*

Deriv. In honor of Olans Celsius, D. D., Professor of Oriental Languages in the University of Upsal.

GEN. CHAR. Calyx 5-partite : corolla flat-rotate; segments slightly unequal : stamens 4 : style compressed and dilated at the apex : capsule globose, ovoid or oblong, dehiscent.

(1) C. COROMANDELIANA. (*Vahl.*)

Ident. Vahl. Symb. III. p. 79.—Dec. prod. X. p. 246.—Roxb. flor. Ind. III. p. 100.

Syn. C. viscosa, *Roth.*

Engrav. Hook. Journ. Bot. I. t. 129.

SPEC. CHAR. Herbaceous, pubescent, viscid : radical leaves lyrate : upper ones oblong, ovate or orbiculate, toothed : racemes subpanicled: peduncles longer than the calyx: calycine segments ovate or oblong, entire or serrated: flowers largish, yellow: filaments bearded with purple hairs.

Waste places in the Deccan. Banks of rivers and still waters in the Peninsula. Flowering in the cold season.

GENUS III. LINARIA.

Didynamia Angiospermia. *Sex: Syst:*

Deriv. From *Linum*, flax, on account of the similarity in the leaves.

GEN. CHAR. Calyx deeply 5-parted: corolla personate; tube spurred at the base; upper lip erect; palate prominent, sometimes closing the throat, sometimes depressed, with the throat pervious: capsules ovoid or globose: cells often nearly equal, dehiscing by a single or many-valved pore: seeds sometimes ovoid, wingless, angular or wrinkled, sometimes discoid and surrounded by a membranaceous wing.—Herbs, rarely undershrubs; lower leaves opposite or verticilled; upper ones alternate.

(1) L. RAMOSISSIMA. (*Wall.*)

Ident. Wall. Cat. No. 3911.—Dec. prod. X. p. 270.
Engrav. Wall. Pl. As. Rar. II. t. 153.

SPEC. CHAR. Glabrous: branches elongated, very slender: leaves alternate, petioled, smooth, triangular-hastate, lower ones sometimes 5–7-lobed: pedicels longer than the petiole: flowers yellow, with a short curved spur: capsule glabrous.

Patna. Delhi. Deccan. Flowering in the rainy season.

(2) L. INCANA. (*Wall.*)

Ident. Wall. Pl. As. Rar. II. p. 43,—Dec. prod. X. p. 270.

SPEC. CHAR. Villous: stems prostrate, woody at the base: leaves ovate or cordate-orbiculate, rarely hastate, lobed or toothed: pedicels filiform, rigid, longer than the petiole: calycine segments lanceolate, very acute, membranaceous at the margin: spur of the corolla much shorter than the tube.

Growing on walls in the Deccan.

GENUS IV. SUTERA.

Didynamia Angiospermia. *Ser.; Syst.*

GEN. CHAR. Calyx 5-parted, segments linear, not membranaceous: tube of the deciduous corolla nearly equal, limb small, scarcely spreading, segments entire, two upper ones a little less: fertile stamens four, included: style shortly 2-lobed at the apex: capsule septicidally 2-valved, valves shortly 2-cleft.

(1) S. GLANDULOSA. (*Roth.*)

Ident. Dec. prod. X. p. 362.
Syn. S. dissecta, *Walp. Rep.* III. p. 271.
Engrav. Wight's Icon. t. 856.
SPEC. CHAR. Diffuse, much branched, viscous with short glutinous hairs: lowest leaves opposite, upper ones alternate, outline ovate, petioled, cut and pinnatifid, or dissected, segments cut or toothed, floral leaves decrescent: pedicels axillary: flowers sometimes subsessile, upper ones racemose: corolla slender, a half shorter than the calyx: capsule viscidly puberulous or glabrate.

On dark soil in Central India.

GENUS V. MAZUS.

Didynamia Angiospermia. *Ser.: Syst.*

Derio. From *Mazus*, a teat; in allusion to the mouth of the corolla being closed by tubercles.

GEN. CHAR. Calyx broad campanulate, 5-cleft: upper lip of the corolla erect, ovate, shortly 2-cleft, lower much longer, spreading, 3-cleft, bigibbous at the throat: stamens four: cells of the anthers contiguous, divaricate: style bilamellate at the apex, segments ovate, equal: capsule globose, compressed, obtuse, loculicidally 2-valved, valves entire: placenta thick, somewhat fleshy.

(1) M. RUGOSUS. (*Lour.*)

Ident. Lour. flor. Coch. p. 385.—Dec. prod. X. p. 375.
Syn. Lindernia Japonica, *Thunb.*—Columnea tomentosa, *Roxb. flor. Ind.* III. p. 98. 1—Stemodia tomentosa, *Don's Gard. Dict.*—Hornemannia bicolor, *Willd.*
Engrav. Sweet. Brit. fl. Gard. t. 36.

358

SPEC. CHAR. Annual: leaves obovate or oblong-cuneate, coarsely toothed, glabrous: calycine segments ovate-lanceolate, acute; corolla twice as long as the calyx, pale blue with white and yellow throat.

Common in the Peninsula. Silhet. Saharunpore. Flowering in the rainy season.

GENUS VI. LINDENBERGIA.

Didynamia Angiospermia. *Bar: Syst:*

Deriv. In honor of J. D. Lindenberg, author of a Synopsis of European Hepaticæ.

GEN. CHAR. Calyx campanulate, half 5-cleft: upper lip of the corolla erect, emarginate, lower one longer, 3-lobed, palate prominent, 2-plicate: stamens fertile, four, anther-cells disjoined, stalked, polliniferous; style subclavate at the apex, capitately stigmatose: capsules oblong or ovoid, 2-furrowed, loculicidally 2-valved, valves entire.—Decumbent or Scrumosely scandent herbs, rarely suffrutescent; flowers subsessile, axillary or arranged in terminal spikes, bracts leafy.

(1) L. URTICÆFOLIA. (*Lehm.*)

Ident. Dec. prod. X. p. 377.—Dals. Bomb. flor. p. 176.

Syn. Stemodia ruderalis, *Vahl.*—*Roxb. fl. Ind.* III. p. 94.—S. muraria, *Roxb. in Don. prod. fl. Nep.*

SPEC. CHAR. Annual, erect or ascending: leaves ovate, serrated: flowers axillary, yellow: corolla three times the length of the calyx.

Bombay Fort, and similar places in the Peninsula. Flowering all the year.

GENUS VII. PTEROSTIGMA.

Didynamia Angiospermia. *Bar: Syst:*

Deriv. From *Pteron*, a wing, and *Stigma*; in reference to its being girded by a wing.

GEN. CHAR. Calyx 5-parted, posterior segment larger: upper lip of the corolla emarginate, lower one 3-lobed, throat not plicate: antheriferous stamens four, cells disjoined, stalked: style dilated at the apex, entire or shortly bilamellate, extremity broad substigmatose and more or less 2-winged below the stigma: capsules ovate, subrostrate, 2-furrowed, loculicidally dehiscing, valves

2-cleft : seeds small, striated.—Villous herbs, usually aromatic : leaves opposite, wrinkled, crenulated : flowers axillary or arranged in leafy terminal spikes : pedicels short, solitary, often bracteolate at the apex.

(1) P. OVATUM. *(Dec.)*

Ident. Dec. prod. X. p. 280.

Syn. Herpestis ovata, *Wall. in Benth. Scroph. Ind. p. 30.*

SPEC. CHAR. Pubescent or afterwards glabrate : leaves short-petioled, ovate, crenated : flowers axillary, subsessile.

Peninsula.

(2) P. VILLOSUM. *(Benth.)*

Ident. Benth. Scroph. Ind. p. 21.—Dec. prod. X. p. 380.

Syn. Adenosma villosum, *Wall. Cat. No. 3851.*—Stemodia camphorata, *Vahl.*

SPEC. CHAR. Leaves ovate, acuminated or ovate-lanceolate, crenated, wrinkled, villous : lower flowers axillary, upper ones racemose.

Pundua. Silhet.

(3) P. CAPITATUM. *(Benth.)*

Ident. Benth. Scroph. Ind. p. 21.—Dec. prod. X. p. 380.

Syn. Stemodia capitata, *Benth. in Bot. Reg. No. 1470.*—P. spicatum, *Benth. Scroph. l. c.*—Erinus labiatus, *Roxb. flor. Ind. III. p. 92.*

SPEC. CHAR. Erect, branched : leaves ovate-oblong or lanceolate, serrated : flowers congested into an oblong or globose spike : corolla exceeding the calyx.

Peninsula. Assam.

GENUS VIII. STEMODIA.

Didynamia Angiospermia. *Sex. Syst.*

Deriv. From *Stemon*, a stamen, and *Dis*, double ; in allusion to each stamen bearing two anthers.

GEN. CHAR. Calyx 5-parted, segments narrow, nearly equal or the posterior one slightly larger : upper lip of the corolla emarginate or rarely 2-cleft, lower 3-lobed, throat not plicate : stamens four, included, anther-cells disjoined, stalked, all pollen-bearing: style dilated at the apex, often 2-lobed, at last shortly deflexed, without wings: capsule globose, ovate or oblong, septicidally dehiscing or parting, valves at last 2-cleft, margin of the carpels bent inwards: seeds small, numerous.

(1) S. viscosa. (Roxb.)

Ident. Roxb. flor. Ind. III. p. 04.—Dec. prod. X. p. 381.
Engrav. Roxb. Cor. II. t. 163.—Wight's Icon. t. 1408.

Spec. Char. Herbaceous, erect, pubescent, viscous: leaves sessile, ovate-oblong or lanceolate, acute, narrowed towards the base, at the base dilated and cordate, stem-clasping: flowers axillary, solitary, uppermost ones racemose; pedicels longer than the calyx: calycine segments lanceolate, acute: flowers small, violet.

Coromandel. Bombay. Flowering in the cold season. The plant has a pleasant aromatic smell.

GENUS IX. LIMNOPHILA.

Didynamia Angiospermia. *Sar: Syst:*

Deriv. From *Limne*, a pool, and *Phileo*, to love; alluding to the habitat of the species.

Gen. Char. Calyx deeply 5-cleft or 5-parted, nearly equal or with the posterior segment larger: upper lip of the corolla emarginate or 2-lobed, lower 3-lobed, throat not plicate: stamens four, included, cells of the anthers disjoined, oblong, very often stipitate: style deflexed at the apex, spathulate-dilated, entire or shortly bilamellate, often auriculately 2-winged at the joint: capsule ovate, globose or compressed, loculicidally 2-valved, valves at last 2-parted, margins of the carpels slightly inflexed, exposing a broad placentiferous dissepiment.—Herbs, growing in swampy or watery places, often dotted with pellucid glands, leaves opposite, or 3-4 times verticillate, lowest ones in the aquatic species submersed in the water, often cut and divided like threads: flowers solitary, axillary, or upper ones arranged in a foliated raceme, often bibracteolate in the calyx itself.

(1) L. menthastrum. (*Benth.*)

Ident. Dec. prod. X. p. 386.—Dals. Bomb. flor. p. 177.
Syn. Stemodia menthastrum, *Benth.*—Herpestis rugosa, *Roth.*

Spec. Char. Erect, smoothish: leaves subsessile, ovate-oblong, narrowed at both ends: floral leaves of the same shape, longer than the calyx: calyx deeply 5-cleft: segments lanceolate, subulate: flowers axillary, sessile, clustered in globose heads.

Tulkat ghaut. Silhet. Flowering in September.

(2) L. CONFERTA. (*Benth.*)

Ident. Dec. prod. X. p. 387.—Benth. Scroph. Ind. p. 23.

Syn. Stemodia sessilis, *Benth.*

SPEC. CHAR. Procumbent, glabrous: leaves sessile, oblong, rather obtuse, serrato-crenated, narrow at the base, floral ones similar: flowers axillary, subsessile, solitary or aggregated on little branches: calyx glabrous or ciliated: segments lanceolate-subulate: flowers blue, scarcely longer than the calyx.

Goalparah. Malwan. Flowering in September.

(3) L. MICRANTHA. (*Benth.*)

Ident. Dec. prod. X. p. 387.

Syn. Stemodia micrantha, *Benth.* Scroph. Ind. p. 23.

SPEC. CHAR. Very small, diffuse, much branched, glabrous: leaves sessile, oblong-lanceolate, acute, narrow at the base, floral ones of the same shape: flowers sessile, solitary, axillary: calycine segments lanceolate-subulate, recurved, spreading: corolla twice as long as the calyx: flowers numerous.

Silhet.

(4) L. REPENS. (*Benth.*)

Ident. Dec. prod. X. p. 387.

Syn. Stemodia repens, *Benth.* Scroph. Ind. p. 21.

SPEC. CHAR. Glabrous, somewhat clammy, creeping: leaves sessile, oblong or lanceolate, narrowed at the base, acute, serrated: floral ones of the same shape: flowers axillary, solitary or glomerate: pedicels shorter than the calyx: calycine segments lanceolate, subulate at the apex, recurved, spreading: corolla twice as long as the calyx.

Peninsula.

(5) L. HIRSUTA. (*Benth.*)

Ident. Dec. prod. X. p. 388.

Syn. Stemodia hirsuta, *Benth.* Scroph. Ind. p. 24.

SPEC. CHAR. Stem ascendent, hirsute: leaves subsessile, narrow at the base, uppermost ones scarcely stem-clasping, glabrous or hirsute: floral ones shorter than the flower: raceme terminal, with the flowers opposite: calycine segments lanceolate-subulate.

Peninsula. Silhet. Assam.

(6) L. hyssopifolia. *(Roth.)*

Ident. Roth. Nov. Sp. p. 297.—Benth. in Dec. prod. X. p. 388.

Spec. Char. Glabrous, somewhat erect; leaves linear-lanceolate, remotely denticulate, upper ones half stem-clasping: floral ones of the same shape: flowers axillary: fructiferous pedicels longer than the leaf: calycine segments short, lanceolate.

Peninsula.

(7) L. punctata. *(Blume.)*
Var. subracemosa.

Ident. Dec. prod. X. p. 388.
Syn. L. gratissima, *Blume.*

Spec. Char. Glabrous, somewhat erect: leaves oblong-lanceolate, serrated, round or narrow at the base, stem-clasping: floral ones small: racemes axillary and terminal: flowers axillary.

Peninsula.

(8) L. Roxburghii. *(G. Don.)*

Ident. Don. Gen. Syst. IV. p. 543.—Dec. prod. X. p. 388.
Syn. Capraria gratissima, *Roxb. flor. Ind.* III. p. 92.

Spec. Char. Stem glabrous, thickish: leaves emersed, upper ones opposite, lanceolate, stem-clasping, 3-nerved: pedicels in the upper axils shorter than the calyx: uppermost ones somewhat racemose: flowers purple.

Peninsula. Sircar mountains. Deccan. Flowering nearly all the year.

(9) L. gratioloides. *(R. Br.)*

Ident. R. Br. prod. p. 442.—Dec. prod. X. p. 389.

Syn. Gratiola virginiana, *Linn.*—G. trifida, *Willd.*—Hottonia Indica, *Linn.*—Hydropityon pedunculatum, *Ser. in Dec. prod.* I. p. 422.—Columnea balsamea, *Roxb. fl. Ind.* III. p. 97.—L. trifida, *Spreng.*

Engrav. Rheede Mal. IX. t. 85.—Burm. Zeyl. t. 55, f. 1.

Spec. Char. Stem glabrous, thin: leaves emersed, opposite, lowest ones 3-parted, cut or many-cleft like threads, upper ones entire, serrulate: pedicels in the upper axils much longer than the calyx: flowers greenish-white, streaked with purple.

Coromandel. Bombay. Common on the margins of tanks. Flowering nearly all the year.

(10) L. SESSILIFLORA. (*Blume.*)

Ident. Dec. prod. X. p. 389.

Spec. Char. Stem pubescent or hirsute: all the leaves 3-partite: segments of the lower ones cut into many narrow divisions, of the upper ones cut or toothed: flowers sessile, axillary.

Goalparah. Silhet.

(11) L. HETEROPHYLLA. (*Benth.*)

Ident. Benth. Scroph. Ind. p. 25.—Dec. prod. X. p. 300.
Syn. Columnea heterophylla, *Roxb. flor. Ind.* III. p. 97.

Spec. Char. Glabrous: submerged leaves, partite, capillary, many-cleft, emerged ones quite entire or serrated, oblong: flowers subsessile, axillary: calycine segments very acute: capsule subglobose.

Bengal. Hindostan. Flowering in February and March.

(12) L. RACEMOSA. (*Benth.*)

Ident. Benth. Scroph. Ind. p. 26.—Dec. prod. X. p. 300.
Syn. Cyrilla aquatica, *Roxb. flor. Ind.* III. p. 115.
Engrav. Roxb. Cor. II. t. 189.—Wight's Icon. t. 861.

Spec. Char. Stem pubescent: submerged leaves partite, capillaceously cut, emerged ones broad-lanceolate, crenulated, stem-clasping, 5-7-nerved: floral ones bract-shaped: raceme terminal, dense, many-flowered: flowers bluish.

Concans. Coromandel. Bengal. Flowering nearly all the year.

(13) L. POLYSTACHYA. (*Benth.*)

Ident. Benth. Scroph. Ind. p. 26.—Dec. prod. X. p. 300.
Syn. Stemodia aquatica, *Willd.*—Conobea Indica, *Spreng.*
Engrav. Wight's Icon. t. 860.

Spec. Char. Glabrous or minutely puberulous: submerged leaves partite, cut in many threads, emerged ones lanceolate, acute, serrated, stem-clasping, 3-5-nerved, floral ones bract-shaped: racemes subpanicled: flowers sessile.

Peninsula.

(14) L. HYPERICIFOLIA. (*Benth.*)

Ident. Dec. prod. X. p. 380.

Syn. Stemodia hypericifolia, *Benth. Scroph. Ind. p. 23.*—Herpestis connata, *Spreng.*—Cybianthera cuneata, *Ham. in Don. prod. flor. Nep. p. 86.*

Engrav. Wight's Icon. t. 1409.

Spec. Char. Glabrous, rooting at the base, ascending: leaves sessile, ovate-oblong, obtuse, cordately half-stem-clasping at the base, floral ones smaller: racemes terminal or axillary: flowers sessile; calycine segments lanceolate.

Kotagherry, Neilgherries, in swampy ground, flowering in August.

GENUS X. HERPESTIS.

Didynamia Angiospermia. *Sex: Syst:*

Deriv. From *Herpestis*, any thing that creeps ; in allusion to the habit of the plants.

Gen. Char. Calyx 5-parted, posterior segment broader, often the largest, lateral ones often very narrow : upper lip of the corolla emarginate or 2-lobed, lower 3-lobed, lobes arranged in two lips or all nearly equal : anthers 2-celled, cells distinct, contiguous, parallel or divaricated ; style concave-dilated at the apex or shortly 2-lobed : capsule 2-furrowed, dehiscing by two bipartite or entire valves, margins of the carpels inflexed, showing the entire placentiferous column: seeds numerous, small.—Herbs, often found in swamps: leaves opposite, entire, toothed, or capillary : peduncles axillary, or oppositely subracemose at the tops of the branchlets, 1-flowered, without bracts or furnished with two minute bristly bracts under the calyx.

(1) H. HAMILTONIANA. (*Benth.*)

Ident. Benth. Scroph. Ind. p. 30.—Dec. prod. X. p. 400.

Spec. Char. Erect or decumbent at the base ; leaves lanceolate, quite entire, narrow at the base : flowers subsessile, opposite, solitary, bibracteate; posterior segments of the calyx very broad, cordate.

Malwan. Assam.

(2) H. FLORIBUNDA. (R. Br.)

Ident. R. Br. prod. p. 442.—Dec. prod. X. p. 400.
Syn. Lindernia sessamoides, *Spreng.*—H. linearis, *do.*—H. lanceolata, *Wight. in Benth. Scroph. Ind. p. 30.*

SPEC. CHAR. Erect: leaves lanceolate-linear, quite entire, narrow at the base, 1-nerved: pedicels 1–3, shorter than the calyx, somewhat bibracteolate at the apex: posterior segments of the calyx ovate, equalling the corolla.

Peninsula.

(3) H. MONNIERIA. (*H. B. & Kth.*)

Ident. H. B. et Kunth. Nov. Gen. et Sp. II. p. 366.—Dec. prod. X. p. 400.
Syn. Gratiola monniera, *Linn.*—Limosella calycina, *Forsk.*—Septas repens, *Lour.*—Bramia Indica, *Lam.*—H. procumbens, *Spreng.*—H. spathulata, *Blume.*
Engrav. Bot. Mag. t. 2557.—Roxb. Cor. II. t. 178.—Rheede Mal. X. t. 14.

SPEC. CHAR. Creeping: leaves obovate-cuneiform, quite entire or with a few crenatures, nerveless or obscurely 1–3-nerved: pedicels bibracteolate at the apex: posterior segments of the calyx ovate: flowers smallish, pale blue.

Margins of tanks all over India, flowering nearly all the year.

GENUS XI. DOPATRIUM.

Didynamia Angiospermia. *Ser: Syst.*

GEN. CHAR. Calyx deeply 5-cleft: throat of the corolla dilated, upper lip short, 2-cleft, lower large, broadly 3-lobed: two posterior stamens fertile, included, cells of the anthers parallel, distinct, equal, two anterior stamens small, sterile: style bilamellate at the apex: capsule 4-valved, margins of the carpels not inflexed: seeds tubercled.—Swampy herbs: leaves opposite, approximated at the base of the stem, obovate or oblong, quite entire, upper ones remote, minute: pedicels filiform, opposite (or alternate by abortion), 1-flowered, ebracteate.

(1) D. NUDICAULE. (*Benth.*)

Ident. Benth. Scroph. Ind. p. 31.—Dec. prod. X. p. 407.
Syn. Gratiola nudicaulis, *Willd.*—G. aphylla, *Roth.*—G. concifera, *Roxb. flor. Ind. Ed. Car.* I. p. 142.

Spec. Char. Low, branched: floral leaves minute, very acute: capsule oblong: radical leaves subrosulate, obovate: pedicels divaricate: flowers small, blue.
Peninsula. Flowering in the rainy season.

(2) D. LOBELIOIDES. (*Benth.*)

Ident. Benth. Scroph. Ind. p. 31.—Dec. prod. X. p. 407.
Syn. Gratiola lobelioides, *Retz.*
Engrav. Wight's Icon. t. 859.

Spec. Char. Stem elongated, slightly branched: floral leaves minute, obtuse, radical ones oblong: calycine segments obtuse: capsule globose: corolla much longer than the calyx.
Peninsula.

(3) D. JUNCEUM. (*Ham.*)

Ident. Ham. in Benth. Scroph. Ind. p. 31.—Dec. prod. X. p. 407.—Roxb. flor. Ind. I. p. 142.
Syn. Gratiola juncea, *Roxb.*—Morgania juncea, *Spreng.*
Engrav. Roxb. Cor. II. t. 129.

Spec. Char. Stem elongated, scarcely branched: floral leaves minute, obtuse; capsule globose: corolla scarcely three times longer than the calyx: flowers small, rosé-coloured.
Coromandel. Saharunpore. Western Coast. Flowering in the rainy season.

GENUS XII. ARTANEMA.

Didynamia. Gymnospermia. *Bar : Syst*

Deriv. From *Artao*, to support, and *Nema*, a filament; a tooth is borne on one side of each of the longer filaments.

Gen. Char. Calyx 5-parted, segments somewhat leafy, much imbricated, nearly equal: upper lip of the funnel-shaped corolla broad, emarginate, lower scarcely longer, 3-parted: stamens four, fertile: filaments decurrent into the tube, increased by a small scale at the base, anterior one furnished at the throat with a short obtuse appendage: capsules largish, subglobose.

(1) A. SESAMOIDES. (*Benth.*)

Ident. Benth. Scroph. Ind. p. 39.—Dec. prod. X. p. 408.
Syn. Columnea longifolia, *Linn.*—Achimenes sesamoides, *Vahl.*—Diceros longifolius, *Pers.*
Engrav. Wight's Icon. t. 1410.

SPEC. CHAR. Herbaceous, erect; stem acutely 4-sided; leaves opposite, oblong or ovate-lanceolate, entire or serrated; pedicels shorter than the calyx; corolla subcampanulate; flowers in terminal racemes, blue.

Concans. Travancore. Peninsula. Flowering nearly all the year.

GENUS XIII. CURANGA.

Didymamia Angiospermia. *Sex. Syst.*

GEN. CHAR. Calyx flat, 4-parted, posterior segment entire and with the anterior one entire or 2-cleft, lateral ones narrow, covered; corolla ringent, upper lip forked, emarginate, lower spreading, 3-cleft; posterior stamens fertile, anthers coherent, cells diverging, distinct, anterior stamens sterile, clavate; style bilamellate at the apex; capsule ovate, enclosed by the calyx.

(1) C. AMARA. (*Juss.*)

Ideal. Juss. Ann. Par. 9. p. 319.—Dec. prod. X. p. 408.

Syn. Gratiola amara, *Roxb. flor. Ind.* I. p. 136.—Herpestis amara, *Benth.*

Engrav. Rumph. Amb. V. t. 170. fig. 11.

SPEC. CHAR. Leaves petioled, ovate, crenated; racemes short; pedicels without bracts; flowers small, white, or purple-dotted.

Assam, flowering in the cold season.

GENUS XIV. TORENIA.

Didymamia Angiospermia. *Sex. Syst.*

Deriv. In honor of Olof Toren, a Swedish Botanist.

GEN. CHAR. Calyx tubular, plicate or winged, obliquely 5-toothed at the apex or 2-lipped; corolla ringent, upper lip emarginate or 2-cleft, lower 3-cleft, larger; posterior stamens fertile, anterior ones arched, antheriferous, increased at the base by a tooth-shaped or filiform appendage; anthers approximate in pairs or coherent; style slightly bilamellate at the apex; capsule oblong, not exceeding the calyx.—Herbs with opposite leaves; racemes short, few-flowered, fascicle-shaped or rarely elongated, terminal or axillary from a false branch or placed in the fork of the branches.

(1) T. CORDIFOLIA. (*Roxb.*)

Ident. Dec. prod. X. p. 409.—Roxb. flor. Ind. III. p. 95.

Engrav. Rheede Mal. IX. t. 66.—Bot. Mag. t. 3715.—Roxb. Cor. II. t. 161.

SPEC. CHAR. Glabrous or roughish with few hairs: leaves petioled, ovate, serrate-crenated: calyx broadly 3-winged, ovate, rounded at the base: anterior filaments minutely toothed: flowers bluish-purple.

Circars. South Concan. Courtallum. Mussooree. Flowering in the rainy season.

(2) T. EDENTULA. (*Griff.*)

Ident. Dec. prod. X. p. 140.—Griff. in Gen. Scroph. p. 4.
Syn. T. Asiatica, (*ex parte*) Benth. Scroph. Ind. p. 38.

SPEC. CHAR. Decumbent or slightly erect, glabrous, or hirsute with soft hairs: leaves petioled, ovate, serrate-crenated: wings of the calyx three, broadish, acute at the base, decurrent: corolla scarcely longer than the calyx: anterior filaments without teeth.

Nathpur, Assam.

(3) T. DIFFUSA. (*Don.*)

Ident. Don. prod. flor. Nep. p. 86.—Dec. prod. X. p. 410.

SPEC. CHAR. Diffuse: leaves petioled, ovate, serrate-crenated, rounded at the base: calyx oblong, 3-winged, wings acutely decurrent: corolla a half longer than the calyx: appendage of the anterior filaments subulate.

Goalpara.

(4) T. ASIATICA. (*Linn.*)

Ident. Linn. Spec. p. 862.—Dec. prod. X. p. 410.
Syn. T. vagans et T. hians, Roxb. fl. Ind. III. p. 96.
Engrav. Wight's Icon. t. 862.—Lam. III. t. 523. fig. 1.

SPEC. CHAR. Diffuse, glabrous or slightly hairy: leaves petioled, ovate or ovate-lanceolate, serrate-crenated: calyx elongated, acute at the base, ribs five, nearly equal or 3 narrow-winged: corolla more than twice as long as the calyx: appendage of the anterior filaments subulate: flowers violet.

Peninsula. Silhet. Concana. Chittagong. Flowering in the rainy season.

(5) T. HIRSUTA. (*Lam.*)

Ident. Dec. prod. X. p. 410.

Syn. T. cordifolia, Benth. in Wall. Cat. (*not Roxb.*)

SPEC. CHAR.. Hirsute: diffuse: leaves petioled, ovate, serrate-crenate, subcordate at the base: calyx elongated, 5-ribbed, wingless, obtuse at the base: corolla scarcely twice as long as the calyx: appendage of the anterior filaments subulate.

Neilgherries.

(6) T. PARVIFLORA. (*Ham.*)

Ident. Dec. prod. X. p. 410.—Benth. Scroph. Ind. p. 39.

SPEC. CHAR. Diffuse or suberect, glabrous: calyx in flower linear, slightly 5-ribbed, acute at the base: corolla scarcely exceeding the calyx: appendage of the filaments short, subulate.

Peninsula.

(7) T. FLAVA. (*Ham.*)

Ident. Dec. prod. X. p. 411.

SPEC. CHAR. Suberect, hirsute: leaves ovate or oblong, serrate-crenated, cuneate at the base: calyx incurved, 5-ribbed, wingless, acute at the base: corolla scarcely exceeding the calyx: appendage of the filaments tooth-shaped.

Goalpara. Assam.

(8) T. BICOLOR. (*Dalz.*)

Ident. Dalz. Bomb. flor. p. 181.—Hook. Journ. Bot. III, p. 38.

SPEC. CHAR. Stem creeping and rooting; leaves petioled, triangular, scarcely cordate at the base, crenate-serrated: calyx linear, incurved, equally 5-ribbed: corolla curved: flowers axillary, in twos or threes; under lip of the corolla white, the rest of a deep violet.

Vingorla.

GENUS XV. VANDELLIA.

Didynamia Angiospermia. *Sex. Syst.*

Deriv. In honor of Dominico Vandelli, Professor of Botany at Lisbon.

GEN. CHAR. Segments of the calyx nearly equal, scarcely imbricated in æstivation, almost free from the base or more or less combined into a 5-toothed not plicate calyx: upper lip of the corolla erect, shortly 2-cleft, lower larger, spreading, 3-cleft: stamens four, fertile, filaments of the anterior ones increased by a tooth-shaped and filiform appendage at the base, arched, anthers cohering between themselves under the upper lip; style dilated at the apex, often bilamellate: capsule globose, oblong or linear. —Herbs with opposite toothed leaves: flowers opposite or solitary by abortion, axillary or racemose at the end of the branches, racemes often contracted into false umbels.

(1) V. CRUSTACEA. (*Benth.*)

Ident. Benth. Scroph. Ind. p. 35.—Dec. prod. X. p. 413.

Syn. Capraria crustacea, *Linn. Mant.*—Torenia crustacea, *Cham. et Schlect. in Linn.*—Gratiola lucida, *Vahl.*—*Rosb. flor. Ind.* I. p. 138.—Morgania lucida, *Spreng.*—T. flaccida, *R. Br.*—T. varians, *Roxb.*—V. varians, *G. Don.*—G. aspera, *Roth.*—M. aspera, *Spreng.*—Hornemannia ovata, *Link. et Otto.*—Tittmannia ovata, *Reichb. Ic. ex.* 1. p. 27.—V. alba, *Benth. l. c.*—Antirrhinum hexandrum, *Forsk.*

Engrav. Wight's Icon. t. 863.—Roxb. Cor. II. t. 202.—Rumph. Amb. V. t. 170. fig. 3.

SPEC. CHAR. Diffuse, glabrous or roughish with few hairs: leaves short-petioled, ovate: peduncles axillary or subracemose, rarely sub-fascicled, 2-3 times longer than the calyx: capsules ovate-oblong, shorter than the calyx: flowers small, indigo-coloured.

Coromandel. Bombay. Bengal. Flowering in the rainy season.

(2) V. MULTIFLORA. (*G. Don.*)

Ident. Don's. Gard. Dict. IV. p. 549.—Dec. prod. X. p. 414.

Syn. Tittmannia trichotoma, *Benth.*—Torenia multiflora, *Roxb. flor. Ind.* III. p. 90. ?

SPEC. CHAR. Glabrous: suberect, much-branched: leaves ovate, obscurely crenated, lower ones long-attenuated into the petiole, upper ones sessile or half-stem-clasping, floral ones minute: racemes elongated: pedicels almost twice as long as the calyx: segments of the calyx lanceolate-subulate: capsule sub-globose: flowers small, bluish-white.

Monghyr, flowering in the rainy season.

(3) V. HIRSUTA. *(Ham.)*

Ident. Ham. in Benth. Scroph. Ind. p, 36.—Dec. prod X. p. 414.

Syn. Tittmannia ovata, *Benth.*—Gratiola viscosa, *Hornem. in Hort. Hafn.*—Hornemannia viscosa, *Willd.*—T. viscosa, *Reichb. Icon. ex. L p. 26, t. 39.*

SPEC. CHAR. Hirsute, erect or much-branched: leaves ovate, cuneated, lower ones narrowed into the petiole, upper ones sessile or half-stem-clasping: floral ones minute: racemes elongated: pedicels 2-3 times longer than the calyx: segments of the calyx lanceolate-subulate: capsule subglobose.

Peninsula. Vingorla.

(4) V. SCABRA. *(Benth.)*

Ident. Benth. Ind. Scroph. p. 36.—Dec. prod. X. p. 414.

Syn. Torenia hirta, *Cham. & Schlect. in Linn.*—Gratiola pusilla, *Willd.*—Bonnaya Vahlei, *G. Don.*—Columnea minuta, *Rasb. for. Ind. III. p. 98.*—Stemodia minuta, *G. Don.*

SPEC. CHAR. Diffuse, scabrously hairy or nearly glabrous: leaves subsessile, broad-ovate, rounded or sub-cordate at the base: peduncles axillary or very shortly fasciculately racemose, many times longer than the calyx: segments of the calyx linear-lanceolate, shortly and stiffly hairy: capsule subglobose.

Peninsula.

(5) V. LAXA. *(Benth.)*

Ident. Benth. Scroph. Ind. p. 36.—Dec. prod. X. p. 414.

SPEC. CHAR. Diffuse, hirsute with few long hairs: leaves subsessile, broad-ovate, rounded or subcordate at the base: pedicels axillary and loosely fasciculately racemose, many times longer than the calyx: segments of the calyx subulate, very hirsute: capsule globose: flowers largish, white.

Peninsula. Vingorla.

(6) V. MOLLIS. *(Benth.)*

Ident. Benth. Scroph. Ind. p. 37.—Dec. prod. X. p. 414.

SPEC. CHAR. Hirsute: stem elongated, procumbent: leaves short-petioled, ovate-lanceolate, serrated, softly villous on both sides: pedicels axillary or shortly racemose, many times longer than the calyx: segments of the calyx subulate, very hirsute: capsule ovate-oblong.

Silhet. Lower Assam.

(7) V. ERECTA. (*Benth.*)

Ident. Benth. Scroph. Ind. p. 36.—Dec. prod. X. p. 415.

SPEC. CHAR. Erect: much branched or somewhat diffuse, glabrous: lowest leaves ovate, narrowed into a short petiole, floral ones sessile or half-stem-clasping, ovate, oblong or lanceolate; peduncles all axillary: calycine segments linear-lanceolate: capsules ovate-globose.

Common in the Peninsula.

(8) V. SESSILIFLORA. (*Benth.*)

Ident. Benth. Scroph. Ind. p. 37.—Dec. prod. X. p. 416.

Syn. V. minima, *Benth. l. c.*

SPEC. CHAR. Erect: glabrous or scarcely puberulous: leaves subsessile, broad ovate, serrated, sub-cordate at the base: flowers sessile, somewhat fascicled.

Khasia Hills.

(9) V. PEDUNCULATA. (*Benth.*)

Ident. Benth. Scroph. Ind. p. 37.—Dec. prod. X. p. 416.

Syn. Torenia diffusa, *Roxb.*—V. Roxburghii, *G. Don.*—Gratiola cordifolia, *Vahl.*—Bonnaya cordifolia, *Spreng.*—V. cordifolia, *G. Don.*

SPEC. CHAR. Glabrous: procumbent or diffuse, lax: leaves very shortly petioled, ovate or ovate-lanceolate, crenated, and quite entire, upper ones subcordate: pedicels axillary: calyx somewhat 5-parted: capsule oblong-linear.

Vingorla. Silhet. Assam.

(10) V. ANGUSTIFOLIA. (*Benth.*)

Ident. Benth. Scroph. Ind. p. 37.—Dec. prod. X. p. 417.

Syn. Lindernia macrantha, *Hon. prod. flor. Nep.*

SPEC. CHAR. Glabrous: diffuse: leaves oblong-linear, quite entire or few-toothed: pedicels axillary: capsule linear: calyx 5-parted.

Assam. Upper Provinces.

GENUS XVI. ILYSANTHES.

Didynamia Angiospermia. *Sex. Syst.*

GEN. CHAR. Calyx 5-parted, segments scarcely imbricated in æstivation: upper lip of the corolla short, erect, 2-cleft, lower one longer, spreading, 3-cleft: posterior stamens fertile, rudiments of the anterior ones 2-lobed, one lobe glandulose, obtuse, the other glabrous, thin, acute, sometimes very short, tooth-shaped, sometimes elongated, rarely bearing a 1-celled anther: capsule ovate or oblong, equalling or exceeding the calyx.—Annual glabrous herbs inhabiting swamps: flowers usually axillary: fructiferous peduncles often reflexed.

(1) I. HYSSOPIOIDES. (*Benth.*)

Ident. Dec. prod. X. p. 419.—Dalz. Bomb. flor. p. 179.

Syn. Gratiola hyssopioides, *Linn.*—Morgania hyssopioides, *Spreng.*—Bonnaya hyssopioides, *Benth. Scroph. Ind.* p. 34.

Engrav. Wight's Icon. t. 857.—Roxb. Cor. III. t. 203.

SPEC. CHAR. Stem diffuse or elongated, lax: leaves oblong or lanceolate, remote, narrowed at the base: upper ones small, linear: peduncles elongated, filiform: anterior filaments glandulosely hispid, increased under the apex by a minute glabrous lobelet.

Common in the rains. Peninsula. Western Coast.

(2) I. PARVIFLORA. (*Benth.*)

Ident. Dec. prod. X. p. 419.

Syn. Gratiola parviflora, *Roxb.*—Bonnaya parviflora, *Benth. Scroph. Ind.* p. 31.

Engrav. Roxb. Cor. III. t. 204.

SPEC. CHAR. Much branched, slender: leaves oblong or lanceolate, lower ones narrowed into the petiole, upper ones sessile or half-stem-clasping: upper flowers subracemose, corolla about twice as long as the calyx: anterior filament lightly glandulose, increased by a very minute glabrous lobelet below the apex.

Common in the Peninsula.

(3) I. ROTUNDIFOLIA. (*Benth.*)

Ident. Dec. prod. X. p. 420.

Syn. Gratiola rotundifolia, *Linn.*—Morgania rotundifolia, *Spreng.* —Bonnaya rotundifolia, *Benth. Scroph. Ind.* p. 34.—G. integrifolia, *Roxb.*

SPEC. CHAR. Diffuse: leaves sessile, broad-ovate or orbiculate, rounded or cordate at the base: peduncles axillary: anterior filaments glandulose, increased by a very minute glabrous lobelet under the apex: capsule subglobose.

Peninsula.

(4) I. MINIMA. *(Benth.)*

Ident. Dec. prod. X. p. 420.
Syn. Bonnaya minima, *R. W.* (not *G. Don.*)
Engrav. Wight's Icon. t. 856.

SPEC. CHAR. Very small, erect: leaves few, lower ones ovate, upper ones oblong; peduncles axillary, elongated: corolla three times longer than the calyx: anterior filaments glandulose, very obtuse, increased by an elongated arched glabrous lobelet under the apex.

Peninsula.

GENUS XVII. BONNAYA.

Diandria Monogynia. *Sm. Syst.*

Deriv. In memory of Bonnay, a German Botanist.

GEN. CHAR. Calyx 5-parted, segments nearly equal, scarcely imbricated in æstivation: upper lip of the corolla short, erect 2-cleft, lower larger, spreading, 3-cleft: posterior stamens fertile, rudiments of the anterior ones glandulose, linear, obtuse or clavate, entire: style usually bilamellate at the apex: capsule linear, longer than the calyx.—Glabrous or rarely hairy herbs, creeping or slightly erect: leaves opposite, quite entire or often toothed: flowers opposite or alternate by abortion, axillary or arranged in terminal racemes: anthers free.

(1) B. PEDUNCULARIS. *(Benth.)*

Ident. Benth. Scroph. Ind. p. 34.—Dec. prod. X. p. 421.

SPEC. CHAR. Stems slender, elongated; leaves remote, oblong-lanceolate, nearly quite entire; flowers axillary, solitary, long-peduncled; capsules linear, spreading or nodding, scarcely twice longer than the calyx.

Silhet. Probably a mere variety of *B. grandiflora*, (*Spreng.*)

(2) B. TENUIFOLIA. *(Spreng.)*

Ident. Spreng. Syst. 1, p. 42.—Dec. prod. X. p. 422.
Syn. Gratiola tenuifolia, *Vahl. En.* p. 96.

SPEC. CHAR. Low plant, erect or diffusely much branched: leaves linear, somewhat quite entire: lower pedicels leaf-opposed, fructiferous ones reflexed, uppermost ones subracemose: capsules linear: flowers small, blue.

Peninsula, flowering in the rainy season.

(3) B. BRACHIATA. (*Link. & Otto.*)

Ident. Link. & Otto. Abbild. II. p. 23. t. 9.—Dec. prod. X, p. 420.

Syn. Gratiola serrata, Roxb.—B. pusilla, Benth.

Spec. Char. Erect, much branched: leaves sessile, oblong or obovate, sharply serrated: flowers racemose, pink: capsules spreading, twice the length of the calyx.

Common in pastures during the rainy season.

(4) B. REPTANS. (*Spreng.*)

Ident. Spreng. Syst. I. p. 41.—Dec. prod. X. p. 420.

Syn. Gratiola reptans, Roxb.—G. ruelloides, Koen. Vahl. En. —Henckelia Roxburghiana, Link.—G. ciliata, Vahl.—B. ciliata, Spreng.

Spec. Char. Stem creeping and rooting: leaves rounded or obovate-oblong, sharply serrated, narrowing into the petiole: capsules spreading, 2 to 3 times longer than the calyx: flowers bluish-pink.

Southern Concan. Khasia Hills. Flowering in the rainy season.

(5) B. VERONICÆFOLIA. (*Spreng.*)

Ident. Spreng. Syst. I. p. 41.—Dec. prod. X. p. 421.

Syn. Gratiola veronicæfolia, Retz.—G. marginata, Vahl.—B. marginata, Spreng.—G. grandiflora, Retz.—G. racemosa, Roth.— B. Rothii, Link.—B. rigida et B. procumbens, Benth.

Engrav. Wight's Icon. t. 1411.—Roxb. Cor. II. t. 134.

Spec. Char. Stem decumbent or creeping at the base: flower-bearing branches ascending: leaves subsessile, narrow at the base, oblong, rather thick, sharply serrated, lower ones nearly entire: flowers racemose: capsules 2 to 3, longer than the calyx: flowers violet.

Coromandel. Mahableshwur. Bengal. Silhet. Flowering in the rainy season.

(6) B. VERBENÆFOLIA. (*Spreng.*)

Ident. Spreng. Syst. I. p. 42.—Dec. prod. X. p. 421.

Syn. Gratiola verbenæfolia, Vahl.—G. racemosa, Roxb.—G. Roxburghii, Link.—Tittmannia Colsmanni et Torenia gracilis, Benth.

Engrav. Wight's Icon. t. 1412.

SPEC. CHAR. Leaves subsessile, oblong-lanceolate, or nearly linear, entire or serrated: flowers racemose, blue: capsules 2 to 3 times longer than the calyx: leaves narrower and not so much serrated as in the preceding.

Southern Concan. Nilhet. Hindostan. Flowering in the rainy season.

(7) B. GRANDIFLORA. (*Spreng.*)

Ident. Spreng. Syst. 1. p. 41.—Dec. prod. X. p. 421.

Syn. Gratiola grandiflora, *Roxb.*—Henckelia grandiflora, *Link.* —G. pulegiifolia, *Vahl.*—B. pulegiifolia, *Spreng.*—B. Wightii, *Benth.*

Engrav. Roxb. Cor. II. t. 179.

SPEC. CHAR. Stem diffuse, rather creeping: ovate-oblong or lanceolate, sessile, serrated: flowers axillary, uppermost ones racemose: capsules linear, scarcely twice as long as the calyx: flowers light blue.

South Concan. Coromandel. Flowering in the rainy season.

(8) B. OPPOSITIFOLIA. (*Spreng.*)

Ident. Spreng. Syst. I. p. 41.—Dec. prod. X. p. 421.

Syn. Gratiola oppositifolia, *Roxb.*—Henckelia oppositifolia, *Link.*—G. minima, *Roth.*—B. minima, *G. Don.*

SPEC. CHAR. Erect: leaves oblong or lanceolate, slightly serrated: flowers axillary, uppermost ones racemose: capsules linear, scarcely twice as long as the calyx: flowers small, blue.

Concans. Coromandel. Flowering in the rainy season.

GENUS XVIII. PEPLIDIUM.

Diandria Monogynia. *Ses: Syst.*

Deriv. From *Peplis,* purslane; plants with the habit of Purslane.

GEN. CHAR. Calyx tubular, 5-sided, very shortly 5-toothed: limb of the corolla 5-cleft, lowest segments much longer: posterior stamens none, filaments of the anterior ones dilated at the base, somewhat appendiculate: anthers small, 2-celled, cells parallel: style spathulately dilated at the apex: capsule globose, thinly membranaceous, fragile.

(1) P. HUMIFUSUM. (*Delile.*)

Ident. Del. flor. Æg. p. 148.—Dec. prod. X. p. 422.

Syn. Hedyotis maritima, *Linn. A. Suppl.*—Oldenlandia maritima, *Roth.*—Pæderota cochlearifolia, *Koen.*—Microcarpæa cochlearifolia, *Smith.*

Engrav. Sm. in Bot. Misc. III. Suppl. t. 29.

SPEC. CHAR. Creeping, much branched, quite glabrous: leaves opposite, spathulate or obovate, thickish, fleshy, quite entire, nerveless, narrowed at the base into the petiole: flowers sessile in the axils or short pedicelled, solitary, opposite.

Peninsula, in moist watery places.

GENUS XLIX. GLOSSOSTIGMA.

Diandria Monogynia. *Sex: Syst:*

Deriv. From *Glossa*, a tongue, and *Stigma*; in allusion to the shape of the latter organ.

GEN. CHAR. Calyx campanulate, short, very obtusely 3-lobed, posterior lobe very broad, sometimes 2-3-toothed: corolla small, limb 5-cleft: stamens 2 or 4: cells of the anthers parallel, confluent at the apex: style spathulately dilated at the apex, shortly bilamellate: capsule subglobose, loculicidally 2-valved, valves septiferous in the middle, exposing an entire placentiferous column.

(1) G. SPATHULATUM. (*Arn.*)

Ident. Arn. in Nov. Act. Nat. Cur. V. 18. par. I. p. 355.—Dec. prod. X. p. 526.

Syn. Limosella diandra, *Linn. Mant.*—Microcarpæa spathulata, *Hook. Bot. Misc.* p. 101. & t. 4.—Pæderota minima, *Reis.*

SPEC. CHAR. Cæspitose, creeping: stems stoloniferous and rooting at the nodes: leaves fascicled, linear-spathulate, entire, very small: pedicels solitary in the axils of the leaves: calyx and corolla very small.

Margins of tanks and other moist places.

GENUS XX. MICROCARPÆA.

Diandria Monogynia. *Sex: Syst:*

Deriv. From *Micros*, small, and *Karpos*, fruit; in allusion to the capsules.

GEN. CHAR. Calyx tubular-campanulate, angled, equal, 5-cleft: corolla subcampanulate, limb 5-cleft, segments uppermost and lowest larger: stamens fertile, included, cells of the anthers divaricate, confluent, no vestige of sterile ones: style short, dilated at the apex, somewhat capitately stigmatose: capsule ovoid, included in the calyx, loculicidally 2-valved, valves concave, exposing the dissepiment: seeds oblong, with a membranaceous testa.

(1) M. MUSCOSA. (R. Br.)

Ident. R. Br. prod. p. 435.—Dec. prod. X. p. 433.
Syn. Pædorota minima, Koen.
SPEC. CHAR. Herbaceous, much branched, glabrous: leaves subsessile, opposite, oblong or lanceolate-linear, quite entire, 1-nerved: flowers solitary in the axils, sessile, small, sterile in the other axil: calycine teeth spreading, ciliated with few hairs: corolla shorter than the calyx.

Peninsula.

GENUS XXI. BUDDLEIA.

Tetrandria Monogynia. *Sea: Syst.*

Deriv. In memory of Adam Buddle, an English Botanist.

GEN. CHAR. Calyx 4-toothed or half 4-cleft: tube of the corolla short, subcampanulate or elongated, limb spreading or rarely slightly erect, segments short: stamens inserted at the throat with subsessile anthers, or in the middle of the tube with the anthers nearly equalling the throat or included: style entire, clavate at the apex, the stigmatose portion thick, capitate: capsule septicidally 2-valved, valves 2-cleft or somewhat entire, margins inflexed, exposing the placentiferous column: seeds numerous, small, compressedly fusiform or discoid, testa rather loose, often slightly expanded into a membranaceous wing, reticulated. —Trees, shrubs or herbs, often clothed with down or wool: leaves opposite: peduncles cymosely many-flowered, axillary or usually arranged in a thyrse or terminal panicle.

(1) B. CRISPA. (*Benth.*)

Ident. Benth. Scroph. Ind. p. 43.—Dec. prod. X. p. 444.
SPEC. CHAR. Small tree: tomentum densely ferruginous or hoary: leaves usually petioled, crenated at the margin, lower ones cordate at the base, often cut and crenated or crisp, upper ones rounded at the base, more entire, all thick, wrinkled, tomentose on both sides: heads of flowers dense, many-flowered, peduncled: panicle oblong or somewhat racemose.

Silhet.

(2) B. MISSIONIS. (*Wall.*)

Ident. Dec. prod. X. p. 444.—Benth. Scroph. Ind. p. 43.

SPEC. CHAR. Tree: branches pubescent: leaves oblong, narrowed at both ends, quite entire, coriaceous, glabrous above, smooth, pubescent beneath: panicle ovate, pyramidal, many-flowered, with its branchlets opposite: glomerules somewhat sessile.

Wynaad.

(3) B. NEEMDA. (*Ham.*)

Ident. Roxb. flor. Ind. Ed. Car. I. p. 411.—Dec. prod. X. p. 446.

Syn. B. serrulata, *Roth.*—B. subserrata, *Don. prod. fl. Nep.* p. 93.

SPEC. CHAR. Arboreous: tomentum thick, ferruginous or white: leaves lanceolate, acuminate, entire or serrulate, narrowed at the base into a short petiole, above very thinly puberulous: thyrses spiciform, long, interrupted: glomerules many-flowered, sessile: flowers pure white, fragrant.

Chittagong. Mountain rivulets in Northern India. Flowering in November and December.

(4) B. MACROSTACHYA. (*Benth.*)

Ident. Benth. Scroph. Ind. p. 42.—Dec. prod. X. p. 447.

SPEC. CHAR. Tomentum thick, ferruginous: branches winged-tetragonal: leaves sessile, oblong-lanceolate, acuminated, serrated at the apex, narrowed at the base and auriculately connate, slightly glabrous above, tomentose beneath: thyrse dense, spike-shaped: glomerules sessile, few-flowered, close-packed.

Silhet. Khasia hills.

(5) B. ASIATICA. (*Lour.*)

Ident. Lour. flor. Coch. p. 72.—Dec. prod. X. p. 116.

Syn. B. discolor, *Roth. & Benth.*—B. Salicina, *Lam.*—B. acuminatissima, *Blume.*

Engrav. Wight's Icon. t. 894.

SPEC. CHAR. Arboreous: tomentum thin, adpressed: leaves sub-petioled, lanceolate, acuminate, entire or serrulate, glabrous above: thyrses spike-shaped, long, slender: glomerules somewhat 3-flowered, sessile: capsules reflexed: flowers white.

Hills near Penn. Slopes of the Neilgherries.

GENUS XXII. BUCHNERA.

Didynamia Angiospermia. *Sex Syst:*

Deriv. Named after J. G. Buchner, a German Naturalist.

GEN. CHAR. Calyx tubular, obscurely nerved, shortly 5-toothed: tube of the cup-shaped corolla thin, straight or a little incurved, limb spreading, somewhat equally 5-cleft, segments oblong or obovate: capsule straight, valves subcoriaceous, elastically dehiscing when ripe.—Herbs, often scabrous and stiff: lower leaves opposite, upper ones alternate, lower ones broader, often toothed, upper often narrower, remote, usually quite entire, floral ones bract-shaped, very often shorter than the calyx: flowers solitary in the axils of the bracts, bibracteolate, arranged in a terminal thick or interrupted spike.

(1) B. HISPIDA. *(Ham.)*

Ident. Ham. in Don. prod. flor. Nep. p. 91.—Dec. prod. X. p. 499.

Engrav. Wight's Icon. t. 1413.

SPEC. CHAR. Small plant, 1 to 2 feet high, scarcely branched, leafy at the base; leaves oblong or lanceolate, toothed, upper ones linear: spike terminal, slender, interrupted, many-flowered; flowers light-purple.

Island of Caranjah. Coorg. Flowering in the rainy season.

GENUS XXIII. STRIGA.

Didynamia Angiospermia. *Sex: Syst:*

Deriv. So called from the strigose nature of the plants.

GEN. CHAR. Calyx tubular, ribbed, 5-toothed or cleft: tube of the corolla thin, abruptly incurved at the middle or above the middle, limb 2-lipped, upper lip often shorter, entire, emarginate or 2-cleft, lower 3-cleft: capsule straight, valves coriaceous, elastically bursting when ripe.—Scabrous herbs, sometimes parasitic; lowest leaves opposite, upper ones alternate, commonly linear, quite entire, very rarely toothed, sometimes all scale-shaped; floral ones agreeing with the cauline ones or gradually less: flowers solitary in the axils of the leaves or floral bracts, sessile, arranged in terminal spikes, often minutely bibracteate.

(1) S. OROBANCHOIDES. *(Benth.)*

Ident. Benth. in Comp. Bot. Mag. I. p. 361.—Dec. prod. X. t. 501.
Syn. Buchnera gmerioides, *Willd.*—B. Hydrabadensis, *Roth.* —Orobanche Indica, *Spreng. (not Roxb.)*
Engrav. Wight's Icon. t. 1414.—Benth. l. c. t. 19.
Spec. Char. Glabrous or puberulous, branched: leaves minute, scale-like: stems rigid, erect, 6 to 12 inches high, of a reddish hue: flowers numerous, rose-coloured: parasitic on the roots of different species of Lepidagathis and Euphorbium.

Common in rocky ground in the Northern Concan, and hilly parts of the Deccan.

(2) S. DENSIFLORA. *(Benth.)*

Ident. Benth. l. c. p. 363.—Dec. prod. X. p. 502.
Syn. Buchnera Asiatica, *Vahl.*—B. densiflora, *Benth. Scroph. Ind. p. 41.*
Spec. Char. Glabrous, very rough: leaves lanceolate-linear, floral ones scale-like: spikes rather thick when young, densely-flowered, afterwards elongated, interrupted: flowers white, with 5 striæ up the centre of the segments.

About Surat. Peninsula. Bengal. Flowering towards the end of the year.

(3) S. EUPHRASIOIDES. *(Benth.)*

Ident. Benth. l. c. p. 363.—Dec. prod. X. p. 503.
Syn. Buchnera Euphrasioides, *Vahl.*—B. angustifolia, *G. Don.* —S. glabrata, *Benth. l. c.*
Spec. Char. Glabrous, rough: leaves linear, entire or few toothed, elongated: spikes slender, interrupted: calyx with 15 striæ, which all reach to the apex of the segments: flowers white.

Peninsula, very common.

(4) S. HIRSUTA. *(Benth.)*

Ident. Benth. in Dec. prod. X. p. 502.
Syn. S. lutea, *Lour.*—Buchnera Asiatica, *Linn.*—Campuleia coccinea, *Hook. ex. for.* III. t. 203.—B. coccinea, *Benth. Scroph. Ind. p. 40.*
Spec. Char. Very rough: leaves linear, elongated, or lower ones lanceolate: calyx with 10 striæ, of which 5 run into the sinuses between the segments: flowers red, white, or yellow.

Very common. Peninsula. Bengal. Flowering in the cold season.

(5) S. SULPHUREA. (*Dalz.*)

Ident. Dalz. Bomb. flor. p. 182.

SPEC. CHAR. All scabrous: stem slender, quadrangular: leaves very narrow, linear, acute, flowers very shortly-pedicelled with two subulate bracts: calyx prominently 15-nerved, divided to the middle, divisions linear, strap-shaped: corolla yellow: tube as long as calyx, and pubescent towards the apex: upper lip broad, almost truncate, lower 3-lobed: lobes obovate, all ciliated.

Wet rocks on Sewaree Hill fort.

GENUS XXIV. RHAMPHICARPA.

Didynamia Angiospermia. *Ser: Syst.*

Deriv. From *Ramphis*, a beak, and *Karpos*, fruit.

GEN. CHAR. Calyx campanulate, 5-cleft: tube of the corolla thin, long-exserted, straight or incurved, segments of the limb five, obovate, nearly equal or the upper ones broad connate: anthers obtuse: capsule ovate, compressed from the side, obliquely mucronate or rostrate, valves coriaceous.—Erect, branched, glabrous herbs; lower leaves opposite, upper alternate, narrow, entire or pinnatisect: flowers short-peduncled, racemose, usually without bracts: corolla white or sulphur-coloured.

(1) R. LONGIFLORA. (*Benth.*)

Ident. Benth. in Comp. Bot. Mag. 1. p. 368.—Dec. prod. X. p. 504.

Engrav. Wight's Icon. t. 1415.

SPEC. CHAR. A low, much-branched plant: leaves pinnately divided into linear segments: flowers white, with a very long slender tube, and regular limb: capsule furnished with an incurved oblique beak.

Tellicherry and Cannanore, in moist soil. Common in ghaut pastures. Flowering in the rainy season.

GENUS XXV. MICRARGERIA.

Didynamia Angiospermia. *Ser: Syst.*

GEN. CHAR. Calyx campanulate, 5-toothed, teeth very obtuse: corolla tubular-campanulate, lobes nearly equal, entire: stamens

Included: anthers free, cells nearly equal, affixed at the apex, obtuse at the base: style slightly thickened at the top, obtuse: capsule subglobose, very obtuse, loculicidally dehiscing: seeds numerous, oblong-cuneate.

(1) M. WIGHTII. *(Benth.)*

Ident. Dec. prod. X. p. 509.
Engrav. Wight's Icon. t. 1417.
SPEC. CHAR. Herbaceous, stiff, branched plant: leaves linear, acute or 3-cleft: flowers subsessile: bracts on the short pedicels oblong: capsule longer than the calyx.

Courtallum, flowering in August and September.

GENUS XXVI. SOPUBIA.

Didynamia Angiospermia. *Ser: Syst.*

Deriv. The native name Latinised.

(GEN. CHAR. Calyx campanulate, 5-toothed, teeth valvate in æstivation, short or narrow: corolla infundibuliform or subrotate-campanulate, lobes spreading, entire: anthers two, or all cohering by pairs: style thickened at the apex, obtuse, somewhat tongue-shaped: capsule ovate or oblong, round or compressed at the apex, retuse or emarginate: valves entire or finally septicidally 2-cleft: seeds numerous.—Erect, branched, scabrous herbs: leaves narrow, often dissected, opposite or the upper ones alternate: flowers racemose or somewhat spiked at the tops of the branches, peduncles 1-flowered, with two bracts above the middle: corolla purple or rose-coloured.

(1) S. DELPHINIFOLIA. *(G. Don.)*

Ident. Dec. prod. X. p. 522.
Syn. Gerardia delphinifolia, *Linn.*—Euphrasia Coromandeliana, *Roth. in Spreng. Syst.*
SPEC. CHAR. Annual, erect: leaves opposite, irregularly pinnatifid, with filiform segments: flowers axillary, solitary, short-peduncled, large, rose-coloured.

Common in moist places in the Peninsula, flowering in the rains.

GENUS XXVII. CENTRANTHERA.

Didynamia Angiospermia. *Sea: Syst:*

Deriv. From *Kentron*, a spur, and *Anthera*, an anther.

GEN. CHAR. Calyx leafy, compressed, entire or finally shortly 2-5-cleft: corolla funnel-shaped, tubular; tube ventricose below the throat; limb obscurely 2-lipped: lobes broad, entire: stamens included: cells of the anthers spurred or mucronate, one less or narrower, often wanting: style flatly dilated at the apex, lanceolately tongue-shaped, acute: capsule obtuse; valves entire: seeds very numerous, oblong-cuneate.—Scabrous, stiff herbs: leaves opposite, or the top ones alternate, oblong, often narrow, entire or few-toothed: flowers solitary, in the axils of the falso leaves, lower ones opposite or all alternate: pedicels very short, with two bracts.

(1) C. GRANDIFLORA. (*Benth.*)

Ident. Benth. Scroph. Ind. p. 49.—Dec. prod. X. p. 525.

SPEC. CHAR. Herbaceous, erect, tuberculosely scabrous: calyx oblong-inflated, much acuminated.

Silhet and Assam.

(2) C. PROCUMBENS. (*Benth.*)

Ident. Benth. in Dec. prod. X. p. 525.

SPEC. CHAR. Diffuse, hispid: calyx ovate-oblong, acuminate.

Peninsula.

(3) C. HUMIFUSA. (*Wall.*)

Ident. Dec. prod. X. p. 525.

Syn. Rusumovia Tranquebarica, *Spreng.*—Torenia lepidota, *Roth.*

SPEC. CHAR. Diffuse, low plant, glabrous or scaly-scabrous and sparingly hispid: calyces ovate, acutish: capsule globose, somewhat membranaceous.

Peninsula.

(4) C. HISPIDA. (*R. Br.*)

Ident. R. Br. prod. p. 438.—Dec. prod. X. p. 525.
Syn. C. Nepaulensis, *D. Don.*—Digitalis stricta, *Roxb.*
Engrav. Wall. Pl. As. Rar. 1. t. 45.

SPEC. CHAR. Erect, hispid, about one foot in height: leaves opposite, sessile, linear, almost entire, very rough: flowers axillary, solitary, subsessile, somewhat trumpet-shaped, of a deep purplish red.

Hilly parts of the Concan, Coromandel. Hindostan. Flowering nearly all the year.

GENUS XXVIII. PEDICULARIS.

Didynamia Angiospermia. *Sm: Syst:*

Derie. From *Pediculus*, a louse. The supposed effect on sheep eating it.

GEN. CHAR. Calyx tubular or campanulate, in front and sometimes at the back more or less cleft, 2-5-toothed at the apex; teeth rarely equal, lateral ones connate or free, cristato-dentate or entire, posterior one very often less, entire or wanting: tube of the corolla cylindric or slightly amplified at the throat; helmet compressed, entire or increased in front under the apex by a tooth on either side or produced into a truncated or 2-toothed beak: lower lip somewhat erect, 2-crested above: lobes 3, erect, spreading or deflexed, lateral ones rounded: filaments usually hairy: anthers transverse, in pairs, or all approximate: cells equal, mutic: capsule compressed, ovate or lanceolate, more or less falcate or oblique, especially at the apex, loculicidally dehiscing behind from the apex towards the base and in front often more shortly: seeds laterally attached in the lower part of the capsule, ovoid, largish.—Herbs, usually mountainous: leaves alternate or verticillate, one or many times pinnately divided or rarely simply toothed: flowers spiked, rarely racemose, without bracts: floral leaves bract-shaped, entire or cut, rarely agreeing in shape with the cauline ones.

(1) P. PEROTTETTII. *(Benth.)*

Ident. Benth. In Dec. prod. X. p. 563.
Engrav. Wight's Icon. t. 1418.

SPEC. CHAR. Small: sparingly pilose: branches simple: leaves deeply pinnatifid, lobes ovate or oblong, crenate: flowers axillary, pedicelled: calyx tubular with the limb, crested: corolla many times longer than the calyx, with a slender tube: capsules not beaked.

Koundah hills, Neilgherries.

(2) P. ZEYLANICA. (*Benth.*)

Ident. Benth. Scroph. Ind. p. 54.—Dec. prod. X. p. 580.
Engrav. Wight's Icon. t. 1419.—Spicil. II. t. 185.

SPEC. CHAR. Furfuraceo-pubescent, or rarely nearly glabrous: loosely ramous at the base: branches ascending or erect: leaves petioled, oblong, obtuse, doubly crenate: racemes capitate or elongated: calyx cleft along one side, cristately 2-3-toothed behind: tube of the corolla shortly exserted: helmet incurved, obtuse, erostrate: flowers pink.

Neilgherries, flowering in the rainy season.

(3) P. FLAGELLARIS. (*Benth.*)

Ident. Benth. In Dec. prod. X. p. 581.

SPEC. CHAR. Perennial? stems elongated, procumbent, villous: leaves pinnately parted, segments oblong, cut and toothed: racemes elongated from the base, leafy, compact at the apex: calyx tubular, occasionally cleft, 2-5-toothed, teeth created: corolla tube exserted, helmet arched above, narrowed, beaked: capsule ovate-lanceolate.

Upper Assam.

ORDER CXIX. OROBANCHACEÆ.

Flowers irregular: calyx free, persistent, 4-5-sepaled: sepals cohering into a 4-5-cleft calyx, or united by pairs: corolla monopetalous, hypogynous, pentamerous, or by union of the upper pairs, tetramerous, persistent, æstivation imbricated, tube more or less curved, limb more or less 2-lipped: stamens 4, didynamous, inserted on the base, dehiscing by a longitudinal slit or oblong pore: ovary free, bound at the base by a fleshy disk, 1-celled: placentæ parietal, paired on each side the ovary, either distinct or geminately-connate, or with two broadly 2-lobed placentæ extending from the parietes: placentæ lateral as regards the axis of inflorescence: ovules usually numerous, anatropous, with sometimes a longish funiculus: style terminal, simple: stigma large, capitate, 2-lobed, lobes either over the placentæ, or placed anterior and posterior sometimes obscurely sulcated in the middle, rarely sub-clavate, undivided: capsule 1-celled, two valvate at the apex or through its whole length: valves bearing on the middle or oftener towards the middle, solitary or paired, filiform or broad placentas: seeds numerous, rarely few, minute, globose, oblong, or pear-shaped: testa thick, spongy, scrobiculate or tubercled: albumen copious: embryo minute, oboroid.—Herbaceous leafless plants, growing parasitically on the roots of other plants, often forming dense masses of great extent round the base of plants suited for their support: stems erect, more or less covered with brown, yellowish, or colourless scales in place of leaves.

GENUS L. PHELIPÆA.

Didynamia Angiospermia. *Ser. Syst.*

Deriv. Named by Tournefort, in honor of the family of Phelipeau, Patrons of Natural Sciences.

GEN. CHAR. Flowers hermaphrodite, bibracteolate: calyx tubular, 4-5-cleft or toothed: corolla hypogynous, ringent, upper lip erect, 2-cleft, lower 3-cleft, spreading: stamens four, inserted on the tube of the corolla, included, filaments flattened at the base: anthers 2-celled, cells divaricate at the base, mucronate; ovary 1-celled, placentæ parietal, four, approximated in pairs: ovules many: style simple: stigma capitately 2-lobed: capsule 1-celled, 2-valved at the apex, valves cohering by the base, bearing twin placenta near the axis: seeds many, subglobose, testa fungous, thick.—Leafless herbs, scape simple or branched, parasitic on the roots of other stems.

(1) P. INDICA. *(G. Don.)*

Ident. Don. Gen. Syst. IV. p. 632.—Dec. prod. XI. p. 8.
Syn. Orobanche Indica, *Roxb.*

SPEC. CHAR. Scape simple or branched, with scales here and there: calyx 4-toothed, teeth lanceolate-subulate from a broad base: corolla tubular, infundibuliform, curved, purple, widened in the throat: flowers spiked, terminal.

Parasitic on tobacco plants in Guzerat and the Deccan. Coromandel. Hindostan. Oude. Flowering in January.

GENUS II. ÆGINETIA.

Didynamia Angiospermia. *Ser. Syst.*

Deriv. In honor of Paul Æginetta, a Physician of the 7th century.

GEN. CHAR. Flowers hermaphrodite, without bracts: calyx spathaceous, loose, cleft in front, acute behind: corolla hypogynous, tube cylindric, slightly exceeding the calyx, somewhat incurved, limb somewhat equally 5-cleft, 2-lipped: stamens four, inserted at the bottom of the tube, included: filaments torate, converging: anthers cohering by pairs, cells pendulous from the apex of the thickened connectivum, truncated at the top, dehiscing from the acute base, those of the lower stamens produced behind into a conical obtuse spur: ovary 2-celled: style simple: stigma large, fleshy, peltately-cordate: capsule 2-celled, irregularly 2-valved, valves placentiferous: seeds many, small.—Parasitical herbs: scape scaly, abbreviated, throwing out scapiform naked 1-flowered peduncles, solitary or collected together on turf, flowers largish, showy.

(1) Æ. ABBREVIATA. (*Ham.*)

Ident. Ham. Mss. in Wall. Cat. No. 3905.—Dec. prod. XI. p. 43.

SPEC. CHAR. Scape very short, branched or simple, scaly: limb of the corolla 2-lipped, lower lip enlarged, 3-lobed.

Peninsula. Silhet.

(2) Æ. PEDUNCULATA. (*Wall.*)

Ident. Dec. prod. XI. p. 43.

Syn. Orobanche pedunculata, *Roxb.*

Engrav. Wall. Pl. As. Rar. III. t. 219.—Wight's Icon. t. 1421.

SPEC. CHAR. Glabrous: scape simple, scales few, attenuated, elongated: floral scales triangular: calyx spathaceous, 1-leaved, inner side cleft: tube of the corolla inflated, equalling the calyx: limb 5-partite, nearly equal, segments reniform, denticulate: filaments glabrous: stigma large cordate-peltate: tube of corolla yellow, limb dark violet.

Bengal, parasitic on roots of the Andropogum Muricatum and many Bamboos. Courtallum. Flowering in the rainy season.

(3) *Æ. Indica.* (*Roxb.*)

Ident. Dec. prod. XI. p. 43.
Syn. Orobanche Æginetia, *Linn.*
Engrav. Roxb. Cor. I. t. 91.—Wight's Icon. t. 805.—Rheede Mal. XI. t. 47.

Spec. Char. Scape simple, elongated, naked, bearing at its apex a large curved purple flower, something like a tobacco-pipe: calyx spathe-like, lax, split in front, acute.

Parasitic on the roots of bushes and grass. Circars. Concans. Travancore. Silhet. Flowering in the hot season.

GENUS III. CHRISTISONIA.

Didynamia Angiospermia. *Sts: Syst:*

Deriv. Named after Dr. Robert Christison of Edinburgh.

Gen. Char. Calyx tubular, quinquifid, equal or sublabiate; corolla hypogynous, tube funnel-shaped, limb 5-lobed, bilabiate; stamens didynamous, inserted on the tube of the corolla, all fertile, incluse, or rarely exserted: anthers 2-celled, one polleniferous, dehiscing at the apex by an oblique pore, the other sterile, prolonged into an acute spur: disk one: ovary ovate-oblong, 1-celled: placentiferous margins deeply inflexed, revolute within the cell: ovules numerous: style filiform, simple: stigma bilabiate, or orbicular: capsule enclosed in the calyx, subglobose, 1-celled, 2-valved, dehiscing loculicidally, and bearing the placenta on the middle of the valves: seeds numerous, oblong, obtuse, supported on a short thick funiculus: outer seed-coat loose, membranaceous, reticulated, or sub-scrobiculate (pitted like a thimble); embryo enclosed in copious albumen, orthotropous: cotyledons short, obtuse; radicle thick, blunt.—Herbaceous plants growing parasitically on the roots of other plants; stems short, simple or ramous, scaly below, floriferous towards the apex: flowers large, rose-coloured, or yellow, or deep purplish-blue: pedicels racemose.

(1) C. Stockii. (*Hook.*)

Ident. Hook. Ic. Plant. vol. IX. & t. 836.—Dalz. Bomb. flor. p. 202.
Syn. C. calcarata, *Wight's Icon.* t. 1426.

Spec. Char. Scape thick, fleshy, imbricately scaly: scales broadly ovate, concave, obtuse: flowers racemose; pedicels along gated, erect, without bracts: calyx tubular, cylindric: limb 5-divided, lobes triangular, rather obtuse: corolla pubescent, of a bluish-white colour, the lobes spreading and rounded.

Salsette. Parasitic on the roots of Strobilanthes; flowering in the rains.

(2) C. Lawii. *(R. W.)*

Ident. Dalz. Bomb. flor. p. 202.

Engrav. Wight's Icon. t. 1427.

Spec. Char. Scape thick, fleshy, irregularly shaped: base of the subsessile flowers cliractenlate, embraced by a few loose scales: calyx tubular, 5-toothed, regular; corolla tubular, twice the length of the calyx, lobes suborbicular: flowers large, pale-purple, with yellow spots.

Salsette. Between Ram Ghaut and Belgaum. Flowering in August.

(3) C. Aurantiaca. *(R. W.)*

Ident. Wight's Icon. vol. IV. & t. 1480.

Spec. Char. Erect, few scales, pilose: scales ovate, glabrous: flowers corymbose, long-peduncled, dark yellow: peduncles bibracteolate near the middle: calyx tubular, pilose, with 5 mucronate teeth, reddish-orange.

Neilgherries in long grass between Neddawuttum and Goodaloor.

(4) C. Subacaulis. *(Gardner.)*

Ident. Wight's Icon. vol. IV. & t. 1423.

Syn. Phelipœa subacaulis, *Benth. Scroph. Ind.*—*Dec. prod.* XI. p. 11.

Spec. Char. Stems very short, thick, scaly; peduncles 3-4, shorter than the scales, 1-flowered; corolla tube slender, shortly exserted beyond the calyx, thin, expanding into a large sub-bilabiate, 5-lobed limb; anthers glabrous, cells calcarate; stigma capitate.

Peninsula, but the exact locality not given.

GENUS IV. CAMPBELLIA.

Didymammia Angiospermia. *Sex. Syst.*

Deriv. Named after Dr. W. H. Campbell, Secretary to the Edinb. Bot. Society, and his brother Captain J. Campbell, of the Madras Establishment, an active collector of plants.

GEN. CHAR. Calyx tubular, 5-lobed, bibracteolate : corolla sub-infundibuliform, bilabiate : the upper lip more or less deeply 2-lobed, the under 3-lobed : stamens didynamous, incluse : anthers 1-celled, pendulous, opening by a pore at the apex : ovary spuriously 2-celled at the base, 1-celled at the apex, carpels deeply inflexed : placentiferous margins revolute : style simple : stigma capitate : capsule, like the ovary, imperfectly 2-celled : seed oblong : testa loose, reticulate, produced at the ends into a wing : albumen copious : embryo minute.—Herbaceous plants, parasitic on the roots of others : stems simple, scaly : flowers axillary, peduncled, aggregated towards the apex of the stem, each furnished with two bracteoles : stamens shorter than the corolla : style hooked at the apex : stigma clavate, drooping.

(1) C. AURANTIACA. (R. W.)

Icon. Wight's Icon. vol. IV. & t. 1424.

SPEC. CHAR. Stems simple, covered on all sides with closely appressed sub-orbicular scales : floral ones or bracts broad, obovate, bracteoles lanceolate, entire : flowers sessile, stipulate : corolla scarcely exceeding the calyx, pubescent within, 5-lobed : stamens scarcely didynamous : filaments pilose : style the length of the stamens, pilose : stigma clavate : flowers pale yellow.

Neilgherries, near Neddawuttum, flowering in August and September.

(2) C. CYTINOIDES. (R. W.)

Icon. Wight's Icon. vol. IV. & t. 1425.

Syn. Phelipæa cytinoides, *Dec. prod.* XI. p. 14.—Christisonia Neilgherries, *Gardn. in Calc. Journ.* V. 8. p. 157.

SPEC. CHAR. Erect, glabrous, covered with appressed, broad, ovate, obtuse scales : flowers pedicelled, bright yellow : bracts sub-orbicular, shorter than the lanceolate bracteoles : calyx tubular, irregularly 5-7-toothed : corolla 2-lipped : upper lip emarginate, under broadly 3-lobed : stamens the length of the corolla : filaments glabrous : anthers deflexed, 1-celled : style hooked at the apex : stigma clavate : testa of the seed reticulately scrobiculate.

Neilgherries, parasitic on the roots of Strobilanthes, in woods near Pycarrah, flowering in May.

GENUS V. OLIGOPHOLIS.

Didynamia Angiospermia. *Sm. Syst.*

Deriv. From *Oligos*, few, and *Pholis*, a scale.

Gen. Char. Hermaphrodite, ebracteolate: calyx tubular, 5-toothed: corolla infundibuliform, sub-ringent, 5-lobed: stamens didynamous, incluse: anthers 2-celled, embraced at the base by a cup-shaped disk: placentæ 2, large, fleshy, nearly filling the whole cavity, covered on all sides with minute ovules: style sub-clavate: stigma peltate.

(1) O. tubulosa. *(R. W.)*

Ident. Wight's Icon. vol. IV. & t. 1422.

Spec. Char. Herbaceous, parasitical, with erect, slightly ramous, nearly naked stems, only furnished with a few scales: peduncles axillary, longer than the floral scale, ebracteolate: corolla tubular, ventricose above, more than twice the length of the calyx: filaments thickened below with a ring of hairs at the base.

Courtallum, parasitic on roots of Bamboos, flowering in September.

ORDER CXX. ACANTHACEÆ.

Calyx pentamerous, the odd sepal posterior, sometimes the two anterior ones united, hence 4 or 5-divided, sometimes, but rarely, nearly obsolete, entire or several-toothed: corolla monopetalous, hypogynous, 5-cleft, the segments alternate with the sepals: limb usually bilabiate, but sometimes regular, 5-lobed, contorted in æstivation: stamens inserted on the tube at different heights, sometimes near the base, about the middle, or on the throat, either didynamous, the 5th rudimentary or altogether wanting, or often only two antheriferous: filaments filiform, sometimes united by pairs at the base, or even monadelphous, two or one-celled, cells contiguous, parallel or superposed, or variously divaricated; occasionally one of them sterile, dehiscing longitudinally: ovary free, dicarpellary, two-celled, the septum formed of the inflexed margins of the carpels, either complete (meeting in the axis) or somewhat incomplete: cells anticous and posticous with respect to the axis of inflorescence, often spuriously stipitate from the obliteration of the lower half of the cells, sometimes rostrate at the apex: ovules 1-2 or several in each cell, sessile or borne on processes of of the parietal placenta: style terminal, filiform, simple; stigma entire or 2-lobed: capsule 2-celled, of various consistence, unguiculate or rostrate, bursting elastically: dissepiment opposite the valves, separating in two pieces through the axis (the middle sometimes open), usually adnate to the valves, but sometimes separating from them: seeds usually compressed, 1–2 or several in each cell, attached to cup-shaped, subulate or hooked processes (retinacula) of the placenta: testa coriaceous, fibrous or loose, often tuberculate, sometimes pilose; albumen none: embryo curved or straight: cotyledons large, roundish: radicle taper, descending; and at the same time centripetal, curved or straight.—Herbaceous plants or shrubs: stem and branches nodosely jointed: leaves often beset with white hair-like lines under the epidermis which, after breaking the cuticle, effervesce on the application of an acid: leaves opposite or rarely in fours, exstipulate, entire or serrated, rarely showing a tendency to become lobed, sometimes in unequal pairs: inflorescence terminal or axillary, in spikes, racemes, fascicles, or panicles: flowers usually opposite on the spikes or sometimes alternate, furnished with three bracts of which the lateral pair are now and then deficient: bracts often large and foliaceous, and then the calyx is usually much diminished in size.

GENUS I. THUNBERGIA.

Didynamia Angiospermia. *Sec, Syst*

Deriv. In honor of Charles P. Thunberg, F. R. S., a celebrated Traveller and Botanist.

GEN. CHAR. Calyx short, cup-shaped, truncated or manytoothed, teeth as many as ten when present: bracteoles two at the base of the calyx, larger than the calyx and covering the flower valvately before its expansion: corolla campanulately funnel-shaped, throat inflated, limb 5-cleft, spreading, nearly equal: stamens four, didynamous: anthers erect, adnate, cells parallel, ciliately bearded at the margin, one a little shorter at the base and the same produced into an awn-shaped spur: stigma funnel-shaped, transversely emarginate, somewhat two-lipped: ring nectariferous, thick, lobed, surrounding the ovary: capsule globose at the base, 2-celled, 2-4-seeded, beaked towards the upper part, depressed: dissepiment membranaceous, cohering in the centre, loosening from the valves; seeds globose, callous at the base, perforate.—Shrubs or herbs, scandent; leaves commonly angular, and often, with the bracteoles, hairy: flowers axillary, peduncled, solitary, or arranged in a raceme: corolla showy.

(1) T. GRANDIFLORA. *(Roxb.)*

Ident. Roxb. flor. Ind. III. p. 34.—Dec. prod. XI. p. 54.

Syn. Flemingia grandiflora, *Rottler et Willd.*

Engrav. Roxb. Cor. t. 67.—Wight's Icon. t. 872.—Bot. Reg. t. 493.

Spec. Char. Climbing: leaves cordate, angled, acuminate, hispid; limb of the calyx truncated, quite entire; flowers very large, light blue, with a white tube.

Silhet, Assam, Travancore, Peninsula. Flowering nearly all the year.

(2) T. CORDIFOLIA. *(Nees.)*

Ident. Dec. prod. XI. p. 55.

Spec. Char. Scandent, hirsute: leaves deeply cordate, acuminate, quite entire; limb of the calyx repandly lobed; flowers axillary and terminal, racemose, large.

Assam.

(3) T. LÆVIS. *(Wall.)*

Ident. Dec. prod. XI. p. 56.—Wall. Pl. As. Rar. III. p. 77.

Spec. Char. Scandent: leaves hastato-acutate, subangulate, obtuse, glabrous: calyx 12-toothed; flowers yellow (?)

Dindigul hills.

(4) T. FRAGRANS. *(Roxb.)*

Ident. Roxb. flor. Ind. III. p. 33.—Dec. prod. XI. p. 57.

Engrav. Roxb. Cor. I. t. 67.—Bot. Mag. t. 1881.

Spec. Char. Climbing: leaves oblong, acute, cordate, angular and somewhat hastate at the base, slightly scabrous: calyx 12-

cleft; peduncles axillary, solitary, 1-flowered, round; downy: capsule flat and beaked; flowers large, pure white, or white with sulphur-coloured bottom.

Banks of rivers near Samulcottah, Tanjore, and other parts of the Peninsula. Hindostan and Concans.

Var. 1. Leaves broader, acutely bitentate or even hastate at the base. Travancore.

Var. 2. Leaves narrower, obsoletely repand, somewhat hastate. Hindostan.

Var. 3. Leaves as in *Var.* 1, or almost quite entire, and with the stem densely silky and velvety. Neilgherries.

(5) T. ALATA. (*Bojer.*)

Ident. Dec. prod. XI. p. 58.
Engrav. Hook. Exot. flor. t. 17.—Bot. Mag. t. 2591.

SPEC. CHAR. Twining, silky villous: leaves cordate-sagittate, acute: wing petioled: calyx 12-cleft; bracteoles repand: flowers yellow, with a deep purple bottom.

Assam.

(6) T. ROXBURGHII. (*Nees.*)

Ident. N. ab. E. in Wall. Pl. As. Rar. III. p. 78.—Dec. prod. XI. p. 58.

SPEC. CHAR. Scandent, hirsute: leaves cordate, coarsely toothed at the base: calyx 12-15-cleft.

Neddawuttum. Coonoor, Neilgherries.

(7) T. TOMENTOSA. (*Wall.*)

Ident. Wall. Pl. As. Rar. III. p. 78.—Dec. prod. XI. p. 58.

SPEC. CHAR. Scandent, hirsute: leaves cordate or triangular-hastate, acute: calyx with many bristles.

Peninsula. Neilgherries.

GENUS II. MEYENIA.

Didynamia Angiospermia. *See Syst:*

Deriv. Named after F. J. Meyen, a German Botanical Author.

GEN. CHAR. Calyx small, 5-lobed, enclosed between two large bracteoles. corolla funnel-shaped, throat large, tube short, closed within with a ring of hairs; limb nearly equal: stamens 4, didynamous, anthers bearded at the apex, 2-celled: cells of the longer pair unequal, upper ones diverging, tomentose on the margin of the lower ones, parallel, about equal, both muticous at the base: stigma membranous, dilated, bilabiately 2-lobed: capsule turned at the base, above tapering to a point, 2-celled, 4-seeded, partition persistent, adhering to the axis of the woody valves: seeds globose, attached to spongy cup-shaped processes.—Procumbent or twining undershrubs; leaves opposite, entire; flowers axillary, peduncled; limb of the corolla deep blue, tube brownish-yellow.

(1) W. HAWTAYNEANA. *(Nees.)*

Ident. Nees in Wall. Pl. As. Rar. III. p. 74.—Dec. prod. XI. p. 80.
Syn. Thunbergia Hawtaynii, *Wall. tent. flor. Nep.* 1. p. 49.
Engrav. Wall. L c. II. t. 164.—Wight's Icon. t. 1487.— Spicil. II. t. 160.
Spec. Char. Shrubby, procumbent, glabrous: leaves sessile, cordate, acute; flowers deep purplish blue with a yellow throat.

Neilgherries. Iyamully hills near Coimbatore. Flowering all the year.

GENUS III. HEXACENTRIS.

Didynamia Angiospermia. *Nex; Nyet:*

Deriv. From *Hex*, six, and *Kentron*, a spur; the upper anthers have each one, the lower two spurs.

Gen. Char. Calyx disk-shaped, small, limb unequally toothed or repand: bracts two, connate at one side, debiscing at the other, valvate, covering the calyx: corolla infundibuliform-campanulate, tube very short, limb somewhat unequal, obliquely 5-cleft: stamens inserted at the throat of the tube where there is a bearded ring: anthers erect, 2-celled, glabrous, cells parallel, contiguous, one of the upper stamens very long-spurred at the base, the other shortly mucronulate, each cell of the lower ones very long-spurred, spurs flexuose: rudiment of a fifth sterile stamen short, subulate: stigma 2-forked, segments truncated, folded and channelled, narrower at the base: capsule 2-celled, 4-seeded, beaked: seeds crested: dissepiment loosening from the valves, entire, separable at the axis in two parts.—Scandent shrubs: leaves toothed, glabrous: racemes axillary and terminal, many-flowered: common bracts small: flowers opposite, solitary or fascicled: bracteoles caducous.

(1) H. Mysorensis. *(R. W.)*

Ident. Wight's Icon. vol. III.—Walp. Ann. I. p. 530.
Engrav. Wight's Icon. t. 871.—Bot. Mag. t. 4786.

Spec. Char. Climbing: leaves elliptic-lanceolate, acuminate, 5-nerved, often sub-hastate: raceme very long, pendulous, all the bracts lanceolate, small: corolla flat, upper lip large, somewhat undivided: anthers long-bearded at the base: flowers large, showy, either of a golden-yellow throughout, or with the limb of an orange or blood-red-colour.

Nuggur in Mysore. Dharwar district. Travancore mountains. Flowering in December.

(2) H. coccinea. *(Nees.)*

Ident. N. ab. E. in Wall. Pl. As. Rar. III. p. 78.—Dec. prod. XI. p. 61.
Syn. Thunbergia coccinea, *Wall.* test. *for. Nep.* p. 49.
Engrav. Hook. Exot. flor. t. 195.—Lond. Bot. Cab. t. 1198.

Spec. Char. Climbing: leaves cordate, sagitto-cordate or obovate, repandly toothed; bracteoles ovate: calyx 12-toothed: racemes terminal and axillary: flowers largish, deep red, with a yellow throat.

Pundua, Silhet, flowering in the cold season.

(3) H. dentata. *(Nees.)*

Ident. N. ab. E. in Wall. l. c.—Dec. prod. XI. p. 61.
Syn. Thunburgia coccinea, *Wall. Cat.*

Spec. Char. Climbing: leaves ovate, truncate at the base, coarsely toothed; bracteoles oblong: calyx repand, 5-lobed.

Assam.

(4) H. acuminata. *(Nees.)*

Ident. N. ab. E. in Wall. l. c.—Dec. prod. XI. p. 61.

Spec. Char. Scandent: leaves ovate-oblong, long-acuminate, remotely toothed or quite entire.

Silhet mountains.

GENUS IV. SCHMIDIA.

Didynamia Angiospermia. *Ser. Syst.*

Deriv. In honor of Dr. Bernard Schmid, an Indian Botanist.

GEN. CHAR. Bracts 2, free to the base, calyx entire, very short: corolla tubular, opening obliquely: limb 5-lobed, reflexed: stamens sub-didynamous, inserted near the middle of the tube, included: anthers 2-celled, straight, cells contiguous, parallel, prolonged below the point of attachment and each ending in a longish subulate spur, no rudimentary filament: ovary 2-celled, with two ovules in each: stigma entire, truncated: capsule globose at the base, ending in a conical beak, 2-celled: seed sub-globose, flattened next the partition.

(1) S. BICOLOR. *(R. W.)*

Icon. Wight's Icon. vol. IV. & t. 1848.

SPEC. CHAR. Twining shrub: leaves opposite, broad ovate-lanceolate, acuminate, subcrenato-dentate, 3-5 nerved, glabrous: racemes axillary, long, pendulous, many-flowered: bracts small, subulate: bracteoles large, suborbicular, reniform at the base, mucronate (nearly an inch in diameter) when fresh one-half of a dark brownish-purple, the other pale-yellowish, or cream-coloured; corolla tubular, exceeding the bracteoles, light blue, the lobes of the limb acutely turned back on the apex of the tube.

Neilgherries, below Sisparah, flowering in November and December.

GENUS V. ELYTRARIA.

Didynamia Gymnospermia. *Ser: Syst.*

Deriv. From *Elytron*, an envelope; alluding to the scaly stem.

GEN. CHAR. Calyx either 5-parted with the upper segment broader, the twin lower ones connected a little deeper at the base, or 4-parted with the upper and lower segment broader: corolla 2-lipped or ringent, lower lip 3-cleft, (segments 2-cleft): stamens two fertile, two sterile, included: cells of the anthers parallel: capsule cells many-seeded from the base: retinacula none.—Stemless herbs: leaves radical, entire, toothed, or repand: scapes or peduncles covered in four rows by small scale-shaped leaves: bracts opposite, 1-flowered; bracteoles two, narrower; flowers small.

(1) E. CRENATA. (*Vahl.*)

Ident. Vahl. En. I. p. 206.—Dec. prod. XI p. 63

Syn. E. Indica, *Pers.*—Justicia acaulis, *Linn.* Suppl. p. 84.—
Roxb. *flor. Ind.* I. p. 130.

Engrav. Roxb. Cor. II. t. 127.

SPEC. CHAR. Stemless: leaves obovate-oblong, crenated, villous on the nerves beneath; scape long, slender, simple; bracts ovate, ciliated; flowers spiked, white.

Koondianu, in the Broach Collectorate. Circar mountains. Madura. Gimpie hills. Banks of the Jumna. Flowering nearly all the year, and usually found in pastures under the shade of trees.

GENUS VI. NELSONIA.

Diandria Monogynia. *Ser. Syst.*

Deriv. In honor of D. Nelson, the Botanist who accompanied the Circumnavigator Captain Cook.

GEN. CHAR. Calyx 4-parted, unequal, upper segment larger, lower one 2-cleft: corolla 2-lipped, lower one 3-cleft: stamens two, none sterile: anthers included; connectivum obliquely lanceolate at the apex, one cell placed over the other, lower one mutinous, equal; stigma 2-cleft, segments ovate: capsule attenuated from an ovate base, cells 8-seeded, seed-bearing from the base? retinacula none.—Diffuse herbs, often tomentose, growing in damp places: leaves middle-sized or small, broadish: spikes terminal or axillary: flowers solitary, small, covered by a large bract: lateral bracts small or none.

(1) N. TOMENTOSA. (*Willd.*)

Ident. Willd. Sp. Pl. I. p. 419.—Dec. prod. XI. p. 65.

Syn. Justicia tomentosa, *Roxb.*—N. origanoides, *Roem. & Schult.*—J. origanoides, *Vahl.*—J. vestita, *Roem. & Schult.*—J. Bengalensis, *Spreng.*—J. lamiifolia, *Koen. in Roxb.*

SPEC. CHAR. Herbaceous, prostrate, villous: lower leaves petioled, elliptic, obtuse, upper submessile, smaller: spikes ovate: bracts rounded, elliptic, obtuse, mucronulate: flowers purple, a little longer than the calyx.

Common in the Warree jungles. Peninsula. Assam. There is a variety which grows in Silhet and the Coromandel Coast, the flowers of which are variegated with deep and light purple, flowering in the cold season.

GENUS VII. ADENOSMA.

Didynamia Angiospermia, *Lin. Syst.*

Deriv. From *Aden*, a gland, and *Osme*, smell; from the fragrant glands of the leaves, like Mint.

GEN. CHAR. Calyx 5-parted, equal or the upper segment larger: corolla ringent: stamens 4, didynamous: anthers 2-celled: cells parallel: capsule narrow, beaked, many-seeded: retinacula none. —Herbs or undershrubs, growing in swampy places and by the sea-shore, erect or diffuse: leaves serrated or crenated, scattered with glands, pubescent or glabrous: flowers middle-sized, purplish, sessile in the axils of the smaller upper leaves, single or in threes, opposite, forming a leafy spike.

(1) A. BALSAMEA. (*Spreng.*)

Ident. Spreng. Syst. II. p. 829.—Dec. prod. XI. p. 68.
Syn. Ruellia balsamea, *Linn.*
Engrav. Wight's Icon. t. 446.
SPEC. CHAR. Stem erect: leaves lanceolate, callosely-toothed, glutinous, glabrous: flowers axillary, verticilled: upper segment of the calyx oblong, the rest linear, obtuse.

Very common in rice fields after the harvest. Neilgherries. Courtallum.

(2) A. TRIFLORA. (*Nees.*)

Ident. N. ab. E. in Wall. Pl. As. III. p. 79.—Dec. prod. XI. p. 68.
Syn. Ruellia triflora, *Roxb.*—Cardanthera triflora, *Ham.*
SPEC. CHAR. Stem ascendent and with the ovate crenate-serrated leaves glandulosely pubescent: all the segments of the corolla retuse: flowers axillary, verticilled, in threes, deep and light blue, with a white reddish tube and a yellow-stained throat.

Ditches in Bengal. Silhet. Coromandel. Flowering in the cold season.

(3) A. VILLOSA. (*Benth.*)

Ident. Dec. prod. XI. p. 68.
SPEC. CHAR. Stem erect, villous: leaves ovate, acute at both ends, sharply serrated, slightly villous: flowers axillary.

Khasia hills.

(4) A. VERTICILLATA. (Nees.)

Ident. N. ab. E. in Wall. Pl. As. Rar. III. p. 79.—Dec. prod. XI. p. 69.

Engrav. Wight's Icon. t. 1524.

Spec. Char. Stem ascending and like the oval oblong serrulate-crenate leaves, hairy: segments of the calyx linear-spathulate, obtuse: flowers verticilled, purplish.

Mysore and Coorg.

(5) A. ULIGINOSA. (R. Br.)

Ident. R. Br. prod. I. p. 398.—Dec. prod. XI. p. 69.

Syn. Ruellia uliginosa, *Linn.*—Roxb. flor. Ind. III. p. 52.— R. ringens, *Linn.*

Spec. Char. Stem creeping, geniculate, much branched, hairy above: leaves oval, crenate or quite entire, subsessile, pubescently scabrous: flowers axillary, opposite, somewhat solitary; small, blue, arranged in a terminal tetragonal spike.

Very common in rice fields near Tranquebar and other parts of the Peninsula, flowering in the cold season.

(6) A. THYMUS. (Nees.)

Ident. N. ab. E. in Wall. l. c. III. p. 79.—Dec. prod. XI. p. 69.

Syn. Ruellia cernua, Roxb. flor. Ind. III. p. 45.

Spec. Char. Stem creeping, much branched: leaves petioled, drooping, ovate-lanceolate, quite entire: flowers axillary, sessile, opposite, small, pale pink.

Mysore, flowering in the rainy season.

GENUS VIII. ERYTHRACANTHUS.

Didynamia Angiospermia. *Ses. Spl.*

Deriv. From *Erythros*, red; alluding to the colour of the leaves underneath.

Gen. Char. Calyx deeply 5-parted, segments nearly equal, lanceolate-subulate: corolla funnel-shaped, limb equal, 5-cleft, obtuse: stamens four, didynamous: anthers 2-celled, cells motionous, obliquely diverging into the hatchet-shaped connectivum: stigma bilamellate, lamellæ parallel, narrow, upper one shorter, truncated: ovary girt with a ring at the base, 2-celled above, cells many-ovuled, ovules in dense rows, horizontally spreading, obtuse: capsules ovato or oblong, depressed, 2-celled from the base, many-seeded, dissepiment bipartite: seeds small, compressed: retinacula none.—Undershrubs: leaves generally red below: raceme terminal, composite, flowers and bracts somewhat alternate: bracts and bracteoles small, bristly, nearly equal, shorter than the calyx: flowers middle-sized.

(1) E. ELONGATUS. *(Nees.)*

Ident. Dec. prod. XI. p. 78.—Dalz. Bomb. flor. p. 184.
Syn. Adenosma elongatum, *Blume. Bidjr.* p. 757.
SPEC. CHAR. Herbaceous: leaves oblong-oval or oblong, pubescent, rough, obtuse, red on the under side: racemes axillary and terminal, simple or compound, elongated, loose: common peduncle very short: bracts lanceolate: capsule oblong.

Warree jungles.

GENUS IX. EBERMAIERA.

Didymaxula Angiospermia. *Sex: Syst.*

Deriv. Named after a German Botanist and Author.

GEN. CHAR. Calyx 5-parted, upper segment broader, middle ones narrower: corolla funnel-shaped, limb obliquely 5-cleft, two upper segments shorter; stamens inserted on the tube; lower ones occasionally sterile, sometimes with the filament of a fifth barren stamen, all included: cells of the anthers ovate, transversely spreading, retrorsely and extrorsely dehiscent: stigma shortly bilamellate, upper segment truncate, or 2-cleft, prominent from either side to the base of the lower lamella which embraces the upper one, hence the stigma when its two divisions are incumbent on each other is in appearance 8-toothed: capsule oblong or oval, obtuse, many-seeded, seed-bearing from the base: valves depressed at the back: retinacula none.—Herbs or undershrubs: stem often hirsutely tomentose or puberulons at the base: leaves entire, often puberulous: flowers pale or yellowish, subsessile or short-pedicelled in the axil of a floral leaf: pedicel often adhering to the bract, forming a spike or spiciform racemes on the stem and branches.

(1) E. GLAUCA. *(Nees.)*

Ident. Dec. prod. XI. p. 73.—Dalz. Bomb. flor. p. 114.
Engrav. Wight's Icon. t. 1488.

SPEC. CHAR. Stem erect, a foot high, pubescent and rough: leaves oblong, attenuated into the petiole, glabrous, entire: calyx pubescent and glandular: capsule oblong, glabrous: spikes leafy, axillary and terminal: corolla tubular.

South Concans. Mysore. Coorg.

(2) E. STAUROGYNE. *(Nees.)*

Ident. Dec. prod. XI. p. 75.
Syn. Staurogyne argentea, *Wall. Cat. No.* 4905. *(partly.)*
Engrav. Wall. Pl. As. Rar. II. t. 186.

SPEC. CHAR. Perennial: leaves oblong, paler beneath, sharply attenuated into the petiole, glabrous: petiole and stem rough tomentose at the base: raceme terminal, short: bracts oblong or lanceolate: bristly at the apex: bracteoles lanceolate, attenuated at the base, scabrous: segments of the calyx with long bristles at the apex, upper segment broader.

Silhet.

(3) E. ARGENTEA. *(Nees.)*

Ident. Dec. prod. XI. p. 76.
Syn. Staurogyne argentea, *Wall. (partly.)*

SPEC. CHAR. Perennial: leaves lanceolate-oblong, silvery-white below, sharply attenuated into the petiole, glabrous: petiole and stem strigosely hirsute at the base: racemes axillary and terminal, short, erect, hirsute: bracts and upper segment of the calyx lanceolate, the other segments of the calyx and all the bracteoles, subulate, stiff.

Silhet.

GENUS X. HEMIADELPHUS.

Diandria Monogynia. *Sax. Syst.*

Deriv. From *Hemi*, half, and *Adelphia*, a fraternity; there are only two stamens.

GEN. CHAR. Calyx 5-cleft, segments equal: corolla ringent, closed, upper lip 2-toothed, lower 3-lobed: stamens two: filaments dilated at the base, furnished above the middle with a bristly tooth (the rudiment of an abortive filament): anthers 2-celled, cells parallel, muticous, at length twisted: capsules lanceolate, compressed, many-seeded: seeds subtended with retinacula, convex-concave, inarginate.

(1) H. POLYSPERMA. (*Nees.*)

Ident. Dec. prod. XI. p. 80.—Wall. Pl. As. Rar. III. p. 80.

Syn. Justicia polysperma, *Roxb.*—Ruellia polysperma, *Roth.*—Adenosma polysperma, *Spreng.*

Engrav. Wight's Icon. t. 1492.

SPEC. CHAR. Herbaceous, creeping, glabrous: leaves elliptic-oblong, somewhat crenate: spikes short, terminal in the branches and branchlets: common bracts obovate or oval, roughish: flowers small, pale yellow.

Bengal. Oude. Silhet. Assam. Flowering in the cold season.

GENUS XI. PHYSICHILUS.

Didynamia Angiospermia. *Sex. Syst.*

Deriv. From *Physis*, a bladder, and *Cheilos*, a lip; in allusion to the corolla.

GEN. CHAR. Calyx deeply 5-parted, equal, segments narrow: corolla personate; palate densely blistered; upper lip 2-cleft or entire, lower 3-cleft, of the middle segment smaller: stamens joined in pairs at the base: filaments inflexed: anthers attached above the base, linear-oblong; cells contiguous, parallel, muticous, never twisted, those of the shorter stamens smaller: capsules oblong, depressed, 8-seeded from the base; seeds subtended with scale-shaped retinacula, orbiculate, convexo-concave, marginate.—Herbs or undershrubs: leaves narrow: flowers violet, either axillary, subsessile and collected into a somewhat secund terminal spike or verticillate.

(1) P. AERPHYLLUM. (*Nees.*)

Ident. Dec. prod. X. p. 525.—Dalz. Bomb. flor. p. 184.

Engrav. Wight's Icon. t. 1493.

SPEC. CHAR. Diffuse, branched, creeping: leaves strigose and hirsute, those on the stem nearly orbicular: floral ones oblong or oblong-lanceolate: flowers subsessile, collected into a terminal spike, axillary, purple.

Common in rice fields. Bombay. Mysore. Malwan.

GENUS XII. GYMNOSTACHYUM.

Diandria Monogynia. *Ser : Syst :*

Deriv. From *Gymnos*, naked, and *Stachys*, a spike.

GEN. CHAR. Calyx 5-parted, upper segment usually shorter: corolla 2-lipped, upper lip narrower, 2-toothed, lower 3-cleft: stamens two, inserted below the middle, included, no rudiments of sterile ones: anthers 2-celled, cells equal, parallel, contiguous, one or both sides mucronate at the base, when 1-mucronate then usually the anthers are 1-celled : stigma 2-cleft, segments compressed : capsule tetragonal, 2-celled and seed-bearing from the base to the apex, 12-seeded.—Herbs: raceme spiciform, slender, subsecund, simple or 2-3-cleft, flowers distant, very shortly pedicelled : bracts small, subulate, sub-opposite, one sterile a little inferior.: bracteoles none, except when the raceme has the flowers fascicled in threes when the lateral ones are bibracteolate.

(1) G. ALATUM. (*R. W.*)

Icon. Wight's Icon. vol. IV. & t. 1525.

SPEC. CHAR. Stemless: leaves glabrous, all radical, humifuse, long petioled, cordately sub-orbicular, entire : petioles winged : spikes ascending : flowers sessile, solitary, sub-alternate : calyx segments all equal, sub-pubescent, acute: corolla many times longer than the calyx, limb large, ventricose, upper lip emarginate, under 3-toothed : anthers 2-celled, pubescent.

Coorg.

(2) G. POLYANTHUM. (*R. W.*)

Icon. Wight's Icon. vol. IV. & t. 1494.

SPEC. CHAR. Glabrous: flowers fascicled on slender glabrous racemes : fascicles short-peduncled, sub-aggregate : fascicles furnished with minute subulate bracts, glabrous: leaves cordately orbicular, somewhat cuspidately acuminate : corolla cylindrical : anthers 2-celled, cells parallel, distinct, except at the apex, ecalcarate : capsules slender, many-seeded.

Coorg.

GENUS XIII. NOMAPHILA.

Didynamia Angiospermia. *Ser : Syst :*

Deriv. From *Nomos*, a pasture, and *Phileo*, to love ; alluding to the habitat of the plants.

GEN. CHAR. Calyx 5-parted, segments narrow, upper one a little longer and broader: corolla subpersonate, upper lip broad-concave, 2-lobed, lower 3-lobed, disk inflated, keeled, hirsute: stamens four, shorter than the upper lip, connate in pairs at the base: anthers nearly equal, ovate, cordate (violet), cells parallel, rusticous, antrorsely contiguous, separated behind from the oblong connectivum; stigma simple: capsule narrow, 6-striated, slightly tapering, marked with a furrow at the back; 2-celled and seed-bearing even to the base, many-seeded: seeds small, compressed, sub-orbiculate: retinacula middle-sized, obtuse.—Herbs growing in pastures and fields: cymelets axillary, short-peduncled, 2-cleft, 3-many-flowered and then capituliform, flowers second upwards: bracteoles small, shorter than the calyx: corolla purple, the palate of a deeper colour.

(1) N. PINNATIFIDA. *(Dals.)*

Ident. Dalz. in Hook. Journ. Bot. III. p. 38.—Bomb. flor. p. 164.

SPEC. CHAR. All glandular and pubescent: leaves petioled, deeply pinnatifid, linear-lanceolate, the segments linear, oblong-obtuse, serrulated: flowers in the opposite axils of the leaves, sessile, or clustered in terminal heads: flowers purple.

The river-banks of the Southern Concan, flowering in January and March.

(2) N. STRICTA. *(Nees.)*

Ident. Dec. prod. XI. p. 84.

Syn. Justicia stricta, *Vahl.*—Justicia pubescens, *Lam.*

SPEC. CHAR. Perennial: leaves oval, or oval-oblong, repandly subcrenate, decurrent into a long petiole, attenuated into an obtuse acumen: cymes axillary, dichotomous and with the calyx glandulosely hairy.

Malabar.

GENUS XIV. HYGROPHILA.

Didynamia Angiospermia. *Sex: Syst:*

Deriv. From *Hygros*, moist, and *Phileo*, to love; alluding to the habitat of the plants.

GEN. CHAR. Calyx tubular, 5-cleft to about the middle, segments equal: corolla ringent, lower lip convex in the middle, wrinkled, 3-cleft: stamens not exserted, frequently connate in pairs at the base: anthers 2-celled, cells parallel, divergently sagittate at the base, muticous or somewhat mucronate, violet: stigma simple, subulate, incurved: capsules narrow, 6-striated, 2-celled and seed-bearing to the base, many-seeded: seeds small, orbiculate, compressed, smoothish: retinacula short, obtuse.—Herbs: root stoloniferous: stems erect or procumbent, 4-angled; leaves quite entire, or here and there crenated, densely lineolate above, furnished with often many-ribbed divisions, hairy or glabrous: flowers axillary, cymosely glomerate, forming dimidiate or entire verticils: teeth of the calyx rough or bearded: corolla purple or yellowish.

(1) H. OBOVATA. (*Nees.*)

Ident. Dec. prod. XI. p. 91.
Syn. Ruellia obovata, *Roxb.*
Engrav. Wight's Icon. t. 1489.—Rheede Mal. II. t. 48.
SPEC. CHAR. Stem herbaceous, erect: cauline leaves oblong, those of the branches obovate, obtuse, attenuated into the petiole, entire, slightly hirsute on both sides: flowers blue, half verticilled: calyx 5-fid, the segments and the inferior lip of the corolla bearded.

Courtallum. Chittagong. Malabar. Flowering during the rainy season.

(2) H. SALICIFOLIA. (*Nees.*)

Ident. Dec. prod. XI. p. 92.—Dalz. Bomb. flor. p. 184.
Syn. Ruellia salicifolia, *Vahl.*—*Roxb. flor. Ind.* I. p. 50.—R. longifolia, *Roth.*
Engrav. Wight's Icon. t. 1490.
SPEC. CHAR. Stem herbaceous, erect, rough round the joints: leaves lanceolate, acuminate at both ends, lineolate, hirsute on the veins beneath: whorls dimidiate: segments of the calyx subulate, hairy: flowers axillary, about 7 together, pale blue: capsule quadrangular, narrow-compressed.

Wet places in the Southern Concan. Quilon.

(3) H. PHLOMOIDES. (*Nees.*)

Ident. Nees. ab. Esen. in Wall. Pl. As. Rar. III. p. 80.—Dec. prod. XI. p. 90.
SPEC. CHAR. Herbaceous, creeping at the base: stem quadrangular, hirsute: leaves oblong-elliptic, subcrenated, sessile, hirsute: flowers verticilled: bracts and calyces very hirsute.

Silhet.

(4) H. INCANA. (*Nees.*)

Ident. Dec. prod. XI. p. 91.
Syn. Ruellia hirsuta, *Roxb. fior. Ind.* III. p. 51.
SPEC. CHAR. Stem procumbent or erect, tetragonal, geniculate: branches ascending: leaves lanceolate, quite entire, subsessile, hairy-villous: flowers verticilled: verticils complete: bracts and calyces very hirsute.

Chittagong.

(5) H. RADICANS. (*Nees.*)

Ident. N. ab. E. in Wall. Pl. As. Rar. III. p. 81.—Dec. prod. XI. p. 92.
SPEC. CHAR. Herbaceous, procumbent, creeping: leaves oblong, acute at both ends, repandly subcrenate, lineolate, glabrous: verticills dimidiate; segments of the calyx subulate, hirsute.

Silhet.

GENUS XV. CRYPTOPHRAGMIUM.

Diandria Monogynia. *Sea: Syst.*

Deriv. From *Kryptos*, concealed, and *Phragmion*, a partition; alluding to the division of the cells of the anthers.

GEN. CHAR. Calyx 5-parted, equal: corolla either 2-lipped or ringent, tube straight or incurved: stamens two, lower side inserted on the tube: anthers 2-celled, especially contiguous in front and debiscing with a simple cleft, the bottom of the cell divided by a common bilamellate partition: capsule tetragonal, 2-celled with an adnate partition, many-seeded, seed-bearing from the base: seeds subtonied with retinacula.—Herbs, often much branched: leaves toothed, or repandly toothed: flowers nodding, purplish, arranged in small cymes: bracts and bracteoles small, nearly equal.

(1) C. CANESCENS. (*Nees.*)

Ident. Dec. prod. XI. p. 95.
Syn. Justicia pubescens, *Lam.*
Engrav. Wight's Icon. t. 1405.
SPEC. CHAR. Spikes axillary, passing into terminal, bifid, secund-flowered, glandulosely-hirsute: leaves ovate, acutish, cunei-

form at the base, repand, pubescent: capsule twice the length of the setaceous calyx.

Courtallum, flowering in August and September.

(2) C. LATIFOLIUM. (Dalz.)

Ident. Dalz. in Hook. Journ. Bot. III. p. 37.—Bomb. flor. p. 185.

Syn. Phlogauthus latifolius, *R. W.*

Engrav. Wight's Icon. t. 1537.

SPEC. CHAR. Suffruticose, glabrous: leaves very long-petioled, rounded-ovate, acuminate, truncate at the base, crenulate, 1 foot long: spikes axillary, short trichotomous: flowers yellowish-white: capsule 4 times longer than the calyx.

Chorla Ghaut.

(3) C. GLABRUM. (Dalz.)

Ident. Dalz. l. c. p. 138.

SPEC. CHAR. Herbaceous: leaves elliptic-acuminate, denticulate, glabrous, running down with a wing into the petiole: spike terminal, compound: the branches opposite, nearly a foot long.

In shady places in the Southern Concan.

(4) C. VENUSTUM. (Nees.)

Ident. N. ab. E. in Wall. Pl. As. Rar. III. p. 100.—Dec. prod. XI. p. 94.

Syn. Justicia venusta, *Wall. Pl. As. Rar.* III. p. 53.

Engrav. Wall. l. c. t. 60.

SPEC. CHAR. Stem compressed-tetragonal, canescent, scabrous: leaves elliptic, wingedly decurrent into the petiole, denticulate, scabrous: corolla much longer than the calyx, pubescent, purple: spike terminal, compound, brachiate.

Silhet.

(5) C. ELONGATUM. (Nees.)

Ident. Dec. prod. XI. p. 95.

Syn. Justicia elongata, *Vahl.*—C. cordifolia, *N. ab. E. in Wall. Pl. As. Rar.* III. p. 100.

SPEC. CHAR. Stems many from a single root, quadrangular: leaves ovate, acute, dotted, scabrous above, lower ones subcordate: racemes axillary, secund: flowers distant, purplish.

Tranquebar. Courtallum. Foot of the Neilgherries.

GENUS XVI. PHLEBOPHYLLUM.

Diandria Monogynia. *Sex. Syst.*

Deriv. From *Phleps*, a vein, and *Phyllon*, a leaf.

GEN. CHAR. Calyx deeply 4-parted, upper segment deeply 2-cleft: corolla funnel-shaped, limb 5-cleft, segments emarginate and repand: stamens two, not exserted: anthers 2-celled, cells parallel, muticous: stigma simple, subulate: capsule compressed, without seeds at the base, 2-celled towards the apex, 4-seeded.—Spike with broad imbricated bracts, bracteoles twin, narrow.

(1) P. KUNTHIANUM. (*Nees.*)

Ident. Dec. prod. XI. p. 102.—N. ab. E. in Wall. Pl. As. Rar. III. p. 83.

Engrav. Wight's Icon. t. 448.

SPEC. CHAR. Small erect shrub with obsoletely 4-sided branches and oval, subundulate, acutely serrated leaves, clothed with whitish tomentum beneath: flowers pale-bluish, sometimes nearly white.

On hill pastures at great elevations in the Peninsula.

GENUS XVII. CODONACANTHUS.

Diandria Monogynia. *Sex. Syst.*

Deriv. From *Kodon*, a little bell; alluding to the shape of the flowers.

GEN. CHAR. Calyx 5-parted, equal: corolla campanulate from a short tube, limb equal, 5-cleft, segments ovate, obtuse: stamens two, included: anthers broadish, oval, 2-celled, cells parallel, nearly equal, muticous.—Herbs: racemes terminal on the stem and branches, compound, second-flowered, with small bracts and bracteoles: flowers nodding, middle-sized, blue.

(1) C. PAUCIFLORUS. (*Nees.*)

Ident. Dec. prod. XI. p. 103.

Syn. Asystasia pauciflora, *N. ab. E. in Wall. Pl. As. Rar.* III. p. 90.

SPEC. CHAR. Root and stem creeping below: branches diffuse: leaves oblong, acute at both ends, subrepand, glabrous: raceme terminal, subdivided, few-flowered.

Silhet and the neighbouring mountains.

GENUS XVIII. ENDOPOGON.

Didynamia Angiospermia. *Ser. Syst.*

Deriv. From *Endon*, within, and *Pogon*, a beard; alluding to the throat of the corolla.

GEN. CHAR. Calyx regular, 5-parted: corolla in the bud, often convolutely mucronate, bilabiate, throat inflated, upper lip broad, bifid within, having a decurrent canal bearded on both sides for the reception of the style, lower lip trifid: stamens two: cells of the anthers parallel, equal, muticous: capsule 4-angled, sutures prominent, 2-celled, four-seeded near the base, partition adnate, narrow and incomplete above: seed either depressed, lenticular, mucronulate with a shield-like depression on both sides near the hilum, or ovate, subcordate, carinate on one side, smooth.— Shrubs with serrated leaves: flowers spicate: common bracts opposite, imbricated, broadish, the proper ones narrow: corolla showy, blue.

(1) E. VISCOSUM. *(Nees.)*
Var. HUMILIS.

Ident. Dec. prod. XI. p. 104.
Engrav. Wight's Icon. t. 1498.

SPEC. CHAR. Low branched shrub: bracts oblong, lanceolate (the lower ones sometimes oval), obtusely unguiculato-cuspidate, carinate, as long as the calyx, rigid, and with the rachis, hirsutely glandulose: leaves oval-oblong, or oval, attenuated at both ends, hispid: seed oval: spikes very dense, villous.

Courtallum.

(2) E. VERSICOLOR. *(R. W.)*

Ident. Wight's Icon. vol. IV. & t. 1497.

SPEC. CHAR. Bracts lanceolate, subulately attenuated at the apex, and like the calyx, densely glanduloso-hirsute: calyx 5-cleft, segments lanceolate: leaves long-petioled, broadly ovate, acuminate, crenate, glabrous above, white beneath.

Neilgherries, flowering in March and April.

(3) E. CAPITATUM. *(R. W.)*

Ident. Wight's Icon. vol. IV. & t. 1409.

SPEC. CHAR. Large branched shrub: spikes short, capitate: exterior bracts leaf-like, limb glabrous, the dilated base, calyx, branchlets and petioles thickly covered with rigid glandular hairs: leaves ovate, acuminate, serrated: limb glabrous, densely lineolate.

Neilgherries, flowering in March and April.

(4) E. STROBILANTHES. *(R. W.)*

Ident. Wight's Icon. vol. IV. & t. 1500.

SPEC. CHAR. Spikes elongated, glabrous, exterior bracts foliaceous, oblong, lanceolate, or ovate, acuminate, longer than the calyx; calyx 5-cleft, segments lanceolate: leaves broadly ovate, acuminate, serrated, glabrous, lineolate on both sides.

Neilgherries, flowering in March and April.

(5) E. FOLIOSUM. *(R. W.)*

Ident. Wight's Icon. vol. IV. & t. 1501.

SPEC. CHAR. Shrub: spikes short, capitate, glabrous; exterior bracts leaf-like: limb ovate, acuminate, serrated: flowers diandrous: leaves long-petioled, glabrous.

Neilgherries, flowering in March and April.

(6) E. RHAMNIFOLIUM. *(R. W.)*

Ident. Wight's Icon vol. IV. & t. 1521.

Syn. Buteræa rhamnifolia, *Nees.*

SPEC. CHAR. Shrub: young branches and spikes softly whitish-hirsute: bracts rhomboid-oblong: petioles glabrous: stem glabrous below, subcrenate towards the apex, woolly-hirsute: leaves crenato-serrated, rough above with scattered bristles, glabrous beneath; bracts subcrenate at the base, obtuse or ending in short acumen at the apex.

Locality not specified.

(7) E. INTEGRIFOLIUM. *(Dalz.)*

Ident. Dalz. in Hook. Journ. Bot. II. p. 342.—Bomb. flor. p. 185.

SPEC. CHAR. Shrub: leaves narrow-elliptic or lanceolate-acuminate, running down with a wing into the petiole; bracts and bracteoles linear, ciliated with long hairs, and as long as the calyx: flowers blue, rather large: rachis short, quadrangular, viscous and glandular.

Hills near Panwell.

(8) E. HYPOLEUCUM. *(Nees.)*

Ident. N. ab. E. in Wall. Pl. As. III. p. 99.—Dec. prod. XI. p. 104.

SPEC. CHAR. Bracts bristly-cuspidate, pubescently glandular, leaves broad-ovate, crenated, hoary-tomentose below.

Banks of rivers round the Kolyrgherries.

(9) E. CONSANGUINEUM. *(Nees.)*

Ident. N. ab. E. in Wall. Pl. As. Rar. III, p. 99.—Dec. prod. XI. p. 104.

SPEC. CHAR. Bracts ovate-elliptic, acute, hirsute, glandular: leaves ovate-elliptic, attenuated at both ends, glabrous.

Dindigul hills. Courtallum. Negapatam.

(10) E. KHASSYANUM. *(Nees.)*

Ident. N. ab. E. in Dec. prod. XI. p. 104.

SPEC. CHAR. Hirsute: bracts oblong, spreading, shorter than the calyx: spikes shorter than the leaf: leaves ovate, acute, crenated, sharply passing at the base into a very hirsute petiole.

Khasia hills.

(11) E. MACROSTROIUM. *(Nees.)*

Ident. Nees. in Dec. prod. XI. p. 105.

SPEC. CHAR. Undershrub; glabrous: bracts ovate, cuspidate, shorter than the calyx: leaves oblong-acuminate, sharply passing into a short petiole, somewhat toothed in the middle: spikes axillary, somewhat fascicled.

Assam. Khasia.

(12) E. DECURRENS. *(Nees.)*

Ident. Nees. in Dec. prod. XI. p. 105.

SPEC. CHAR. Undershrub: spikes tetragonal: bracts oval, cuspidate, densely ciliated, decurrent at the base, densely imbricated: leaves oblong, caudately cuspidate, quite entire, scabrous at the margin; seeds cordately orbiculate.

Assam. Khasia.

(13) E. VITELLINUM. *(Nees.)*

Ident. Nees. in Dec. prod. XI. p. 723.

Syn. Justicia vitellina, *Roxb.*

SPEC. CHAR. Shrub, 3 feet: branches 4-sided: angles scabrous: leaves broad-lanceolate, attenuated at both ends, reflexed, glabrous: racemes somewhat cylindric: flowers somewhat fascicled, yellowish-red: upper lip of the corolla 2-cleft.

Chittagong hills.

GENUS XIX. STENOSIPHONIUM.

Didynamia Angiospermia. *Sm. Syst.*

Deriv. From *Stenos*, narrow, and *Siphon*, a tube; alluding to the corolla.

GEN. CHAR. Calyx 5-cleft, segments equal, first joined beyond the middle by membranaceous margins, after flowering and separating at the base: corolla funnel-shaped, tube very slender, limb inflated, campanulate, 5-cleft, segments equal, obtuse: stamens either four, didynamous, or only two: anthers narrow, cells parallel, narrowly contiguous, equal, muticous: stigma membranaceously crested at the back, subulate at the apex: capsule tetragonally columnar, 2-celled even to the base, from the base beyond the middle 8-seeded: partition complete, adnate: seeds lenticular, mucronate, smoothish.—Shrubs: leaves more or less toothed: spikes axillary and terminal: flowers separated, opposite, single or aggregated: bracts broadish, appressed, stiffish: bracteoles narrow, resembling the calyx: flowers spotted in the throat.

(1) S. CORYMBOSUM. (*Nees.*)

Ident. Dec. prod. XI. p. 105.

Engrav. Wight's Icon. t. 1503.

SPEC. CHAR. Leaves broad ovate, dentate, marked beneath: bracts ovate-lanceolate or oblong, obtusely acuminate (shorter than the calyx) and like the calyx, glandulosely hirsute: flowers geminate or ternate.

Pulney mountains, Neilgherries.

(2) S. RUSSELLIANUM. (*Nees.*)

Ident. Dec. prod. XI. p. 105.—N. ab. E. in Wall. Pl. As. Rar. III. p. 84.

Engrav. Wight's Icon. t. 1503.

SPEC. CHAR. Shrub: leaves ovate, dentate, naked beneath: bracts rhomboid, obtuse, cuspidate, as long as the calyx.

Pulney mountains. Neilgherries.

(3) S. DIANDRUM. (*Nees.*)

Ident. Dec. prod. XI. p. 105.

Engrav. Wight's Icon. t. 1502.

SPEC. CHAR. Shrub: leaves ovate, crenato-dentate, glabrous: bracts oblong, somewhat obtuse, and like the calyx, glandulosely-hispid: flowers solitary, diandrous: filaments hairy at the base.

Courtallum, flowering in July and August.

(4) S. TUBAERICUM. *(Nees.)*

Ident. N. ab. E. in Wall. Pl. As. Rar. III. p. 84.—Dec. prod. XI. p. 105.

Syn. Ruellia cordifolia, *Vahl.*

Spec. Char. Shrub: leaves ovate, toothed, younger ones below white tomentose; bracts rhombeo-ovate, long-cuspidate, equalling the calyx: young branches covered with white down.

Hills near Chingleput. Tranquebar.

GENUS XX. DYSCHORISTE.

Didynamia Angiospermia. *Sex: Syst:*

Deriv. From the Greek, meaning *difficult to separate*; probably alluding to its near alliance to other genera.

Gen. Char. Cayx tubular, scarcely 5-cleft to the middle, equal: corolla funnel-shaped, limb oblique, 5-cleft: stamens four, didynamous: anthers 2-celled, cells parallel, equal, mucronate at the base, seldom muticous: capsule oblong, slightly depressed, stiff, 4-seeded almost from the base, without seeds upwards: partition adnate: seeds ovate, obtuse.—Small depressed shrubs, with small leaves: flowers axillary, middle-sized, blue, terminal, solitary, sessile on a very short branchlet, furnished at the base with two bracteoles resembling leaves, surrounded with bracts and narrow bracteoles.

(1) D. LITORALIS. *(Nees.)*

Ident. N. ab. E. in Wall. Pl. As. Rar. III. p. 81.—Dec. prod. XI. p. 108.

Syn. Ruellia litoralis, *Linn. Suppl.*

Engrav. Wight's Icon. t. 447.—Burm. flor. Ind. t. 4. fig. 3.

Spec. Char. Stem fruticose, diffuse: leaves cuneiform, retuse, dentate towards the apex, glabrous.

In dry rocky places in the Peninsula.

(2) D. DEPRESSA. *(Nees.)*

Ident. N. ab. E. l. c. III. p. 81.—Dec. prod. XI. p. 106.

Spec. Char. Stem creeping; leaves obovate-orbiculate, obtuse or retuse, quite entire, mucronate, glabrous: flowers erect,

Tanjore.

(3) D. czanca. *(Nees.)*

Ident. N. ab. E. l. c.—Dec. prod. XI. p. 106.

SPEC. CHAR. Stem procumbent, diffuse, scabrous: leaves spathulate, quite entire, glaucous: flowers and fruit reflexed.

Peninsula, the exact locality not given.

GENUS XXI. CALOPHANES.

Didynamia Angiospermia. *See: Syst.*

Deriv. From *Kalos*, beautiful, and *Phainomai*, to appear; alluding to the appearance of the plants.

GEN. CHAR. Calyx cohering from the base to a fourth part or a little beyond, segments bristly: corolla funnel-shaped, limb 5-cleft, somewhat regular: anthers 2-spurred at the base, seldom muticous, cells parallel, flat, membranaceous: filaments connate in pairs at the base: capsule lanceolate, without cells at the base, hence 4-seeded in the middle: flowers axillary, opposite or 2-5-cymosely aggregate or sessile, bracts and bracteoles narrow.—Herbs or undershrubs, usually low, more or less pubescent: corolla blue, spotted in the throat.

(1) C. VAGANS. *(R. W.)*

Ident. Wight's Icon. vol. IV. & t. 1526.

SPEC. CHAR. Shrubby, diffuse, climbing: leaves oval or subovate, petioled, entire: peduncles axillary, longer than the petioles, cymose, 2-3-flowered and with the calyx somewhat viscosely pubescent: calyx-lobes subulate, about half the length of the bilabiate corolla: seeds hairy.

Coorg, climbing among bushes.

(2) C. NAGCHANA. *(Nees.)*

Ident. Nees. in Dec. prod. XI. p. 109.

Syn. Dipteracanthus Nagchana, *N. ab. E. in Wall. Pl. As. Rar.* III. p. 82.

SPEC. CHAR. Stem procumbent, rooting at the base, scabrous: leaves oblong, obtuse, cuneiform at the base, repandly-crenated, scabrous: peduncles axillary, very short, 3-flowered: bracteoles spathulate: corolla shorter than the subulate-elongated segments of the calyx: spurs of the anthers undivided.

Patna.

GENUS XXII. PETALIDIUM.

Didynamia Angiospermia. *Ser. Syst.*

Deriv. A Greek diminutive, alluding to the small petals.

GEN. CHAR. Calyx equal, deeply 5-parted, valvately enclosed by two bracteoles: corolla funnel-shaped, limb nearly equal, 5-cleft: stamens included; anthers oblong, sagittate, cells parallel, equal, awned at the base: stigma 2-cleft, segments filiform: capsule compressed at the base and seedless, hence 4-seeded in the middle: dissepiment complete, adnate, persistent: seeds furnished with hooked subulate retinacula, ovate, acute, compressed.—Shrubs: leaves crenated: flowers peduncled, axillary, solitary, or showing an axillary fascicle on a short branchlet, blue, showy.

(1) P. BARLERIOIDES. (*Nees.*)

Ident. N. ab. E. in Wall. Pl. As. Rar. III. p. 75. & t. 82.— Dec. prod. XI. p. 114.

Syn. Ruellia barlerioides, *Roth.*—R. bracteata, *Roxb.*—Eranthemum barlerioides, *Roxb.*

Engrav. Bot. Mag. t. 4053.

SPEC. CHAR. Shrubby: leaves oblong, crenate dentate, somewhat glaucous: peduncles axillary, solitary, 1-flowered: flowers an inch long, white or pale-blue: bracteoles large, opposite, covering the calyx.

The Ghauts: hills near Panwell.

(2) D. PATULUM. (*Nees.*)

Ident. N. ab. E. in Dec. prod. XI. p. 126.—Dalz. Bomb. flor. p. 185.

Syn. Dipteracanthus patulus, *Dec. l. c.*—Ruellia patula, *Jacq.* —*Roxb.*—R. erecta, *Roth.*

Engrav. Jacq. Ic. Rar. I. t. 119.

SPEC. CHAR. Stem erect: leaves ovate, oval or oblong-obtuse, hoary and puberulous: flowers fascicled in threes or fives, or solitary, smaller than in the preceding, white: bracteoles oval or oblong, longer than the calyx.

The flowers open in the evening, and fall off in the morning; while in the preceding, the flowers open in the morning. Negapatam. Foot of the Neilgherries. Pondicherry.

GENUS XXIII. DIPTERACANTHUS.

Didynamia Angiospermia. *Ser. Syst.*

Deriv. From *Dis*, double, and *Pteros*, a wing; alluding to the pair of large leafy bracts.

GEN. CHAR. Calyx equal, more or less deeply 5-cleft: corolla funnel-shaped, limb nearly equal, 5-cleft: stamens didynamous, included, filaments contiguous or joined together at the base: anthers linear-sagittate, cells parallel, equal, muticous: stigma bilamellate, knotted at the base: capsule compressed and seedless at the base, usually from the middle, rarely nearer from the base, 2-8-12-16-seeded: dissepiment membranaceous in the middle, afterwards in a great measure evanescent: retinacula hooked, præmorse: seeds orbiculate, compressed, surrounded by a divided tumid margin.—Herbs, creeping or erect, rarely shrubs: flowers either all or the lower ones at least axillary, or solitary or fascicled, sessile or peduncled, upper ones generally collected in a small-bracteated raceme: bracts two, large, leafy, often petioled, placed below the calyx or fascicle, smaller and narrower in the racemose parts: bracteoles either none or very small: capsules by abortion 4 or 2-seeded.

(1) D. CILIATUS. (*Nees.*)

Ident. N. ab. E. in Wall. Pl. As. Rar. III. p. 87.—Dec. prodr. XI. p. 120.

Syn. Ruellia ciliatus, *Spreng.*

SPEC. CHAR. Stem erect, scabrously pubescent: leaves ovate, acute at both ends, repandly crenated, hairy or hirsute: flowers axillary, almost terminal, sessile or capitate, peduncled: common bracts spathulate, petioled: flowers large blue.

Coromandel, flowering in the cold season.

(2) D. SISUA. (*Nees.*)

Ident. N. ab. E. in Wall. Pl. As. Rar. III. p. 81.—Dec. prod. XI. p. 120.

Syn. Ruellia suffruticosa, *Roxb.*

SPEC. CHAR. Stems short, from a creeping rhizome, densely leafy at the apex: leaves elliptic, obtuse, quite entire, hirsute: flowers solitary, subsessile, white: bracts oblong, longer than the calyx.

Nukunagúr. Dinagepore. Flowering in May.

(3) D. PROSTRATUS. (*Nees.*)

Ident. Dec. prod. XI. p. 124.

SPEC. CHAR. Stem herbaceous, prostrate, covered with whitish pubescence: leaves ovate, obtuse, almost quite entire, very small, pubescently hairy, canescent: flowers axillary, solitary, short-peduncled: bracteoles oval, petioled, longer than the calyx: capsule pubescent, many-seeded.

In pastures and hedges at Negapatam. Foot of the Neilgherries, Tranquebar.

(4) D. DEJECTUS. (*Nees.*)

Ident. N. ab. E. in Wall. Pl. As. Rar. III. p. 82.—Dec. prod. XI. p. 125.

Syn. Ruellia repens, *Blume.*—R. ringens, *Roxb.*

SPEC. CHAR. Stem herbaceous, suffrutescent at the base, creeping, procumbent or ascending: leaves long-petioled, ovate-elliptic, acute at both ends, entire, strigose or rough on the veins: capsule half an inch long, oblong, narrow at the base, 12-16-seeded: flowers axillary, sessile, solitary, pale blue, expanding in the evening and drooping the next morning.

Common all over India, flowering nearly all the year.

GENUS XXIV. HEMIGRAPHIS.

Didynamia Angiospermia. *Sex. Syst.*

GEN. CHAR. Calyx 5-parted, segments nearly equal, upper one larger and broader, usually deeply joined together at the base in pairs, namely, two lower ones and two upper ones, (one of these being smaller), the fifth free: corolla funnel-shaped, resupinate, limb 5-lobed, lobes obliquely sub-retuse, two upper ones a little smaller: stamens didynamous, inserted at the apex of the tube, confluent at the base: anthers 1-celled, cell adnate to a narrow keeled connectivum, one cell of the upper stamens changed into a fringe of hairs placed below the cell, one cell of the lower stamens altogether wanting, the perfect cell host-shaped at the base, mucronate: stigma furnished with a blunt tooth at the dorsal base, pubescent, simple: capsule 6-8-seeded below the middle, without cells at the base: seeds furnished with retinacula, echinate.—Perennial, branched, flexuose, softly villous, viscid herbs: leaves oblong, attenuated at both ends, serrated, floral ones (bracts) quite entire: flowers axillary, either solitary, subsessile, bibracteolate, or glomerate, terminal, capitate, bracteate.

(1) H. ELEGANS. (*Nees.*)

Ideal. Dec. prod. XI. p. 722.
Syn. Ruellia elegans, *Sims.*—R. diffusa, *N. ab. E. in Wall. Pl. As. Rar.* III. p. 83.—R. Crossandra, *Steud.*
Engrav. Roxb. Cor. I. t. 67.—Bot. Mag. t. 3389.

SPEC. CHAR. Leaves oblong or oval-oblong, petioled, serrated, hirsute: stem tetragonal, procumbent, branched, hirsute: branches diverging, geniculate: spikes axillary, short-peduncled, capituliform; bracts lanceolate, leafy, longer than the calyx: flowers blue, with a yellow tube, the two lower segments marked with 3 purple lines, three upper ones reddish at the base.

Hurdwar. Chunar.

(2) H. LATEBROSA. (*Roxb.*)

Ideal. Roxb. flor. Ind. III. p. 46.—Dec. prod. XI. p. 723.
Syn. Ruellia latebrosa, *Roth.*
Engrav. Wight's Icon. t. 1504.

SPEC. CHAR. Leaves ovate, coarsely serrated: bracts ovate-lanceolate, equalling the calyx.

Delhi. Hurdwar.

GENUS XXV. RUELLIA.

Didynamia Angiospermia. *Sex. Syst.*

Deriv. In honor of John Ruelle of Soissons, Botanist and Physician to Francis 1st.

GEN. CHAR. Calyx 5-parted at the base, segments either unequal, the upper one being deeper, or nearly equal and linear, somewhat dilated at the apex, or equal and acuminate: corolla funnel-shaped, tube passing gradually into a narrow-campanulate 5-cleft limb, segments equal, obtuse: stamens four, didynamous, inserted at the base of the throat: anthers oblong, 2-celled, cells parallel, contiguous, equal, muticous or mucronulate at the base; stigma subulate, spiral, spongy at the back, channelled, increased at the base by a small tooth: capsule narrow, 4-angled, 2-celled to the base, from the base to beyond the middle 6-8-16-seeded: dissepiment complete, adnate: seeds furnished with middle-sized retinacula.—Herbs and undershrubs, pubescent and hirsute: spikes usually contracted into the shape of capituli, leafy-bracteate and therefore little conspicuous, or when the inflorescence is more perfect, dichotomous and truly cymose: bracteoles none or narrow: flowers middle-sized.

(1) R. HIRTA. (*Vahl.*)

Ident. Vahl. Symb. III. p. 84.—Dec. prod. XI. p. 145.
Syn. R. sarmentosa, *N. ab. E. in Wall. Pl. As. Rar.* III. p. 83.
Engrav. Vahl. l. c. t. 67.

SPEC. CHAR. Perennial: leaves ovate, crenately serrated towards the apex, hirsutely hoary: stem creeping: spikes axillary, peduncled, capituliform: bracts leafy, oblong: bracteoles none: capsule 12-seeded: flowers large, bluish-purple.

Sukanagur. Coromandel. Circars. Flowering in the cold season.

(2) R. TRISTA. (*Linn.*)

Ident. Linn. Sp. p. 80.—Dec. prod. XI. p. 90.—N. ab. E. in Wall. Pl. As. Rar. III. p. 83.

SPEC. CHAR. Stem herbaceous, creeping at the base: leaves ovate, acute at both ends, undulately crenated, lineolate, bispid: capitula axillary, opposite, bracteate, hirsute: peduncle nearly equalling the calyx: bracts obovate, crenated at the apex: corolla short, with a campanulate throat.

Locality not given.

(3) R. ASPERA. (*Nees.*)

Ident. N. ab. E. in Wall. Pl. As. Rar. III. p. 84.—Dec. prod. XI. p. 147.
Syn. Strobilanthes scaber, *N. ab. E. l. c.*

SPEC. CHAR. Leaves oblong, obtuse, repandly crenate, sharply passing into the short petiole, thick-nerved below, loosely hairy, floral ones lanceolate, sessile, quite entire, pilosely scabrous: spikes axillary, collected into a compound spike at the apex of the branchlets, glandulosely hirsute: bracts either linear-spathulate or linear-lanceolate: segments of the calyx linear, obtuse, of the corolla, roundish: tube incurved: capsule 6-seeded.

Silhet.

(4) R. DURA. (*Nees.*)

Ident. N. ab. E. in Dec. prod. X. p. 146.—Dalz. Bomb. flor. p. 186.

SPEC. CHAR. Stem quadrangular, procumbent, bispid and bristly: leaves oblong-obtuse, subcrenate, attenuated into the petiole, bispid: spikes axillary, subsessile and terminal, somewhat capitate, subtended by subovate hirsute bracts: flowers middle-sized, blue: capsule 8-seeded, shorter than the calyx.

About Surat, common. Coromandel. Masulipatam. Central India. Flowering in the cold season.

(5) R. PUNCTATA. *(Nees.)*

Ident. Dec. prod. XI. p. 147.
Engrav. Wight's Icon. t. 1563.

SPEC. CHAR. Leaves oval, attenuated at both ends, entire or sub-repand, glandulose-punctate, and like the four-sided herbaceous stem, hirsute: capitula terminal, bracteate, pubescent, foliaceous, acute: bracteoles linear-oblong, closely ciliate.

Courtallum.

GENUS XXVI. ASYSTASIA.

Didynamia Angiospermia. *Ses: Syst:*

GEN. CHAR. Calyx 5-parted, equal: corolla somewhat funnel-shaped, limb 5-lobed, equal, the upper lobe slightly concave: stamens 4, didynamous, within the tube, approaching by pairs: anthers 2-celled, cells parallel, appendiculate at the base: stigma capitulate, 2-lobed, or 2-toothed: capsule contracted at the base, rough, often 4-angled, 2-celled, 4-seeded; seed attached to processes, discoidly lobed, with a prominent angle at the base.—Suffruticose or herbaceous, diffuse or climbing plants: racemes spike-like, 1-sided, axillary or terminal: bracts small, equal: flowers primrose-blue or lilac, or variously tinged with yellow, sometimes handsome.

(1) A. PLUMBAGINEA. *(Nees.)*

Ident. N. ab. E. in Wall. Pl. As. Rar. III. p. 89.—Dec. prod. XI. p. 164.

SPEC. CHAR. Stem obtusely 4-cornered: leaves trapezoid, repand, acute, rough above: spikes terminal, ternate: lower flowers opposite, upper ones secund: segments of the narrow corolla entire.

Gathered from the Governor's garden in Madras.

(2) A. NEESIANA. *(Wall.)*

Ident. Wall. Pl. As. Rar. III. p. 89.—Dec. prod. XI. p. 164.
Syn. Ruellia Neesiana, *Wall. Pl. As. Rar.* I. p. 83.

Spec. Char. Stem quadrangular, erect: leaves ovate-oblong, acute, sharply ending in a short petiole, quite entire, glabrous: racemes terminal, solitary, 2-cleft at the base: flowers alternate, secund, glandulosely pubescent, rose-coloured.

Silhet mountains, flowering in October.

(3) A. DENTICULATA. *(Nees.)*

Ident. N. ab. E. in Wall. Pl. As. Rar. III. p. 89.—Dec. prod. XI. p. 164.

Spec. Char. Stem 4-furrowed, flexuose, pubescently scabrous; leaves oblong, denticulately crenated, cuspidate, cuneiform at the base, glabrous: spikes axillary, opposite, 3-cleft: flowers opposite, yellowish: segments of the corolla emarginate: filaments very hirsute.

Poodun.

(4) A. CHELONOIDES. *(Nees.)*

Ident. N. ab. E. in Wall. Pl. As. Rar. III. p. 89.—Dec. prod. XI. p. 164.

Spec. Char. Stem dichotomous, somewhat hairy: leaves elliptic-oblong, acuminated, acute at the base, repandly crenated, above and at the margin slightly hairy or glabrous: racemes axillary and terminal, simple or trifid, secund: calyx glandulosely scabrous: pedicels a little shorter than the calyx: flowers purple.

Var. 1. Leaves less, glabrous, cuneately attenuated at the base. Neilgherries in moist wood at the Koondahs, flowering in October.

Var. 2. Leaves more manifestly crenated: corolla white. Conoor, on the Neilgherries, flowering in February.

(5) A. PANICULATA. *(Nees.)*

Ident. Dec. prod. XI. p. 167.

Spec. Char. Stem trichotomous; leaves oval, acuminated, serrated, rusty-tomentose below: panicles axillary, trichotomous; flowers sessile, secund: segments of the corolla lunate.

Silhet.

(6) A. COROMANDELIANA. *(Nees.)*

Ident. Dec. prod. XI. p. 165.

Syn. Justicia gangetica, *Linn. Amæn.* IV. p. 290.—Ruellia secunda, *Vahl.*—R. intrusa, *Vahl.*—R. Zeylanica, *Roen.*—Roxb, fl. Ind. III. p. 42.

Engrav. Rheede Mal. 9. t. 45.—Wight's Icon. t. 1506.

SPEC. CHAR. Stem erect: branches numerous, almost smooth: leaves cordate-ovate or suborbiculate, glabrous; racemes axillary, elongated, secund, straight; flowers large, pale-blue: capsule an inch long.

Very common in most parts of the country. There are varieties with white, purplish-blue and primrose-coloured flowers.

(6) A. VIOLACEA. *(Dalz.)*

Ident. Dalz. Bomb. for. p. 186.

SPEC. CHAR. Stem ascending, jointed, smooth, striated, obtusely quadrangular: leaves ovate or oblong-acute, entire, lower ones attenuated into the petiole, upper rounded at the base, short petioled or subsessile: racemes terminal, secund, solitary or twin: flowers an inch long, of a deep-blue, somewhat 2-lipped, the lower lip of a dark-violet: the throat spotted with purple.

Concans. Travancore.

(7) A. LAWIANA. *(Dalz.)*

Ident. Dalz. in Hook. Journ. Bot. IV. p. 344.—Bomb. for. p. 186.

SPEC. CHAR. Stem herbaceous, erect, quadrangular, knotted, trichotomous: leaves elliptic-oblong, acute, suddenly narrowing into a petiole of 1 inch, roughish above, hispid on the nerves beneath: spikes terminal, solitary, short: flowers approximated, sessile, opposite, decussate: bracts and bracteoles lanceolate, foliaceous, 3-nerved, villous: flowers small, white, shorter than the bracts.

Near Dharwar, flowering in the rains.

GENUS XXVII. ECHINACANTHUS.

Didynamia Angiospermia. *Ser. Syst.*

Deriv. From *Echinus*, a hedge-hog; probably alluding to the bristly anthers.

GEN. CHAR. Calyx deeply 5-cleft, nearly equal, erect when in fruit: corolla funnel-shaped, limb equal: stamens four, didynamous, included, connate by pairs at the base: anthers sagittate, oval, oblong, hairy, cells parallel, at the base either one or two spurred: stigma simple: capsule 2-celled, many-seeded from the base: dissepiment complete, adnate: seeds cordately subround, compressed, subtended by retinacula.—Herbs: leaves more or less denticulate: cymes either axillary, 2-cleft with an intermediate flower, the branches secund-flowered upwards, or arranged in a terminal panicle: bracts narrow: bracteoles none: flowers middle-sized.

(1) E. ATTENUATUS. *(Nees.)*

Ident. N. ab. E. in Wall. Pl. As. Rar. III. p. 90.—Dec. prod. XI. p. 168.

SPEC. CHAR. Herbaceous, erect, simple: stem obtusely 4-cornered : lower leaves oblong-oval, cuneiform at the base, somewhat toothed, glabrous, floral ones lanceolate : calyx densely pubescent : panicle terminal, somewhat naked, glandulosely pubescent.

Assam.

(2) E. CALYCINUS. *(Nees.)*

Ident. Dec. prod. XI. p. 168.
Syn. Asystasia calycina, N. ab. E. in Wall. Pl. As. Rar. III. p. 90.

SPEC. CHAR. Glabrous : leaves oblong, serrated : spikes axillary and terminal, flexuose : calyx large, subulately cuspidate, bibracteate : bracteoles none.

Khasia hills.

GENUS XXVIII. TRIÆNANTHUS.

Didynamia Angiospermia. *Nov. Syst.*

Deriv. From *Triaina*, a trident, and *Anthos*, a flower ; alluding to the upper segment of the calyx.

GEN. CHAR. Calyx 5-parted, upper segment larger, 3-cleft to the middle, middle division longest : corolla funnel-shaped, limb regular, 5-cleft, tube short : stamens four, didynamous, contiguous by pairs at the base : anthers 2-celled, cells parallel, glabrous, sagittate at the base, muticous : stigma truncated : capsule sterile from the base for a short space, hence 4-seeded in the middle : seeds furnished with retinacula.—Herbs : leaves broadish, sub-serrated : spikes axillary, simple, loose, very flexuose : bracts narrow, longish, bracteoles none : flowers alternate, sub-secund.

(1) T. GRIFFITHIANUS. *(Griff.)*

Ident. Griffith in Herb. Hook.—Dec. prod. XI. p. 169.

SPEC. CHAR. Herbaceous : leaves ovate, glabrous : spikes opposite, shorter than the leaf, passing into a 3-cleft terminal panicle : bracts linear, twice as long as the calyx, glandulosely pubescent : segments of the calyx bristly-acuminate.

Khasia.

GENUS XXIX. LEPTACANTHUS.

Didynamia Angiospermia. *Sex : Syst :*

Deriv. From *Leptos*, slender ; in allusion to the appearance of the plants.

GEN. CHAR. Calyx deeply 5-parted, segments narrow, upper one often longer, middle ones shorter, others equal : corolla funnel-shaped, limb 5-lobed, unequal, two upper segments larger, ascending : stamens four, didynamous, not exserted : anthers at first cordate, afterwards half-oval, cells antrorsely parallel, contiguous : stigma subulate, retrorsely hooked at the upper base: ovary 4-ovuled from the base to the middle: ovules orbiculate, subtended with thick retinacula : capsule oblong, a little depressed on the back, 2-celled from the base, 4-seeded in the middle, seeds furnished with retinacula. Herbs : flowers arranged in a terminal panicle, trichotomous and interwoven with small leaves, scund, a small branchlet being opposite to the panicle ; pedicels slender from the axils of the floral leaflets : bracteoles when the calyx is very unequal none, when equal two, narrow, adnate to the base of the calyx : corolla showy, blue : fruit pendulous.

(1) L. WALKERI. (*Nees.*)

Ident. Dec. prod. XI. p. 170.

Engrav. Wight's Icon. t. 1507.

SPEC. CHAR. Panicle densely glanduloso-villous : lobes of the perianth linear-filiform, the upper one a little longer : cauline leaves oval-oblong, pubescent beneath : floral ones, at least the primaries, ovate, acuminate, small : upper branches hairy, leaves acuminate or caudato-cuspidate ; segments of the calyx narrow, very villous, the upper one longer, straight : corolla cylindrical, ventricose, lobes of the limb sub-repand, dark pink, or purplish-coloured.

Neilgherries, flowering in February and March.

(2) L. ALATUS. (*R. W.*)

Ident. Wight's Icon. vol. IV. & t. 1527.

SPEC. CHAR. Shrubby ; panicles racemose, numerous on the naked branches, or in single axillary racemes on the leafy branchlets : leaves oval-oblong, acuminate, entire, decurrent on the petiole, amplexicaul, glabrous : peduncles, bracts and calyx thickly beset with long bristly hairs : lobes of the calyx linear obtuse, much longer than the bracteoles : longer filaments hairy.

Coorg.

GENUS XXX. GOLDFUSSIA.

Didynamia Angiospermia. *Ser: Syst:*

Deriv. Named after Dr. Goldfuss, Professor of Natural History at Bonn upon the Rhine.

GEN. CHAR. Calyx 5-parted, about equal: corolla funnel-shaped, limb 5-cleft, lobes obtuse, equal: stamens incluse, didynamous, the lower ones often very short, reflexed: anthers nodding with the oblique ovate membranaceous cells on a hooked gland, long connective: stigma simple, subulate, irritable, crenate on one side: capsule 6-angled, valves easily separable from the dissepiment, cells 2-seeded: seed discoid, supported on retinacula.—Shrubs with serrated penni-nerved leaves, nerves curved, all tending towards the apex, but not reaching it: flowers few, capitulate, rarely spicate, bibracteolate: bracts deciduous: spikes elongating after the fall of the bracts: capitula peduncled, with the peduncle simple or divided.

(1) G. ZENKERIANA. (*Nees.*)

Ident. Dec. prod. XI. p. 172.

Syn. Strobilanthes ciliatus, *Nees.*

Engrav. Wight's Icon. t. 1517.

SPEC. CHAR. Stem fruticose: leaves ovate, acuminate, acute at the base, callosely serrated, glabrous: spikes axillary, opposite, oblong, slightly involucrate, peduncled: bracts oblong: segments of the calyx subulate-acuminate, glabrous, somewhat ciliated: corolla regular: flowers blue.

Neilgherries, flowering in September.

(2) G. TRISTIS. (*R. W.*)

Ident. Wight's Icon. vol. IV. & t. 1508.

SPEC. CHAR. Shrubby, erect: leaves unequal, elliptic-lanceolate, acuminate, acutely serrated, glabrous on both sides: inflorescence paniculately-spicate: spikes subcapitate, long pedicelled, drooping, few-(above 2)-flowered, involucrate: involucral leaves or bracts? lanceolate, acute: lobes of the calyx long, ciliate at the apex: corolla infundibuliform, limb regular, tube very hairy within: stamens monadelphous at the base: anthers oblong, capsule 4-seeded: seeds near the base, the lower one often aborting, upper oblong, obtuse, subtruncate, pubescent.

Western slopes of the Neilgherries, flowering in Fe...

(3) G. DECURRENS. *(Nees.)*

Ident. Wight's Icon. vol. IV. & t. 1522.
Syn. Strobilanthes decurrens, *Nees.—Dec. prod.* XI. p. 189.
SPEC. CHAR. Herbaceous, stem 4-furrowed, slightly rough: leaves oval, acute at both ends, sub-dentate, glabrous, closely lineolate: spikes axillary or ternate, terminal, peduncled, oblong, drooping; bracts oblong elliptic, obtuse, broadly decurrent, glabrous, lineolate: flowers with the rudiment of the fifth filament.

Courtallum.

(4) G. LESCHENAULTIANA. *(Nees.)*

Ident. Dec. prod. XI. p. 172.
SPEC. CHAR. Stem herbaceous, ascending: leaves ovate, obtusely acuminate, calloso-serrulate, glabrous: spike terminal, subcylindric, short-peduncled, bracts linear-caudate from a subovate base, obtuse or retuse with a recurved acumen, pubescent: segments of the calyx lanceolate, 3-nerved, pubescent, glandulose.

Neilgherries.

(5) G. SESSILIS. *(Nees.)*

Ident. Dec. prod. XI. p. 172.
SPEC. CHAR. Stem ascending, glabrous: leaves very unequal, ovate-orbiculate, serrated towards the apex, cuneate at the base, sessile: peduncles axillary, equalling the leaf, 1-3-spiked: spikes globose, drooping: bracts fugacious: calyx glandular-hirsute: flowers blue.

Assam.

(6) G. DISCOLOR. *(Nees.)*

Ident. Dec. prod. XI. p. 172.
SPEC. CHAR. Herbaceous: leaves elliptic, long cuspidate, acuminate, cuneate at the base, calloso-serrate, glabrous, glaucous or purplish below, unequal: peduncles axillary and terminal, 1-3-spiked, shorter than the leaf: spikes somewhat globose: bracts oblong-lanceolate and with the calyces covered with glandular hairs: flowers blue.

Khasia, flowering in November.

(7) G. GLOMERATA. *(Nees.)*

Ident. N. ab. E. in Wall. Pl. As. Rar. III. p. 88.—Dec. prod. XI. p. 173.
Engrav. Bot. Mag. t. 3881.
Spec. Char. Stem fruticose, hirsute: leaves unequal, one ovate, cuspidate, unequally dentato-crenate, the other much smaller, ovate, subrotund, all obtuse at the base: spikes axillary, opposite, solitary, globose, very shortly peduncled, hirsute: bracts lanceolate, quite entire, inner ones exceeding the capitulum: flowers bluish-violet.

Khasia hills, flowering in the cold season.

(8) G. NUTANS. *(Nees.)*

Ident. N. ab. E. in Wall. Pl. As. Rar. III. p. 88.—Dec. prod. XI. p. 174.
Syn. Ruellia hirta, *Don. flor. Nep.* p. 119.
Spec. Char. Stem fruticose, creeping at the base: leaves oval, acute at both ends, unequal, serrated, hirsute: spikes axillary and terminal, oblong, nodding, fruit-bearing ones often erect: bracts ovate, deciduous.

Khasia hills.

(9) G. ANISOPHYLLA. *(Nees.)*

Ident. N. ab. E. in Wall. Pl. As. Rar. III. p. 88.—Dec. prod. XI. p. 176.
Syn. Ruellia anisophylla, *Hook.*
Engrav. Bot. Mag. t. 3401.—Bot. Reg. XI. t. 955.—Hook. Exot. flor. t. 191.
Spec. Char. Undershrub: leaves oblong, caudate-acuminate, opposite one the smallest: flowers pale-purplish blue, with a yellow base.

Khasia hills, flowering in the cold season.

(10) G. ISOPHYLLA. *(Nees.)*

Ident. N. ab. E. in Wall. Pl. As. Rar. III. p. 88.—Dec. prod. XI. p. 176.
Spec. Char. Undershrub: leaves lanceolate, equal, remotely serrulate: peduncles opposite, usually trifid: flowers pale-purplish blue.

Khasia hills, flowering in the cold season.

(11) G. COLORATA. *(Nees.)*

Ident. N. ab. E. in Wall. Pl. As. Rar. III. p. 89.—Dec. prod. XI. p. 176.

Spec. Char. Herbaceous, erect, glabrous: leaves ovate-oblong, acuminate, callosely crenate-serrate, opposite ones smaller: capitula panicled: branches of the panicle glabrous.

Assam.

(12) G. CHIKITA. *(Nees.)*

Ident. Dec. prod. XI. p. 176.

Spec. Char. Perennial: leaves oblong-oval, acuminate, calloso-serrate, attenuated at the base, glabrous, congruous: peduncles axillary, trichotomous, slender, hirsute, 1-flowered: calyx naked, glabrous.

Assam.

GENUS XXXI. STROBILANTHES.

Didynamia Angiospermia. *Sex. Syst.*

Deriv. From *Strobilos*, a cone, and *Anthos*, a flower; alluding to the inflorescence.

Gen. Char. Calyx about equal, 5-parted to the base: segment linear, somewhat broader towards the apex: corolla funnel-shaped, the tube not passing abruptly into the limb: lobes equal or nearly so, rarely sub-bilabiate: stamens four, didynamous, inserted on the middle of the throat, the filaments united at the base by a membrane, monadelphous: anthers oblong, muticous, cells parallel, equal, contiguous, or in some diverging at the base, whence the anthers are sagittate: stigma subulate, incurved or involute, spongiose on the back: capsule columnar, 4-sided, 2-celled almost to the base, 4-seeded about the middle: partition thin, incomplete towards the apex, adnate, or sometimes separating from the valves: seed discoid, angular, with an areola on both sides, the angles more prominent towards the hilum, attached to hooked retinacula.—Herbs and shrubs, spikes more or less dense, axillary and terminal, erect, cernuous, or drooping: bracts foliaceous or foliaceo-membranous, persistent or cadocous, exposing the flowers, bracteoles small or sometimes wanting: flowers, usually, delicate blue or white.

(1) S. SESSILIS. (*Nees.*)

Ident. N. ab. E. in Wall. Pl. As. Rar. III. p. 85.—Dec. prod. XI. p. 177.

Engrav. Hook. Bot. Mag. t. 3962.—Wight's Icon. t. 151 L.

SPEC. CHAR. Suffruticose, very hirsute: stem erect, quadrangular; leaves sessile, ovate, acuminate, crenate: spikes axillary, opposite and with a terminal one: bracts ovate, cuspidate, flowers pale blue.

Common in woods at Ootacamund, flowering from October to December.

(2) S. BARBATUS. (*Nees.*)

Ident. N. ab. E. in Wall. Pl. As. Rar. III. p. 85.—Dec. prod. XI. p. 179.

SPEC. CHAR. Stem fruticose: leaves oval, acuminated at both ends, crenulate, glabrous: spikes axillary, opposite, ovate, very shortly peduncled: bracts orbiculate, linear-cuspidate, upper lip of the bilabiate corolla bearded.

Courtallum hills.

(3) S. FIMBRIATUS. (*Nees.*)

Ident. N. ab. E. in Wall. Pl. As. Rar. III. p. 86.—Dec. prod. XI. p. 180.

SPEC. CHAR. Stem fruticose: leaves oblong, acuminated at both ends, denticulate, glabrous: spikes axillary, alternate, sessile, hirsute; bracts ovate, cuspidate, cut and serrated, 2 lower ones truncated: corolla regular: flowers purple.

Khasia hills.

(4) S. ECHINATUS. (*Nees.*)

Ident. N. ab. E. in Wall. Pl. As. Rar. III. p. 85.—Dec. prod. XI. p. 181.

SPEC. CHAR. Stem fruticose, scabrous: leaves oblong, acuminated at both ends, mucronately serrated, bristly scabrous: spikes axillary and with a terminal one, short-peduncled: bracts oval, obtuse, serrated, hispid: bracteoles and calyx linear, serrated at the apex.

Khasia hills.

(5) S. GLABRATUS. (*Nees.*)

Ident. N. ab. E. in Wall. Pl. As. Rar. III. p. 85.—Dec. prod. XI. p. 183.

SPEC. CHAR. Fruticose, glabrous: stem swollen at the joints: leaves oblong, attenuated at both ends, denticulate: capitula oblong, peduncled, axillary, opposite, quartern, nodding; bracts oblong, imbricated, covered: segments of the corolla emarginate.

Khasia hills.

(6) S. LUPULINUS. (*Nees.*)

Ident. N. ab. E. in Wall. Pl. As. Rar. III. p. 85.—Dec. prod. XI. p. 184.

SPEC. CHAR. Stem herbaceous, glabrous: leaves elliptic, acute, passing into the petiole, crenate, glabrous, closely lineolate: spikes axillary, fascicled, ovate, hirsute: bracts oval, ventricose, emarginate, lowest ones oblong, more remote, spreading: calycine segments 2-toothed at the apex.

Near Courtallum.

(7) S. HOMOTROPUS. (*Nees.*)

Ident. Dec. prod. XI. p. 187.

SPEC. CHAR. Fruticose, erect: stem above, as well as the branches of the inflorescence pubescent-glandular: leaves oblong, acuminate, repandly subserrate, embracing at the narrower and somewhat cordate base: spikes terminal on the branches and stem, compound, distant-flowered: flowers opposite, secund; calyx equalling the joints, a little longer than the oval bract: flowers blue.

Common on the outskirts of the woods near Ootacamund, flowering from November to March.

(8) S. SCABER. (*Nees.*)

Ident. N. ab. E. in Wall. Pl. As. Rar. III. p. 84.—Dec. prod. XI. p. 177.

SPEC. CHAR. Fruticose: stem hispid: leaves obovate-elliptic, dentato-crenate, lineolate above, rough with distant bristles, pale below: spikes infraterminal, opposite, terminal ones tern, glandulose: bracts lanceolate, with an obtuse acumen: corolla pubescent, purplish.

Khasia hills, flowering in the cold season.

(9) S. Brunonianus. *(Nees.)*

Ident. N. ab. E. in Wall. Pl. As. Rar. III. p. 87.—Dec. prod. XI. p. 188.

Spec. Char. Fruticose: leaves lanceolate, attenuated into the petiole, repand, glabrous: spikes axillary, with a terminal one, compound, secund, rather lax, hairy: bracts oblong: stamens monadelphous: flowers blue.

Assam. Khasia hills.

(10) S. monadelphus. *(Nees.)*

Ident. N. ab. E. in Wall. Pl. As. Rar. III. p. 87.—Dec. prod. XI. p. 188.

Spec. Char. Fruticose: leaves oval, acuminated at both ends, crenato-serrate, hispid above with scattered bristles: spikes axillary and terminal, 2-cleft, slightly secund, hirsute: bracts ovate, recurvedly-spreading: stamens monadelphous: corolla pubescent, purple.

Silhet mountains.

(11) S. petiolaris. *(Nees.)*

Ident. Dec. prod. XI. p. 189.

Spec. Char. Fruticose: leaves ovate, acuminate, crenato-serrate in the middle, cuneately decurrent into a long petiole, hispid above with few bristles: spikes axillary, branched, a little leafy below, confluent into a terminal spike: bracts recurvedly spreading, lower ones oblong-linear, rather remote, upper ones obovate-spathulate, imbricated: segments of the calyx very densely hirsute: stamens monadelphous: corolla pubescent, purple.

Assam. Khasia.

(12) S. Sabinianus. *(Nees.)*

Ident. N. ab. E. in Wall. Pl. As. Rar. III. p. 80.—Dec. prod. XI. p. 190.

Syn. Ruellia Sabiniana, *Wall.*

Engrav. Bot. Mag. t. 3317.

Spec. Char. Herbaceous: leaves ovate, acuminate, attenuated into the petiole, repandly subcrenate, glabrous, opposite one smaller, upper ones cordate, stem-clasping: spikes axillary and terminal, somewhat loose, viscidly pubescent: bracts orbiculate, cuneiform at the base.

Pundua. Khasia.

(13) S. MACULATUS. (*Nees.*)

Ident. Dec. prod. XI. p. 190.
Syn. Ruellia maculata, *Wall. Pl. As. Rar.* III. p. 93. & t. 250.
Spec. Char. Fruticose: leaves oblong-lanceolate, caudate-acuminate, obtusely serrated, attenuated at both ends, alternate, long-petioled, pilose, white-spotted above: spikes axillary and terminal, somewhat loose, pubescent: bracts oblong-cuneate, obtuse, sessile: flowers blue.

Silhet mountains, flowering in July.

(14) S. LANATUS. (*Nees.*)

Ident. Dec. prod. XI. p. 191.
Spec. Char. Stem herbaceous (?) erect, deeply and obtusely quadrangular: spikes and leaves woolly-tomentose: leaves ovate, acuminate, quite entire, glabrous above, petiole tomentose: spikes axillary, opposite, terminal, tern, cylindric, dense: bracts oblong, obtusely cuspidate, glabrous and lineate above.

Neilgherries.

(15) S. MYSURENSIS. (*Nees.*)

Ident. N. ab. E. in Wall. Pl. As. Rar. III. p. 86.—Dec. prod. XI. p. 192.
Syn. Ruellia Mysurensis, *Roth.*
Spec. Char. Stem erect, deeply quadrangular: leaves ovate, crenulate, hirsute above: spike terminal: bracts oblong-lanceolate, acuminate, villous, densely ciliated, deciduous.

Mysore.

(16) S. UROPHYLLUS. (*Nees.*)

Ident. Dec. prod. XI. p. 102.
Syn. Dipteracanthus urophyllus, *N. ab. E. in Wall. Pl. As. Rar.* III. p. 82.
Spec. Char. Fruticose, glabrous: leaves oblong-lanceolate, caudate-cuspidate, serrate: spikes axillary, opposite: bracts and bracteoles leafy, spathulate, petioled: two upper segments of the incurved corolla shorter: corolla pubescent.

Pundua.

(17) S. PLATIPIFOLIUS. (*Nees.*)

Ident. Dec. prod. XI. p. 194.

Spec. Char. Glabrous: stem fruticose, when young deeply quadrangular and slightly scabrous above: leaves oval-oblong, short-acute, long and crenately decurrent into the petiole, here and there serrate towards the apex, glabrous, thin: spikes axillary, shorter than the leaf, peduncled, simple, 6-flowered: bracts oblong, obtuse, attenuated into the petiole, inconstant: segments of the glabrous calyx linear, upper one a little longer: flowers blue.

Assam.

(18) S. EXTENSUS. (*Nees.*)

Ident. Dec. prod. XI. p. 195.
Syn. Goldfussia extensa, *N. ab. E. in Wall. Pl. As. Rar.* III. p. 88.

Spec. Char. Herbaceous; stem erect, tetragonal, purple: leaves cordate, acuminate, coarsely serrate in the middle, hirsute, unequal, upper ones sessile; spikes terminal, simple or 3-cleft, peduncled, few-flowered; peduncles hirsute: bracts at the division, ovate-acuminate, septuplinerved, partial ones linear-lanceolate, falling off: flowers blue, pale at the base.

Silhet mountains. Assam.

(19) S. PEROTTETIANUS. (*Nees.*)

Ident. Dec. prod. XI. p. 179.
Engrav. Wight's Icon. t. 1313.

Spec. Char. Shrubby, branches reddish, hairy: leaves ovate, caudately-cuspidate, undulato-crenate, hairy, very rough above: spikes axillary, opposite, secund, oval, nodding, dense, hairy: bracts ovate, acute, the interior ones larger, thinner, and coloured: stamens monadelphous: corolla pale blue.

Nelgherries on the outskirts of forests near Ootacamund.

(20) S. WIGHTIANUS. (*R. W.*)

Ident. Wight Cat. No. 1980.—Dec. prod. XI. p. 180.
Engrav. Wight's Icon. t. 1514.

Spec. Char. Shrubby, erect, very hairy, obtusely 4-angled or nearly terete: leaves ovate, petioled, undulato-crenate, rugous: spikes axillary, opposite and terminal: bracts foliaceous, ovate: corolla a little longer than the bracts, lobes emarginate: flowers straw-coloured and reticulated with purplish veins.

Locality not given.

(21) S. ASPER. (R. W.)

Ident. Wight's Icon. vol. IV. & t. 1518.

Spec. Char. Shrubby, erect, four-sided, young shoots furrowed on two sides, older branches glabrous, branchlets hirsute: leaves unequal, ovate oblong, long petioled, acuminate, crenato-serrate, rough on both sides, venoso-reticulate: peduncles axillary, shorter than the petioles, trifid: spikes compact, ovate, bracts broad ovate, ventricose, undulate, attenuated below into a winged petiole, cuspidato-acuminate above: bracteoles linear-lanceolate, bristly, as long as the calyx.

Neilgherries in woods above Pycarrah.

(22) S. MICRANTHES. (R. W.)

Ident. Wight's Icon. vol. IV. & t. 1519.

Spec. Char. Suffruticose, or herbaceous, erect, stems 4-angled, glabrous: leaves long petioled, broad ovate, serrated, abruptly acuminate, decurrent into the petiole, somewhat bispid above, reticulato venous and sparingly pubescent beneath: spikes axillary, opposite, drooping: peduncles refract near the apex: bracts ovate, lanceolate, acute, the lower ones foliaceous, pubescent, those above membranous, ciliate: bracteoles linear lanceolate, longer than the calyx.

Neilgherries.

(23) S. GRAHAMIANUS. (R. W.)

Ident. Wight's Icon. vol. IV. & t. 1520.—Dals. Bomb. flor. p. 187.

Spec. Char. Shrubby, ramous, 4-sided, older branches glabrous, tuberculate: leaves broad ovate, cuspidato-acuminate, slightly crenato-dentate, decurrent on the long petiole, stellately-hirsute above, pubescent beneath, reticulately veined: peduncles axillary or from the naked branches, trifid, shorter than the petioles: spikes ovate oblong, glabrous: bracts orbicular, ventricose, the lower one a little more remote, densely lineolate, sometimes bispid: bracteoles none.

Bombay.

(24) S. NERMANA. (R. W.)

Ident. Wight's Icon. vol. IV. & t. 1523.—Dals. Bomb. flor. p. 188.

Spec. Char. Suffruticose, branchlets subterete, glabrous: leaves unequal, elliptic-ovate, acuminate, acute, slightly unequal at the base, coarsely crenate-serrated, stellately hirsute, densely lineolate above, lineolate and sparingly pubescent beneath: peduncles axillary, often trifid, numerous and sub-panicled towards the ends of the branches, bibracteolate about the middle: spikes short, ovate-capitulate; bracts foliaceous, acuminate, retuse at the point, clothed with viscid pubescence: calyx and shorter bracteoles densely pilose: corolla sparingly pubescent without, bristly hirsute within: longer filament hirsute, ovary 4-ovuled.

Neilgherries.

(25) S. CAMPANULATUS. (B. W.)

Ident. Wight's Icon. vol. IV. & t. 1562.

SPEC. CHAR. Herbaceous, erect, ramous, four-sided: angles rounded: leaves broadly ovate, or subcordate at the base, cuspidately acuminate, pilose on both sides: spikes axillary, capitate, peduncled: bracts glabrous, somewhat shining, sub-orbicular: bracteoles about the length of the calyx: flowers scarcely exceeding the bract, tube short, limb campanulate, equally 5-lobed, and like the longer filaments hairy within.

Coorg.

(26) S. LURIDUS. (R. W.)

Ident. Wight's Icon. vol. IV. & t. 1515-6.

SPEC. CHAR. Large ramous shrub: branches virgate, bearing the inflorescence on the lower naked portions: leaves oval, oblong, acuminate, pubescent on both sides, finely serrated: spikes ascending, one or two together, opposite: bracts large, orbicular, emarginate or slightly retuse at the apex, dark-livid brown: bracteoles linear, obtuse, about the length of the calyx.

Neilgherries in woods near Neddiwuttom, flowering in January and February.

(27) S. SESSILOIDES. (R. W.)

Ident. Wight's Icon. vol. IV. & t. 1512.

SPEC. CHAR. Suffruticose, very hairy all over, stem erect, four-sided: leaves sessile, round-cordate, serrate, bullately reticulate, coriaceous: spikes axillary and terminal, bracts broad cordate, cuspidate, entire: flowers deep lilac, very handsome.

Neilgherries, rare.

(28) S. BUDDSYS. (R. W.)

Ident. Wight's Icon. vol. IV. & t. 1619.

Spec. Char. Shrubby, erect, ramous; branchlets sparingly pubescent, 4-sided, furrowed, angles blunt, older branches glabrous: leaves broad, ovate acuminate, coarsely crenato-serrated, decurrent on the petiole, wrinkled, hirsute on both sides: spikes globose, axillary, simple or compound, when compound peduncles trifid or sometimes twice trifid: lower bracts remote, sterile, reflexed, all obovate, round above, glabrous: bracteoles none.

Coonoor, Neilgherries.

(29) S. CALLOSUS. (*Nees.*)

Ident. N. ab. E. in Wall. Pl. As. Rar. III. p. 85.—Dec. prod. XI. p. 185.—Dals. Bomb. flor. p. 188.

Spec. Char. Shrubby: stem verrucose: leaves elliptic-cuspidate, running down into a long petiole, with minute callous teeth on the margin, scabrous and ciliated: spikes axillary, compound, shorter than the leaf: bracts orbicular, ventricose, lower more remote, sterile; branches 4-sided, glabrous, often rough with warts and grey points: flowers deep-blue: seeds quite smooth.

The Ghauts, flowering in August.

(30) S. HEYNEANUS. (*Nees.*)

Ident. Dec. prod. XI. p. 184.—Dals. Bomb. flor. p. 187.

Spec. Char. Stems about a foot high, herbaceous, strigose and hirsute: leaves elliptic-cuspidate, running down into a long petiole, crenate serrate, hirsute: spikes axillary, compound, shorter than the leaf, sub-globose, glabrous: bracts orbicular, ventricose: calyx short, the segments oblong-obtuse, glabrous.

Chorla Ghaut.

(31) S. WARRENSIS. (*Dalz.*)

Ident. Dalz. in Hook. Journ. Bot. II. p. 341.—Bomb. flor. p. 187.

Spec. Char. Stem suffruticose, dichotomously branched, knotty and smooth: leaves oblong-acuminate, running gradually into the petiole, glabrous on both sides, repand-toothed: spikes in the opposite axils, peduncled, simple, solitary, drooping: peduncles jointed in the middle: flowers small, spotted with purple.

Warree country.

(32) S. TETRAPTERUS. *(Dalz.)*

Ident. Dalz. Bomb. flor. p. 187.

SPEC. CHAR. Shrubby, subscandent, glabrous: leaves oval, shortly acuminated, running down the petiole along the stem, crenate, coriaceous, shining above: spikes axillary, opposite and terminal, solitary, peduncled: bracts herbaceous, rhomb-cuneate, long cuspidate, ciliated: corolla somewhat 2-lipped, white.

Warree country.

(33) S. ASPERRIMUS. *(Nees.)*

Ident. Dec. prod. XI. p. 183.

SPEC. CHAR. Stem rigid, rough and tuberculated, hirsute at the joints: leaves elliptic-acute, running down like a wing into the petiole, shorter than the leaf, crenated, hispid and bristly: peduncles axillary, trichotomous: spikes ovate: bracts broadly oval, ventricose, glabrous, lower ones more remote and smaller: rachis hirsute.

The Ghauts.

(34) S. GRACILIS. *(Bedd.)*

Ident. Beddome in Madr. Journ. of Lit. (1804) p. 55.

SPEC. CHAR. Shrub 18 feet high, stems terete, glabrous: leaves sessile, auricled at the base, narrow lanceolate with a long acumination, attenuated towards the base, sharply serrated, glabrous on both sides: panicles terminal or from the upper axils, loose, many-flowered, glanduloso-puberulous: flowers in distant pairs, each furnished with a small lanceolate bract: calycine lobes linear lanceolate: corolla lilac, one-third longer than the calyx.

Annamallays.

(35) S. ANDERSONII. *(Bedd.)*

Ident. Beddome in Madr. Journ. of Lit. (1861) p. 55.

SPEC. CHAR. Shrubby, 12 to 20 feet high, stems terete, hirsute: leaves petioled, ovate acuminate, serrate, hirsutely pubescent on both sides, petioles hirsute: peduncles axillary, much shorter than the leaves: flowers in dense bracteated heads, bracts large, glabrous or slightly ciliated, ovate-obtuse: calycine lobes narrow, lanceolate, ciliate: corolla glabrous, large, pale blue.

Annamallays at 5000 feet.

GENUS XXXII. BARLERIA.

Didynamia Angiospermia. *Sex. Syst.*

Deriv. In honor of Rev. James Barrelier, a Dominican and M. D. of Paris.

GEN. CHAR. Calyx 4-parted, segments cruciately opposite, upper and lower one much broader and usually longer and more exterior, entire or the lower one emarginate, rarely 2-parted, middle and lateral ones narrower: corolla funnel-shaped, tube with respect to the limb either short or very long, limb with the throat conically dilated, segments of the limb five, deeply divided, of which the upper one is shorter: stamens 4, didynamous, inserted round the base of the tube, the greater pair often long, nearly equalling the corolla, the smaller very short, in a few species without anthers, sometimes the four stamens are equal, with short filaments; anthers linear, 2-celled: cells parallel, muticous, shorter than the smaller stamens or imperfect: stigma compressed, funnel-shaped, truncated; limb entire: capsule conical-acuminate, nearly 2-celled round the base, and there 4, or by abortion, 2-seeded: dissepiment entire, adnate: seeds covered with a pellicle, which afterwards become floccose: retinacula thick, concave.—Herbaceous or fruticose plants, of a dissimilar habit, but of a constant type: inflorescence axillary or spiked: bracts in the spikes broad or narrow; bracteoles 2, narrow, and sometimes, as also the bracts, ciliated or spinous: corolla showy, blue, white or yellowish, more or less veined.

(1) B. CÆRULEA. (Roxb.)

Ident. Roxb. flor. Ind. III. p. 39.—Dec. prod. XI. p. 226.

SPEC. CHAR. Undershrub: stem strigose: leaves elliptic-oblong, decurrent into the petiole, lineolate above, glabrous: spikes axillary, subsessile, short, dense: bracts elliptic, somewhat attenuated at the apex, mucronulate-ciliated: upper segment of the calyx emarginate: tube of the corolla elongated: flowers pale-blue.

Samulcottah. Upper Assam. Flowering in the cold season.

(2) B. CILIATA. (Roxb.)

Ident. Roxb. flor. Ind. III. p. 38.—Dec. prod. XI. p. 228.— N. ab. E. in Wall. Pl. As. Rar. III. p. 92.

Syn. B. cristata, *Roth.*

SPEC. CHAR. Fruticose, strigose: leaves lanceolate, acute, short-petioled: flowers axillary, solitary or twin, rarely in threes, subsessile: bracts linear, straight, dentato-ciliate: larger segments of the calyx equal, ovato-oblong, closely ciliato-dentate: flowers purplish-pink, rosy-streaked downwards.

Bengal. Hindostan. Travancore. Flowering in the cold season.

(3) B. NUDA. *(Nees.)*

Ident. N. ab. E. in Wall. Pl. As. Rar. III. p. 92.—Dec. prod. XI. p. 229.

SPEC. CHAR. Herbaceous, sparingly strigose: leaves oblong-elliptic, decurrent into the petiole, acute, sparingly strigose above: peduncles axillary, 3-flowered, very short, also terminal, capitate: bracts linear, ciliated, spreading: larger segments of the calyx very unequal, rhombeo-oblong, almost quite entire: flowers blue.

Silhet.

(4) D. INVOLUCRATA. *(Nees.)*

Ident. N. ab. E. in Wall. Pl. As. Rar. III. p. 92.—Dec. prod. XI. p. 232.

SPEC. CHAR. Somewhat diandrous, herbaceous, slightly strigose, leaves elliptic, attenuated at both ends: peduncles axillary, very short, 3-flowered: bracteoles lanceolate, entire: larger segments of the calyx oblong-lanceolate, obtuse, silky, lower somewhat 2-cleft: flowers whitish-blue.

Neilgherries, among moist rocks, flowering in October.

(5) D. PANICULATA. *(Herb. Madr.)*

Ident. N. ab. E. in Wall. Pl. As. Rar. III. p. 92.—Dec. prod. XI. p. 233.

SPEC. CHAR. Fruticose, hirsute and glandular: leaves elliptic-oblong, very long-acuminated: peduncles axillary, cymosely 2-3-cleft, somewhat 3-flowered: bracteoles linear-lanceolate, reflexed: larger segments of the calyx oblong, reticulated, unequal, upper one acute, lower shorter, 2-cleft, inner ones linear-lanceolate, acuminate, almost twice as short: flowers blue.

Travancore.

(6) B. TOMENTOSA. *(Roth.)*

Ident. Roth. Nov. Sp. p. 314.—N. ab. E. in Wall. Pl. As. Rar. III. p. 92.—Dec. prod. XI. p. 233.

SPEC. CHAR. Fruticose, hirsutely tomentose and glandulose with fascicled hairs, yellowish: leaves elliptic, acute at both ends, strigose above: peduncles axillary, cymosely 2-3-cleft, somewhat

3-flowered: bracteoles linear, reflexed: larger segments of the calyx oblong-elliptic, obtuse, reticulated, inner ones lanceolate, acute, much shorter: flowers blue.

Common in dry sunny places near the sea coast at Negapatam.

(7) B. PILOSA. *(Herb. Madr.)*

Ident. N. ab. E. in Wall. Pl. As. Rar. III. p. 93.—Dec. prod. XI. p. 234.

SPEC. CHAR. Fruticose: leaves ovate, obtuse at the base, and with the younger branches covered with spreading hairs: peduncles axillary, short, 1-flowered: bracteoles linear, reflexed: larger segments of the calyx nearly equal, ovate, orbiculate, obtuse, toothed, ciliated, inner ones linear-lanceolate, acute, twice as short: flowers blue.

Courtallum.

(8) B. LONGIFLORA. *(Linn.)*

Ident. Linn. Suppl. p. 200.—Dec. prod XI. p. 235.—Roxb. fl. Ind. III. p. 40.—N. ab. E. in Wall. Pl. As. Rar. III. p. 93.

SPEC. CHAR. Fruticose, very softly tomentose, hoary: leaves ovate, obtuse at the base: peduncles axillary, very short, 1-flowered: bracteoles linear, reflexed: larger segments of the calyx equal, ovate, obtuse, reticulated, inner ones small, ovate, bristly-mucronate: flowers white.

Malabar. Hills near Vellore. Circars. Travancore.

(9) B. HYSTRIX. *(Linn.)*

Ident. Linn. Mant. p. 89.—Dec. prod. XI. p. 239.—N. ab. E. in Wall. Pl. As. Rar. III. p. 93.

SPEC. CHAR. Fruticose: bracteoles sterile outside the branches, twin, or rarely in fours, thorny, divaricate: leaves elliptic or elliptic-oblong, spinulosely-mucronate, strigose or hirsute at the ribs: flowers axillary, sessile, in threes, upper ones spiked: bracts nervoso-striated, strigilose: proper bracteoles subulato-spinescent: larger segments of the calyx oblong-lanceolate, acuminately spinescent, strigilose: flowers purplish (?)

Pondicherry.

(10) B. NOCTIFLORA. *(Linn.)*

Ident. Linn. Sp. p. 290.—Dec. prod. XI. p. 239.

SYN. B. cristata, *Lam.*—B. Mysorensis, *Roth.*—Justicia lanceolata, *Forsk.*

Spec. Char. Fruticose: bracteoles thorny, ramous near the base, axillary, twin: leaves obovate-oblong, spinously-mucronate, strigose, glabrescent: flowers axillary, sessile, solitary: larger segments of the calyx ovate, reticulate, unequal, upper one larger, spinously ciliated, embracing the lower one which is quite entire or denticulate and ciliated.

Tanjore. Coortallum. Neilgherries.

(11) B. BIAPINOSA. *(Vahl.)*

Ident. Vahl. Symb. I. p. 46.—Dec. prod. XI. p. 243.—N. ab. E. in Wall. Pl. As. Rar. III. p. 91.

Syn. Justicia bispinosa, *Forsk.*

Spec. Char. Fruticose: bracteoles thorny, axillary, twin, divaricate, simple, 1-flowered: leaves obovate, spinoso-mucronulate, strigose: flowers sessile: upper segments of the calyx, which equals the thorns, equal, elliptic, mucronulate, ciliated, very hirsute with strigæ: flowers pale rosy-lilac.

Courtallam hills, flowering in the cold season.

(12) B. ACUMINATA. *(R. W.)*

Ident. Dec. prod. XI. p. 231.—N. ab. E. in Wall. Pl. As. Rar. III. p. 93.

Engrav. Wight's Icon. t. 450.

Spec. Char. Shrubby, tomentose: leaves ovate or cordate, acute, sometimes prolonged into a slender acumen, whitish beneath: peduncles axillary, cymosely 2 or 3-cleft: bracteoles linear-lanceolate, reflexed: larger segments of the calyx oblong, and like the interior shorter lanceolate ones, reticulated.

Travancore. Madura Hills.

(13) B. PRIONITIS. *(Linn.)*

Ident. Linn. Sp. Pl. p. 887.—Dec. prod. XI. p. 237.—Roxb. S. Ind. III. p. 36.

Engrav. Rheede Mal. XI. t. 41.—Wight's Icon. t. 452.

Spec. Char. Shrubby, the sterile spinous bractes and bracteoles in 4-cleft fascicles, the fertile bracts subulate, spinous: leaves elliptic-oblong, attenuated at both ends, glabrous beneath, on the lines and margins slightly hairy: flowers sessile, axillary, verticilled, the terminal ones spicate: larger segments of the calyx ovate, spinously cuspidate, quite entire, glabrous: flowers bright orange.

Peninsula. Bengal. Silhet. Assam. Flowering nearly all the year.

(14) B. CUSPIDATA. *(Herb. Madr.)*

Ident. Dec. prod. XI. p. 239.—N. ab. E. in Wall. Pl. As. Rar. III. p. 93.

Engrav. Wight's Icon. t. 451.

SPEC. CHAR. Shrubby, bracts and bracteoles spinous, fascicled: leaves lanceolate or oblong-lanceolate, spinously-mucronate, sprinkled with a few adpressed hairs: flowers axillary, subsolitary: segments of the calyx entire, spinously acuminate.

Dindigul and Madura hills. Metapollium. Flowering in March.

(15) B. BUXIFOLIA. *(Linn.)*

Ident. Linn. Sp. Pl. p. 887.—Dec. prod. XI. p. 241.—N. ab. E. in Wall. Pl. As. Rar. III. p. 94.

Engrav. Wight's Icon. t. 870.—Rheede Mal. II. t. 47.

SPEC. CHAR. Shrubby; hairs spreading, bracteoles axillary, paired, spinous, opposite, divaricated, alternately one-flowered and sterile; leaves elliptic, acute at the base, spinously-mucronate: segments of the calyx shorter than the spines, the inferior one obtuse, emarginate: flowers white or rose-coloured.

Courtallum hills. Metapollium. Coimbatore.

(16) B. TERMINALIS. *(Nees.)*

Ident. Dec. prod. XI. p. 225.—Dalz. Bomb. flor. p. 188.

SPEC. CHAR. Stem strigose: leaves oval-oblong, running down with a wing into a long petiole: flowers spicate: spikes destitute of bracts, crowded at the apex of the branches: bracteoles bracteate, nearly as long as the calyx: calyx pubescent, ciliated, the larger segments subequal, oval-acute, entire; flowers deep blue.

Western Ghauts, flowering in November and December.

(17) B. NITIDA. *(Nees.)*

Ident. N. ab. E. in Wall. Pl. As. Rar. III. p. 91.—Dec. prod. XI. p. 234.

Engrav. Wight's Icon. t. 454.

SPEC. CHAR. Stem fruticose, strigose: leaves ovate or elliptic, petioled, the younger ones clothed with scattered bristles: flowers spicate, bracteas ovate-elliptic, acute, denticulate, ciliate, shining: the larger segments of the calyx unequal, rhombeo-ovate, somewhat acute, ciliate and strigose.

(18) B. COURTALLICA. *(Nees.)*

Ident. Dec. prod. XI. p. 226.—Dals. Bomb. flor. p. 188.
Engrav. Wight's Icon. t, 1529.

SPEC. CHAR. Stem fruticose ; leaves oblong, attenuated at the base and apex, glabrous, shining ; spikes axillary and terminal, short, glandulose-hirsute ; bracts and bracteoles linear-subulate : larger segments of the calyx about equal, oval-oblong, attenuated at the apex ; anthers of the shorter stamens imperfect, acute at the base.

Courtallum, flowering between July and August.

(19) B. DICHOTOMA. *(Roxb.)*

Ident. Roxb. fl. Ind. III. p. 39.—Dec. prod. XI. p. 227.

SPEC. CHAR. Suffruticose, adpressed, strigose : stem with opposite branches ; leaves elliptic-oblong, attenuated at both ends, petioled : spikes axillary and terminal : flowers white, secund : bracts linear-lanceolate, pectinate, ciliated : larger calycine segments ovate-subulate, serrate.

Near the village of Penn. Patna. Bengal. Flowering in the cold season. Though certainly a native of India, it has been found truly wild. It is a favourite plant of the Brahmins, and is often found planted near temples. *(Dalzell.)*

(20) B. CRISTATA. *(Linn.)*

Ident. Linn. Sp. Pl. p. 887.—Dec. prod. XI. p. 229.—Roxb. flor. Ind. III. p. 37.
Engrav. Wight's Icon. t. 453.—Bot. Mag. t. 4. 1615.—Bot. Rep. X. t. 625.

SPEC. CHAR. Herbaceous, strigose with adpressed hairs : leaves elliptic, attenuated at both ends, petioled : peduncles axillary, very short, few-flowered : bracts linear-subulate, ciliated : larger calycine segments unequal, elliptic-oblong, ciliated and serrated : flowers blue.

Bombay and the Concans.

(21) B. MONTANA. *(Herb. Madr.)*

Ident. Dec. prod. XI p. 232.—N. ab. E. in Wall. Pl. As. Rar. III. p. 92.
Syn. B. purpurea, Lodd. Bot. Cab. IV. t. 344.

SPEC. CHAR. Herbaceous, erect, all quite smooth : leaves oblong-elliptic, attenuated into the petiole, a little scabrous on the margin; flowers axillary, solitary, sessile, opposite: bracteoles linear: larger calycine segments equal, elliptic, herbaceous: flowers of a beautiful rose colour.

Island of Caranjah. Travancore mountains. Flowering in September.

(22) B. CISSONI. *(Dala.)*

Ident. Dals. in Hook. Journ. Bot. II. p 339.—Bomb. flor. p. 189.

SPEC. CHAR. Suffruticose, all quite smooth: leaves elliptic-acute at both ends, glaucous beneath, ciliolated on the margins: flowers spicate: spikes short, terminal, solitary: bracts small, foliaceous, narrow-ovate, obtusely acuminated: bracteoles linear-acute: larger calycine segments oval, subequal, quite entire: flowers of a beautiful pink colour.

The Ghauts, and on the Brahmiowara range.

(23) B. GRANDIFLORA. *(Dals.)*

Ident. Dals. Bomb. flor. p. 189.

SPEC. CHAR. Stem fruticose: leaves elliptic acuminate, attenuated into the petiole, upper ones subsessile, quite smooth on both sides: flowers short-pedicelled, solitary in the opposite axils, very large, pure white: bracts inserted on the middle of the pedicel, short, subulate: large calycine segments equal, herbaceous, ovate-acute, glabrous, 10 to 12-nerved: smaller ones narrow-subulate, half the length.

Mangellee Ghaut.

(24) B. ELATA. *(Dals.)*

Ident. Dals. Bomb. flor. p. 189.

SPEC. CHAR. Shrubby, 6 feet high: stem round-strigose, swollen at the joints: leaves herbaceous, unequal, long-petioled, elliptic-acuminated, suddenly attenuated into the petiole, pubescent on both sides: spikes terminal, and in the upper axils, solitary, short-spreading, stout, 2-3-flowered: flowers very shortly pedicelled, secund, pedicels subtended by lanceolate, foliaceous bracts as long as the calyx: tube of the corolla reddish-purple, limb blue.

Phoonda Ghaut, flowering in November.

GENUS XXXIII. ASTERACANTHA.

Didynamia Angiospermia. *Linn. Syst.*

Deriv. From *Aster*, a star, and *Acanthus*, a spine; alluding to the whorl of spines round the flowers.

Gen. Char. Calyx 4-parted at the base, upper segment a little larger, lower one 2-toothed: corolla deeply 2-lipped, upper lip 2-cleft, lower 3-cleft, bicallose at the origin of the segments: stamens four, didynamous, connate by pairs at the base: anthers 2-celled, equal, cells parallel, glabrous, naked: stigma simple, acuminate: capsule 2-celled, compressed, 8-seeded from the base, dissepiment adnate: seeds furnished with small retinacula, ovate, compressed, truncate at the base, smooth.—Marshy, annual herbs, thorny at the joints, narrow-leaved, more or less hirsute: verticils of axillary, sessile, bracteated and bracteoled flowers, surrounded by a circle of stiff spines: flowers red.

(1) A. LONGIFOLIA. *(Nees.)*

Ideal. N. ab. E. in Wall. Pl. As. Rar. III. p. 90.—Dec. prod. XI. p. 247.

Syn. Barleria longifolia, *Linn.*—Ruellia longifolia, *Roxb.*

Figres. Wight's Icon. t. 419.—Rheede Mal. II. t. 45.

Spec. Char. Herbaceous, erect: stem quadrangular: leaves lanceolate, attenuated at both ends, serrulate, ciliated: flowers sessile in the axils, verticilled, blue, surrounded by rigid spines.

In swampy places, very common, flowering in the rainy season.

GENUS XXXIV. NEURACANTHUS.

Didynamia Angiospermia. *Linn. Syst.*

Deriv. From *Neuron*, a nerve; alluding to the bracts.

Gen. Char. Calyx 2-parted, upper lip 3-cleft, lower 2-cleft: corolla 2-lipped, somewhat ringent, upper lip 2-toothed, lower deeply 3-cleft: stamens perfect, four, didynamous, hidden in the tube: filaments very short: anthers small, 2-celled, cells obliqua, pendulous from the top of the filament, diverging downwards, upper one cristately hairy: the lower cell of the anthers of the shorter stamens smaller, incomplete: stigma lateral, oblong: capsule compressed, 4-sided, 4-seeded from the base: dissepiment complete, adnate: seeds ovate-roundish, compressed, smooth, imbedded with retinacula.—Shrubs: spikes axillary, opposite, unequal, bracts nerved, quadrifarious: flowers solitary, small; calyx ciliated.

(1) N. sphaerostachyus. (*Dalz.*)

Ident. Dalz. in Hook. Journ. Bot. II. p. 140.—Bomb. flor. p. 100.
Syn. Lepidagathis sphaerostachya, *N. ab. E. in Deo. prod.* XI. p. 254.—N. Lawii, *Wight's Icon.* t. 1531.

Spec. Char. Stems many from a perennial root, erect, simple, obtusely quadrangular, pubescent and scabrous: leaves opposite, oblong-truncate or subcordate at the base, obtuse at the apex, pubescent and scabrous on both sides: spikes in the opposite axils sessile, globose, densely silky and tomentose: bracts orbicular, suddenly acuminate, reticulately veined: corolla blue, subentire, the limb ventricose and rotate.

Malabar hill. Island of Caranjah.

(2) N. trinervius. (*R. W.*)

Ident. Wight's Icon. vol. IV. & t. 1532.—Dalz. Bomb. flor. p. 100.

Spec. Char. Branches round, glabrous and shining: leaves shortly petioled, subobovate-mucronate, glabrous: spikes axillary, secund, dense, terminal one as long as the leaves: bracts ovate-acute, coriaceous, densely hairy, 3 to 5-nerved: calycina lobes lanceolate, pubescent: corolla obsoletely 5-lobed; flowers small, blue.

Hills near Alibaug.

GENUS XXXV. LEPIDAGATHIS.

Didynamia Angiospermia. *Ser: Syst:*

Deriv. From *Lepis*, a scale, and *Agathis*, a ball.

Gen. Char. Calyx 5-parted, upper segment largest, two lower ones more deeply joined together, in some almost coalescing into one: corolla 2-lipped, upper lip either entire or 2-toothed, or 2-cleft, the segments then agreeing with those of the lower 3-cleft lip: stamens four, didynamous, included, consociated by pairs: anthers 2-celled, bivalved, 4-seeded near the base: seeds compressed.—Herbs or undershrubs, small or middle-sized, many-flowered: spikes simple or glomerately compound, secund, scorpioid: flowers single in the angle of each fertile bract, small, purplish or whitish.

(1) L. UVATICA. (Nees.)

Ident. N. ab. E. IN. Lepideg. p. 16.—Wall. Pl. As. Rar. III. p. 95.—Dec. prod. XI. p. 253.

SPEC. CHAR. Stem suffruticose, creeping at the base: leaves ovate-elliptic or oblong, attenuated at both ends, repand, scabrous above at the ribs: spikes ovate, axillary and terminal, aggregate; bracts ovate-oblong and with the lanceolate bracteoles and calyx awned, villosely ciliated, transparent at the base: calyx 4-parted; lower segment deeply 2-cleft: flowers white, dotted with brown.

Silhet. Mysore. Amara. Flowering nearly all the year.

(2) L. USTULATA. (Nees.)

Ident. N. ab. E. III. Lepideg. p. 16.—Wall. Pl. As. Rar. III. p. 95.—Dec. prod. XI. p. 253.

SPEC. CHAR. Stem fruticose, creeping at the base: leaves oval or oblong, scabrous, lowest ones smaller, spathulate: spikes ovate, terminal: bracts and bracteoles and the larger segments of the almost 5-parted calyx lanceolate, acuminately awned, nearly equal; alternate ones transparent and coloured: flowers white, dotted with brown.

Goruckpore, flowering in September and October.

(3) L. RUPESTRIS. (Nees.)

Ident. Dec. prod. XI. p. 256.—N. ab. E. in Wall. Pl. As. Rar. III. p. 96.

SPEC. CHAR. Stem suffruticose, diffuse; leaves oblong or lanceolate, sessile: spikes capitate, conglomerate round the root, woolly: bracts and bracteoles ovate-orbiculate or oval, cuspidately mucronate: calyx 4-parted: segments densely woolly, minutely mucronate, lower one deeply 2-cleft.

Among rocks on the hills near Ongole.

(4) L. CUSPIDATA. (Nees.)

Ident. N. ab. E. III. Lepideg. p. 31.—Wall. Pl. As. R..r: III. p. 97.—Dec. prod. XI. p. 258.

SPEC. CHAR. Stem fruticose, pubescent: leaves ovate-elliptic, attenuated at both ends, pubescent: spikes terminal, compound at the base, dense, glandulosu-pubescent: bracts oblong, ciliated at the base, very densely pubescent: bracteoles and calyx spinoso-mucronate: sterile bracts narrower: calyx 5-parted, dorsal segment ovate, lower ones oval, 3-nerved: flowers pale blue.

Hills in the Deccan, flowering in October.

(5) L. SCABIOSA. (*Nees.*)

Ident. N. ab. E. in Wall. Pl. As. Rar. III. p. 95.—Dec. prod. XI. p. 251.

Engrav. Wight's Icon. t. 457.

SPEC. CHAR. Stem shrubby, and like the under surface of the ovate repand leaves pulverulently tomentose; spikes terminal, capitate, involucrate, all the bracteas membranaceous, and like the segments of the 4-cleft calyx, shortly armed, the inferior one deeply bifid, acuminated.

Shady vallies in Madura districts. Palamcottah. Travancore.

(6) L. CRISTATA. (*Willd.*)

Ident. N. ab. E. in Wall. Pl. As. Rar. III. p. 96.—Dec. prod. XI. p. 256.

Engrav. Wight's Icon. t. 455.—Roxb. Cor. II L t. 287.

SPEC. CHAR. Stem suffruticose, diffuse, and with the lower lanceolate leaves, glabrous; spikes capitate-congested, conglomerated near the root, on the branches axillary, woolly; bracteas and bracteoles conformable, oblong, mucronate; calyx 4-parted, segments mucronately aristate, the inferior one bifid; flowers small, rose-coloured.

Coromandel. Bombay. Banks of the Jumna. Malabar. Flowering in the cold season.

(7) L. PUNGENS. (*Nees.*)

Ident. N. ab. E. in Wall. Pl. As. Rar. III. p. 97.—Dec. prod. XI. p. 258.

Engrav. Wight's Icon. t. 456.

SPEC. CHAR. Stem shrubby, very ramous; leaves (small), spinously dentate; spikes binate or ternate, capitately congested, axillary, villous; dorsal bracteæ ovate, and like the fertile ones and bracteoles oblong-lanceolate, rigid, spinous at the apex; calyx 4-parted, segments mucronate, spinulose, the inferior one bifid at the apex.

Plains in Tinnevelly. Neilgherries.

(8) L. WALKERIANA. (*Nees.*)

Ident. Dec. prod. XI. p. 260.

Engrav. Wight's Icon. t. 1530.

SPEC. CHAR. Herbaceous, glabrous; leaves ovate, oblong, acuminate, obtusely dentate; entire at the base and attenuated into the long petiole: spikes axillary, trichotomously compound, crowded: peduncles shorter than the leaves; bracts herbaceo-scarious, about half the length of the calyx, and, like the oblong somewhat obtuse exterior lobes of the calyx, 3-nerved; bracteoles, like the bracts, narrower, somewhat acute, 1-nerved.

Coortallum.

(9) L. GRANDIFLORA. *(Dals.)*

Ident. Dals. in Hook. Journ. Bot. II. p. 138.—Bomb. flor. p. 190.

SPEC. CHAR. Stem erect, suffruticose, quadrangular, glabrous, 3 to 4 feet high; leaves ovate acuminate, entire, glabrous, attenuated into the petiole: spikes axillary and terminal, simple or trifid, long, slender, densely woolly: bracts, bracteoles and upper lip of calyx of the same shape, obtuse, 3-nerved, reticulately veined, woolly: corolla deeply bilabiate, large, blue, with two lines of yellow hairs in the throat.

The Ghauts.

(10) L. PROSTRATA. *(Dals.)*

Ident. Dals. l. c.—Bomb. flor. p. 190.

SPEC. CHAR. Stem shrubby, creeping and rooting, glabrous, obtusely quadrangular: younger branches softly tomentose; leaves small, sessile, opposite or tern, elliptic, spinous-pointed, younger ones tomentose; bracts, bracteoles and calyx segments lanceolate, spinous-pointed: spikes rarely axillary, more frequently terminal and simple at the apex of short ascending branches.

Malwan.

(11) L. LUTEA. *(Dals.)*

Ident. Dals. l. c.—Bomb. flor. p. 190.

SPEC. CHAR. Stems several, erect, filiform, dichotomously branched from the base, velvety and tomentose: leaves linear-folded, 3-nerved, minutely hispid above, glabrous beneath; spikes clustered about the root, velvety and tomentose; bracts ovate-orbicular, with a long spinous point: anterior and posterior segments of calyx rhomb-cuneate, spinous-pointed, lateral ones linear: flowers small, yellow.

Malwan, on rocks.

(12) L. MITIS. *(Dals.)*

Ident. Dals. l. c. III. 226.—Bomb. flor. p. 191.

SPEC. CHAR. Stem branched, diffuse: branches trichotomous, glabrous, almost 4-sided: leaves sessile, linear-oblong, acute, glabrous on both sides, minutely ciliated on the margin: spikes clustered about the root into a ball: bracts acuminated from a broad base, bracteoles and calyx-segments linear-acute, all without points, somewhat cartilaginous, smooth at the base, silky and villous at the lip: flowers white, spotted with pink and yellow.'

On rocks at Phoonda Ghaut, flowering in November.

(13) L. CLAVATA. *(Dalz.)*

Ident. Dalz. l. c. p. 340.—Bomb. flor. p. 191.

SPEC. CHAR. Stems several, from a woody root, simple, ascending, obtusely quadrangular, glabrous, naked at the base: spikes terminal, solitary, simple, oblong, quadrangular: leaves small, sessile, ovate acuminate, spinous-pointed, glabrous, entire, coriaceous, rigid: bracts densely imbricated in four rows, of the same shape as the leaves, along with the bracteoles and calyx, silky and tomentose.

Chorla Ghaut.

(14) L. RIGIDA. *(Dalz.)*

Ident. Dalz. l. c. p. 341.—Bomb. flor. p. 191.

SPEC. CHAR. Stem erect, suffruticose, covered with soft spreading hairs: leaves linear-lanceolate, folded, gradually attenuated into the base, glandular and pubescent on both sides: spikes terminal on the short branchlets, cylindric, compound at the base, glandular and pubescent: bracts linear-subulate, calyx 4-divided: dorsal segments oblong-acute, 3-nerved, anterior divided to the middle, lateral subulate, all spinous-pointed.

Ram Ghaut.

(15) L. GOSSYPII. *(Dalz.)*

Ident. Dalz. l. c. p. 340.—Bomb. flor. p. 191.

SPEC. CHAR. Stem herbaceous, dichotomous, diffuse: leaves broadly ovate-acute, softly pubescent, repand-dentate: spikes terminal or bifid, peduncles lax, few-flowered, bracts broadly ovate, rather obtuse: bracteoles linear, and with the calycine segments densely glandular, pubescent.

The Warree country.

GENUS XXXVI. ÆTHEILEMA.

Didynamia Angiospermia. *See : Syst :*

Deriv. From *Aithos*, shining, and *Eilema*, a wrapper; alluding to the bracts.

GEN. CHAR. Calyx 5-parted, upper segment largest, bract-shaped: corolla 2-lipped, upper lip 3-cleft, lower 3-cleft, or subringent, upper lip entire or minutely 2-toothed: stamens four, didynamous, included, consociated by pairs: anthers 2-celled, cells parallel, contiguous: capsule 2-celled, bivalved, 4-seeded at the base: seeds compressed.—Spikes axillary or terminal, leafy; peduncles very short, alternate in the axils of the floral leaves, and on that account secund, 2-5-flowered: common bract broad: proper bracteoles none.

(1) Æ. RENIFORME. (*Nees.*)

Ident. N. ab. E. in Wall. Pl. As. Rar. III. p. 94.—Dec. prod. XI. p. 261.

Syn. Ruellia imbricata, *Vahl.—Roxb. fl. Ind.* III. p. 48.—R. densiflora, *Retz.—Æ.* parviflorum, *Spreng.—*Phaylopsis parviflora, *Willd.*

Engrav. Wight's Icon. t. 1533.

SPEC. CHAR. Stem herbaceous, and like the ovate, unequal at the base, repand leaves, pubescent: one of the leaves smaller: bracts reniform and with the upper ovate membranaceous laciniæ of the calyx ciliate: flowers small, white.

Paulghaut in hedge rows. Silhet. Flowering nearly all the year.

GENUS XXXVII. BLEPHARIS.

Didynamia Angiospermia. *See : Syst :*

Deriv. A Greek term, signifying the eyelash; alluding to the bracts of the calyx.

GEN. CHAR. Calyx 4-parted, unequal, lower and upper segment broader, lower 2-toothed, bracteated at the base: corolla 1-lipped, lip 3-cleft, throat cartilaginous, upper margin tridenticulate: stamens four, sub-didynamous: upper anthers 1-celled, cell ciliately bearded on the margin: lower ones obliquely 2-celled, common division between the lips, while closed, hidden; anthers of the upper stamens adnate to the filament, of the lower ones lateral on the obtuse filament, somewhat stalked or sessile:

capsule 2-celled, 2-2-seeded at the base: dissepiment adnate: seeds subtended with retinacula.—Suffruticose, creeping, hispid herbs: leaves verticilled, unequal; spike bracteate, imbricated, lower bracts empty, often awnedly ciliated: flower terminal, bibracteolate, bracteoles agreeing with or differing in shape from the bracts.

(1) D. DOERNIAVIÆFOLIA. (*Juss.*)

Ident. Dec. prod. XI. p. 260.—N. ab. E. in Wall. Pl. As. Rar. III. p. 97.

Syn. Acanthus Maderaspatensis, *Linn.*—Roxb. flor. Ind. III. p. 35.

Engrav. Wight's Icon. t. 453.

Spec. Char. Stem creeping: leaves in fours, ovate-rhomboid or oblong, repand-dentate, the two opposite smaller: flowers axillary, sessile or peduncled: bracteoles flat, wedge-shaped, ciliated with bristles at the apex: flowers pale blue, with a yellow spot on the under lip.

Madras. Banks of the Jumna. Travancore. Flowering in the cold season.

(2) D. ASPERRIMA. (*Nees.*)

Ident. Dec. prod. XI. p. 267.

Engrav. Wight's Icon. t. 1331.

Spec. Char. Stem herbaceous, suberect: leaves oblong or ovate, entire or remotely denticulate, opposite: proper bracteoles quaternary, white, reticulated with green lines at the base, alternate, cuneiform, trifid and lanceolate: flowers blue, solitary, or pair ed in the opposite axils.

Mysore. Belgaum. Coorg. Common on the Ghauts.

(3) D. MOLLUGINIFOLIA. (*Juss.*)

Ident. Juss. Pers. Syn. II. p. 180.—Dec. prod. XI. p. 260.— Dals. Bomb. flor. p.

Syn. Blepharis repens, *Roth*—Acanthus repens, *Vahl.*

Spec. Char. Hispid and bristly: stem creeping: leaves in fours, oblong, sublinear, densely serrulate, scabrous on the margin, the two opposite ones half smaller: flowers axillary, attenuate, sessile: bracteoles boat-shaped, pointed with a bristle, and strongly ciliated.

Bengal. Concans. Madras. Monghyr.

GENUS XXXVIII. DILIVARIA.

Didynamia Angiospermia. *Ser. Syst.*

GEN. CHAR. Calyx 4-parted, upper and lower segments larger, entire, callous at the base: corolla 1-lipped, lip 3-lobed, palate convex thickened, upper margin quite entire: stamens four, didynamous: all the anthers 1-celled, barbato-ciliated at the margin, adnate: filaments agreeing with each other, straight: capsule 2-celled, compressed, 4-seeded from the base to the middle, sides chartaceous, dissepiment woody, 1-furrowed, separating from the valves, dilated at the top, obliquely truncated: seeds cordate-ovate, compressed, tubercled: retinacula thick, obtuse, straightish. —Shrubs growing on banks of streams, erect: leaves usually thorny and toothed: spikes leafless: flowers furnished with bracts and occasionally with bracteoles, showy.

(1) D. ILICIFOLIA. *(Juss.)*

Ident. Juss. Gen. p. 103.—Dec. prodr. XI. p. 268.
Syn. Acanthus ilicifolius, *Roxb. flor. Ind.* III, p. 32.
Engrav. Rheede Mal. II. t. 48.—Wight's Icon. t. 439.

SPEC. CHAR. Shrubby, spinous or unarmed, glabrous: leaves elliptic, serrately dentate, spinous: spikes many-flowered: flowers bracteate and bracteolate, large, blue.

Sea shores and backwaters, flowering nearly all the year.

GENUS XXXIX. ACANTHUS.

Didynamia Angiospermia. *Ser. Syst.*

Deriv. From *Akantha*, a spine: some of the species being spiny.

GEN. CHAR. Calyx 4-parted, upper and lower segments much larger, the latter 2-cleft at the apex: corolla 1-lipped, lip 3-cleft or 3-lobed and sometimes auricled at the base, upper margin quite entire, in a few 5-lobed: stamens four, sub-didynamous: lower filaments indexed at the top: all the anthers 1-celled, ciliated, upper ones erect, lower ones transverse in the hook of the filament: capsule 2-celled, compressed, 4-seeded from the base to the middle, sides chartaceous.—Herbs or shrubs: leaves usually pinnatifid, spinous: spike terminal, bracteate, leafless: flowers bluish or white, large, 3-bracteate, common bract bristly-ciliate.

(1) A. LEUCOSTACHYUS. (Wall.)

Ident. Wall. Cat. No. 2512.—N. ab. E. in Wall. Pl. As. Rar. III. p. 98 —Dec. prod. XI. p. 270.

Spec. Char. Herbaceous; leaves oblong, subrepand, spinously toothed; ribs below very prominent and together with the stem tomentose; spike densely pubescent and scabrous; upper and lower segments of the calyx oblong-linear; flowers white.

Khasia hills. Assam.

GENUS XL. CROSSANDRA.

Didynamia Angiospermia. *Ser; Syst;*

Deriv. From *Krossos*, a fringe, and *Aner*, a male, or anther.

Gen. Char. Calyx 5-parted, segments broad, inner ones shorter; corolla with a long tube, limb flat, 5-cleft, divided above as far as the tube: stamens four, didynamous, hidden in the tube; anthers 1-celled, hairy, ciliated at the margin: capsule compressed, 2-valved, 4-seeded from the base; dissepiment adnate.—Shrubs: leaves almost quite entire, flowers showy, red; spike terminal, 4-cornered, bracts opposite, broad, herbaceous, proper ones narrow, membranaceous.

(1) C. OPPOSITIFOLIA. (R. W.)

Ident. Dec. prod. XI. p. 281.—N. ab. E. in Wall. Pl. As. Rar. III. p. 98.

Spec. Char. Suffruticose, glabrous; leaves approximated in fours, oblong-lanceolate, smooth; bracts ciliated; spikes subsessile, terminal, aggregated.

Courtallum hills.

(2) C. INFUNDIBULIFORMIS. (Nees.)

Ident. N. ab. E. in Wall. Pl. As. Rar. III. p. 98.—Dec. prod. XI. p. 280.

Syn. C. undulæfolia, *Salisb.*—Justicia Infundibuliformis, *Linn.* —Ruellia Infundibuliformis, *Andr.*—Roxb. flor. Ind. III. p. 41.

Engrav. Rheede Mal. XI. t. 62.—Bot. Reg. I. t. 69.— Wight's Icon. t. 451.— Bot. Mag. t. 2186.—Bot. Repos. 8. t. 542.

Spec. Char. Stem pubescently rough, leaves in whorls of 3 or 4 obovate oblong, punctulately rough and scabrous, bracteæ ciliate; spikes long-peduncled: flowers large, copper-coloured, or orange.

Madura district, flowering in the cold season.

(3) C. AXILLARIS. (*Nees.*)

Ident. N. ab. E. in Wall. Pl. As. Rar. III. p. 98.—Dec. prod. XI. p. 281.

Engrav. Wight's Icon. t. 460.

SPEC. CHAR. Young stems somewhat scabrous, leaves quarternate, oblong, glabrous, even: spikes axillary, subsessile, attenuate, shorter than the leaves: bracteas pubescently scabrous, margin naked.

Shady valleys in the Madura district. About Dharwar.

GENUS XLI. PHLOGACANTHUS.

Diandria Monogynia. *Sec. Syst.*

Deriv. From *Phlox*, a flame; in allusion to the colour of the flowers.

GEN. CHAR. Calyx 5-parted, equal: corolla obliquely 2-lipped, upper lip broader and longer, 2-cleft, lower 3-cleft, tube 8-cornered: stamens two: anthers 2-celled, cells parallel, contiguous, afterwards hastately diverging, muticous: stigma simple, acute: capsule compressed, 2-celled, cells 4-seeded above.—Leaves minutely pappulose above: raceme terminal or lateral, simple or 3-fold, spiciform: flowers verticillately quatern: common bract and the twin partial ones narrow, elongated: corolla showy, yellow or fulvous.

(1) P. CURVIFLORUS. (*Nees.*)

Ident. N. ab. E. in Wall. Pl. As. Rar. III. p. 99.—Dec. prod. XI. p. 320.

Syn. Justicia curviflora, Wall. l. c. II. p. 9.

Engrav. Bot. Mag. t. 3780.—Wall. l. c. II. t. 112.

SPEC. CHAR. Suffruticose: stem erect, quadrangular, strigilose, tomentose: leaves large, elliptic, acute at both ends, repandly toothed, glabrous: corolla elongated, pubescently tomentose, yellowish-red: tube moderately incurved.

Silhet mountains, flowering in November and December.

(2) P. TUBIFLORUS. (*Nees.*)

Ident. N. ab. E. in Wall. Pl. As. Rar. III. p. 99.—Dec. prod. XI. p. 321.

Spec. Char. Suffruticose: stem erect, tetragonal, pubescently scabrous above: leaves large, elliptic, acute at both ends, somewhat repand, pubescent below: raceme thyrsoid: corolla short, densely tomentose and viscid.

Goalparah. Assam. Khasia.

(3) P. THYRSIFLORUS. (*Nees.*)

Ident. N. ab. E. in Wall. Pl. As. Rar. III. p. 99.—Dec. prod. XI. p. 321.

Syn. Justicia thryslflora, *Roxb.*

Spec. Char. Fruticose: stem erect: leaves oblong-cuneiform, quite entire, glabrous: raceme elongated: capsule quadrangular: flowers deep orange.

Interior of Bengal. Khasia hills. Oude. Assam. Flowering in January and February.

(4) P. GUTTATUS. (*Nees.*)

Ident. N. ab. E. in Wall. Pl. As. Rar. III. p. 99.—Dec. prod. XI. p. 321.

Syn. Justicia guttata, *Wall.*

Engrav. Bot. Reg. t. 1334.—Wall. l. a. I. t. 28.

Spec. Char. Low undershrub: stem procumbent at the base, afterwards erect, quadrangular: leaves oblong, repandly crenate, glabrous: racemes short: flowers opposite, lower ones in threes, upper ones solitary, very pale greenish, stained with blood-red spots.

Khasia hills, flowering in December and January.

(5) P. ASPERULUS. (*Nees.*)

Ident. N. ab. E. in Wall. Pl. As. Rar. III. p. 99.—Dec. prod. XI. p. 321.

Syn. Justicia aspernla, *Wall.*—J. quadrangularis, *Hooker.*

Engrav. Bot. Cab. t. 1681.—Bot. Mag. t. 2845.

Spec. Char. Suffruticose; young stem 4-cornered angles rough with small teeth: leaves elliptic-oblong, long-acuminate, glabrous: spike terminal, leafy at the base: flowers yellow or orange.

Khasia hills, flowering in February.

GENUS XLII. LOXANTHUS.

Didynamia Angiospermia. *Sex. Syst.*

Deriv. From *Loxos*, oblique, and *Anthos*, a flower.

GEN. CHAR. Calyx 5-parted, equal, short; corolla tubulose, incurved, coriaceous, limb obliquely 5-lobed, lower segment more deeply divided; fertile stamens two, anthers 2-celled, cells parallel, barren stamens small, at the base of the fertile ones, subulate: capsule many-seeded from the base.

(1) L. GOMEZII. *(Nees.)*

Ident. N. ab. E. in Wall. Pl. As. Rar. III. p. 99.—Dec. prod. XI. p. 322.

SPEC. CHAR. Shrub; branches 4-cornered: leaves oblong, cuneiform, quite entire, glabrous: thyrse terminal; peduncles 2-cleft, minutely bracteolate: common bract a little longer than the peduncle, linear-subulate, tomentose; bracteoles 2, alternate, very small: corolla incurved, densely tomentose, orange.

Silhet mountains.

GENUS XLIII. THYRSACANTHUS.

Didynamia Angiospermia. *Sex. Syst.*

Deriv. From *Thyrsos*, a thyrse; alluding to the form of inflorescence.

GEN. CHAR. Calyx 5-cleft beyond the middle, equal, short; whole corolla tubular or dilated towards the apex, incurved, soft, limb either 5-lobed, almost regular, or more distinctly 2-lipped, upper lip 2-cleft, lower 3-cleft: fertile stamens two, anthers 2-celled, oval, cells parallel, separate by an oblong, nearly oblique connectivum; sterile stamens subulate, hooked, or capitate at the base of the fertile ones, or none: stigma 2-toothed: capsule depressed and barren from the base to the middle, from thence 2-4-seeded: seeds furnished with retinacula, discoid.—Herbs or shrubs, bark smooth, lax, coloured: leaves large, cuneate-sessile, or attenuated into the petiole: thyrse terminal, sometimes dense, branches short, cymose (fascicles) opposite, bearing verticils, sometimes more lax, passing into a simple raceme: bracts and bracteoles small: flowers longish pedicelled, scarlet.

(1) T. INDICUS. *(Nees.)*

Ident. Dec. prod. XI. p. 325.

SPEC. CHAR. Suffruticose: branches acutely 4-cornered, angles smooth: leaves oblong-lanceolate, glabrous, acuminated, long-decurrent into a short petiole: racemes axillary, short, ending in a more or less terminal, somewhat secund-flowered thyrse: segments of the somewhat 2-lipped corolla short: filaments sterile, straight.

Assam, Khasia.

GENUS XLIV. GRAPTOPHYLLUM.

Diandria Monogynia. *Sex. Syst.*

Derio. From *Grapho*, to write, and *Phyllon*, a leaf; alluding to the appearance of the leaves.

GEN. CHAR. Calyx 5-parted, equal: corolla ringent, upper lip straight, arched, margin reflexed, lower 3-cleft: stamens two: anthers 2-celled, incurved, sagittate, muticous, cells parallel, equal: capsule rostrate, 2-celled at the base, 4-seeded: seeds furnished with retinacula.

(1) G. HORTENSE. *(Nees.)*

Ident. N. ab. E. in Wall. Pl. As. Rar. III. p. 285.—Dec. prod. XI. p. 328.

Syn. Justicia picta, *Linn.*—*Roxb. flor. Ind.* I. p. 117.

Engrav. Bot. Reg. t. 1227.—Rheede Mal. VI. t. 60.—Bot. Mag. t. 1870.

SPEC. CHAR. Fruticose: raceme terminal, short-ovate: bracteoles small: leaves oblong or ovate, cuspidate, glabrous, variegated: tube of the corolla amplified upwards, compressed, segments revolute at the edges, glandular within: flowers crimson.

Patna. Assam. Flowering nearly all the year. There is a variety with the leaves dark blood-coloured. Both are common in gardens, but possibly are not really natives of India.

GENUS XLV. HEMICHORISTE.

Didynamia Angiospermia. *Sex. Syst.*

Derio. From *Hemisus*, half, and *Choristos*, separated.

GEN. CHAR. Calyx 5-parted, equal: corolla ringent, upper lip entire, lower 3-cleft: stamens four, didynamous, inserted on the tube towards the base, anthers of the longer ones 2-celled, cells separated by the dilated connectivum, obliquely diverging, spurred at the base, of the lower or inner ones 1-celled, spurred at the base, or rudimentary: stigma obtuse, bifid, divisions narrowly contiguous: ovary 4-ovuled.

(1) H. MONTANA. *(Nees.)*

Ident. N. ab. E. in Wall. Pl. As. Rar. III. p. 102.—Dec. prod. XI. p. 367.

Engrav. Wight's Icon. t. 1538.

SPEC. CHAR. Shrubby, smooth: leaves large, oblong-entire, attenuated into the petiole: thyrses of whitish flowers terminal.

The Ghauts, pretty common. This may be easily mistaken for an Adhatoda, which it much resembles.

GENUS XLVI. ROSTELLULARIA.

Diandria Monogynia. *Bur. Syst.*

Deriv. From *Rostellum*, a little beak; alluding to the beaked anther.

GEN. CHAR. Calyx 4-5-parted, two upper segments often smaller, the fifth, when present, least: corolla 2-lipped, upper lip flat, truncately 2-toothed, lower 3-lobed, convex, broad: stamens two: connectivum thickened at the top: cells obliquely placed, the lower produced into a beak at the base, the upper smaller: capsule 4-seeded, bearing seeds from the base: retinacula short, lamelliform.—Herbs, usually annual, small, low: spike terminal and axillary, bracteate: flowers decussately solitary, bibracteolate, bracteoles resembling the segment of the calyx and alternate with them, bracts lax in some, in others almost similar to the calycine segments.

(1) R. DIFFUSA. *(Nees.)*

Ident. N. ab. E. in Wall. Pl. As. Rar. III. p. 100.—Dec. prod. XI. p. 371.

Syn. Justicia diffusa, *Willd.*—J. procumbens, *Vahl.*

SPEC. CHAR. Stem procumbent, diffuse: leaves lanceolate-elliptic or rounded, glabrous or sparingly hairy: spikes compressed, slender: calycine segments lanceolate, membranaceous on the margin, minutely ciliated: bracts of the same shape, and shorter than the calyx: flowers small, pale-purple.

Common in pastures, flowering in the cold season.

(2) R. PROCUMBENS. *(Nees.)*

Ident. N. ab. E. l. c. p. 101.—Dec. prod. l. c.
Syn. Justicia procumbens, *Linn.*—*Roxb. flor. Ind.* L p. 132.—
J. ascendens, *R. Br. prod.*
Engrav. Wight's Icon. t. 1539.

SPEC. CHAR. Stem procumbent or ascending: leaves from ovate to lanceolate, ciliated, hairy: spikes subtetragonal: calyx segments and bracts lanceolate, linear, equal, hairy, ciliated: flowers small, rose-coloured.

A common and variable plant liable to be confounded with *R. diffusa* and *Mollissima*.

Ootacamund. Western Ghauts. Flowering in the cold season.

(3) R. PEPLOIDEA. *(Nees.)*

Ident. N. ab. E. l. c. p. 101.—Dec. prod. XI. p. 375.—Dalz. Bomb. flor. p. 193.

SPEC. CHAR. Branches diffuse, spreading: leaves ovate-obtuse, glabrous: spikes dense at the apex, interrupted and leafy at the base: bracts bracteoles and calyx-segments oblong-spathulate, with white margins: whole plant smooth and glaucous.

About water courses in the Deccan.

(4) R. CILIATA. *(Nees.)*

Ident. N. ab. E. l. c. p. 101.—Dec. prod. l. c. p. 373.

SPEC. CHAR. Stem procumbent, ascending, trichotomous, pubescent: leaves elliptic or ovate, hairy: spikes terminal, sessile, short: bracts very rough with spreading hairs: flowers pale-purple, very small: capsule oval, glabrous, white.

Vingorla, flowering in August.

(5) R. GRACILIS. *(R. W.)*

Ident. Wight's Icon. vol. IV. & t. 1541.

SPEC. CHAR. Repent, stems erect, ramous, glabrous: leaves oblong, oval-lanceolate, pointed at both ends, sessile, entire, slightly revolute on the margin: spikes terminal, short: bracts sub-lanceolate, subulate, pointed longer than the calyx: calyx 5-parted, segments subulate, glabrous.

Locality not given. Probably a mere variety of *R. diffusa.*

(6) R. HEDYOTIDIFOLIA. *(Nees.)*

Ident. N. ab. E. l. c. p. 100.—Dec. prod. XI. p. 370.
Expres. Wight's Icon. t. 1540.

SPEC. CHAR. Stem erect, divaricately branched from the base, and like the ovate subcrenate acute leaves roughish: spikes terminal and towards the ends of the branches, axillary, short, interrupted at the base: calyx 4-parted, lobes lanceolate, membranous on the margin, glabrous; bracts equalling bracteoles, shorter than the calyx, setaceous on both sides.

Locality not given.

(7) R. SIMPLEX. *(R. W.)*

Ident. Wight's Icon. vol. IV. & t. 1542.

SPEC. CHAR. Root somewhat repent: stems erect, simple, 4-sided, and with the veins on the under surface of the leaves, more or less thickly beset with rigid bristly hairs: leaves oblong-oval, lanceolate, blunt, glabrous above, but marked with numerous transverse lineoles: spikes terminal, longish: bracts about the length of the calyx, and like it pectinately bristle ciliate on the margin: costa below beset with similar bristles.

Station not known.

(8) R. VAHLII. *(Nees.)*

Ident. N. ab. E. l. c. p. 102.—Dec. prod. XI. p. 376.
Syn. Justicia Vahlii, *Roth.*—J. Vahliana, *Rorm. & Schult.*—J. diffusa, *Vahl.*

SPEC. CHAR. Stem procumbent: leaves lanceolate or linear-lanceolate, scabrous: calyx and bracteoles equal, oblong-lanceolate and with the longer bracts cuspidate, broad-membranaceous at the margin, slightly scabrous and ciliolate at the apex: flowers purple.

Southern India.

(9) R. ROTUNDIFOLIA. *(Nees.)*

Ident. N. ab. E. in Wall. Pl. As. Rar. III. p. 100.—Dec. prod. XI. p. 370.

SPEC. CHAR. Annual: stem creeping, and with the orbiculate leaves very hirsute: segments of the calyx broad-membranaceous at the margin, hairy-ciliate.

Travancore. Coromandel. Flowering in August.

(10) R. ADENOSTACHYA. *(Nees.)*

Ident. N. ab. E. l. c. p. 101.—Dec. prod. XI. p. 373.

Spec. Char. Annual; stem procumbent, diffuse: leaves oblong-lanceolate, lineolate, glabrous: spikes shorter than the peduncle: segments of the calyx, bracts and linear-obtuse bracteoles hairy: margin narrow, membranaceous: corolla white, lower lip purple.

Travancore.

(11) R. MOLLISSIMA. *(Nees.)*

Ident. N. ab. E. c. l. p. 101.—Dec. prod. XI. p. 373.

Spec. Char. Stem procumbent, creeping: leaves from ovate to lanceolate, hirsute: spikes tetragonal: calyx and lanceolate bracteolts and rhombeo-oval bracts equal, membranaceously margined, ciliated, yellowish.

Hills at Billicul, Neilgherries.

(12) R. QUINQUANGULARIS. *(Nees.)*

Ident. N. ab. E. l. c. p. 101.—Dec. prod XI. p. 375.

Syn. Justicia quinquangularis, Koen.—Roxb. fl. Ind. I. p. 134.

Spec. Char. Stem erect or procumbent, scabrous: leaves linear-lanceolate, acute, lineolate, elongated, scabrous: calyx, bracteoles and bracts membranaceous at the margin, smooth or scabrous at the edge: anthers glabrous: corolla small, glabrous, white, the lower lip rose-coloured.

Banks of the Ganges at Rajmahal. Silhet. Moradabad. Rice fields near Samulcottah. Flowering nearly all the year.

GENUS XLVII. LEPTOSTACHYA.

Diandria Monogynia. *See: Syst:*

Deriv. From *Leptos*, slender, and *Stachys*, a spike of flowers.

Gen. Char. Calyx 5-parted, equal, small, furnished with small bracts and bracteoles: corolla ringent, tube longish, upper lip straight, bidenticulate, lower convex, 3-cleft, segments short: stamens two, inserted on the tube, their base and the tube strigosely hairy at the insertion of the stamens: anthers 2-celled, cells somewhat contiguous, one placed over the other, upper one oblique, lower spurred, smaller: stigma 2-toothed: capsule compressed at the base, depressed at the top, 4-seeded; dissepiment adnate: seeds muricated, furnished with 2-toothed retinacula.

(1) L. Wallichii. *(Nees.)*

Ident. N. ab. E. in Wall. Pl. As. Rar. III. p. 103.—Dec. prod. XI. p. 370.

Engrav. Wight's Icon. t. 1513.

Spec. Char. Shrubby, glabrous, terete, smooth: leaves oblong or oval-oblong, rough above, tapering at both ends, thin, submembranous: panicles racemose, branchlets glandulose-pubescent: flowers opposite: bracts and bracteoles shorter than the calyx.

Courtallum.

GENUS XLVIII. ADHATODA.

Diandria Monogynia. *Sex. Syst.*

Deriv. The name in Malabar.

Gen. Char. Calyx deeply 5-cleft, lobes equal: corolla ringent, tube shortish, upper lip concave, lower 3-lobed: stamens two, inserted below the middle of the tube: anthers 2-celled, cells obliqu: on the connective, one somewhat above the other, the lower ones spurred: stigma obtuse, capsule depressed, 4-seeded in the middle: seeds either lenticular or flat.—Herbs or shrubs: flowers various in form: leaves quite entire: spikes either axillary, opposite, or the flowers axillary, or the spikes terminal: bracts and bracteoles often large, longer than the calyx: flowers either opposite, or, by abortion, 1-ranked.

(1) A. Neilgherrica. *(Nees)*

Ident. Dec. prod. XI. p. 380.—N. ab. E. in Wall. Pl. As. Rar. III. p. 103.

Engrav. Wight's Icon. t. 1341.

Spec. Char. Low, procumbent: leaves lanceolate, sessile, glabrous, smooth: spikes terminal, 4-sided: bracts and bracteoles ovate, acuminate, remotely 3-nerved, glabrous.

Neilgherries, frequent in pastures about Ootacamund, flowering nearly all the year.

(2) A. Wynaudensis. *(Nees.)*

Ident. N. ab. E. l. c. p. 104.—Dec. prod. XI. p. 403.

Syn. Gendarussa Wynandensis, *N. ab. E. l. c.*

Engrav. Wight's Icon. t. 1515.

Spec. Char. Shrubby: stems long, slender, terete, smooth: leaves oblong, attenuated at both ends, lower ones crenate-dentate:

spikes axillary, spreading and drooping, glandular and pubescent : flowers solitary, opposite, pubescent: bracts ovate, deciduous : bracteoles linear-subulate, shorter than the calyx.

Jungly parts of the Concan. Eastern slopes of the Neilgherries on the Banks of the stream near Burliar.

(3) A. RAMOSISSIMA. *(Nees.)*

Ident. N. ab. E. in Dec. prod. XI. p. 385.—Wall. Pl. As. Rar. III. p. 103.

Syn. Justicia ramosissima, Roxb.

SPEC. CHAR. Shrubby, creeping · leaves broadly ovate, obtusely acuminated : glabrous : spikes axillary and ten.inal, secund, rather lax : bracts and bracteoles ovate-lanceolate, acuminate, glabrous, white, reticulated with green veins : flowers of a dull-white colour.

Common in the higher Ghauts. Coromandel. Flowering in the cold season.

(4) A. TRINERVIA. *(Nees.)*

Ident. N. ab. F. l. c. p. 103.—Dec. prod. XI. p. 386.—Dalz. Bomb. flor. p. 194.

Syn. Justicia trinervia, Vahl.—Dicliptera trinervia, Roem. & Schult.

SPEC. CHAR. Suffruticose : stem procumbent : leaves lanceolate or oval, obtuse, sessile, glabrous : spikes terminal, secund, slender : bracts and bracteoles oblong-lanceolate, acuminate, reticulately veined.

On Wag Donger, near Vingorla.

(5) A. VASICA. *(Nees.)*

Ident. N. ab. F. l. c. p. 103.—Dec. prod. XI. p. 387.

Syn. Justicia adhatoda, Linn.

Engrav. Rheede Mal. IX. t. 43.—Bot. Mag. t. 861.

SPEC. CHAR. Shrub 4 to 5 feet high : leaves elliptic-oblong, attenuated at both ends, glabrous : spikes axillary, opposite, ovate, long peduncled : bracts herbaceous, glabrous, ovate : flowers rather large, white, with brown spots.

On the Ghauts, pretty common. Bengal. Travancore. Silhet. Flowering in the cold season.

(6) A. BETONICA. (*Nees.*)

Ident. N. ab. E. in Wall. Pl. As. Rar. III. p. 103.—Dec. prod. XI. p. 395.

Syn. Justicia Betonica, *Linn.—Roxb. flor. Ind.* I. p. 128. J. pseud.—Betonica, *Roth.—J.* ochroleuca, *Blume.*

Engrav. Rheede Mal. II. t. 21.

Spec. Char. Annual : leaves ovate, acute at both ends, linear-late above, slightly pubescent, quite-entire, repand or toothed : spike terminal, secund : bracts and bracteoles ovate-elliptic, acuminated, ciliated, venoso-reticulate : flowers whitish, tinged with pale-rose and purple.

Coromandel. Bombay. Concans. Flowering in the cold season.

(7) A. ARENARIA. (*Nees.*)

Ident. N. ab. E. in Wall. Pl. As. Rar. III. p. 103.—Dec. prod. XI. p. 387.

Spec. Char. Suffruticose : stem procumbent, diffuse : leaves linear, obtuse, emarginate, glabrous, sessile : spikes terminal, secund : bracts and bracteoles oblong, herbaceous : flowers solitary, in the axils of the larger bracts, whitish, with a purple, trifid spot at the insertion of the stamens.

In sandy moist places at Negapatam and Tanjore.

(8) A. QUADRIFARIA. (*Nees.*)

Ident. Dec. prod. III. p. 398.

Syn. Gendarussa quadrifaria, *N. ab. E. in Wall. Pl. As. Rar.* III. p. 105.

Spec. Char. Herbaceous : stem erect, obtusely tetragonal, somewhat tomentose : leaves oblong-lanceolate, puberulous on both sides, sessile : bracts obcordate-spathulate ; bracteoles very short, scale-shaped ; flowers verticilled, pale, streaked with white and purple.

Khasia hills, flowering in the cold season.

(9) A. NEESIANA. (*Nees.*)

Ident. Dec. prod. XI. p. 397.

Syn. Gendarussa Neesiana, *N. ab. E. in Wall. Pl. As. Rar.* III. p. 105.—Justicia Neesiana, *Wall.*

SPEC. CHAR. Fruticulose, erect, leafless below, many-jointed, densely leafy above, ... : leaves linear-lanceolate, glabrous: bracteole small, spathulate ; upper lip narrow, entire ; stamens reflexed : flowers axillary, twin, somewhat verticilled, pale greenish rose.

Khasia hills, flowering in the cold season.

(10) A. TRANQUEBARIENSIS. (*Nees.*)

Ident. Dec. prod. XI. p. 399.

Syn. Gendarussa Tranquebariensis, *N. ab. E. in Wall. Pl. As. Rar.* III. p. 103.—Justicia Tranquebariensis, *Linn*.—J. parvifolia, *Lam.*

Engrav. Wight's Icon. t. 462.

SPEC. CHAR. Fruticulose, hoary-pubescent : leaves roundish, small : bracts orbiculate, retuse ; bracteoles equalling the calyx, linear : flowers axillary, solitary, ascending in a terminal spike, yellowish, purple-dotted.

Tranquebar. Pondicherry. Flowering in February and March.

(11) A. ORIXENSIS. (*Nees.*)

Ident. Dec. prod. XI. p. 400.

Syn. Gend.rnssa Orixensis, *N. ab. E. in Wall. Pl. As. Rar.* III. p. 104—Justicia Orixensis, *Kœn in Roxb. fl Ind.* I. p. 132—T. Tranquebariensis, *Roxb.*—J. brachiata, *Roth.*—J. brachionodes, *Spreng.*

SPEC. CHAR. Herbaceous : stem pubescently scabrous ; bracteoles opposite, linear-subulate : leaves ovate, densely line late : spikes terminal and axillary, somewhat round ; bracts orbiculato-spathulate, ciliated, upper ones sterile : flowers small, pale-yellowish-red.

Coromandel. Samalcottah. Orissa. Flowering in February and March.

(12) A. VENTRICOSA. (*Nees.*)

Ident. Dec. prod. XI. p. 407.

Syn. Justicia ventricosa, *Wall. Pl. As. Rar.* I. p. 83, & t. 93.

Engrav. Bot. Mag. t. 2700.

SPEC. CHAR. Fruticose, jointed : leaves oblong, glabrous ; spikes on the stem and terminal branches somewhat verticill-fl-weral : bracts decussate, oval or orbiculate, ciliate ; proper bracts subulate, small : corolla pubescent, greenish white, the limb inside-sprinkled with purple.

Khasia hills, flowering in the cold season.

(13) A. vasculosa. (*Nees.*)

Ident. Dec. prod. XI. p. 407.

Spec. Char. Fruticulose: stem creeping below, with an ascending apex: leaves elliptic and oblong-elliptic, attenuated at both ends, laeolate above: spikes terminal, compound: branches secund: bracts and bracteoles a little shorter than the calyx, ovate-lanceolate: flowers opposite, pubescent, white.

Khasia hills. Assam.

GENUS XLIX. GENDARUSSA.

Diandria Monogynia. *Sex: Syst.*

Deriv. . An alteration of the native name.

Gen. Char. Calyx regular, 5-parted, furnished with small bracts at the base: corolla 2-lipped, upper lip arched, lower transversely obliquely folded, tube short: stamens two, inserted below the throat: connectivum rhomboid-lanceolate, oblique: cells placed one above the other obliquely, semiovate, lower one spurred: capsule narrow, depressed from base to apex, 4-seeded, stiff at the base, thin above.

(1) G. vulgaris. (*Nees.*)

Ident. N. ab E., in Wall. Pl. As. Rar. III. p. 101.—Dec. prod. XI. p. 410.

Syn. Justicia gendarussa, *Linn.*—*Roxb. fl. Ind.* I p. 128.

Icones. Wight's Icon. t. 468.—Bot. Reg. s. t. 635.—Rheede. Mal. XI. t. 42.

Spec. Char. Shrubby: spikes terminal, flowers somewhat whorled, leafy at the base: branches small: leaves lanceolate, glabrous: flowers pale greenish white, sparingly stained with purple.

Prasania. Concans. Bengal. Silhet. Flowering in the cold season.

GENUS L. JUSTICIA.

Diandria Monogynia. *Sex: Syst.*

Deriv. Named in honor of J. Justice, an eminent Scotch Horticulturist and Botanist.

GEN. CHAR. Calyx 5-parted almost to the base, small, segments equal; corolla bilabiately hypocrateriform, tube long, upper lip narrow, reflexed, lower 3-cleft, segments equal: stamens two; anthers 2-celled, cells parallel, somewhat unequal at the base, mutitous; capsule compressed at the base, seedless, dilated at the apex, depressed, ovate, cuspidate, 2-celled, 2-seeded; seeds deeply cordate, girt with an elevated margin, compressed, tubercled; retinacula hooked, strong.—Shrubs : leaves firm ; spike terminal ; bracts torbaceous, afterwards deciduous, broad, flowers opposite, solitary, reddish ; bracteoles small, subulate.

(1) J. ECBOLIUM. *(Linn.)*

Ident. Linn. flor. Zeyl. p. 17.—Sp. Pl. p. 85.—Dec. prod. XI. p. 420.

Engrav. Wight's Icon. t. 463. (not Roxb.)—Bot. Mag. t. 1847.

SPEC. CHAR. Shrubby: spikes terminal, 4-sided ; bracteae oval, entire, ciliate, mucronate, equalling the fruit ; leaves ellipticoblong, attenuated at both ends, pubescent ; upper lip of the corolla linear, reflexed : flowers large, greenish livid.

Hills throughout the Concans. Island of Bombay.

(2) J. LIVIDA. *(R. W.)*

Ident. Wight's Icon. vol. IV. & t. 1546.

SPEC. CHAR. Shrubby : leaves oblong, acuminate at both ends, glabrous, shining : petioles obtusely margined : spikes terminal, 4-sided : bracts oval, long, cuspidate, repando-subdentate, ciliate, somewhat shorter than the capsule : upper lip of the corolla linear, reflexed.

Courtallum.

(3) J. ROTUNDIFOLIA. *(Nees.)*

Ident. N. ab. E. in Wall. Pl. As. Rar. III. p. 108.—Dec. prod. XI. p. 427.

SPEC. CHAR. Suffruticose : leaves sessile, elliptic-ovate, obtuse, glabrous : spike terminal, tetragonal : bracts roundish, subdenticulate, shorter than the fruit : upper lip of the corolla subulate : lower segments obovate.

Tanjore, in moist places. Foot of the Neilgherries. Flowering all the year.

(4) J. DENTATA. *(Klein.)*

Ident. Dec. prod. XI. p. 427.—N. ab. E. in Wall. Pl. As. Rar. III. p. 108.

Spec. Char. Fruticose: leaves elliptic-oblong, attenuated at both ends, glabrous; petioles with leafy margins; spike terminal, tetragonal; bracts ovate, long-cuspidate, erose-dentate, pubescent, equalling the capsule; upper lip of the corolla linear-retracted; tube very long: flowers liver-verdigris with a white tube.

Banks of the Ganges. Upper Assam. Common in most parts of India. Flowering nearly all the year.

GENUS LI. RHINACANTHUS.

Diandria Monogynia. *Ges, Syst.*

Deriv. From *Rhis*, a snout; alluding to the elongation of the capsule.

Gen. Char. Calyx regularly 4-parted, bracts and bracteoles small, subulate: corolla hypocrateriform, 2-lipped, tube long, slender, upper lip narrow, lower 3-cleft, segments equal: stamens two, inserted in the throat; anthers 2-celled, muticous, one cell placed above the other in an almost linear row: capsule clavate, compressed at the base with a long continuation, commissure of the valvelets contiguous, seedless, in the upper part 4-ovuled, 4 or by abortion, 2-seeded: dissepiment complete, adnate: seeds ovate, biconvex, furnished with concave, obtuse retinacula.—Loose shrubs, somewhat scandent: panicles axillary, passing into terminal trichotomous branches, 2-cleft: flowers agglomerate or short-spiked round the tops of the branchlets, white or bluish.

(1) R. communis. *(Nees.)*

Ident. N. ab. E. in Wall. Pl. As. Rar. III. p. 109.—Dec. prod. XI. p. 442.

Syn. Justicia nasuta, *Linn.—Roxb. flor. Ind.* I. p. 120.

Engrav. Wight's Icon. t. 464.—Rheede Mal. IX. t. 09.—Bot. Mag. X. t. 325.

Spec. Char. Shrubby, 4 to 5 feet high: leaves oblong or ovate-oblong: panicles axillary and terminal, bitrichotomous, spreading: flowers small, white: corolla with a long slender compressed tube.

Mahableshwur; generally to be found in gardens. Paulghaut. Travancore. Flowering nearly all the year.

(2) R. calcaratus. *(Nees.)*

Ident. Dec. prod. XI. p. 444.

Syn. Justicia calcarata, *Wall. Pl. As. Rar.* II. p. 0. & t. 113.

SPEC. CHAR. Fruticose: leaves oval-oblong: panicle terminal, trichotomous: calyx glandulosely pubescent: segments lanceolate-acute: corolla pubescent: upper lip linear-attenuate, bifid, very acutely 2-cleft at the apex: bracts subulate from a triangular base, spreading: flowers white with a tinge of sulphur.

Khasia hills, flowering in March.

GENUS LII. ERANTHEMUM.

Didynamia Angiospermia. *Sex. Syst.*

Deriv. From *Ear*, spring, and *Anthos*, a flower; applied by the ancients to their *Anthemis*.

GEN. CHAR. Calyx 5-cleft, equal: corolla hypocrateriform or long-funnel-shaped, tube long, slender, limb nearly equal: stamens two, fertile, inserted the mouth of the tube, long decurrent, two sterile ones very short, filaments of the longer ones connected at the base: anthers exserted, 2-celled, muticous, cells parallel contiguous, of a thicker texture: capsule depressed below, valvelets contiguous, seedless, upper one 2-celled, 4-seeded: dissepiment a looso: seeds discoid, furnished with retinacula. —Shrubs or undershrubs, usually inhabitants of mountainous districts, flowers showy, blue, rose-coloured, or variegated white: leaves entire or serrated: flowers spiked: common bracts larger or smaller, all the bracteoles small opposite.

(1) E. NERVOSUM. (R. Br.)

Ident. R. Br. prod. p. 333.—Dec. prod. XI. p. 445.—Dal. Bomb. flor. p. 103.

Syn. E. pulchellum, *Roxb.* Justicia nervosa, *Vahl.*

Engrav. Bot. Mag. t. 1358—Roxb. Cor. t. 177.

SPEC. CHAR. Stem quadrangular: leaves ovate or elliptic, acuminated at both ends, subserrate or entire, glabrous: spikes axillary, opposite, imbricated: bracts eliptic, long and acutely cuspidate, reticulated with veins: diameter of the limb of the corolla as long as the tube: flowers blue.

Assam. Coucana. Hardwar. Silhet. Flowering in February.

(2) E. MONTANUM. (Roxb.)

Ident. Roxb. flor. Ind. I. p 110.—Dec prod. XI. p. 448.

Syn. E. capense, *Linn. fl. Zeyl.*—E. fastigiatum, *Spreng.*—Justicia fastigiata. *Lam.*

Engrav. Wight's Icon. t. 466.—Bot. Mag. t. 4031.

SPEC. CHAR. Stem quadrangular: leaves oblong, attenuated at both ends, repand-crenate, glabrous: peduncles terminal, trichotomous, and with the spikes pubescent and viscid: bracts lanceolate, attenuated, ciliated.

The Ghauts near Dharwar. Circars. Travancore mountains. Dindigul & Courtallum hills. Flowering in the cold season.

(3) R. CRENULATUM. *(W. & A.)*

Ident. Dec. prod. XI. p. 453.—Dals. Bomb. flor. p. 105.

Syn. Justicia latifolia, *Vahl*.—R. dianthemum, *Blume.*

Engrav. Wall. in Bot. Reg. t. 870.

SPEC. CHAR. Shrubby, erect: leaves oblong, acuminated at both ends, repand-crenate, glabrous: raceme terminal, simple or compound, or several axillary aggregated, simple: flowers somewhat fascicled, subverticelled or secund, white: bracts and bracteoles subulate, short, and with the calyx glandular and scabrous.

Warree Jungles. Silhet.

(4) E. ROSEUM. *(Roem. & Schult.)*

Ident. Dec. prod. XI. p. 447.—Roem. & Schult. Syst. I. p. 175.—Dalz. Bomb. flor. p. 105.

SPEC. CHAR. Suffruticose, leaves elliptic, glabrous, scabrous on the veins beneath: spikes axillary, peduncled, imbricated: bracts oval, somewhat wedge-shaped, acute, ciliated with long hairs, reticulately veined: flowers rose-coloured.

Around Bombay. Assam. Purindar hills. Flowering from October to December.

(5) E. PURPURASCENS. *(R. W.)*

Ident. Dec. prod. XI. p. 447.

SPEC. CHAR. Fruticose: stem tetragonal: leaves broad-ovate, cuspidate-acuminate, repando-crenate, glabrous; upper ones subcordate, very shortly petioled: spikes axillary, opposite, imbricated, lower ones very long-peduncled: bracts ovate-rhomboid, rostrate-attenuate, ciliated: flowers purple: segments of the corolla equal, obovate-truncate.

Mongbyr. Dindigul and Courtallum hills, flowering in October.

(6) E. STRICTUM. *(Colebr.)*

Ident. Colebr. in Roxb. fl. Ind. (Ed. Car.) I. p. 115.—Dec. prod. XI. p. 448.

Engrav. Bot. Reg. t. 867.—Bot. Mag. t. 3068.

SPEC. CHAR. Stem suffruticose, scabrous; leaves elliptic-oblong, acuminated at both ends, pubescently scabrous below: spikes on the stem and terminal branches elongated: bracts quadrifarious, spreading, oblong, ciliated: anthers included, violet: tube of the corolla incurved, pubescent, segments obliquely obovate: flowers blue.

Khasia hills, flowering from January to March.

(7) E. WIGHTIANUM. (*Wall.*)

Ident. N. ab. E. in Wall. Pl. As. Rar. III. p. 107.—Dec. prod. XI. p. 449.

SPEC. CHAR. Fruticose: stem tetragonal, pubescently tomentose: leaves ovate, acute at both ends, glabrous: floral ones subrotund: spikes terminal, corymbose: bracts linear-lanceolate, acute, straight, pubescently scabrous; corolla-tube long, slender, segments roundish.

Courtallum and Dindigul hills.

(8) E. PALATIFERUM. (*Nees.*)

Ident. N. ab. E. in Wall. Pl. As. Rar. III. p. 108.—Dec. prod. XI. p. 457.

Syn. Justicia palatifera, *Wall. Pl. As. Rar.* I. p. 60. & t. 92.

SPEC. CHAR. Fruticose, erect: leaves oblong, acuminate, quite entire, short-petioled, glabrous: racemes terminal, somewhat in threes, spiciform, recurved: bracts and bracteoles subulate, short: flowers erecto-second, bilabiate.

Hills near Silhet, flowering in March and April.

GENUS LIII. RUNGIA.

Diandria Monogynia. *Ser: Syst*.

GEN. CHAR. Calyx 5-parted, regular: corolla 2-lipped, upper lips 2-toothed, lower 3-lobed, palate 2-folded: stamens two: cells of the anthers obliquely placed one above the other, lower one with a lamellar orbiculate appendage: capsule 2-valved, 2-celled, 4-seeded, seed-bearing from the base; dissepiment membranaceous separating together with the lateral partitions of the capsule from the back of the dehiscing valves: seeds concentrically wrinkled, compressed, furnished with retinacula, of which the lower one is bent back from the base of the dissepiment.—Herbs, often creeping, remarkable for the white-margined mucronate bracts of the spikes: flowers and capsules small: proper bracts two, opposite, linear, scarcely longer than the calyx: segments of the calyx linear, ciliate, membranaceous.

(1) R. PARVIFLORA. (*Nees.*)

Ident. N. ab. E. in Wall. Pl. As. Rar. III. p. 110.—Dec. prod. XI. p. 469.

Syn. Justicia parviflora, *Retz.*—J. pectinata, *Roxb.* (*not Linn.*) Dicliptera cærulea, *Blume.*

Engrav. Roxb. Cor. t. 153.

Spec. Char. Stem diffuse or creeping: leaves oval or lanceolate, rather obtuse: fertile bracts suborbicular, mucronate or unpointed, nerved and veined, glabrous, ciliated, with membranous margin, sterile ones oval or oblong, margined on one or both sides, acute, ciliated: bracteoles membranous margined, emarginate mucronate: corolla small, upper lip acute: flowers of a fine blue.

Peninsula. Bengal. Khasia hills. Flowering in the cold season.

(2) R. POLYGONOIDES. (*Nees.*)

Ident. N. ab. E. in Wall. Pl. As. Rar. III. p. 110.—Dec. prod. XI. p. 471.

Spec. Char. Stem creeping at the base: common bracts suborbicular, mucronate, 3-nerved: bracteoles boat-shaped, ciliated and with a broad, membranous margin: leaves unequal, obtuse, lower oval, upper lanceolate: spikes axillary, glomeruliform, size of a pea.

Bombay.

(3) R. REPENS. (*Nees.*)

Ident. N. ab. E. in Wall. Pl. As. Rar. III. p. 110.—Dec. prod. XI. p. 472.

Syn. Justicia repens, *Linn.*—*Roxb. flor. Ind.* I. p. 132.—Dicliptera retusa, *Juss.*—D. repens, *Roem. & Schult.*

Engrav. Wight's Icon. t. 465.—Roxb. Cor. t. 152.—Burm. Zeyl. 111. fig. 2.

Spec. Char. Leaves oblong, lanceolate-acute: stem creeping: bracts ovate-cuspidate, without nerves, with a broad, white margin, subciliated: bracteoles lanceolate: flowers small, pink.

Very common. Everywhere. Flowering nearly all the year.

(4) R. ELEGANS. (*Dalz.*)

Ident. Dalz. Bomb. flor. p. 196.

Spec. Char. Stem somewhat angular, covered with soft, white hairs: leaves sessile, ovate or ovate-lanceolate, acuminate, puberulous on the upper surface, pale beneath, with prominent nerves: flowers in a sessile terminal spike, 1 inch long: bracts all broad ovate-cuspidate, ciliate, with a broad, white, scariuus margin: flowers of a beautiful blue: capsule ovoid, glabrous, 4-seeded.

High hills around Joonere, flowering in August.

(5) R. LATIOR. *(Nees.)*

Ident. Dec. prod. XI. p. 472.
Engrav. Wight's Icon. t. 1548.

Spec. Char. Leaves obovate or oval, moderately attenuated at both ends, somewhat obtuse: stem diffusely repent: corolla longer than the bracts, upper lip acute: bracts uniform, obovate, retuse, shortly mucronate, 3-nerved, ciliate, margin membranaceous.

Courtallum. Ootacamund. Flowering during the autumnal months.

(6) R. WIGHTIANA. *(Nees.)*

Ident. N. ab. E. in Wall. Pl. As. Rar. III. p. 110.—Dec. prod. XI. p. 472.
Engrav. Wight's Icon. t. 1540-50.

Spec. Char. Suffruticose, erect: leaves ovate oblong, much attenuated at the apex: bracts veined, the margins hyaline towards the apex, most delicately ciliate, sterile ones oblong, acute, the fertile ones rhombeo-cuneiform, obtuse: bracteoles membranaceous, oval, mucronulate: spikes lax, terminal: flowers rose-coloured.

Courtallum, flowering during the rainy months.

(7) R. MURALIS. *(Royle.)*

Ident. Dec. prod. XI. p. 470.

Spec. Char. Stem low, woody, much branched: bracts rough, mucronate, ciliated, fertile ones oval, with a membranaceous margin, sterile ones lanceolate-oblong, margin naked, very narrowly membranaceous: bracteoles emarginate, mucronate: leaves somewhat round-spathulate.

On walls at Berampore. Courtallum. Coromandel.

(8) R. FUNDUANA. *(Nees.)*

Ident. N. ab. E. in. Wall. Pl. As. Rar. III. p. 110.—Dec. prod. XI. p. 473.

Spec. Char. Stem fruticose: leaves oblong, attenuated at both ends: spikes axillary, broadish: bracts rhombeo-spathulate, acute, veined, margin of the fertile ones membranaceous, ciliated: bracteoles lanceolate.

Pundua.

GENUS LIV. DICLIPTERA.

Diandria Monogynia. *Ser. Syst.*

Deris. From *Dis*, double, and *Kleio*, to shut: in allusion to the 2-valved fruit.

Gen. Char. Calyx equal, 5-parted, sessile on bracteated usually bivalved capitolum: corolla resupinate, 2-lipped, lips flat or concave, upper one 3-toothed, lower entire or 2-toothed: anthers 2-celled, cells placed one behind the other, muticous, half-oval, membranaceous after the ejection of the pollen, undulated: capsule 2-valved, 2-celled, compressed at the base for a short way unguiculate, seedless, depressed towards the apex, 4-seeded: seeds furnished with hooked retinacula, discoid.—Herbs: stem usually sexangular: capitule enclosed by involucral bracts, outer ones twin opposite and longer, arranged like a fan in axillary and afterwards terminal umbellets.

(1) D. BIVALVIS. (Jass.)

Ident. Jass. in Ann. Mus. 9. p. 268.—Dec. prod XI. p. 478.
Syn. Justicia bivalvis, *Linn.*
Engrav. Wight's Icon. t. 1551.

Spec. Char. Leaves ovate oblong, acuminate, acute at the base, lineolate, hispido-scabrous: peduncles axillary, longer than the petiole, trifid: capitule 2 or 3-flowered: bracts broad ovato-roundish, aristato-mucronate, 5-nerved, hispid, margin naked.

Courtallum.

(2) D. CUNEATA. (Nees.)

Ident. N. ab. E. in Wall. Pl. As. Rar. III. p. III.—Dec. prod. XI. p. 481.
Engrav. Wight's Icon. t. 1552.

Spec. Char. Leaves ovate, obtuse or acute at the base, and with the stem, minutely lineolate, glabrous: peduncles axillary, longer than the petioles, 3-5-cleft: common involucrum shorter than the umbel, subulate: partial involucrum diphyllous: leaves cuneiform, mucronate, pubescent scabrous.

Courtallum, flowering during the autumnal rains.

(3) D. ROXBURGHIANA. *(Nees.)*

Ident. N. ab. E. in Wall. Pl. As. Rar. III. p. 111.—Dec. prod. XI. p. 483.

Syn. D. chinensis, *Roem. & Schult.*—Justicia chinensis, *Roxb.*

SPEC. CHAR. Leaves ovate, acute at both ends: umbels axillary, in fours or fives, 3 to 5 divided: leaflets of proper involucre unequal, obovate, mucronulate, ciliated, 3-nerved, veined: capsule oval, somewhat rounded, compressed, hirsute: flowers pale rose, stained with red.

Coromandel. Silhet. Assam. Bombay. Flowering nearly all the year.

(4) D. BURMANNI. *(Nees.)*

Ident. N. ab. E. in Wall. Pl. As. Rar. III. p. 112.—Dec. prod. XI. p. 483.

Syn. Justicia chinensis, *Burm. fl. Ind. t. 4 fig. I.*

SPEC. CHAR. Stem obsoletely quadrangular: branches pubescent and scabrous: leaves oval or lanceolate, acute at both ends, mucronate: umbels axillary, simply or doubly in fours or fives, very shortly-peduncled: proper involucre 2-leaved: leaflets unequal, spathulate, lanceolate, pointed with a bristle, ciliated: capsule orbicular.

Peninsula.

(5) D. MICRANTHES. *(Nees.)*

Ident. N. ab. E. in Wall. Pl. As. Rar. III. p. 112.—Dec. prod. XI. p. 364.

Syn. Justicia chinensis, *Vahl.*—J. cuspidata, *Vahl.*

SPEC. CHAR. Leaves ovate-acuminated: umbels axillary, subsessile, 3 to 5-divided: flowers in heads of three: leaflets of proper involucre unequal, sessile, oblong, partial ones in fours, lanceolate pointed, ciliated: capsule sessile, oblong, tetragonal: seeds glochidiate.

Gujarat: on Sagurghur, near Alibaug. Deccan.

(6) D. BUPLEUROIDES. *(Nees.)*

Ident. N. ab. E. in Wall. Pl. As. Rar. III. p. 111.—Dec. prod. XI. p. 485.

Syn. Justicia latebrosa, *Korn. in Roxb. fl. Ind.* I. p. 128.

SPEC. CHAR. Herbaceous; stem obsoletely hexagonal: leaves ovate, acuminate, acute at the base, lineolate above, glabrous, quite entire: peduncles axillary, umbelliferous, simple or compound: flowers capitate, twin: common involucre, subulate or linear, shorter than the umbel: leaflets of the proper 2-leaved involucre unequal, oblong-lanceolate, glabrous, mucronate, loosely ciliated or scabrous at the margin: bracteoles acuminated with hairs: flowers pale rose.

Silhet. Assam. Flowering nearly all the year.

(7) D. PARVIBRACTEATA. (*Nees.*)

Ident. N. ab. E. in Wall. Pl. As. Rar. III. p. 117.—Dec. prod. XI. p. 488.

Syn. Justicia retorta, *Vahl.*

SPEC. CHAR. Herbaceous: leaves ovate, acuminate, acute at the base, lineolate above, scattered with bristles: peduncles axillary, somewhat twin: rays simple or 3-cleft: leaflets of the 2-leaved involucre lanceolate, scarcely exceeding the calyx: flowers capitate, tern, pale.

Rajamundry.

GENUS LV. PERISTROPHE.

Diandria Monogynia. *Sex. Syst.*

Derio. From *Peri*, around, and *Strophe*, a turning; alluding to the twisted anthers.

GEN. CHAR. Calyx equal, 5-cleft or parted, sessile on a bracteate 2-valved capitulum; corolla resupinate, 2-lipped, lips flat, upper one 3-toothed, lower entire or 2-toothed; stamens two; anthers narrow, 2-celled, cells obliquely placed one behind the other, direction parallel, at length twisted, muticous: capsule 2-valved, 2-celled, cells 2-seeded: dissepiment adnate, persistent: seeds discoid: retinacula hooked.—Herbaceous plants with showy purple flowers with a long tube: capitals of flowers enclosed in a 2-valved involucre and arranged in axillary and finally terminal umbels, simple or compound; stem usually sexangular, fleshy at the knots.

(1) P. MONTANA. (*Nees.*)

Ident. N. ab. E. in Wall. Pl. As. Rar. III. p. 113.—Dec. prod. XI. p. 493.

Engrav. Wight's Icon. t. 1563.

Spec. Char. Leaves oblong, attenuated at both ends, linrolate and like the stems, glabrous: umbels axillary and terminal, five-cleft; capitula 3-5 flowered: involucrum diphyllous: leaflets equal ovato-elliptic, obtuse mucronulate, glabrous.

Courtallum flowering during the rainy months.

(2) P. BICALYCULATA. *(Nees.)*

Ident. N. ab. E. in Wall. Pl. As. Rar. III. p. 113.—Dec. prod. XI. p. 490.

Syn. Justicia bicalyculata, *Vahl.*—J. ligulata, *Lam.*—J. Malabarica, *Ait.*—Dianthera bicalyculata, *Retz.*—D. Malabarica, *Linn.*

Engrav. Lam. Ill. t. 12. fig. 2. Cav. ic. I. t. 71.

Spec. Char. Stem hexagonal, rough and hairy; leaves ovate-acuminate, glabrous or puberulous; peduncles axillary, bifrifid, their branches dichotomous: flowers solitary, pale-rose: common involucra of one leaf linear, double the length of the flower-head: calyx small, membranaceous: corolla pubescent, nearly half an inch long.

Peninsula. Hindostan. Flowering in the cold season.

(3) P. TINCTORIA. *(Nees.)*

Ident. N. ab. E. in Wall. Pl. As. Rar. III. p. 113.—Dec. prod. XI. p. 493.

Syn. Justicia tinctoria, *Roxb.*—J. Roxburghiana, *Roem. et Schult.*—J. purpurea, *Lour.*—Dianthera Japonica, *Thunb.*—J. crenata, *Vahl.*

Engrav. Rumph. Amb. VI. t. 22. f. 1.—Thunb. flor. Jap. t. 4.

Spec. Char. Herbaceous: leaves ovate, obtuse, lineolate, and with the stem pubescently scabrous below; umbels axillary and terminal, 3-cleft; rays compound and decompound; leaflets of the 2-leaved involucro unequal, subenrdate-ovate, pubescently ciliated: flowers deep rose with a white tube and a purple throat.

Rungpore. Bengal. Assam. Flowering nearly all the year.

(4) P. SPECIOSA. *(Nees.)*

Ident. N. ab. E. in Wall. Pl. As. Rar. III. p. 113.—Dec. prod. XI. p. 495.

Syn. Justicia speciosa, *Roxb.*

Engrav. Bot. Mag. t. 1722.

Spec. Char. Fruticose: leaves ovate, acute at the base, lanceolate above, glabrous; stem obtusely hexagonal; peduncles axillary, trifid or trichotomous: flowers umbellately capitate, involucrated; outer bracts cuneiform-linear, obtuse: proper ones lanceolate, acute: flowers crimson, with the upper lip at the base dark purple spotted.

Interior of Bengal, flowering at the end of the cold season.

(5) P. UNDULATA. *(Nees.)*

Ident. Dec. prod. XI. p. 496.

Syn. Justicia undulata, *Vahl.*

Spec. Char. Herbaceous: leaves lanceolate, undulated, repandly-toothed; umbels terminal, simple and 3-cleft.

Malabar.

(6) P. LANCEOLARIA. *(Nees.)*

Ident. N. ab. E. in Wall. Pl. As. Rar. III. p. 114.—Dec. prod. XI. p. 496.

Syn. Justicia lanceolaria, *Roxb. fl. Ind.* I. p. 121.

Spec. Char. Suffruticose: leaves oblong-lanceolate, acuminated at both ends, lineolate, slightly scabrous; panicle terminal, trichotomous, viscidly hairy: leaflets of the involucre tern, lanceolate-acuminate, viscidly hairy: flowers in heads of threes, rose-coloured, with the upper lip towards the base cream-coloured and dark purple-dotted.

Khasia hills, flowering in December and January.

GENUS LVI. RHAPHIDASPORA.

Diandria Monogynia. *Sus. Syst.*

Deriv. From *Rhaphis*, a needle, and *Spora*, a seed.

Gen. Char. Calyx small, 5-cleft: corolla usually resupinate tube recurved, upper lip (presently the lower) concave, uppermost broader 3-cleft: stamens two, exserted: cells of the anthers placed one above the other, upper shorter, lower spurred at the base: capsule at the base compressed seedless narrower, 4-seeded at the apex: dissepiment complete: seeds subtended with hooked retinacula, discoid, cohinately hispid.—Herbs, loose and more or less diffuse: leaves broadish, inflorescence expanded: panicles axillary, trichotomous, sometimes passing into terminal: flowers capitately tern, usually solitary by abortion of the lateral flowers, girt in appearance by 8-9 bracts, namely 2 or 3 a little smaller, and twin bracteoles of each flower, some of which are often obliterated: bracts and bracteoles small bristly.

(1) R. GLABRA. (Nees.)

Ident. N. ab. E. in Wall. Pl. As. Rar. III. p. 115.—Dec prod. XI. p. 499.
Syn. Justicia glabra, Koen. Roxb. flor. Ind. I. p. 130.
Engrav. Wight's Icon. t. 1554.
SPEC. CHAR. Peduncles axillary, many-flowered: leaves ovate, attenuated at the apex, glabrous.

Coromandel, flowering in the rainy season.

GENUS LVII. HYPOESTES.

Diandria Monogynia. *Gar. Syst.*

Deriv. A Greek term signifying *an under garment*; alluding to the covering of the involucrum.

GEN. CHAR. Calyx 5-cleft or parted, equal, enclosed in a somewhat 1-flowered capitulum, the involucre 4-leaved and leaflets free or cohering at the base: corolla 2-lipped, lower lip deeply 3-cleft: stamens two: anthers 1-celled, cohering before evolution, cells lateral in a narrow connectivum: stigma 2-cleft: capsule 2-celled towards the apex, 4-seeded: dissepiment adnate, complete: seeds furnished with subulate retinacula, ovato-subrotund, bi-convex, compressed, tubercled;—Herbs, shrubs, or small trees, often with showy purple or rose-coloured flowers; leaves entire or crenato-dentate.

(1) H. LANATA. (Dalz.)

Ident. Dalz. in Hook. journ. Bot. II. p. 343.—Bomb. flor. p. 197.

SPEC. CHAR. Suffruticose; stem glabrous, ascending, geniculate: leaves lanceolate-acuminate, entire, slightly hispid above, glabrous beneath: branches of the inflorescence trichotomous, covered with white wool: heads few-flowered, 1 to 3, sessile in the opposite axils of the floral leaves: flowers light-purple.

Northern Concan near Rohe.

(2) H. PURPUREA. (R. Br.)

Ident. R. Br. prod. I. p. 474.—Dec. prod. XI. p. 509.—N. ab. E. in Wall. Pl. As. Rar. III. p. 114.
Syn. Justicia purpurea, Vahl.

SPEC. CHAR. Herbaceous: leaves oval, acuminated at both ends, quite entire, below together with the branches pubescently scabrous: thyrses axillary and with a terminal one, spiciform, narrow: floral leaves ovate, and with the segments of the tubular involucre mucronate: flowers purple.

Assam.

GENUS LVIII. HAPLANTHUS.

Diandria Monogynia. *Sex. Syst.*

Deriv. From *Haplos*, single, and *Anthos*, a flower; some of the species are 1-flowered.

GEN. CHAR. Calyx equal, 5-parted: corolla somewhat funnel-shaped, tube incurved, limb 5-cleft, somewhat 2-lipped: stamens included: anthers 1-celled, connectivum and abortive cell tomentose or villous; filaments dilated at the base inwardly: capsule linear, depressed, 8-16 seeded from the base: seeds small, angled.—Erect branched herbs: flowering branches often dissimilar, spiciform, few-flowered, either a somewhat spiked terminal branchlet with small subulate bracts, or flowers solitary at the base of the fascicled somewhat verticillate branchlets.

(1) H. VERTICILLARIS. (*Nees.*)

Ident. Don. prod. XI. p. 513.—Dalz. Bomb. flor. p. 197.

Syn. Justicia verticillata, Roxb.

SPEC. CHAR. Stem herbaceous, simple, erect, naked and smooth at the base: leaves ovate-oblong, attenuated at both ends; branches assuming the form of short, rigid spines, which are bifid at the apex: flowers of a pale-lilac half an inch long.

Hills in the Deccan. Neilgherries. Assam. Flowering from September till December.

(2) H. TENTACULATUS. (*Nees.*)

Ident. Don. prod. XI. p. 513.—Dalz. Bomb. flor. p. 197.

Syn. Ruellia tentaculata, *Linn.*—R. aciculata, *Roth.*

Engrav. Burm. flor. Ind. t. 40. fig. 1.

SPEC. CHAR. Leaves oval-obtuse, smaller than in the preceding: axillary branches verticelled, bifid, longer than the leaves.

Jungles in the Concan. Central India. Flowering in January.

H. NEILGHERRENSIS. (R. W.)

Ident. Wight's Icon. vol. IV. & t. 1556.

Spec. Char. Herbaceous, ramous, declining; branches axillary, opposite, shorter than the leaves: flowers racemose on the ends of the branches and stem: leaves hispid, elliptic-oblong, acuminate, long petioled: petioles winged: flowers opposite from the axil of a minute leaf: calyx 5-parted, small, and like the numerous bracts, setaceo-hispid: bracts linear, 2-3-toothed at the apex: anthers two-celled, both polleniferous with a dense tuft of woolly pubescence on the back.

Neilgherries and Koorg jungles.

GENUS LIX. ANDROGRAPHIS.

Diandria Monogynia. *Bar: Syst:*

Gen. Char. Calyx deeply 5-parted, equal, lobes narrow: corolla 2-lipped, upper lip entire or bifid, inferior trifid, unless when resupinate, when the contrary is the case: stamens two, anthers two-celled, cells parallel, bearded at the base: capsule ovate, or lanceolate, depressed, 2-celled to the base, 4 or many-seeded: partition attached to the valves: seeds oval, obtuse, roundish, obliquely truncated at the base, pitted thimble-like, with a deep bilum.—Herbaceous annuals or under-shrubs, decumbent or erect: stem and branches acutely 4-angled; racemes axillary or terminal, simple or forked: flowers opposite or all turned to one side: bracts opposite, shorter than the calyx, bracteoles wanting, or two, minute at the base of the pedicel: flowers more or less rough or glandular, white or variously purple: lobes of the calyx linear or filiform; capsule linear, oblong, flattened.

(1) A. PANICULATA. (*Nees.*)

Ident. N. ab. E. in Wall. Pl. As. Rar. III. p. 116.—Dec. prod. XI. p. 515.

Syn. Justicia paniculata, Burm. Roxb. for. Ind. I. p. 117.

Engrav. Rheede Mal. IX. t. 56.—Wight's Icon. t. 518.

Spec. Char. Herbaceous, glabrous: stem 4-sided: leaves lanceolate, attenuated into the petiole entire: racemes axillary, horizontal, long, second, bifid, or dichotomous: capsule many-seeded: flowers remote, white, purple-dotted, long-pedicelled.

Peninsula. Most parts of India, flowering nearly all the year.

(2.) A. WIGHTIANA. (*Arn.*)

Ident. Dec. prod. XI. p. 517.

Engrav. Wight's Icon. t. 1358.—Rheede Mal. IX t. 44.

SPEC. CHAR. Herbaceous; stem and branches; leaves sessile, sub-cordate, attenuated towards the point or ovate, short petioled, glabrous, rough on the margin: racemes axillary and terminal, simple or bifid: flowers short pedicelled: laciniæ of the calyx subulate: fruit about 12-seeded.

Malabar.

(3) A. ECHIOIDES. *(Nees.)*

Ident. N. ab. E. in Wall. Pl. As. Rar. III. p. 117.—Dec. prod. XI. p. 518.

Syn. Justicia ciliaris, *Lam.*—J. Echioides, *Linn. fl. Zeyl.—Roxb. flor. Ind.* 1. p. 118.

Engrav. Wight's Icon. t. 467.—Rheede Mal. IX. t. 46.

SPEC. CHAR. Herbaceous, hairy; leaves oblong, subsessile, somewhat crenated: racemes rigid, reflexed: capsules 4-seeded; flowers whitish with dark purple spots.

Ravines in the Deccan. Southern Peninsula. Flowering nearly all the year.

(4) A. VISCOSULA. *(Nees.)*

Ident. N. ab. E. in Wall. Pl. As. Rar. III. p. 116.—Dec. prod. XI. p. 517.

Engrav. Wight's Icon. t. 1559.

SPEC. CHAR. Suffruticose, diffuse, ramose: and like the oblong-lanceolate leaves glabrous; racemes terminal, trifid, glanduloso-pubescent: flowers secund: capsula oval, 8-seeded.

Courtallum.

(5) A. CEYLANICA. *(Nees.)*

Ident. Dec. prod. XI. p. 518.

Engrav. Wight's Icon. t. 1560.

SPEC. CHAR. Herbaceous, stem birsutulous: leaves oblong, lanceolate, usually short petioled, strigose above, pubescent beneath: racemes axillary and terminal, secund, glanduloso-pubescent: flowers pedicelled: laciniæ of the calyx subulata; capsule oblong-linear, hairy, 10-seeded.

Courtallum, flowering during the rainy months.

(6) A. LOBELIOIDES. *(R. W.)*

Ident. Wight's Icon. vol. IV. & t. 1557.

Syn. Erianthera lobelioides, *Nees.*

Spec. Char. Herbaceous, diffuse, procumbent: leaves subovato-orbicular, mucronulate; flowers terminal, racemose, purple.

Neilgherries. Metapollium. Dry pastures at Kaity. Flowering in February.

(7) A. Neesiana. (*R. W.*)

Ident. Wight's Icon. vol. IV & t. 1561.

Spec. Char. Herbaceous, erect, nearly simple, acutely 4-angled, glabrous except round the joints, where it is furnished with a ring of short brown hair: leaves hirsute, elliptic-oblong, acute at both ends, short petioled: panicles terminal, contracted: branches trifid: calyx and corolla glanduloso-pubescent, capsule linear, hirsute, about 8-seeded.

Pulney mountains.

(8) A. serpyllifolia. (*Nees.*)

Ident. N. ab. E. in Wall. Pl. As. Rar. III. p. 115.—Dec. prod. XI. p. 514.

Syn. Justicia serpyllifolia, *Rotll.*—Erianthera serpyllifolia, *Nees.*

Engrav. Wight's Icon. t. 517.

Spec. Char. Stem procumbent: leaves suborbicular, subsessile: pedicels from 1 to 3 flowered: flowers axillary, lower lip of the corolla 2-lobed, lobes painted with three purple dotted lines.

Courtallum. Palamcottah. Bangalore. Mysore.

(9) A. gracilis. (*Nees.*)

Ident. Dec. prod. XI. p. 516.

Spec. Char. Herbaceous: stem glabrous: leaves ovate-lanceolate, glabrous, lineolate-dotted, muriculate-scabrous at the margin, lower ones ending in a short petiole, upper ones subsessile, narrower: racemes axillary, simple, very slender, few-flowered: flowers purple.

Courtallum.

(10) A. lineata. (*Nees.*)

Ident. N. ab. E. in Wall. Pl. As. Rar. III. p. 116.—Dec. prod. XI. p. 516.

Spec. Char. Herbaceous: stem glabrous: leaves ovate, oblong, sessile: racemes axillary, 2-3-chotomous, glandular-pubescent: segments of the calyx subulate, very densely pubescent; corolla pubescent: segments purple.

Neilgherries.

(11) A. AFFINIS. *(Nees.)*

Ident. N. ab. E. in Wall. Pl. As. Rar. III. p. 116.—Dec. prod. XI. p. 517.

SPEC. CHAR. Herbaceous: leaves oval-oblong, obtuse, very shortly petioled, strigose above, pubescent below; racemes axillary, dimidiately cymose or 3-cleft, densely glandular-pubescent: bracts and calycine segments lanceolate-subulate: flowers pale, marked with purple, anthers violet, distinctly bearded at the base.

Neilgherries.

(12) A. GLANDULOSA. (*R. W.*)

Ident. N. ab. E. in Wall. Pl. As. Rar. III. p. 115.—Dec. prod. XI. p. 518.

Syn. Justicia glandulosa, *Roth.*

SPEC. CHAR. Fruticose: leaves oval, petioled, glandulosely villous: racemes axillary, bifid, subcymose, glandulosely villous; stem softly pubescent.

Mysore.

ORDER CXXI. VERBENACEÆ.

Flowers hermaphrodite, rarely polygamo-dioicous, 4-5-merous, rarely more: regular or irregular or bilabiate, unibracteate: bracts sometimes enlarging after blooming: calyx free, monosepalous, 4-5-rarely 6-8-toothed, persistent, more or less enlarging with the fruit: corolla hypogynous, monopetalous, tubular, deciduous; limb 4-5-sometimes 6-12-lobed, usually unequal, secund or more or less perfectly bilabiate, rarely equal: æstivation imbricate (in Symphorema, inflexed): stamens inserted on the tube of the corolla, incluse or exserted, 4-5, rarely more: usually didynamous, and all fertile, or with a superior exantheriferous or rudimentary one: anthers 2-celled: cells generally opening longitudinally, parallel, divaricate or vertically superposed: connectivum sometimes produced beyond the cells: ovary free, seated on an annular disk composed of 2 or 4 carpels : 2–4-celled with the margins of the carpels forming the primary partitions, or by these partitions sometimes splitting within the cavity and introflexed, 4–8-celled : ovules usually solitary in the cells, rarely geminate, collateral, or two opposite, erect from the base, anatropous in verbenaceæ, in the other tribes pendulous from an ascending parietal spermophore in the central angle of the cell ; style terminal, simple : stigma undivided, capitate or bifid : fruit either capsular of 2–4, or rarely 6, one-seeded cocci, separating at maturity, or drupaceous, 1–2-celled : seeds erect, exalbuminous : embryo straight, cotyledons thick, oily : radicle inferior, short.—Herbs, shrubs, or large trees, sometimes scandent, often furnished with resinous glands, thence aromatic or fetid : branchlets 4-sided : leaves opposite, whorled or alternate, simple or pinnate, incised, divided or digitate : stipules none, inflorescence either indefinite (centripetal), racemoso-spicate, capitate or definite (centrifugal) di-or trichotomously cymose : cymes axillary or forming terminal panicles : calyx and bracts often coloured, enlarging with the fruit : corolla variously coloured, white, red, blue, yellow, often small and inconspicuous.

GENUS I. PRIVA.

Didynamia Angiospermia. *Ser : Syst :*

Gen. Char. Calyx tubular, 5-folded, shortly 5-toothed : corolla almost hypocrateriform, tube cylindric, limb 5-cleft, somewhat unequal, oblique ; stamens 4, didynamous, inserted on the tube of the corolla, included : anthers erect, 2-cleft at the base, 2-celled, cells dehiscing by a gaping longitudinal cleft : ovary 4-celled, cells 1-ovuled : style equalling the lower stamens, stigma lateral, somewhat laminar, often curved back suddenly : capsule enclosed by the enlarged membranaceous calyx which is often twisted at the apex, when ripe septicidally separating into two cocci sometimes 2, or by abortion, 1-celled : pericarp hard and dry, tuberculed or angled at the back, angles muricated, prickly or echinate, rarely smooth.

(1) P. LEPTOSTACHYA. (Juss.)

Ident. Juss. Ann. Mus. 7. p. 70.—Dec. prod XI. p. 533.—Dalz. Bomb. flor. p. 198.

Syn. Tortula aspera, *Roxb. in Willd. Sp. III. p.* 359.—Streptium asperum, *Roxb.*

Engrav. Roxb. Cor. II. t. 146.—Wight in Hook. Journ. bot. I. t. 130.

SPEC. CHAR. Perennial: stem and branches puberulous: leaves subcordate, ovate-acuminate, coarsely crenate-serrate, hispid on both sides, pale beneath: fruit-bearing calyx subglobose, hoary with hooked pubescence: capsules obcordate: flowers small, in terminal racemes, white.

On old walls at Dapoorie. Coromandel. Flowering in the cold season.

GENUS II. BOUCHEA.

Didynamia Angiospermia. *Sar: Syst:*

GEN. CHAR. Calyx long-tubular, plicately 5-angled, somewhat equally 5-toothed, truncated between the teeth: corolla infundibular-hypocrateriform: stamens inserted at the throat of the corolla, four anther-bearing, didynamous, included: anthers sub-didymous with contiguous cells, adnate to the dorsal connectivum: cells 1-ovuled: style a little thickened upwards: stigma dilated into an oblong somewhat 2-lobed lamina: capsule enclosed by or exceeding the calyx, dicoccous: pericarp hard and dry, filled with seed.

(1) B. HYDRABADENSIS.. *(Walpers.)*

Ident. Walp. Repert. 4. p. 12.—Dec. prod. XI. p. 559.

Engrav. Wight's Icon. t. 1462.

SPEC. CHAR. Suffruticose, sparingly pubescent, branches obsoletely 4-angled: leaves ovate-elliptical, cuneately narrowing into the petiole, acutely and coarsely serrated, glaucous beneath; spikes terminal, peduncled; pedicels short, minutely bracteolate: bracts lanceolato-subulate, margin membranaceous, roughly ciliate, two or 3-times shorter than the calyx: calyx plicately five-angled, truncated, with five subulate, unequal teeth: corolla large: capsule the length of the calyx, linear compressed at the apex, smooth.

Serramallie hills, near Dindigul. Mysore, in shady jungles.

GENUS III. LIPPIA.

Didynamia Angiospermia. *Sex: Syst:*

Derio. Named after Augustus Lippi, a French Physician and traveller in Abyssinia.

GEN. CHAR. Calyx small, membranaceous, tubular, 2-winged, 2-keeled or without angles, 2-cleft, lobes more or less manifestly 2-toothed, and at last often 2-valved and adherent to the capsule, or almost equal and herbaceous-4-toothed: corolla tubular-subinfundibuliform, tube amplified upwards, limb obliqua, flat or bending, somewhat 2-lipped, upper lip entire or 2-cleft, lower 3-cleft: stamens didynamous inserted on the tube, enclosed: anthers 2-celled, cells dehiscing by a gaping cleft: ovary 2-celled, cells 1-ovuled: style terminal, short, filiform: stigma lateral, linear: capsules dicoccous, cocci easily separating or cohering when ripe, pericarp very hard, back smooth.

(1) L. NODIFLORA. (*Rich.*)

Ident. Rich. in Michx. flor. Bor. Am. II. p. 15.—Deb. prod XI. p. 585.

Syn. Verbena nodiflora, *Linn.*—V. capitata, *Forsk.*—Blairia nodiflora, *Gærtn.*—Zapania nodiflora, *Lam.*

Engrav. Lam. Ill. t. 17. fig. 3.—Wight's Icon. t. 1463.—Burm. Ind. t. 6. f. 1.—Rheede Mal. X. t. 47.

SPEC. CHAR. Creeping, all strigose with adpressed hairs: stems filiform: leaves cuneate, spathulate, sharply serrated in the upper half: peduncles axillary, solitary: heads of flowers ovoid, and afterwards cylindric, small, white.

Common in grassy and sandy places, flowering nearly all the year.

GENUS IV. LANTANA.

Didynamia Angiospermia. *Sex: Syst:*

Derio. An ancient name of *Viburnum*, and applied by Linnæus, because of its affinity.

Calyx membranaceous, small, obsoletely 3-4-toothed, ciliate, covering the fruit, and with its increase, becoming greatly extended and translucent, at length withering away: corolla tubuloso-infundibuliform, slightly swelling upwards: limb obliqua, flat, or inclined, somewhat bilabiate, the upper lip entire or bifid, the lower one lobed: stamens 4, inserted within the tube of the corolla, didynamous : anthers 2-celled, opening longitudinally : ovary 2-celled, cells with a single erect ovule : style terminal, short : stigma linear or obliquely capitate : drupe fleshy or succulent with 2-nuts, shell hard, rough, and tuberculate, or rarely smooth : cotyledons thick radicle interior, short.—Shrubs or under shrubs, stems 4-sided : leaves opposite or verticelled, simple or feather-nerved, rugous : peduncles axillary usually single : capitula compact, usually elongating during flowering : calyx pubescent : corolla variously coloured, white, orange, red, purple and often changeable.

(1) L. ALBA. (*Miller.*)

Ident. Miller ex Link enum. Pl. Hort. Berol. II. p. 126.— Dec. prod. XI. p. 606.

Syn. L. Indica, *Roxb.*—L. dubia, *Royle.*

Engrav. Royle Himal. Bot. t. 73. fig. 3.—Wight's Icon. t. 1464.

SPEC. CHAR. Shrubby, erect, straight : branches twiggy, 4-sided, strigose and hairy : leaves opposite, short petioled, elliptic or rounded-ovate or subcordate, acuminate at the ends, coarsely serrate-crenate, much wrinkled, scabrous above, hoary and villous beneath : peduncles axillary, spreading, thickened upwards : heads of flowers hemispherical, light-purple : throat yellow, scentless : fruit dark-violet, of the size of a pea.

About Dharwar, and other parts of Deccan. Mysore and most parts of the country. Flowering all the year. There are several varieties with white, orange, and pale violet flowers.

GENUS V. SYMPHOREMA.

Octandria Monogynia. *Scr: Syst.*

Deriv. From *Symphoreo*, to accumulate ; in reference to the involucre.

GEN. CHAR. Involucre 6-8-leaved, spreading, subtending a contracted cyme : calyx cup-shaped-campanulate, 6-8-toothed, persistent : corolla tubular, limb 6-18-cleft, segments linear reflexed : stamens the number of the lobes of the corolla, inserted at the summit of the tube, equal, long exserted : filaments capillary : anthers inserted by the back, 2-celled, cells dehiscing by a gaping cleft : ovary 2-celled, cells binovulate, ovules collateral : style filiform, exserted beyond the stamens : stigma 2-cleft : capsule coriaceous, indehiscent, enclosed by the calyx, 1-seeded by abortion : seed erect.—Climbing shrubs.

(1) S. INVOLUCRATUM. *(Roxb.)*

Ident. Roxb. flor. Ind. II. p. 62.—Dec. prod. XI. p. 621.
Engrav. Roxb. Cor. II. t. 186.—Wight's Icon. t. 362.

SPEC. CHAR. Stem woody, climbing: branches inflorescence, and underside of the leaf covered with soft tomentum: leaves opposite, short-petioled, oval or rounded-elliptic, obtuse at the base, 3-nerved, with a short, obtuse, acumen, the margin almost entire, or irregularly repand-toothed or serrate: inflorescence terminal, panicled, consisting of long-peduncled bifid cymes: involucre 7 to 9-flowered; flowers white.

The Concans, between Nagotna and Alibaug. Forests of Coromandel. Flowering in the hot season.

(2) S. POLYANDRUM. *(R. W.)*

Ident. Dec. prod. XI. p. 621.
Engrav. Wight's Icon. t. 363.

SPEC. CHAR. Corolla many (14–18) cleft: stamens equalling the number of segments: leaves from broadly ovate, sub-acuminate to nearly orbicular, stellately hairy above, thickly tomentose beneath, flowers pure white.

Ballaghaut hills, near Madras, flowering in April.

GENUS VI. SPHENODESMA.

Pentandria Monogynia. *Ser: Syst:*

Deriv. From *Sphen*, a wedge, and *Desme*, a fascicle; alluding to the flowers.

GEN. CHAR. Involucre 5-leaved, spreading, veiling a sessile 6-flowered glomerule: calyx cup-shaped, 5-cleft, persistent: corolla hypocrateriform or infundibular, somewhat regular, tube equalling the calyx, throat more or less pubescent, limb 5-cleft: stamens five, inserted at the throat, slightly enclosed or exserted, equal, anthers 2-celled: ovary 2-celled, cells 1-(2)-ovuled: style very short, stigma 2-cleft: ripe capsule coriaceous, indehiscent, completely filled up by a single seed.—Twining shrubs: leaves simple, opposite, quite entire: panicles axillary and terminal, rarely solitary: glomerules consisting of two 3-flowered cymes: involucral leaves usually unequal, often coloured.

(1) S. Wallichiana. *(Schauer.)*

Ident. Dec. prod. XI. p. 622.
Syn. Roscoea pentandra, *Roxb. fl. Ind.* III. p. 54.
Engrav. Wight's Icon. t. 1475.

Spec. Char. Suffruticose, climbing: branchlets pubescently tomentose: leaves coriaceous, short petioled, ovate, oblong, obtuse at the base, narrow acuminate, glabrous, shining above, beneath bearded in the axils of the veins: panicles large, bracteolate, leafy below: bracts ovate: peduncles filiform, as long as the involucre, and like it and the flowers, glabrous: leaflets of the involucre linear oblong, sessile obtuse: calyx cup-shaped, 5-nerved, very shortly 5-cleft, truncated ciliate: flowers purple.

Pundua, flowering in the hot season.

(2) S. Jackiana. *(Schauer.)*

Ident. Dec. prod. XI. p. 622.
Syn. S. pentandra, *Jack. Mal. Misc.—Hook. Bot. Misc.* I. p 283.
Engrav. Wight's Icon. t. 1476.

Spec. Char. Climbing: branchlets pubescently tomentose: leaves coriaceous, short petioled, oblong, obtuse at the base, attenuate-acuminate at the apex, glabrous, shining, beneath pubescent or becoming glabrous, bearded in the axils of the veins: panicles large, brachiate, leafy below, bracts oblong: peduncles filiform, about equal to, or a little shorter than the involucre, and like it thinly sprinkled with hairs: leaflets of the involucre linear, lanceolate, sessile, obtuse, scarcely mucronulate: calyx glabrous, tubuloso-campanulate, 10-nerved, 10-toothed: five teeth lanceolate, reflexed, 5 broad, triangular, acute, erect, flowers purple.

Silhet.

(3) S. Angulculata. *(Nees.)*

Ident. Dec. prod. XI. p. 623.

Spec. Char. Fruticose, climbing: branchlets panicle and young leaves yellowish with stellate tomentum: leaves coriaceous, short-petioled, lanceolate or oblong, obtuse at the base, acuminated: panicle large, chequered with leaves: peduncles filiform, long, much spreading: old leaves of the involucre acutish, spathulate, long-attenuated at the base: calyx cup-shaped, 4-5-cleft, lobes triangular, acute.

Pundua.

GENUS VII. CONGEA.

Didynamia Angiospermia. *Sex: Syst:*

GEN. CHAR. Involucre 3-leaved, spreading, veiling a 6–9-flowered, sessile glomerule: calyx tubular, amplified upwards, 5-cleft, persistent: corolla 2-lipped, tube equalling the calyx, limb very unequal, upper lip elongated, erect, 2-cleft, lower spreading, shortly 3-lobed: stamens inserted at the throat, long exserted, filaments capillary, gyrose: anthers 2-celled: ovary 2-celled, cells binovulate: style capillary, nearly equalling the stamens: stigma 2-cleft: ripe capsule coriaceous, indehiscent, completely filled by a single seed.

(1) C. TOMENTOSA. *(Roxb.)*

Ident. Dec. prod. XI. p. 623.
Syn. Roscoea tomentosa, *Roxb.*—R. villosa, *Roxb. fer. Ind.* III. p. 36.
Engrav. Wight's Icon. t. 1479-2.

SPEC. CHAR. Suffruticose, climbing: leaves ovate, slightly cordate, acute or sub-acuminate, hispid above, tomentose beneath: leaflets of the involucre oval, obtuse at both ends, tomentose above, softly pubescent beneath: umbels 7-flowered; calyx teeth short, blunt: flowers small, white.

Chittagong. Coromandel. Flowering in March.

GENUS VIII. TECTONA.

Pentandria Monogynia. *Sex. Syst.*

Deriv. From *Tekka*, the Malabar name of the tree.

GEN. CHAR. Calyx campanulate, 5-cleft: corolla funnel-shaped, tube short nearly equalling the calyx, limb 5-cleft nearly equal, spreading, throat hairy; stamens 5-6, inserted on the tube, exserted, nearly equal: anthers cordate, 2-celled, cells parallel dehiscing by a longitudinal cleft: ovary minute at the bottom of the corolla, 4-celled, cells 1-ovuled: style terminal, terete, the length of the stamens, stigma acutely 2-cleft: drupe 1-stoned, sarcocarp spongioso-suberous, or noramentaceous, altogether devoid of flesh, enclosed in the increased or a bladder-inflated ample calyx, stone 4-celled, putamen and dissepiment bony: seed thick, oily.

(1) T. GRANDIS. (*Linn.*)

Ident. Linn. Sl. suppl. p. 151.—Dec. prod. XI. p. 629.—Roxb. flor. Ind. I. p. 600.

Engrav. Rheede Mal. IV. L. 24.—Rumph. Amb. III. t. 18. —Lam. flor. L. 136.—Roxb. Cor. I. t. 6.

SPEC. CHAR. Tree: branches quadrangular: leaves opposite, large, ovate or subelliptic-acuminated, short-petioled, shining above: cymes axillary, dichotomous, or collected in a terminal panicle: flowers numerous, small, white: drupes enclosed in the inflated calyx: nut 4-celled, one-seed in each: seeds thick, oily.

Banks of rivers, in Malabar and the Western coast. Bundelkund. Flowering in the rains. This is the Teak tree so well known for its valuable timber.

GENUS IX. PREMNA.

Didynamia Angiospermia. *Sex; Syst;*

Deris. From *Premnon*, the stump of a tree, the trees are of dwarf size.

GEN. CHAR. Calyx capsule-shaped, cup-shaped or subcampanulate, somewhat bilabiately 4–5-cleft or toothed or bilabiate, one or the other lip at least entire, persistent, at length increasing: corolla tubular, tube somewhat funnel-shaped, short: limb sometimes bilabiately 4-cleft, spreading, upper lip half bifid or emarginate, lower 3-cleft or 3-parted, lobes nearly equal, sometimes somewhat regular, reflexed: throat villous, often long bearded: stamens didynamous or nearly equal, equalling the corolla or exserted, equidistant; anthers subrotund, diverging from the base of the cell: ovary 4-celled, cells 1-ovuled : style filiform, nearly equalling the stamens: stigma 2-cleft, feet divaricate : drupe pea-shaped, fleshy, 1-stoned, putamens perforated at the axis, bard, wrinkled or warty-tuburcled, 4-celled or by abortion 2–3-celled : seed erect.—Shrubs, undershrubs or trees, glabrous, or pubescent : leaves opposite, simple, quite entire or toothed : flowers small, arranged in terminal trichotomously cymose, corymbiform, bractiately-pyramidal or anthuroideous panicles, often polygamous, the primary axils alone being fertile.

(1) P. INTEGRIFOLIA. (*Linn.*)

Ident. Linn. Mant. II. p. 253.

Syn. P. serratifolia, *Linn.*—Cornutia corymbosa, *Burm.*—P. corymbosa, *Rottl. et Willd.*—P. spinosa, *Roxb.*

Engrav. Burm. Ind. t. 41. fig. 1.—Wight's Icon. t. 1469.

Spec. Char. Arboreous, the trunk and older branches armed with opposite spines; unarmed ramuli, panicles and petioles pubescent: leaves short-petioled, ovate or oval, shortly and obtusely acuminate, rounded towards the base, entire, or crenato-dentate, the adult ones glabrous on both sides, shining above, dull, opaque beneath: panicles terminal, loosely corymbose: calyx bilabiate, the upper acutely bidentate, inferior often entire: tube of the corolla cylindrical, twice the length of the calyx.

Bengal, flowering in the rains.

(2) P. CORDIFOLIA. *(Roxb.)*

Ident. Roxb. flor. Ind. III. p. 78.—Dec. prod. XI. p. 632.
Engrav. Wight's Icon. t. 1493.
Spec. Char. Shrubby: branchlets, cymes and petioles of the younger leaves, villous: leaves short-petioled, cordate, or cordato-ovate, acuminate, entire, bullate, glabrous on both sides, shining above, dull and pale beneath: panicles terminal, small, contracted-corymbose: flowers greenish white.

Khandalla, flowering in May and June.

(3) P. SCANDENS. *(Roxb.)*

Ident. Roxb. flor. Ind. III. p. 82.—Dec. prod. XI. p. 632.
Spec. Char. Large climbing shrub: branches and cymes pubescent: leaves ovate-oblong, or subcordate, cuspidate acuminate, quite entire, glabrous, shining above: panicle terminal, corymbose, rather large; flowers very small, greenish white: drupe like a pea, black, smooth.

Kandalla. Silhet. Flowering in the hot season.

(4) P. TOMENTOSA. *(Willd.)*

Ident. Willd. Spec. III. p. 314.—Dec. prod. XI. p. 634.
Syn. Cornutia corymbosa, *Lam.*—P. flarescens, *Juss.*
Engrav. Wight's Icon. t. 1468.
Spec. Char. Shrub or small tree: branchlets young leaves and cymes everywhere tomentose: leaves petioled, ovate, or ovate-oblong, long-acuminate, entire, venoso-rugous, stellately pubescent on both sides, sparingly above, copiously beneath: panicles large, terminal, many-flowered, compact: flowers whitish.

Coimbatore. Pondicherry. Flowering during the hot season.

(5) P. LATIFOLIA. (*Roxb.*)

Ident. Roxb. flor. Ind. III. p. 76.—Dec. prod. XI. p. 635.

SPEC. CHAR. Shrub erect, branched: leaves petioled, rounded-cordate or oval, quite entire, or obsoletely repand in the upper part: panicles corymbose, terminal and axillary: flowers small, greenish.

Very common in hedges in the Concan. Coromandel. Flowering in April.

(6) P. WIGHTIANA. (*Schauer.*)

Ident. Dec. prod. XI. p. 635.

Engrav. Wight's Icon. t. 1485.

SPEC. CHAR. Shrub or small tree: cymes and petioles puberulous; leaves petioled, ovate, abruptly acuminate, rounded or slightly produced at the base, entire, or shortly toothed anteriorly, subulate, glabrous, altidulous on both sides: panicles terminal, thyrsoid: calyx sub-bilabiate, unequally 5-toothed; tube of the bilabiate corolla twice the length of the calyx.

Courtallum. Dindigul. Serramallay. Travancore.

(7) P. GLABERRIMA. (*R. W.*)

Ident. Wight's Icon. vol. IV. & t. 1484.

SPEC. CHAR. Everywhere glabrous, except a slight villosity on the inflorescence: leaves oborate, oblong, abruptly acuminate, acute, rigid, entire, somewhat shining above, pale-whitish beneath: panicles terminal, lax, corymbose, ultimate divisions dichotomously cymose: calyx campanulate, obscurely 5-toothed, shortly villous: corolla bilabiate: upper lip emarginate, under 3-lobed, middle lobe the largest, throat hairy.

Courtallum, flowering in August.

(8) P. LONGIFOLIA. (*Roxb.*)

Ident. Roxb. fl. Ind. III. p. 79.— Dec. prod. XI. p. 634.

SPEC. CHAR. Arboreous: branchlets, panicle and petioles covered with mealy, tomentose, densely stellate pubescence: leaves petioled, ovate-oblong or oblong, short-acuminate, afterwards glabrate on both sides, shining above, pale below: panicles terminal, spreading-corymbose, many-flowered: calyx campanulate, bilabiate, 5-toothed: flowers white, fragrant.

Forests in Bengal. Khasia hills. Assam. Flowering nearly all the year.

(9) P. MUCRONATA. (*Roxb.*)

Ident. Roxb. fl. Ind. III. p. 80.—Dec. prod. XI. p. 635.

SPEC. CHAR. Small tree: branches panicle petioles and nerves of the leaves hairy pubescent: leaves petioled, broad-ovate, mucronulate into a point, attenuated, quite entire, pubescent on both sides, soft below: panicle divaricately corymbose, loose, many-flowered: calyx somewhat equally 5-cleft: lower lip of the somewhat equally 4-cleft funnel-shaped corolla together with the throat, villously bearded.

Silhet.

(10) P. ESCULENTA. (*Roxb.*)

Ident. Roxb. fl. Ind. III. p. 80.—Dec. prod. XI. p. 636.

SPEC. CHAR. Shrub: branchlets and cymes thinly mealy-puberulous: leaves very shortly petioled, oblong, acuminate, somewhat narrowed at the base, glabrous, coarsely and sharply toothed above, glaucescent below: cyme terminal, small, contracted: calyx 2-lipped, unequally 5-toothed: tube of the corolla greenish-yellow.

Chittagong.

(11) P. HERBACEA. (*Roxb.*)

Ident. Roxb. fl. Ind. III. p. 80.—Dec. prod. XI. p. 637.

SPEC. CHAR. Suffruticose, depressed or herbaceous, without a stem: branchlets, cymes and network of the leaves hairy-pubescent: leaves very shortly petioled, obovate or rotund, subacuminate, serrato-dentate above, villosely ciliated; upper surface at last glabrate, lower somewhat hoary and densely glandular-dotted: cymes terminal and axillary, shorter than the leaf: calyx 5-cleft, round-lobed, ciliated: flowers small, pale whitish yellow.

Interior of Bengal. Goruckpore. Flowering in February and March.

GENUS X. CALLICARPA.

Tetandria Monogynia. *Sex. Syst.*

Deriv. From *Kalos*, beautiful, and *Karpos*, fruit.

GEN. CHAR. Calyx cupula-shaped cup-shaped, or more rarely tubular, 4-5-ribbed, and often angular or plicate, 4-5-toothed or rarely 4-cleft, persistent: corolla sub-campanulato-tubular, tube short, limb 4-5-cleft equal: stamens 4, rarely 5, inserted on the tube of the corolla, equal, exserted: anthers inserted at the back above the base, glandulosely dotted in front and at first at the back, 2-celled, cells parallel, dehiscing above by a deep gaping cleft or oblong pore at the apex: ovary 4-celled, cells 1-ovuled; style filiform, nearly equalling the stamens or longer, clavately thickened towards the top: stigma capitate, very shortly 2-lobed: drupa baccate, 4-stoned, stones when ripe distinct, 1-celled, putamen hard; seed erect.—Shrubs or undershrubs, rarely trees, more or less hoary or rusty with stellate, mealy or scurfy down, usually beset with copious oil-bearing glands: leaves opposite, simple, quite entire at the base; cymes axillary, dichotomous: flowers sometimes polygamous.

(1) C. CANA. (*Linn.*)

Ident. Linn. Mant. II. p. 190.—Dalz. Bomb. flor. p. 200.—Roxb. flor. Ind. I. p. 392.

Syn. C. Wallichiana, *Walp.*—Dec. prod. XI. p. 641.—C. Heyniii *Roth.*—C. tomentosa, *Lam.*—C. acuminata, *Roxb.*

Engrav. Wight's Icon. t. 1480.—Bot. Mag. t. 2107.

SPEC. CHAR. Shrub or small tree: branchlets cymes and petioles densely rusty-tomentose: leaves membranaceous, broadly ovate, roundish, or narrow, obtuse, or even acuminate at the base, long petioled, attenuato-acuminate, entire, or slightly repand, and minutely denticulate, reticulate, rugous: adult ones, except on the veins, glabrous above, densely woolly and whitish tomentose beneath: cymes many-flowered, bipartite, divaricately-dichotomous, corymbose: peduncles of the length of the petiole: calyx truncate or slightly 4-lobed: flower pale red.

Travancore frequent among low jungles. Silhet, flowering in February and March.

(2) C. ARBOREA. (*Roxb.*)

Ident. Roxb. fl. Ind. I. p. 390.—Dec. prod. XI. p. 641.

SPEC. CHAR. Tree: branches tetragonal, together with the cymes and petioles hoary with dense, dusty tomentum: leaves coriaceous, elliptic, obovate or oblong-elliptic, cuneate at the base, narrowed into a long petiole, entire or sharply repand-toothed, adult ones glabrate above, except at the nerves, shining, reticulate below, white with stellate, mealy tomentum: cymes solitary or tern, corymbose: peduncle angled, nearly equalling the petiole: calyx pulverulent, very small, very shortly 4-petioled: flowers small, purple lilac.

Chittagong. Goalpara. Oude. Flowering in the cold season.

(3) C. MACROPHYLLA. (*Vahl.*)

Ident. Vahl. syml. III. p. 13.—Dec. prod. XI. p. 644.—Roxb. fl. Ind. I. p. 393.

SPEC. CHAR. Shrub: branchlets, peduncles, together with the ramification of the inflorescence and petioles white with woolly tomentum: leaves lanceolate-elliptic, short-petioled, obtuse or rounded at the base, attenuated into a point, coarsely mucronate-crenate, venoso-rugous, adult ones scabrous above, below softly ashy tomentose: cymes many-flowered, dichotomously much-branched, divaricately corymbose: peduncle shorter than the petiole: calyx somewhat villous, very small, 4-ribbed, mucronately 4-toothed: flowers small rose-coloured.

Chittagong. Silhet. Assam. Flowering in the rainy season.

(4) C. RUBELLA. (*Lindl.*)

Ident. Dec. prod. XI. p. 645.

Engrav. Lindl. Bot. Reg. t. 883.

SPEC. CHAR. Suffruticose: dotted below with tomentum: branches together with the inflorescence hoary with dense flocose tomentum: leaves lanceolate, very shortly petioled, cordate at the base, acuminated, serrato-denticulate, pubescent above, hoary-tomentose below: cymes many-flowered, bipartite, divaricately dichotomous, loose, corymbose: peduncle three times longer than the petiole: calyx minute, 4-ribbed, truncated, 4-mucronulate.

Pundua.

(5.) C. LONGIFOLIA. (*Lam.*)

Ident. Lam. dict. I. p. 562.—Dec. prod. XI. p. 645.

Syn. C. lanceolaria, *Roxb.*—C. acuminata, *do.*—C. Roxburghiana, *Roem. et Schult.*—C. adenanthera, *R. Br.*—C. albida, *Blume.*

Engrav. Lam. Ill. t. 69. f. 2.—Bot. Reg. t. 864.

SPEC. CHAR. Tree: branchlets, with the branches of the inflorescence, calyx and network of the leaves, floccoso-tomentose: leaves membranaceous, lanceolato-oblong or lanceolate, attenuated at both ends, short-petioled, long-acuminate, serrato-denticulate, adult ones glabrate on the upper surface, clothed on the lower with more or less scattered, stellate wool: cymes many-flowered, divaricately dichotomous, slightly compact: peduncle nearly equalling the petiole: calyx short, 4-ribbed, truncated at the mouth, very shortly 4-mucronulate: flowers small pale purplish lilac.

Khasia hills, flowering in the cold season.

GENUS XI. CLERODENDRON.

Didynamia Angiospermia. *Sex ; Syst ;*

Deriv. From *Kleros*, a lot, and *Dendron*, a tree ; in allusion to the uncertain medical properties of the species.

GEN. CHAR. Calyx campanulate, rarely tubular, sometimes 5-angled, or somewhat salver-shaped, tube usually conspicuously exceeding the calyx, sometimes very long : limb five-parted, the two upper divisions a little larger : stamens 4, inserted on the tube of the corolla, much exserted, sub-didynamous : anthers 2-celled, cells parallel opening longitudinally : ovary 4-celled, cells with one pendulous ovule : style filiform, exserted ; stigma 2-cleft, acute, drupe within the enlarged, persistent calyx, baccate, 4 or by abortion 1-seeded, usually 2-4-lobed, nuts woody, smooth : seed solitary, pendulous, cotyledons oily, radicle short inferior.—Shrubs, or small trees, leaves opposite or ternate, simple, entire or rarely lobed : cymes trichotomous, axillary or collected into a terminal panicle.

(1) C. INFORTUNATUM. (*Linn.*)

Ident. Linn. flor. Zeyl. p. 232.—Dec. prod. XI. p. 667.

Syn. Volkameria infortunata, *Roxb.*—C. viscosum, *Vent.*

Engrav. Wight's Icon. t. 1471.—Bot. Reg. t. 629.—Burm. flor. Zeyl. t. 29.—Rumph. Amb. IV. t. 49.—Bot. Mag. t. 1805.—Rheede Mal. II. t. 29.

SPEC. CHAR. Under-shrub, 2 to 3 feet high : branchlets quadrangular : leaves long-petioled, rounded or ovate-cordate, the upper ones ovate-entire or dentate-strigose, and hairy on both sides : panicle terminal, large, spreading, naked : flowers white tinged with rose inside : the calyxes increasing and turning red after the flower withers : drupe black, within the increased calyx.

Common at Vingorla. Belgaum. Oude. Bengal. Flowering in February and March.

(2) C. SERRATUM. (*Spreng.*)

Ident. Spreng. Syst. II. p. 758.—Dec. prod. XI. p. 664.

Syn. Volkameria serrata, *Linn. Roxb.*—C. Macrophyllum? *Sims.*—C. Javanicum, *Walpers.*—V. farinosa, *Roxb.*—C. ternifolium, *Don. prod fl. Nep.*

Engrav. Bot. Mag. t. 2526—Wight's Icon. t. 1472.

SPEC. CHAR. Suffruticose: branchlets quadrangular, furrowed, glabrous: leaves opposite or in threes, papery, obovate-oblong or lanceolate, remotely serrate-toothed: panicles terminal, raceme-like, hoary and farinaceous: flowers pale blue, lower lip indigo.

Courtallum. Khandala. Assam. Western Ghauts. Flowering in May and June.

(3) C. INERME. (R. Br.)

Ident. R. Br. in Ait. Hort. Kew. IV. p. 65.—Dec. prod. XI. p. 660.—Roxb. flor. Ind. III. p. 58.

Syn. Volkameria inermis, *Linn.*—C. buxifolium, *Spreng.*

Engrav. Rheede Mal. V. t. 49.—Jacq. coll. suppl. t. 4. fig. 1. —Rumph. Amb. V. t. 46.

SPEC. CHAR. Climbing branched shrub: leaves small, smooth, shining, oval or elliptic: cymes axillary, as long as the leaf, 3-flowered, collected into a terminal corymbose panicle: flowers white: the tubes of the corolla long and slender, greenish-white.

Common along the coast, near the sea. Soonderbuns, flowering nearly all the year.

(4) C. PHLOMOIDES. (Linn.)

Ident. Linn. Suppl. p. 292.—Dec. prod. XI. p. 663.—Roxb. flor. III. p. 57.

Syn. Volkameria multiflora, *Burm.*

Engrav. Burm. Ind. t. 45. fig. 1.—Wight's Icon. t. 1473.

SPEC. CHAR. Shrub or small tree, branchlets terete, and like the petioles and peduncles, whitish tomentose: leaves membranaceous, opposite, petioled, ovate, or ovato-rhomboid, acuminate, somewhat obtuse, entire at both ends, irregularly and bluntly serrated in the middle: glabrous above, puberulous beneath: panicles terminal, large, fastigiate, leafy below: cymes tricholomous, lax: bracteoles oblong: calyx glabrous, campanulato-ventricose, half 5-cleft, segments sub-ovate, acute: tube of the corolla sub-glandulose, thrice the length of the calyx: flowers white.

Coromandel. Deccan. Bengal. Flowering nearly all the year. There is a variety with red-flowers.

(5) C. NUTANS. (Wall.)

Ident. Wall. cal. No. 1793.—Dec. prod. XI. p. 663.

Engrav. Bot. Mag. t. 3049.

Spec. Char. Shrubby: branchlets acutely quadrangular: leaves opposite or in threes, membranaceous, oblong-lanceolate, long attenuated at the base, subsessile, long-acuminate, quite entire, shining above, paler below: panicle terminal, naked, very lax, oblong-raceme-shaped, nodding: cymes short-peduncled, somewhat 3-flowered: calyx ventricose, 5-angled, half 5-cleft; segments oval, much spreading: flowers largish, white.

Silhet, flowering in the cold season.

(6) C. BRACTEATUM. (*Wall.*)

Ident. Wall. cat. No. 1800.—Dec. prod. XI. p. 665.

Spec. Char. Arboreous (?) branchlets obsoletely tetragonal and with the branches of the panicle and petioles hoary with downy wool: leaves long-petioled, ovate-oblong, acuminated at both ends, quite entire, strigosely hairy above, canescent with soft hair below and glandular-dotted: panicle terminal, somewhat fastigiate; compact: cymes axillary, shorter than the leaf, capitate; bracts leafy, lanceolate, nearly equalling the calyx: calyx pubescent and glandular-dotted, very large, 5-angled, half 5-cleft; segments spreading, ovate: tube of the corolla slender, pubescently hairy, twice exceeding the calyx.

Pundua.

(7) C. SIPHONANTHUS. (*R. Br.*)

Ident. R. Br. in Ait. Hort. Kew. IV. 65.—Dec. prod. XI. p. 670.

Syn. Siphonanthus Indica, and S. angustifolia, *Willd.*—Ovieda mitis, *Burm.*

Engrav. Lam. Ill. t. 79 f. 1.—Burm. flor. Ind. t. 43. f. 1 & 2.

Spec. Char. Shrub: stem and branches obtusely angled, furrowed: leaves verticilled in threes or fours, long or rarely oblong-lanceolate, thin-acuminated at both ends, subsessile, quite entire, or somewhat repand, subrevolute at the margin: cymes once-thrice 3-cleft, spreading, forming a pyramidal panicle, lower ones axillary, shorter than the leaf: calyx coloured, large, 5-parted; segments lanceolate or ovate: tube of the corolla filiform, very long, funnel-shaped above: flowers greenish white.

Bengal. Silhet. Flowering nearly all the year.

(8) C. COLEBROOKIANUM. (*Walp.*)

Ident. Walp. Repert. IV. p. 114—Dec. prod. XI. p. 672.

Spec. Char. Branches tetragonal: leaves long-petioled, broad-ovate, subacuminate, quite glabrous above, puberulous below: panicles axillary and terminal, dichotomous, fastigiate; peduncles, pedicels and calyx very minutely, puberulous: calyx sharply 5-toothed, often furnished with a callous gland at the base of the segments: tube of the corolla long; segments elliptic-ovate: flowers purple (?).

Silhet.

(9) C. HASTATUM. (*Wall.*)

Ident. Wall. cat. No. 1786.—Dec. prod. XI. p. 671.
Syn. Siphonanthus hastatus, *Roxb.*
Engrav. Bot. Reg. t. 1307.—Bot. Mag. t. 3308.

Spec. Char. Shrubby: branchlets obsoletely tetragonal and with the leaves and peduncles villous: leaves opposite, petioled, subhastato-cordate: lobes acute, quite entire or subrepand, lower ones divaricate, terminal one the largest, oblong: panicle terminal, trichotomously corymbose, few-flowered: calyx campanulate, spreading; segments lanceolate-acute: tube of the corolla villous, filiform, very long: flowers greenish-white, with the mouth of the throat marked with 5 purple dots.

Silhet, flowering in April and May.

(10) C. DENTATUM. (*Wall.*)

Ident. Walp. repert. IV. p. 114.—Dec. prod. XI. p. 674.
Syn. Volkameria dentata, *Roxb. flor. Ind.* III. p. 61.

Spec. Char. Shrubby: leaves round-cordate, acute, sharply toothed: panicle terminal, brachiate, coloured, flowers scarlet.

Silhet, flowering in May and August.

(11) C. NERIIFOLIUM. (*Wall.*)

Ident. Wall. cat. No. 1789.—Dec. prod. XI. p. 660.
Syn. Volkameria neriifolia, *Roxb.*

Spec. Char. Shrub: young branches and cymes strigosely puberulous; leaves coriaceous, opposite and in threes: lanceolate or oblong or subelliptic, acuminated at both ends, quite entire, revolute at the margin, shining above pale below: cymes axillary 3-flowered, twice as short as the leaf: calyx obconical, campanulate, limb spreading, shortly and sharply 5-toothed; tube of the corolla long, filiform; flowers white.

Chittagong, flowering in the rainy season.

(12) C. FRAGRANS. *(Vent.)*

Ident. Vent. Jard. de Malm. t. 70.—Dec. prod. XI. p. 666.
Syn. Volkameria Japonica, *Jacq.*
Engrav. Bot. Reg. t. 41.—Bot. Mag. t. 1834.—Jacq. Hort. Schoenbr. III. t. 338.

Spec. Char. Shrub: branchlets somewhat tomentose-pubescent: leaves long-petioled, broad-ovate or subrotund, short-acuminate, irregularly repand-toothed, a little hairy above, below especially in the network and scattered round the insertion of the petioles with a few glands: panicle terminal, subsessile, compact, many-flowered: bracts lanceolate, exceeding the calyxes, and with them beset at the back with few, pelviform glands: calyx slightly puberulous, segments subulate-lanceolate, reflexed: flowers deep rose-coloured, fragrant.

Chittagong.

(13) C. GLANDULOSUM. *(Lindl.)*

Ident. Lindl. Bot. Reg. 1844. not. ad. t. 19.—Dec. prod. XI. p. 672.

Spec. Char. Tree (?) leaves somewhat round-ovate, truncated at the base or slightly cordate, hairyish, without scales, subdentate: panicle dense, capitate: bracts linear-lanceolate, longer than the calyx, marked with glands at the back: calycine segments acuminate: petals oblong, reflexed: style very long: flowers large, whitish.

Khasia hills, flowering in the cold season.

GENUS XII. GMELINA.

Didynamia Angiospermia. *Sex: Syst:*

Deriv. Named in honour to *George Gmelin*, a German naturalist and traveller.

Gen. Char. Calyx cup-shaped, 4-5-toothed, persistent, somewhat enlarged with the fruit: corolla tubular at the base, greatly enlarged at the throat, ventricosely bell shaped, limb spreading, bilabiately 4-5-lobed, the anterior one larger, inflexed in æstivation: stamens 4, didynamous, ascending, scarcely exserted: anthers 2-celled, attached by the middle, cells distinct, opening longitudinally: ovary 2-4-celled, cells 1-ovuled: style filiform: stigma equally bifid, drupe baccate, nut solitary, berry smooth, 4-celled, perforated at the base: seed pendulous: radicle inferior.—Shrubs or sometimes large trees, branches usually thorny: leaves simple, opposite, entire or lobed: inflorescence cymoso-paniculate, panicles raceme-like or composed of short few-flowered decussating cymules, or simply racemed: bracts often caducous; corolla conspicuous, drupe large, oblong.

(1) G. ARBOREA. *(Roxb.)*

Ident. Roxb. flor. Ind. III. p. 84.—Dec. prod. XI. p. 680.
Syn. Premna arborea, *Roth.*
Engrav. Rheede Mal. I. t. 41.—Roxb. Cor. t. 346.—Wight's Icon. t. 1470.
SPEC. CHAR. Arboreous, unarmed, branchlets and young leaves covered with a greyish, powdery tomentum: leaves long-petioled, cordate or somewhat produced and acute at the base, acuminate, the adult ones glabrous above, greyish tomentose beneath, with 2-4 glands at the base: panicles tomentose, axillary and terminal, raceme-like: cymules decussate, trichotomous, few-flowered: bracts lanceolate, deciduous: the acutely dentate calyx, eglandulose: flowers large, sulphur, slightly tinged red outside the tube.

Coromandel. Neilgherries. Concans. Oude. Goruckpore. Flowering in the hot season.

(2) G. PARVIFOLIA. *(Roxb.)*

Ident. Roxb. H. B. p. 46. flor. Ind. III. p. 87.—Dec. prod. XI. p. 670.
Syn. G. Coromandeliana, *Burm.*—Premna parvifolia, *Roth.*
Engrav. Roxb. Cor. II. t. 162.—Pluk. t. 14. fig. 4.
SPEC. CHAR. Thorny shrub: leaves petioled, sub-rhomboid ovate, obtuse or emarginate, quite entire or 3-5-lobed, lobes triangular rather obtuse, glabrate, shining above, glaucous below: panicles racemiform, terminal, mealy tomentose, few-flowered; bracts caducous: calyx pedicelled, very shortly 4-toothed: flowers large, bright sulphur.

Coromandel coast. Pondicherry. Mirzapore. Flowering in the hot season.

(3) G. ASIATICA. *(Linn.)*

Ident. Linn. Sp. p. 873.—Dec. prod. XI. p. 679.—Roxb. flor. Ind. III. p. 85.
SPEC. CHAR. Shrub, thorny or unarmed: leaves oval or sub-rhombeo-oval, triangular acute, quite entire or with a lateral lobe at her side, young ones slightly tomentose below, adult ones shining ove, glaucous beneath: racemes terminal, and axillary, tomentose:

bracts leafy, cuspidate, longer than the calyx : calyx pedicelled, very shortly 4-toothed furnished in front with many discoid glands ; flowers large, bright sulphur.

Coromandel. Travancore. Flowering nearly all the year.

(4) G. OBLONGIFOLIA. *(Rarb.)*

Ident. Roxb. flor. Ind. III. p. 83.—Dec. prod. XI. p. 679.

SPEC. CHAR. Arboreous: leaves oblong or oval, entire, somewhat wrinkled, glandularly impressed on both sides at the base of the nerves: panicles terminal, solitary, decussate : bracts small, caducous : calyx entire : flowers numerous, large, rose, fragrant.

East of Bengal.

GENUS XIII. VITEX.

Didynamia Angiospermia. *Ser. Syst.*

Deriv. From *Vieo*, to bind ; in allusion to the flexible branches.

GEN. CHAR. Calyx cup-shaped, campanulate or tubular, funnel-shaped, 5-toothed, or cleft, teeth or segments a little unequal : corolla 2-lipped, upper lip 2-cleft, lower 3-cleft, lateral segments a little longer than the upper ones, the middle one larger than the rest and stretched out, throat often campanulately inflated : stamens four, didynamous, inserted on the tube, ascending, exserted : anthers obcordate, cells separate at the base, dehiscing longitudinally ; ovary 4-celled, cells 1-ovuled ; style terminal filiform, 2-cleft at the apex with acute foot, acute : drupe seated on the enlarged and often broken calyx, juicy, 1-stoned, 4-celled, putamen woody : seed erect.—Trees or shrubs ; leaves opposite, usually digitate, very rarely simple by the abortion of the lateral leaves : cymes trichotomous or simple, axillary or panicled.

(1) V. BICOLOR. *(Willd.)*

Ident. Willd. En. Hort. Ber. p. 606.—Dec. prod. XI. p. 683. —Dalz. Bomb. flor. p. 201.

SPEC. CHAR. Shrub: branchlets, panicle, and underside of the leaves white with a fine tomentum ; leaves petioled, 3 to 5-foliolate : leaflets lanceolate, long acuminated, entire or coarsely cut and crenated : panicle terminal, pyramidal ; flowers light-blue : berry black, size of a pea.

Bombay. Northern India.

(2) V. ALATA. *(Heyne.)*

Ident. Heyne in Roth. Nov. Spec. 316—Dec. prod. XI. p. 685.
—Roxb. flor. Ind. III. p. 72.
Engrav. Rheede Mal. V. t. 1.
SPEC. CHAR. Small tree; branchlets with obtuse angles, densely tomentose: leaves trifoliolate, the petiole with a broad wing: leaflets ovate or elliptic-oblong, narrow at both ends, acuminate quite entire, subcoriaceous, shining above, pubescent or hoary and glandulad-dotted beneath: panicle terminal, compound, spreading, pyramidal: flowers pale-yellow, tinged with blue: petioles with a broad veined wing.

Southern Maratta Country. Chittagong. Silhet. Travancore. Flowering in April and May.

(3) V. ALTISSIMA *(Linn.)*

Ident. Linn. fil. suppl. p. 294.—Dec. prod. XI. p. 685.—Roxb. flor. Ind. III. p. 21.
SPEC. CHAR. Large tree with the branchlets quadrangular, compressed, and channelled: petioles and back of the leaf white, with a short, woolly pubescence: leaves long petioled, trifoliolate: leaflets elliptic or elliptic-oblong, acuminate at both ends, entire: panicle hoary with a dense tomentum, terminal, compound, spreading, pyramidal: cymes interruptedly verticilled: corolla small, lower lip woolly: flowers white, tinged with blue.

Ravines near Nagatna. Forests of Coromandel. Goalpara. Flowering in April and May.

(4) V. LEUCOXYLON. *(Linn.)*

Ident. Linn. fil. suppl. p. 293.—Dec. prod. XI. p. 692.—Roxb. flor. Ind. III. p. 74.
Engrav. Rheede Mal. IV. t. 36.—Wight's Icon. t. 1467.
SPEC. CHAR. Small tree: leaves long-petioled, 3 to 5-foliolate: leaflets elliptic or ovate-oblong, shortly and obtusely acuminated: attenuated into the petiole, entire, subcoriaceous, shining and glabrous above: cymes axillary, long peduncled, corymbose, many-flowered: lower lip of corolla densely woolly: drupe large, obovate, black when ripe: flowers whitish-yellow.

Warren Country. Peninsula. Chittagong. Silhet. Assam. Flowering from February to April.

(5) V. PUBESCENS. (*Vahl.*)

Ident. Vahl. Symb. III. p. 85.—Dec. prod. XI. p. 685.
Syn. V. arborea, Roxb. H. B.—Wallrothia articulata, Roth.
Engrav. Wight's Icon. t. 1463.

Spec. Char. Large tree: branchlets 4-sided, channelled, and with the petioles and young leaves pubescent or slightly tomentose: leaves long-petioled, 3-5-foliolate; leaflets elliptic or ovate-oblong, attenuate-acuminate, rounded at the base, subsessile, coriaceous, penninerved, glabrous, shining above, pale and finely puberulous beneath: panicles whitish, powdery tomentose, terminal, compound, ovato-pyramidal, compact: cymes interspersed with foliaceous bracts longer than the calyx.

Forests in Malabar.

(6) V. NEGUNDO. (*Linn.*)

Ident. Linn. Spec. p. 890.—Dec. prod. XI. p. 684.
Syn. V. paniculata, Lam.—Roxb. flor. Ind. III. p 71.
Engrav. Rheede Mal. II. t. 12.—Wight's Icon. t. 519.

Spec. Char. Tree: branchlets, petioles and rachis of the panicle pubescently tomentose: leaves long-petioled, 3-5-foliolate, leaflets oblong, attenuate-acuminate, acute at the base and with the middle ones petioled, quite entire or coarsely serrated, puberulous above, canescent below, or afterwards glabrous; panicle terminal, compound, spreading: cymes very shortly peduncled, divaricately-dichotomous, bracteolate, white with velvety tomentum: calyx shortly 5-toothed: lower lip of the corolla somewhat downy at the base, flowers smallish, blue.

Peninsula. Bengal. Rajmahal. Flowering nearly all the Year.

(7) V. TRIFOLIA. (*Linn.*)

Ident. Linn. Suppl. 293.—Dec. prod. XI. p. 683.—Roxb. for. Ind. III. p. 69.
Engrav. Bot. Mag. t. 2187.—Rheede Mal. II. t. 11-12.

Spec. Char. Tree: branchlets and face of the leaves canescent with long powdery pubescence: leaves trifoliolate or simple, leaflets obovate-oblong or obovate, acute and obtuse, usually long attenuated at the base, sessile, quite entire, at last glabrous above: panicle terminal, cymes peduncled, erect: calyx repand-toothed; flowers small, bluish-white.

Coromandel. Concans. Deccan. Patna. Flowering in April and May.

(8) V. PEDUNCULARIS (*Wall.*)

Ident. Wall. cat. No. 1733.—Dec. prod. XI. p. 687.

SPEC. CHAR. Tree: branchlets glaucous-pruinose; petioles and panicles powdery-tomentose: leaves long-petioled (petioles sometimes winged) trifoliate; leaflets lanceolate-oblong, thin acuminate, narrowed at the base into the petiole, quite entire, glandular-dotted beneath: panicles axillary, spreading, peduncled, racemiform, lax, without bracts: cymes peduncled, much spreading, few-flowered, divaricately dichotomous: calyx short, sessile, minutely repand-toothed, and with the corolla densely glandular-dotted.

Silhet.

(9) V. HETEROPHYLLA. (*Roxb.*)

Ident. Roxb. H. B. p. 46.—Dec. prod. XI. p. 686.
Engrav. Wall. Pl. As. Rar. III. t. 220.

SPEC. CHAR. Tree: branchlets quadrangular and with the petioles and glabrous leaves subpruinose; leaves 3-5-foliolate, petioled; leaflets elliptic-oblong, attenuated or somewhat cuspidately acuminate, narrowed into the petiole, quite entire, undulated, shining, slightly scabrous above, scarcely paler below, densely glandular-dotted: panicle terminal, compound, many-flowered, pyramidal, hoary with dense, powdery tomentum: cymes corymbose, spreading, minutely bracteolate: calyx shortly repand-toothed: corolla white-tomentose, beardless; lips nearly equal: flowers yellowish.

Tipperah. Goalpara. Flowering in the hot season.

(10) V. SALIGNA. (*Roxb.*)

Ident. Dec. prod. XI. p. 692.

SPEC. CHAR. Tree: branchlets together with the petioles powdery pubescent: cymes and younger leaves pruinose: leaves long-petioled, trifoliolate; leaflets lanceolate, acuminate at both ends, very unequal at the base, quite entire, pale below, downy as far as the rib, adult ones afterwards glabrous: cymes axillary, long-peduncled, corymbose, divaricately dichotomous, many-flowered: calyx cut and toothed: lower lip of the corolla densely woolly at the base; tube tawny-tomentose outside, flowers pure white.

Coromandel, flowering in April.

GENUS XIV. HOLMSKIOLDIA.

Didynamia Angiospermia. *Sex : Syst :*

Deriv. Named in Honour of *Theodore Holmskiold*, a Danish botanical author.

GEN. CHAR. Calyx membranaceous, coloured, dilated from a very short tube with a large spreading subrotate campanulate entire limb : corolla 2-lipped, tube elongated, slightly incurved, glabrous within, throat somewhat dilated, upper lip of the limb 2-cleft, segments erecto-patent, lower 3-cleft, lateral segments small, reflexed, middle one ovate, spreading : stamens four, exserted, ascendent, lower ones longer, upper filaments dilated : anthers 2-celled, cells parallel, adnate to the connectivum by the back : style nearly equalling the stamens, stigma with one of the feet very short, acutish : ovary 4-celled, cells 1-ovuled : ovule pendulous : drupe (?) seated at the bottom of the calyx, somewhat fleshy, 4-lobed, 2-stoned, stones 2-celled.

(1) H. SANGUINEA. (*Retz.*)

Ident. Retz. obs. 6. p. 31.—Dec. prod. XI. p. 696.

Syn. H. rubra, *Pers.*—Hastingia coccinea, *Smith.*—Platunium rubrum, *Juss.*

Engrav. Smith's Exot. Bot. II. t. 80.—Decaisne in Voy. Jacquem. part. bot. t. 140.

SPEC. CHAR. Shrub : branches brachiately divaricated, tetragonal : leaves opposite, ovate, acuminate, rounded from the base, almost quite entire or slightly crenato-serrate, the teeth very shortly apiculated, pubescent on the nerves beneath : panicles axillary, nearly equalling the petiole, loose : cymelets 3-flowered, occasionally reduced to one flower : calyx and corolla scarlet.

Hindostan. Silhet.

GENUS XV. HEMIGYMNIA.

Pentandria Monogynia. *Sex. Syst :*

Deriv. From *Hemisus*, half, and *Gymnos*, naked.

GEN. CHAR. Calyx funnel-shaped, striated, 5-toothed : tube of the corolla funnel-shaped, segments five, narrow, twice as long as the tube : stamens five, equal, enclosed : ovary 4-celled, 4-ovuled :. ovules solitary, ascending : style 2-cleft deciduous deeply 2-parted, inner face stigmatose : fruit (immature) drupaceous, rostrato-cuspidate, half girt by the cupuliform calyx.

(1) H. MACLEODII. *(Griff.)*

Ident. Griffith in M'Clelland's Calc. Journ. Nat. Hist. III. p. 363.—Dec. prod. XI. p. 697.

SPEC. CHAR. Middle-sized tree: young parts tomentose with pubescence: leaves opposite, cordate or cordate-rotund: inflorescence terminal, cymosely-corymbose: flowers congested at the top of short pedicels, pointed, white?

Forests near Juppulpore.

GENUS XVI. VERBENA.

Didynamia Angiospermia. *Ser. Syst.*

Deriv. Said to be from the Celtic name *Ferfaen.*

GEN. CHAR. Calyx tubular, plicately 5-ribbed: corolla sub-hypocrateriform, tube cylindric, enlarged upwards, straight or curved, villous at the insertion of the stamens, throat bearded, limb sublabiate-oblique, 5-cleft, segments more or less unequal, emarginate: stamens inserted on the upper part of the tube, enclosed: anthers ovate, 2-celled, cells dehiscent: ovary 4-celled, cells 1-ovuled, ovule erect: style equalling the stamens, 2-cleft or lobed at the apex: capsule enclosed by the calyx, septicidally splitting when ripe into four cocci: pericarp dry, striated at the back—.Herbs or undershrubs: leaves opposite, in threes or very rarely alternate, entire or tripartitely laciniate: flowers collected on spikes or terminal heads, each subtended by a bract.

(1) V. OFFICINALIS. *(Linn.)*

Ident. Linn. Sp. p. 29.—Dec. prod. XI. p. 547.

Engrav. Engl. Bot. t. 767.—Sweet Brit. fl. gard. III. t. 202.

SPEC. CHAR. Herbaceous: stem 4-cornered, erect, striated, slightly rough at the angles: leaves oblong-lanceolate oblong or lanceolate, cuneate-attenuate at the base, sessile, subpinnatifid or trifid, coarsely incise-dentate, shining above, glabrous or scabrid, strigose below in the network, segments and teeth ovate, acutish or obtuse: spikes terminal and axillary, paniculate, filiform, strigosely pubescent: bracts ovate, acuminate, shorter than the mucronately toothed calyx: flowers small, pale-rose.

Rajmahal. Oude. Flowering in the cold season.

GENUS XVII. AVICENNIA.

Didynamia Angiospermia. *Ser : Syst :*

Deriv. Named after *Avicenna*, a celebrated Persian philosopher and physician.

GEN. CHAR. Calyx deeply 5-parted, almost 5-leaved, equal; leaves concave, obtuse, imbricated: tube of the corolla short, campanulate, limb 5-cleft, posterior lobe usually broader and somewhat dissimilar; stamens four, inserted above on the tube, glabrous, shortly exserted, somewhat unequal, alternate with the lobes: anthers 2-celled, cells distinct, collateral, dehiscing lengthwise: ovary sessile, silky, 2-celled: ovules pendulous, two or one by abortion: style, when present, enclosed: stigmas two, short, diverging after flowering: fruit obliquely ovate, compressed, apiculated with the rudiment of the style, supported by the calyx and bracts; pericarp coriaceous, very smooth within: albumen scarcely any.—Trees: leaves opposite, quite entire, glabrescent above, white below with a dense covering: peduncles single, axillary, torn at the tops of the branches, often 3-headed in the middle: bracts and bracteoles similar, and with the calyx and back of the silky corolla villous-ciliated at the margin: corolla small.

(1) A. OFFICINALIS. *(Linn.)*

Ideal. Linn. Sp. I. p. 110.—Dec. prod. XI. p. 700.

Syn.—A. resinifera, *Forst.*—A. tomentosa, *R. Br.*—*Roxb. flor. Ind.* III. p. 88.—A. alba, *Blume.*—Sceura marina, *Forst.*—Mangium album, *Rumph.*—Oepata, *Rheede.*

Engrav. Rheede Mal. IV. t. 45.—Wight's Icon. t. 1481-2.—Wall. Pl. As. Rar. III. t. 271.—Rumph. Amb. III. t. 76.

SPEC. CHAR. Tree: leaves oblong-lanceolate, subelliptic or obovate, acuminated, acute or obtuse, attenuated into the petiole, at last shining above, white below: capitules globose, dense: lobes of the corolla half-exserted from the calyx, nearly equal, erect-spreading, recurved at the apex, ovate, acute or obtuse or emarginate, silky above, shining below; ovary enclosed, obovate, glabrous at the base, villous at the apex: stigmas subsessile: flowers white or dingy-yellow.

Malabar. Soonderbuns. In damp marshy localities by the sea-shore. Flowering in the hot season.

ORDER CXXII LABIATÆ.

Flowers hermaphodite, usually irregular, calyx free, persistent, 5-(rarely 4-) merous, monosepalous: corolla monopetalous, hypogynous, deciduous, 5-merous, or through the union of the upper lobes, 4-merous, irregular: æstivation bilabistely imbricate, the upper lip exterior, middle lobe of the lower inmost lateral ones intermediate: stamens inserted on the corolla, alternate with its lobes, the upper one, and sometimes even the upper or lateral pair aborting or altogether wanting: anthers various: ovary free, seated on a gynophore or thick disk, 4-parted or rarely 4-cleft, lobes erect, attached transversely, or obliquely by the interior side towards the base: style central, erect, between the lobes, usually bifid at the apex, the divisions anterior and posterior: ovules solitary in each lobe of the ovary, erect, anatropous ; fruit conformable to the ovary, sometimes by abortion 3 or 1-lobed, with one erect seed in each: testa thin: albumen sparing or wanting: embryo straight or rarely (in Scutellaria,) incurved with the seed: radicle short next the hilum: cotyledons fleshy, parallel to the axis of the fruit.—Herbs or undershrubs or shrubs with opposite or whorld 4-sided branches: leaves opposite or whorld, exstipulate, entire or divided, reticulately penninerved: leaves and calyx in many, and in some the stems and corolla, covered with globose glands, filled with fragrant very aromatic essential oil: inflorescence (called a thyrsus) formed of axillary, opposite, centrifugal flowering cymes, with a terminal flower, the rest unilateral on the branches: bracts two, opposite, under the branches of the cymes, with solitary ones opposite the flowers on the branches: cymes heteromorphous, namely, 1st, normally loose, ramose, many-flowered, with the flowers unilateral along the branches: 2nd, condensed on the apex of a common peduncle, and then called capitate: 3rd, condensed into 2 opposite sessile fascicles, forming a false verticil or verticillaster;

with the interior bracts often aborting, the exterior ones forming an involucrum or likewise aborting: 4th, reduced to a single flower, and then the flowers are opposite and solitary.

GENUS I. OCIMUM.

Didynamia Gymnospermia. *Bar: Syst:*

Deriv. From *Oxo*, to smell; alluding to the fragrance of the plants.

Gen. Char. Calyx ovate or campanulate, 5-toothed, deflexed after flowering, throat rarely hairy within: the uppermost tooth with decurrent margins: tube of the corolla very seldom exserted, without a ring inside, throat usually campanulate, limb 2-lipped, upper 4-cleft, lower scarcely longer, declinate, quite entire, flat or slightly concave: stamens four: filaments free, upper ones at the base often appendiculate with a tooth or fascicle of hairs: style shortly 2-cleft at the apex: disk hypogynous, swelling into 1-4 glands which sometimes equal the lobes of the ovary: nuts ovoid or subglobose, smooth, or sometimes very slightly wrinkled, moist when ripe, more often densely mucilaginous.—Herbs, shrubs or undershrubs: floral leaves bract-shaped, often petioled, quite entire: with commonly deciduous verticillasters, 6-flowered, arranged in terminal racemes, pedicels erect, recurved at the top.

(1) O. CANUM. *(Sims.)*

Ident. Dec. prod. XII. p. 32.

Syn. O. americanum, *Linn.*—O. album, *Roxb.*—O. stamineum, *Sims.*

Engrav. Sims. Bot. Mag. t. 2452.

Spec. Char. Stem herbaceous, erect, pubescent: leaves petioled, ovate, rounded at both ends, denticulate or entire, rather hoary beneath: petioles ciliated: verticils of the fruit-bearing raceme, numerous, approximated: calyx small, a little ciliated: flowers white.

Common. Everywhere, flowering at all seasons.

(2) O. BASILICUM. *(Linn.)*

Ident. Linn. Sp. p. 883.—Dec. prod. XII. p. 32.

Spec. Char. Stem erect or ascending: leaves petioled, ovate or oblong, rounded at the base, slightly toothed, glabrous: verticils of the fruit-bearing raceme separated by a space longer than the calyx or more rarely loosely approximated in a branched raceme: calyx ciliated: corolla twice its length, white.

Common. Everywhere, flowering at all seasons.

Var. 1. *pilosum.*—Stem much branched, ascendent: leaves small, oblong, quite entire: petioles and verticils very hairy: racemes elongated: corolla usually glabrous, small, white.—O. pilosum, *Willd.*—*Roxb.*—O. minimum, *Burm.*—O. Basilicum, *Burm.*—O. hispidum, *Lam.*—O. ciliatum, *Horn.*—O. hispidulum, *Schum.*—Basilicum Indicum, *Rumph. Amb.* V. *t.* 92. fig. 1. Oude. Peninsula. Hindostan. Assam. Flowering nearly all the year. Very aromatic and fragrant.

Var. 2. *glabratum.*—Stem erect: petioles and calyx sparingly ciliated: leaves scarcely toothed: racemes long, simple: flowers white.—O. caryophyllatum, *Roxb.*—O. integerrimum, *Willd.*—O. lanceolatum, *Schum.* Peninsula. Bengal. Assam. Flowering nearly all the year. Whole plant very fragrant and aromatic.

(3) O. GRATISSIMUM. (*Linn.*)

Ident. Linn. Spec. p. 832.—Dec. prod. XII. p. 34.

Syn. O. Zeylanicum, *Burm.*—O. frutescens, *Mill. dict.*—O. petiolare, *Lam.*—O. gratissimum, *Jacq.*

Engrav. Burm. Thes. Zeyl. t. 80. fig. 1.—Rheede Mal. X. t. 86.

SPEC. CHAR. Stem rather glabrous: leaves petioled, ovate-acute, crenated or coarsely toothed, narrowed at the base, glabrous or pubescent along the veins: floral leaves like bracts, lanceolate-acuminate, hastate at the base: racemes simple or slightly branched, pubescent: calyx pedicelled: lateral teeth minute, upper united into a bimucronate lip: corolla scarcely longer than the calyx; stamens exserted: flowers white or pale-yellow.

Common in gardens, flowering all the year.

(4) O. ASCENDENS. (*Willd.*)

Ident. Willd. Sp. III p. 166 —Dec. prod. XII. p. 35.

Syn. O. Indicum, *Roth.*—Plectranthus Indicus, *Spreng.*—O. cristatum, *Roxb. H. B. flor. Ind.* III. p. 19.

SPEC. CHAR. Stem prostrate: branches pubescent: leaves petioled, ovate-oblong, obtuse, slightly toothed, narrowed at the base, pubescent: floral leaves like bracts, deciduous: racemes simple: calyx in fruit drooping, the tube striated: wings of the upper tooth reaching to the middle of the calyx: lateral teeth truncated, lower very shortly setaceous, acuminated: corolla twice the length of the calyx: stamens much exserted: flowers pale-rose.

Common all over India, flowering in the rains. This species is scentless.

(5) O. SANCTUM. (*Linn.*)

Ident. Linn. Mant. p. 85.—Drc. prod. XII. p. 38.

Syn. Basilicum agreste, *Rumph.*—O. frutescens, *Burm.*—O. inodorum, *do.*—O. monachorum, *Linn.*—Plectranthus monachorum, *Spreng.*—O. tenuiflorum, *Lam.*—O. villosum, *Roxb. H. B.—I.* tenuiflorum, *Linn. in Dec. prod.* XII. p. 39. ?—Lumnitzera tenuiflora, *Lam.*—O. tomentosum, *Lam.*—O. hirsutum, *Benth. in Wall. Pl. As. Rar.* II. p. 14.

Engrav. Rumph. Amb. V. t. 92. fig. 1.—Burm. Zeyl. t. 80. fig. 1.

SPEC. CHAR. Stems hairy: leaves petioled, oval-obtuse, dentate, pubescent: floral leaves like bracts, sessile, shorter than the pedicels: racemes slender, simple or slightly branched: calyx shorter than the pedicel, drooping, glabrous, throat naked within, upper tooth obovate, concave, shortly decurrent: corolla scarcely longer than the calyx: flowers pale-purple.

Common everywhere, flowering all the year. The plant being held sacred among the Hindoos is often found planted near temples.

GENUS II. GENIOSPORUM.

Didynamia Gymnospermia. *See Syn:*

GEN. CHAR. Calyx ovate while flowering, when fruit-bearing declinate, tubular, 5-toothed, uppermost tooth broader, not decurrent, lateral ones free or cohering with the uppermost one in the upper lip: lowest ones free or shortly cohering in the lower lip: corolla tube short: throat campanulate; upper lip broad, shortly 4-cleft; lower one scarcely longer, declinate, quite entire: stamens 4: filaments free, without teeth: style shortly 2-cleft at the apex: nuts ovoid or oblong, smooth.—Erect or procumbent herbs: verticils many-flowered, loose, arranged in racemes or terminal, simple or branched spikes: floral leaves like bracts, usually coloured at the base: fruit-bearing calyx striated and usually transversely wrinkled, teeth coloured: flowers small, usually pedicelled.

(1) **G. STROBILIFERUM.** (*Wall.*)

Ident. Wall. Pl. As. Rar. II. p. 18.—Dec. prod. XII. p. 45.
Syn. Plectranthus coloratus, *Don. prod. flor. Nep.* p. 116.
Engrav. Hook. Ic. Pl. t. 462.

SPEC. CHAR. Stem erect, branched: leaves subsessile, ovate-oblong, narrow at both ends, dentate, rough, hispid above: verticils many-flowered, spicate at the apex of the branches: floral leaves ovate, acuminate, exceeding the flowers: mouth of the calyx irregularly 5-toothed: flowers small, blue.

Khasia hills, flowering in the cold season.

(2) **G. GRACILE.** (*Benth.*)

Ident. Benth. Lab. p. 21.—Dec. prod. XII. p. 45.

SPEC. CHAR. Stem prostrate, much branched: leaves oblong-linear, slightly serrated, long narrowed at the base: verticils racemose: mouth of the calyx acutely 5-toothed, somewhat 2-lipped.

Coromandel.

(3) **G. PROSTRATUM.** (*Benth.*)

Ident. Benth. in Wall. Pl. As. Rar. II. p. 18.—Dec. prod. XII. p. 45.
Syn. Mentha Zeylanica, *Burm.*—Ocimum menthoides, *Burm.*—O. prostratum, *Linn.*—Lumnitzera prostrata, *Spreng.*—M. arimoides, *Lam.*—Elsholtzia ocimoides, *Pers.*—O. macrostachyum, *Petr.*
Engrav. Burm. Zeyl. t. 70. f. 2.—Rheede Mal. X. t. 92.

SPEC. CHAR. Stem prostrate; branches hispid: leaves petioled, oblong-lanceolate; lower ones subovate, upper ones lanceolate-linear, serrated, narrowed at the base: verticils spiked: fructiferous calyx striated at the base: mouth irregularly 2-lipped; upper lip 3-toothed, middle tooth the larger, lower lip erect, 2-toothed.

Sea-shores of Coromandel.

GENUS III. MESONA.

Didymamia Gymnospermia. *Linn. Syst.*

Deriv. From *Mesos*, middle, because the Genus was supposed to be intermediate between *Ocimum* and *Scutellaria*.

GEN. CHAR. Flowering calyx campanulate, fruit-bearing one, tubular, declinate, 2-lipped, lips membranaceous, upper one 3-cleft, lower entire, truncated: corolla tube very short, throat campanulate, upper lip broad, truncated or 4-toothed, lower one a little longer, oblong: stamens four: filaments free, upper ones with a tooth at the base; style 2-cleft: lobes unequal.—Herbs: racemes terminal: verticils many-flowered: floral leaves bract-shaped, caducous.

(1) M. WALLICHIANA. *(Benth.*

Ident. Dec. prod. XII. p. 46.

Syn. Geniosporum parviflorum, *Benth. Lab.* p. 20.

SPEC. CHAR. Stem ascendent or erect: leaves ovate or oblong-lanceolate, acute at the base, floral ones ovate-acuminate.

Khasia hills. Assam.

GENUS IV. ACROCEPHALUS.

Didynamia Gymnospermia. *Lin: Syst:*

Deriv. From *Akros*, summit, and *Kephale*, head; the flowers are at the top of the branches.

GEN. CHAR. Calyx ovate, tubular, gibbous at the base, 2-lipped, upper lip entire, lower entire or 4-toothed, throat naked: corolla sub-bilabiate, upper lip 4-toothed, lower entire, all the lobes nearly equal: stamens 4, filaments free, toothless: style shortly 2-cleft at the top: nuts smooth.—Herbs: flowers small, densely imbricated in terminal heads or axillary whorls.

(1) A. CAPITATUS. (*Benth.*)

Ident. Benth. in Wall. Pl. As. Rar. II. p. 18.—Dec. prod XII. p. 47.

Syn. Prunella Indica, *Burm.*—Ocimum capitellatum, *Linn. Mant.*—O. capitatum, *Roth.*—Lumnitzera capitata, *Spreng.*

Engrav. Hook. Ic. Pl. t. 436.

SPEC. CHAR. Stem procumbent, much branched at the base: leaves ovate or lanceolate, sub-glabrous: heads of flowers terminal: lower lip of the calyx 4-toothed: leaves remotely serrate, narrowed at the base, floral ones 2–4 under the capitula, ovate, subsessile, longer than the capitulum: flowers white.

Peninsula. Assam. S. Maratta Country. Khasia.

(2) A. AXILLARIS. *(Benth.)*

Ident. Benth. in Dec. prod. XII. p. 48.

SPEC. CHAR. Branches elongated: leaves obovate or oblong, hairy: verticils axillary, many-flowered, remote: lower lip of the calyx 4-toothed: flowers small, white.

Amara.

GENUS V. MOSCHOSMA.

Didynamia Gymnospermia. *Sex Syst:*

Deriv. From *Moschos*, musk, and *Osme*, smell.

GEN. CHAR. Calyx ovate or campanulate, 5-toothed, uppermost tooth larger, margins scarcely decurrent, lower ones nearly equal, throat naked: corolla tube included, limb sub-bilabiate, upper lip shortly 4-cleft, lower quite entire, all the lobes nearly equal: stamens 4: filaments free, toothless: style clavato-capitate at the top, very shortly 2-cleft; nuts ovate, compressed, smooth.—Herbs: whorls 6-10-flowered, secund, loosely racemose, racemes axillary and terminal, sub-paniculate: flowers very small.

(1) M. POLYSTACHYUM. *(Benth.)*

Ident. Benth in Wall. Pl. As. Rar. II. p. 15.—Dec. prod. XII. p. 48.

Syn. Ocimum tenuiflorum, *Burm.*—O. polystachyon, *Linn. Mant.*—Lumnitzera polystachya, *Jacq.*—Plectranthus parviflorus, *R. Br. prod.*—P. micranthus, *Spreng.*—O. polycladon, *Link.*—L. ocimoides, *Jacq.*

SPEC. CHAR. Herbaceous: stem acutely quadrangular, angles smooth or scarcely rough: leaves long-petioled, ovate, rather acute, crenate, rounded or cuneate at the base: racemes numerous, slender: flowers minute, purplish: verticils 6 to 10-flowered, lax, approximated.

Concans. Circars. Travancore. Flowering in the rainy season.

GENUS VI. ORTHOSIPHON.

Didynamia Gymnospermia. *Sex Syst:*

Deriv. From *Orthos*, straight, and *Siphon*, a tube; alluding to the corolla.

GEN. CHAR. Calyx ovate-tubular, 5-toothed, upper tooth ovate, membranaceous, margins often decurrent, deflexed after flowering: tube of the corolla exserted, straight or incurved, throat equal or rarely inflated, upper lip 3–4 cleft, lower quite entire, concave: stamens 4, filament free, without teeth : style clavato-capitate at the top or obtuse : nut very minutely wrinkled.—Perennial herbs or undershrubs : racemes simple, often elongated : verticils 6-flowered, rarely 2–4 flowered, distant, loose : floral leaves bracteiform, ovate, acuminate, reflexed, often shorter than the pedicels : fructiferous pedicels recurved.

(1) O. PALLIDUM. *(Royle.)*

Ident. Benth. Lab. p. 708.—Dec. prod. XI. p. 50.

SPEC. CHAR. Smooth ; stem ascending : leaves petioled, ovate, obtuse, coarsely cut and toothed, entire and cuneate at the base : tube of the small white corolla as long as the calyx : spike short, terminal.

Common in the Deccan, flowering from June to October.

(2) O. GLABRATUM. *(Benth.)*

Ident. Benth. in Wall. Pl. As. Rar. II. p. 14.—Dec. prod. XII. p. 50.

Syn. Ocimum thymiflorum, *Roth.*—Plectranthus thymiflorus, *Spreng.*

SPEC. CHAR. Stems ascending, branched, glabrous : leaves long-petioled, ovate, acute, toothed, rounded or subcordate at the base, glabrous, subincurved ; tube of the corolla twice the length of the calyx : flowers light-purple.

Peninsula. Western coast. Flowering in the rainy season.

(3) O. DIFFUSUM. *(Benth.)*

Ident. Dec. prod. XII. p. 50.

SPEC. CHAR. Stem diffusely much branched, villous : leaves petioled, ovate, tomentosely villous on both sides : racemes slender, few-flowered : corolla scarcely twice as long as the calyx : tube straight, thin.

Dry rocky places in the Peninsula.

(4) O. RUFIDUM. *(Benth.)*

Ident. Dec. prod. XII. p. 50.

SPEC. CHAR. Stem much branched, very hirsute : leaves petioled, ovate-oblong, crenated, tomentosely hirsute on both sides : racemes slender, few-flowered: corolla more than twice as long as the calyx; tube straight, large, throat somewhat dilated.

Peninsula.

(5) O. BRACTEATUM. (*R. W.*)

Ident. Wight's Icon. Vol. IV. & t. 1428.

SPEC. CHAR. Suffruticose, erect, branched, tomentose towards the ends of the branches: leaves sessile, obovate-oblong, obtuse, crenato-serrated, pubescent on both sides: racemes short, terminal: verticils about 3-flowered, covered before expansion with a large leafy deciduous bract: tube of the corolla about thrice the length of the calyx, upper lip larger, somewhat 3-lobed, middle one emarginate, under lip entire, inflexed at the apex.

Sevagherry hills, flowering in August and September.

(6) O. TOMENTOSUM. (*Benth.*)

Ident. Benth. in Wall. Pl. As. Rar. II. p. 14.—Dec. prod. XII. p. 51.

Syn. Ocimum triste, *Roth.*—Plectranthus tristis, *Spreng.*

SPEC. CHAR. Suffruticose; branches ascending, tomentosely pubescent: leaves petioled, ovate, cuneate, rounded or subcordate at the base, thickish, with very short tomentose pubescence on both sides: petioles and axils naked: racemes elongated: corolla thrice longer than the calyx, tube somewhat incurved, throat rather enlarged.

Palaveram near Madras.

(7) O. VISCOSUM. (*Benth.*)

Ident. Benth. in Wall. Pl. As. Rar. II. p. 14.—Dec. prod. XII. p. 51.

SPEC. CHAR. Herbaceous: stem erect, branched: leaves petioled, ovate, crenated, rounded at the base or cordate, viscously pubescent, glandular-dotted: petioles and axils hairy: corolla tube straight twice longer than the calyx.

Dindigul hills.

(8) O. RUBICUNDUM. (*Benth.*)

Ident. Benth. in Wall. Pl. As. Rar. II. p. 14.—Dec. prod. XII. p. 51.

Syn. Plectranthes rubicunda, *D. Don,*—Lumnitzera rubicunda, *Spreng.*

Engrav. Hook. Icon. pl. t. 150.

Spec. Char. Herbaceous: stems cæspitose, leafy at the base, branched: leaves oblong, ovate, coarsely toothed, narrowed at the base: lowest ones petioled, upper ones sessile; corolla twice longer than the calyx, tube straight: flowers white or pale-purplish.
Mountains of Orissa. Monghyr.

(9) O. VIRGATUM. (*Benth.*)

Ident. Benth. in Wall. Pl. As. Rar. II. p. 14.—Dec. prod. XII. p. 52.
Syn. Plectranthes virgata, *D. Don.*—Lumnitzera virgata, *Spreng.*
Spec. Char. Stems cæspitose at the base, branches erect, straight: leaves oblong-lanceolate, sub-dentate, narrowed at the base, sessile, or lowest ones short-petioled; corolla twice longer than the calyx, tube straightish: calyx coloured.
Goruckpore.

(10) O. COMOSUM. (*R. W.*)

Ident. Wight's Cat. No. 2567.—Dec. prod. XII. p. 52.
Spec. Char. Suffruticose: stem tomentosely villous: leaves sessile, oblong, serrato-crenate, wrinkled, tomentose, floral ones petioled, oblong, quite entire, coloured, younger ones comose, deciduous while flowering: calyx long-tubular: corolla tube twice longer than the calyx; verticils 6-flowered: calyx and corolla pubescent.
Peninsula.

(11) O. INCURVUM. (*Benth.*)

Ident. Benth. in Wall. Pl. As. Rar. II. p. 15.—Dec. prod. XII. p. 52.
Engrav. Bot. Mag. t. 3847.
Spec. Char. Herbaceous: stem procumbent at the base, ascending: leaves petioled, ovate or oblong, crenated, narrowed in both ends, very thinly pubescent: verticils subsecund: corolla villous, incurved, somewhat thrice longer than the calyx: flowers pale-rose.
Silhet flowering nearly all the year.

(12) O. STAMINEUM. (*Benth.*)

Ident. Benth. in Wall. Pl. As. Rar. II. p. 15.—Dec. prod. XII. p. 52.
Syn. Ocimum grandiflorum, *Blume.*

Spec. Char. Herbaceous: stem erect, branched: leaves petioled, ovate acuminate, coarsely toothed, cuneate at the base, or uppermost ones subcordate: racemes loose: corolla thrice longer than the calyx, tube straight, upper lip dilated: stamens long-exserted: flowers white or lilac.

Lower Assam, flowering in the rainy season.

GENUS VII. PLECTRANTHUS.

Didynamia Angiospermia. *Sm: Syst.*

Deriv. From *Plektron*, a cock's spur, and *Anthos*, a flower.

Gen. Char. Calyx campanulate, 5-toothed, teeth equal or the upper one larger; enlarging with the seed and then declining, straight, incurved, or inflated with the teeth, equal or variously 2-lipped, sometimes erect, tubular, or campanulate, equally 5-toothed: tube of the corolla exserted, gibbous above the base, or calcarate, then abruptly declining or nearly straight; throat equal or rarely inflated, the upperlip 3-4-cleft, the lower one entire, often longer, concavo: stamens declinate, didynamous, the lower ones longer: filaments free, edentulate: anthers ovate, 2uniform, cells confluent, or rarely somewhat distinct, divaricate: style 2-cleft at the apex, lobes about equal, subulate with minute terminal stigmas.—Herbs, undershrubs, or shrubs: racemes terminal, simple or ramous: verticils lax, many-flowered, usually producing cymes on each side, rarely contracted into dense verticils.

(1) P. Wightii. *(Benth.)*

Ident. Benth. Lab. p. 41.—Dec. prod. XII. p. 56.
Engrav. Wight's Icon. t. 1470.

Spec. Char. Stem herbaceous, erect, branched: leaves petioled, broadly ovate or rounded-acuminate, cordate at the base: lower floral leaves like them, upper and the bracts membranaceous, rounded spathulate, shorter than the peduncle and pedicels: calyx oblong, incurved, striated; mouth oblique, bilabiate: corolla inflated, declinate.

Ram Ghaut.

(2) P. rotundifolius. *(Spr.)*

Ident. Spr. Syst. II. p. 690.—Dec. prod. XII. p. 65.
Syn. Germanea rotundifolia, *Poir.*—Nepeta Madagascariensis, *Lam.*—Coleus rugosus, *Benth. in Wall. Pl. As. Rar. II. p. 15.*
Engrav. Rheede Mal. XI. t. 25.

SPEC. CHAR. Stem procumbent at the base, rooting: branches erect, thick, fleshy: leaves petioled, ovate-rounded or cuneate, running into the petiole, smooth, thick; floral leaves bract-like: racemes simple: verticils rather lax, many-flowered, approximated; corolla three times longer than the calyx, declinate.

The Concans.

(3) P. CORDIFOLIUS. (Don.)

Ident. Don. prod. flor. Nep. p. 116.—Dec. prod. XII. p. 65.
Syn. Ocimum Maypurense, *Roth.*—P. Maypurensis, *Spreng.*—P. secundus, *Roxb.*—P. incanus, *Link.*—O. molle, *Ait.*—P. mollis, *Spreng.*
Engrav. Rheede Mal. X. t. 84.
SPEC. CHAR. Pubescent or tomentose, hoary: stem herbaceous, erect: leaves petioled, broadly ovate-crenate, cordate at the base; floral leaves bract-like, ovate-cuneate: racemes lax, panicled: verticils secund, few-flowered: corolla scarcely twice as long as calyx, tube bent in the middle, the throat dilated: flowers small, pale-blue, or yellow.

About Kandalla. Sawunt Warree. Mysore. Flowering in October and November.

(4) P. BULLATUS. (*Benth.*)

Ident. Dec. prod. XII. p. 56.
SPEC. CHAR. Stem erect, scarcely branched, pubescent or villous: leaves very shortly petioled, ovate-orbiculate, cordate at the base, bullately wrinkled, a little hispid beneath, floral ones small: panicles much branched, many-flowered: calyx incurved, striated, hispidulous: corolla-tube somewhat inflated, scarcely twice as long as the calyx: stamens exserted.

Moist woods on the Neilgherries.

(5) P. GRACILIFLORUS. (*Benth.*)

Ident. Dec. prod. XII. p. 50.
SPEC. CHAR. Stem glabrous: leaves petioled, or the upper ones subsessile, ovate-oblong, long-acuminate, serrated, long-narrowed at the base, scabrously hispid above, glabrous below, floral ones lanceolate-subulate: peduncles and branches of the cymes slender: calyx incurved, striated: corolla-tube more than twice as long as the calyx: stamens scarcely exceeding the lower lip.

Silhet mountains. Assam.

(6) P. STRIATUS. *(Benth.)*

Ident. Benth. in Wall. Pl. As. Rar. II. p. 17. (partly) Dec. prod. XII. p. 56.

SPEC. CHAR. Stem pubescent: leaves petioled, ovate, subacuminate, coarsely serrato-crenate, rounded at the base, upper ones subcordate, more sessile, above sparingly, below only at the veins, roughish; floral ones ovate: peduncle, branches and pedicels of the cymes slender, a little longer than the fructiferous calyx; calyx incurved, striated, pubescent: tube of the corolla more than twice as long as the calyx; stamens shortly exceeding the lower lip.

Khasia.

(7) P. HISPIDUS. *(Benth.)*

Ident. Benth. l. c.—Dec. prod. XII. p. 57.

SPEC. CHAR. Stem densely pubescent; leaves subsessile, ovate, acuminate, dentate, cuneate or rounded at the base or upper ones subcordate, hispid above, pubescent below, floral ones oblong or lanceolate: peduncle, branches and pedicels of the cyme slender: calyx incurved, striated, very hirsute; tube of the corolla twice longer than the calyx; stamens shortly exceeding the lower lip.

Silhet and Khasia hills.

(8) P. Nilgherricus. *(Benth.)*

Ident. Benth. in Dec. prod. XII. p. 57.

SPEC. CHAR. Stem villous: leaves petioled, or upper ones sessile, broad cordate-ovate, acuminate, thick, wrinkled, very villous on both sides, floral ones small, ovate; cymes loose: fructiferous calyx incurved, striated, very hirsute: corolla tube more than twice as long as the calyx; stamens shortly exceeding the lower lip.

Neilgherries.

(9) P. NEPETAEFOLIUS. *(Benth.)*

Ident. Benth. in Dec. prod. XII. p. 57.

SPEC. CHAR. Stem hirsute with soft hairs: leaves petioled or upper ones sessile, broad cordate-ovate, crenated, membranaceous, hirsute on both sides, floral ones ovate: cymes loose; fructiferous calyx somewhat inflated more than twice longer than the calyx; stamens scarcely exceeding the lower lip.

Peninsula.

(10) P. NIANA. (*Benth.*)

Ident. Benth. in Dec. prod. XII. p. 57.

Spec. Char. Stem pubescent: leaves petioled, ovate, subacuminate, coarsely crenated, truncate at the base, hispidulous above, below pubescent at the veins, floral ones ovate or lanceolate: peduncle of the cymes short, branches elongated and with the short pedicels stiff: fructiferous calyx incurved, distinctly 2-lipped: tube of the corolla more than twice as long as the calyx: stamens scarcely exceeding the lower lip.

Neilgherries.

(11) P. MENTHOIDES. (*Benth.*)

Ident. Benth. in Wall. Pl. As. Rar. II. p. 17.—Dec. prod. XII. p. 59.

Spec. Char. Stem herbaceous, erect, branched, hoary pubescent: leaves petioled, ovate, narrowed at both ends, coarsely toothed, wrinkled, tomentosely pubescent on both sides: panicles much branched, many-flowered: calyx declinate, 2-lipped, teeth acutish, fructiferous one oblong, striated, incurved, hoary pubescent.

Southern Peninsula.

(12) P. MACRÆI. (*Benth.*)

Ident. Benth. Lab. p. 42.—Dec. prod. XII. p. 59.

Syn. P. rugosus, *Var.* tomentosus, *Benth. in Wall. Pl. As. Rar.* II. p. 17.

Spec. Char. Stem herbaceous, branched, densely villous: leaves petioled, villosely ovate, acute, toothed, rounded or narrowed at the base, softly pubescent on both sides, floral ones conformable: panicles branched, many-flowered: calyx declinate, oblong, sub-bilabiate, teeth nearly equal, ovate: fructiferous calyx incurved, striated, villous.

Mountains of the Peninsula.

(13) P. MONTANUS. *Benth.*

Ident. Benth. in Wall. Pl. As. Rar. II. p. 17.—Dec. prod. XII. p. 60.

Spec. Char. Stem herbaceous, erect, branched above, tomentosely villous: leaves short petioled, ovate, obtuse, thick, wrinkled, densely tomentose: racemes branched, pyramidally paniculate, tomentose: verticils somewhat distinct, dense: fructiferous calyx declinate, incurved, tomentose, obtusely 5-toothed.

Near Nundidroog in the Peninsula.

(14) P. TERNIFOLIUM. *(D. Don.)*

Ident. Don. prod. flor. Nep. p. 117.—Dec. prod. XII. p. 61.
Syn. Ocimum ternifolium, *Spreng.*
Engrav. Hook. Ic. pl. t. 460.

SPEC. CHAR. Tomentosely villous, stem erect, somewhat branched: leaves ternately verticilled, submamils, lanceolate-oblong, acuminate, serrated, cuneate at the base, wrinkled: panicles branched, densely pyramidal, many-flowered: fructiferous calyx cylindric, erect, striated, teeth, equal, obtuse.

Assam. Silhet.

(15) P. MELISSOIDES. *(Benth.)*

Ident. Benth. Lab. p. 30.—Dec. prod. XII. p. 62.

SPEC. CHAR. Stem herbaceous, angles rough: leaves petioled, broad-ovate, cuneate at the base, rough: floral ones conformable, decrescent: verticils loose, remote: peduncles 3-cleft on both sides: calyx campanulate, glabrous: teeth ovate, acute; corolla bent downwards, obtusely spurred above.

Silhet. Assam.

(16) P. COLEOIDES. *(Benth.)*

Ident. Dec. prod. XII. p. 64.

SPEC. CHAR. Herbaceous: stem erect, somewhat fleshy, puberulous: leaves petioled, ovate, crenate, subcordate at the base, thickish, puberulous, floral ones deciduous: raceme panicled: cymes many-flowered: uppermost tooth of the fructiferous calyx ovate, decurrent, upper ones lanceolate, acute: corolla four times longer than the calyx, tube bent downwards at the middle, throat dilated: flowers lilac.

Moist places on the Neilgherries.

(17) P. SUBINCISUS. *(Benth.)*

Ident. Benth. in Wall. Pl. As. Rar. II. p. 16.—Dec. prod. XII. p. 66.

SPEC. CHAR. Stem herbaceous, erect, scarcely fleshy: leaves petioled, broad ovate, inciso-dentate, cordate at the base, floral ones bract-shaped, ovate: racemes base, sub-panicled: verticils secund: fructiferous calyx somewhat nodding, ovate, uppermost tooth ovate, decurrent, lower ones ovate-lanceolate: corolla four times longer than the calyx, tube bent downwards at the middle, sub-gibbous at the base.

Shady places near Courtallum and Dindigul.

GENUS VIII. COLEUS.

Didynamia Gymnospermia. *Linn's Syst.*

Deriv. From *Koleos*, a sheath; alluding to the manner in which the stamens are united.

GEN. CHAR. Calyx ovate-campanulate, throat naked or hispid within, 5 toothed or 2-lipped, uppermost lip ovate, membranaceous, margins rarely decurrent, lower ones narrower, all acute or the lateral ones ovate-truncate, two lowest ones often connate: corolla tube exserted, declinate, curved downwards or often defracted, throat inflated or equal, limb 2.lipped, upper lip abbreviated, obtusely 3-4-cleft, lower one entire, elongated, concave, often boat-shaped: stamens 4: filaments without teeth connected at the base into a tube sheathing the style: style subulate at the top, equally 2-cleft: nuts roundish compressed, smooth.—Annual herbs rarely shrubs: verticils 6 or many-flowered, sometimes very dense, sometimes loose, boat-shaped: floral leaves bracteiform, more or less comose before flowering at the apex of the branches, deciduous while in flower, reflexed.

(1) C. BARBATUS. *(Benth.*

Ident. Benth. in Wall. Pl. As. Rar. II. p. 15.—Dec. prod XII, p. 71.

Syn. Plectranthus Forskolaei, *Willd.*—Germanea Forskolaei, *Poir.*—P. barbatus, *Andr.*—Ocimum asperum, *Roth.*—P. asper, *Spreng.*—P. monadelphus, *Roxb.*

Engrav. Wight's Icon. t. 1432.—Bot. Mag. t. 2036 & 2318.—Andr. Bot. Rep. t. 594.

SPEC. CHAR. Stem fruticose at the base, ascending, tomentose and hispid: leaves petioled, ovate-crenate, softly tomentose, younger ones strigosely hispid: floral leaves membranaceous, broadly ovate-acuminate, in flowering deciduous: verticils distant, 6-flowered: calyx in fruit deflexed, hispid: flowers light-purple.

Carenjah hill. Deccan hills. Bangalore. Dindigul. Flowering in the cold season.

(2) C. APICATUS. *(Benth.)*

Ident. Benth. in Wall. Pl. As. Rar. II. p. 15.—Dec. prod. XII, p. 71.

Syn. Plectranthus caninus, *Roth.*

Engrav. Wight's Icon. t. 1431.

Spec. Char. Stem procumbent at the base: branches ascending, hairy: leaves petioled, obovate, narrowed at the base, fleshy, floral ones membranaceous, concave, exceeding the flowers, at length deciduous: spikes simple, elongated, dense: verticils 6-10-flowered, approximate: fructiferous calyx, deflexed, hispid; throat villous within, uppermost tooth round, somewhat decurrent, lower ones lanceolate, acute, nearly equal; tube of the corolla bent downwards, lower lip stipitate, boat-shaped.

Dindigul mountains.

(3) C. AROMATICUS. *(Benth.)*

Ident. Benth. in Wall. Pl. As. Rar. II. p. 15.—Dec. prod. XII. p. 72.

Syn. Plectranthus aromaticus, *Roxb.*—P. amboynensis, *Spreng.* —C. amboynicus, *Lour.*—C. crassifolius, *Benth. l. c.*

Engrav. Rumph. Amb. V. t. 102. f. 3.—Bot. Reg. t. 1520.

Spec. Char. Stem fruticose at the base: branches tomentosely pubescent or hispid: leaves petioled, broad-ovate, crenated, round or cuneate at the base, very thick, hispid or hoary-villous on both sides, floral ones scarcely equalling the calyx: racemes simple: verticils remote, globose, many-flowered: calyx tomentose: throat naked within: upper tooth ovate-oblong, membranaceous, not decurrent, lower ones shorter, bristly: flowers pale-blue, very aromatic.

Patna, flowering in April. Is this indigenous? It is very common in gardens in many parts of India, but rarely flowers.

(4) C. PARVIFLORUS. *(Benth.)*

Ident. Dec. prod. XII. p. 72.

Spec. Char. Slightly fleshy: leaves petioled, broad-ovate or orbiculate, crenated, round at the base or decurrent in the petiole, floral ones at length deciduous, shorter than the verticil: racemes long, slender: verticils loosely many-flowered: peduncls of the cymes scarcely any: branches and pedicels short: uppermost tooth of the calyx ovate, lowest ones connate beyond the middle.

Peninsula.

(5) C. OVATUS. *(Benth.)*

Ident. Benth. Lab. p. 57.—Dec. prod. XII. p. 76.

SPEC. CHAR. Stem viscously pubescent above: leaves petioled, ovate, scarcely acuminate, coarsely crenated, round or cordate at the base, hispidulous above, glabrous beneath, floral ones deciduous: racemes somewhat branched: verticils loose: common peduncle and branches scarcely any: corolla-tube thin; throat dilated; lower lip long, broad, concave.

Peninsula.

(6) C. MALABARICUS. *(Benth.)*

Ident. Benth. in Wall. Pl. As. Rar. II. p. 16.—Dec. prod. XII. p. 76.

SPEC. CHAR. Very slightly puberulous; leaves petioled, large, very broadly ovate, shortly acuminate, crenated, round-truncate at the base, floral ones deciduous: racemes loosely branched: verticils somewhat 10-flowered, loose: common peduncle scarcely any: corolla-tube thin, exserted; throat enlarged; lower lip long, stretched out.

Peninsula.

(7) C. MACRII. *(Benth.)*

Var. macrophyllus.

Ident. Benth. Lab. p. 58.—Dec. prod. XII. p. 77.

SPEC. CHAR. Slightly pubescent: leaves long-petioled, ovate, acuminate, crenated, round-truncate or subcuneate at the base, glabrous, floral ones deciduous: racemes paniculately branched: verticils cymose in a somewhat fourfold manner, common peduncle scarcely any, branches long: pedicels very short: corolla-tube exserted; throat very large; lower lip long, stretched out.

Peninsula.

(8) C. GLABRATUS. *(Benth.)*

Ident. Benth. Lab. p. 58.—Dec. prod. XII. p. 78.

SPEC. CHAR. Stem somewhat fleshy, glabrous: leaves petioled, ovate-rotund, serrato-crenate, stiff, glabrous; floral ones minute, deciduous: raceme terminal, pubescent, simple, panicle-shaped: common peduncle, branches and pedicels of the cymes elongated; fructiferous calyx glabrate, long: lower teeth lanceolate acute.

Peninsula near Madura.

(9) C. ORYCIFOLIUS. *(Benth.)*

Ident. Dec. prod. XII. p. 78.

Spec. Char. Glabrous or very slightly puberulous: leaves petioled, large, broad-ovate, coarsely duplicato-crenated, truncated at the base, floral ones deciduous: raceme loose, somewhat simple, panicle-shaped; cymes longish, peduncled: lower teeth of the pubescent calyx lanceolate.

Courtallum.

(10) C. FRUTICOSUS. (R. W.)

Ident. Wight. Cat. No. 2514.—Dec. prod. XII. p. 78.

Spec. Char. Branches pubescent or afterwards glabrata: leaves petioled, orbiculate, scarcely acuminate, coarsely somewhat duplicato-crenate or cut, truncated or sub-cordate at the base, pubescent above, tomentose-woolly below: racemes branched: peduncle of the cymes short, branches at length much elongated: tube of the puberulous corolla broadish, curved downwards; lower lip a half longer than the upper one.

Peninsula.

(11) C. WIGHTII. (Benth.)

Ident. Benth. Lab. p. 68.—Dec. prod. XII. p. 78.

Spec. Char. Stem pubescent: leaves petioled, ovate, crenated, round or subcordate at the base, thick, wrinkled, hispid on both sides, floral ones deciduous: raceme terminal, simple, panicle-shaped: peduncle and branches of the cyme elongated: calyx somewhat longer than the pedicel, fructiferous ones pubescent, lower teeth lanceolate, acute: lower lip of the corolla somewhat four times longer than the upper one.

Neilgherries.

(12) C. PANICULATUS. (Benth.)

Ident. Benth. in Wall. Pl. As. Rar. II. p. 16.—Dec. prod. XII. p. 70.

Spec. Char. Stem procumbent at the base, pubescent: leaves petioled, broad-ovate, cut and toothed, rounded or cuneate at the base, thick, fleshy, hispid, floral ones deciduous: raceme terminal, simple, panicle-shaped: peduncle, branches and pedicels of the cymes elongated: fructiferous calyx pubescent: lower teeth lanceolate-subulate; lower lip of the corolla scarcely twice as long as the upper one.

Dindigul hills.

GENUS IX. ANISOCHILUS.

Didynamia Gymnospermia. *Burr Syd.*

Deriv. From *Anisos,* unequal, and *Cheilos,* a lip.

GEN. CHAR. Fructiferous calyx ovate, subercct, the base or middle inflated, contracted above: limb either bilabiate, the upper lip incumbent on the truncated lower one, closing the calyx, or obliquely 5-toothed, the upper one longer, incurved or incumbent: tube of the corolla slender, abruptly bent beyond the calyx, throat dilated, upper lip short, obtuse, 3–4-cleft, the lower one elongated, concave; stamens 4 : filaments free, edentulate: style subulate at the apex, equally bifid: hypogynous disk lobed, the posterior lobe often higher than the ovaries.—Herbs (or undershrubs?) verticillasters densely imbricated, forming ovate, oblong, or cylindrical spikes: floral leaves bract-like, caducous, shorter than the flowers, or rarely the upper ones longer forming a terminal tuft.

(1) A. CARNOSUM. (*Wall.*)

Ident. Wall. Pl. As. Rar. II. p. 18.—Dec. prod. XII. p. 81.
Syn. Lavandula carnosa, *Linn.*—Plectranthus dubius, *Spreng.*
P. strobiliferus, *Roxb.*
Engrav. Rheede Mal. X. t. 90.

SPEC. CHAR. Stem erect, tetragonal; leaves petioled, ovateroundcd, obtuse-crenated, cordate at the base, or rounded: thick, fleshy, hoary and tomentose, or villous on both sides: spikes longpeduncled, at length cylindric: floral leaves ovate-obtuse; upper lip of calyx acute, glabrous, membranaceous, ciliated on the margins: flowers bluish-purple.

Clefts of rocks in the Circar mountains. Western Ghauts. Flowering nearly all the year.

(2) A. DECUSSATUM. (*Dalz.*)

Ident. Dalz. Bomb. flor. p. 206.

SPEC. CHAR. Stem round, coloured, smooth below, hoary above: leaves on longish petioles, broad ovate, acute, truncate or cordate at the base, crenated, shortly tomentose beneath, sprinkled on both sides with ruby-coloured glands: spikes shortly-cylindric, pointed on long-naked peduncles, brachiately disposed: floral leaves cordate-acute; calyx densely woolly, upper lip deflexed, rounded with a sudden point, lower truncate : corolla bluish-purple, velvetly and villous : anthers 4, perfect, blue: stigmas 2, filiform, acute.

On the highest Ghauts opposite Bombay, in rocky places; flowering in August.

(3) A. ADENANTHUM. (*Dalz.*)

Ident. Dalz. Bomb. flor. p. 206.

SPEC. CHAR. Spikes dense, pyramidal : floral leaves lanceolate-acuminate, pubescent, 3-nerved : calyx minute, oblique, truncate, scarcely toothed, tomentose on the outside : corolla glandular-dotted, tomentose outside, lower lip long, entire, boat-shaped, upper 3 to 4-lobed, rounded, obtuse, short.

Near Dharwar; Bababooden hills.

(4) A. CRASSUM. (*Benth.*)

Ident. Dec. prod. XII. p. 81.

SPEC. CHAR. Stem erect : leaves petioled, ovate, round, obtuse, crenated, subcordate at the base, thick, hoary-tomentose on both sides: spikes long-peduncled, thick, at length cylindric; floral leaves acute, upper ones comose at the top of the spike ; upper lip of the pubescent calyx acute, margin longish-ciliated.

Peninsula.

(5) A. SCABRUM. (*Benth.*)

Ident. Dec. prod. XII. p. 81.

SPEC. CHAR. Stem suberect: leaves petioled, oval, obtuse, crenulate, rounded or narrowed at the base, above and together with the stem rough-pubescent, beneath more densely and rigidly white-tomentose: spikes at length cylindric : floral leaves obtuse : upper lip of the glabrous calyx acute, margin ciliated.

Courtallum,

(6) A. ERIOCEPHALUM. (*Benth.*)

Ident. Dec. prod. XII. p. 81.

SPEC. CHAR. Stem ascendent, branched : leaves orbiculate, obtuse, crenated, subcordate at the base, scarcely canescent on both sides: fructiferous spikes ovoid or globose ; floral leaves small ; upper lip of the woolly much inflated fructiferous calyx acuminate, ciliated.

Peninsula.

(7) A. DYSOPHYLLOIDES. (*Benth.*)

Ident. Benth. in Wall. Pl. As. Rar. II. p. 19.—Dec. prod. XII. p. 82.

Engrav. Wight's Icon. t. 1434.

SPEC. CHAR. Stem procumbent at the base: branches ascending, silky-villous: leaves short-petioled, oblong-lanceolate, obtuse, quite entire, narrowed at the base, thick, densely silky-villous: spikes axillary and terminal: upper lip of the woolly calyx small, acute or slightly obtuse: flowers purplish.

Neilgherries, flowering in January and February.

(8) A. SERICEUM. (*Benth.*)

Ident. Dec. prod. XII. p. 82.

SPEC. CHAR. Stem ascendent (?) silky-tomentose: leaves sessile, oblong, obtuse, slightly narrowed at the base, thick, densely silky-tomentose: spikes axillary and terminal, scarcely peduncled; lower teeth of the calyx short, upper one longer, incumbent.

Peninsula.

(9) A. POLYSTACHYUM. (*Benth.*)

Ident. Benth. in Wall. Pl. As. Rar. p. 19.—Dec. prod. XII. p. 62.

SPEC. CHAR. Stem straight, branched: leaves subsessile, ovate-oblong, acuminate, serrated, cuneate at the base, very thinly pubescent, nerved below: spikes panicled: calyx incurved above; mouth oblique, 5-toothed, uppermost tooth scarcely longer.

Assam.

(10) A. PURPUREUM. (*R. W.*)

Ident. Wight's Icon. Vol. IV. & t. 1345.

SPEC. CHAR. Stem procumbent at the base: branches ascending or erect: leaves petioled, obovate-spathulate, obtuse or suborbicular, entire, fleshy: spikes axillary and terminal, peduncled: bracts lanceolate, acute, hairy, about the length of the calyx: corolla withering, tubular, upper lip 4-lobed, erect, under entire, deflexed; under lip of the fructiferous calyx minute, upper larger, deflexed, 3-toothed: flowers purple.

Eastern slopes of the Neilgherries about Conoor, flowering in February and March.

(11) A. ALBIDUM. (*R. W.*)

Ident. Wight's Icon. Vol. IV. & t. 1346.

SPEC. CHAR. Stem decumbent: branches ascending, silky-villous: leaves sessile, obovate-spathulate, tapering at the base: spikes axillary and terminal; bracts lanceolate-acute about the length of the calyx: corolla deciduous, tubular; tube hairy within: upper lip of the fructiferous calyx deflexed, 3-toothed; teeth reflexed at the point: flowers white or pale, straw colour.

Neilgherries about Concor and Kaity, flowering in February and March.

(12) A. SUFFRUTICOSUM. *(B. W.)*

Ident. Wight's Icon. Vol. IV. & t. 1437.

SPEC. CHAR. Suffruticose, erect, branched: young shoots and leaves densely villous: leaves short-petioled, ovate-lanceolate: spikes numerous, long-peduncled, congested on the ends of the branches: corolla tubular, deflexed from the base; middle lobe of the upper lip larger, under entire, obtuse: upper lip of the fructiferous calyx much larger, entire, round at the apex, deflexed.

Western slopes of the Neilgherries, in rocky places among long grass, flowering in December and January.

GENUS X. LAVANDULA.

Didynamia Gymnospermia. *See Syst:*

Deriv. From *Lavo*, to wash; in allusion to the use made of its distilled water.

GEN. CHAR. Calyx ovate-tubular, 13 (rarely 15) nerved, shortly 5-toothed, 4 lower teeth nearly equal or two lowest ones narrower, uppermost one sometimes a little broader than the lateral ones, sometimes produced at the top into a dilated appendage: corolla tube exserted; throat somewhat dilated, limb oblique, 2-lipped, upper lip 2-lobed, lower 3-lobed, all nearly equal; spreading: stamens 4, enclosed, declinate: filament free, without teeth: style shortly 2-cleft at the apex: disc hollow furnished at the margin with 4 fleshy scales: nuts smooth.—Perennial herbs, shrubs or undershrubs: floral leaves bracteiform, 1-5-flowered on each side, opposite or rarely alternate: flowers approximate on terminal spikes, simple or branched at the base: bracts small, bristly, or none.

(1) L. PEROTTETII. *(Benth.)*

Ident. Benth. Lab. p. 151.—Dec. prod. XII, p. 147.
Engrav. Wight's Icon. t. 1439.

SPEC. CHAR. Softly villous: stems leafy; leaves deeply pinnatifid; lobes oblong or linear-toothed, green on both sides, villous: floral leaves broadly ovate, acute, as long as the calyx; spikes dense, villous: flowers solitary, alternate.

Hills at Sattara, flowering in November.

(2) L. BURMANNI. (*Benth.*)

Ident. Benth. l. c.—Dec. prod. XII. p. 147.

Syn. L. multifida, *Burm.*—Bysteropogon bipinnatum, *Roth*,— Chætostachys multifida, *Benth. in Wall. Pl. As. Rar.* II. p. 19.

Engrav. Wight's Icon. t. 1438.

SPEC. CHAR. Slightly pubescent: leaves bipinnatifid, segments linear-entire; floral leaves membranaceous, dilated at the base, acuminated and setaceous at the apex, longer than the calyx: spikes short, dense; flowers solitary, approximated, either white or of a beautiful deep-blue.

Common in the Deccan. Mysore. Coorg. Bellary.

GENUS XI. POGOSTEMON.

Didynamia Gymnospermia. *Sex! Syst:*

Deriv. From *Pogon*, a beard, and *Stemon*, a stamen.

GEN. CHAR. Calyx ovato-tubular, equal, 5-toothed, throat naked within: tube of the corolla incluse, limb 4-cleft, sub-bilabiate, the upper lip trifid, the inferior one entire, all the lobes quite entire, about equal, spreading: stamens 4, exserted, straight or somewhat declining: filaments bearded about the middle or naked: anthers terminal, one-celled, opening transversely at to equally bifid at the point, lobes subulate.—Herbs (or under shrubs?): leaves opposite, petioled or entire, dentate or somewhat lobed: verticillasters many-flowered, equal or somewhat secund, sometimes glomerato-spicate supported by bracts, the spikelets racemosely-paniculated, sometimes loosely approximate in spike-like racemes.

(1) P. PANICULATUM. (*Benth.*)

Ident. Benth. in Wall. Pl. As. Rar. I. p. 30.—Dec. prod. XII. p. 151.

Syn. Elsholsia paniculata, *Willd.*—Hyssopus cristatus, *Law.*

Engrav. Rheede Mal. X. t. 05.

SPEC. CHAR. Stem erect, pubescent; leaves unequal-ovate, cut and serrated, narrow at the base; verticils globose, secund, remote; racemes terminal; bracts broadly ovate, membranaceous, as long as the calyx; calyx pubescent; teeth lanceolate.

South Concan. Malabar. Peninsula.

(2) P. PLECTRANTHOIDES. *(Desp.)*

Ident. Desp. Ann. Mus. II. p. 154.—Dec. prod. XII. p. 151.

Syn. Origanum Bengbalense, *Burm.*—Mentha secunda, *Roxb.*

Engrav. Bot. Mag. t. 3236.

SPEC. CHAR. Covered with hoary pubescence: stem erect; leaves ovate-cuneate or rounded at the base, doubly serrated; flowers subsecund, clustered, spicate, ovate, cylindric, peduncled, panicled; bracts broad, ovate, glandular-dotted, longer than the calyx; calyx hirsute, glandular; teeth broad, lanceolate-acute; flowers tinged with deep rose.

Near Chicklee, Surat Collectorate. Peninsula, Hindoostan. Flowering in the cold season.

(3) P. HEYNEANUM. *(Benth.)*

Ident. Benth. in Wall. Pl. As. Rar. I. p. 31.—Dec. prod. XII. p. 153.

Syn. Origanum Indicum. *Roth.*

Engrav. Rheede Mal. X. t. 77.—Wight's Icon. t. 1440.

SPEC. CHAR. Stem ascending, pubescent: leaves subglabrous, ovate, narrow at the base, irregularly crenated: verticils many-flowered, subsecund, interruptedly spicate: spikes panicled: bracts ovate or lanceolate, equal to the calyx or a little shorter.

Between the Ram Ghaut and Belgaum. Neilgherries, flowering from June to November. Probably a mere variety of P. Patchouly.

(4) P. VILLOSUM. *(Benth.)*

Ident. Benth. Lab. p. 153.—Dec. prod. XII. p. 152.

Syn. Elsholzia villosa, *Roxb.*

SPEC. CHAR. Villous: stem suberect: leaves ovate, rounded at the base, narrowed at the petiole, crenated: flowers subsecund, glomerately spiked, spikelets cylindric, submessile, panicled: bracts ovate, striated, pubescent, equalling the calyx; teeth of the villous calyx lanceolate, acute: filaments bearded.

Silhet, flowering in the rainy season.

(5) P. PURPURICAULE. (*Dalz.*)

Ident. Dalz. in Hook. Journ. Bot. II. p. 336.—Bomb. flor. p. 207.

SPEC. CHAR. Stem erect, suffruticose, purple, shining : leaves broadly ovate-acuminate, coarsely double-toothed, attenuated into the petiole, subglabrous; verticils denudate, approximated : panicles axillary and terminal, lax, pyramidal : bracts ovate and lanceolate, equal to the calyx.

Very common in the hilly parts of the Concan, and on the Ghauts.

(6) P. PURPURASCENS. (*Dalz.*)

Ident. Dalz. l. c. p. 337.—Bomb. flor. p. 207.

SPEC. CHAR. Stem herbaceous, quadrangular, 4-furrowed, softly tomentose with spreading hairs : leaves broadly ovate-acute, cuneate at the base, doubly serrated, wrinkled, softly villous on both sides : lowest verticils sessile on the axils of the upper leaves, upper terminal, simply spicate, approximated : bracts under the calyx ovate-acute, leafy, reticulately veined, equal to the calyx : calyx pentagonal, villous : segments triangular, subulate, 3-nerved.

Common in shady woods in the Concan.

(7) P. ELSHOLZIOIDES. (*Benth.*)

Ident. Dec. prod. XII. p. 153.

SPEC. CHAR. Very slight hoary-tomentose, soon glabrate : leaves ovate or oblong-lanceolate, long acuminated, serrated, narrowed at the base : verticils nearly equal, interruptedly spiked, panicled : bracts scarcely any : teeth of the hoary tomentose calyx short, lanceolate.

Khasia hills.

(8) P. PALUDOSUM. (*Benth.*)

Ident. Dec. prod. XII. p. 154.

SPEC. CHAR. Stem ascendent or suberect, slightly pubescent : leaves petioled, ovate, coarsely toothed, round-cuneate at the base, sparingly hispidulous, floral ones and bracts minute : racemes simple : verticils remote : teeth of the villous calyx short, acute ; filaments bearded.

Marshy places near Ootacamund.

(9) P. PETIOLARE. (*Benth.*)

Ident. Dec. prod. XII. p. 154.

SPEC. CHAR. Stem ascendent? sparingly hairy; leaves long-petioled, ovate, coarsely cut and serrated, above sparingly, below hairy at the veins, membranaceous, floral upper ones shorter than the calyx: racemes simple; verticils nearly equal, distinct; bracts linear-subulate: calyx campanulate, glabrous at the base, mouth hairy, oblique, teeth short, acute: filaments bearded.

Peninsula.

(10) P. WIGHTII. (*Benth.*)

Ident. Benth. Lab. p. 156.—Dec. prod. XII. p. 154.

SPEC. CHAR. Pilosely hispid: stem erect? leaves petioled, ovate, acute, dup.icato-dentate or cut, round-cuneate at the base: racemes simple: verticils equal, approximate, or lowest ones remote: bracts linear-subulate: teeth of the glabrous calyx lanceolate bristly, hispid; filaments lightly bearded.

Pulney hills.

(11) P. PARVIFLORUM. (*Benth.*)

Ident. Benth. in Wall. Pl. As. Rar. I. p. 31.—Dec. prod. XII. p. 152.

SPEC. CHAR. Stem ascendent, slightly pubescent: leaves ovate-oblong, narrowed at the base, duplicato-crenate, floral ones subsecund, glomerately spiked, spikelets subsessile, racemose: racemes axillary and terminal: bracts ovate, striated, pubescent exceeding the flowers: teeth of the villous membranaceous calyx narrow lanceolate: filaments bearded.

Silhet. Upper Assam.

(12) P. PUBESCENS. (*Benth.*)

Ident. Dec. prod. XII. p. 152.

SPEC. CHAR. Younger stem pubescent: leaves ovate, coarsely few-crenated, rounded at the base, hispidulous above: verticils subsecund, globose, distinct, pubescent: racemes panicled: bracts shorter than the calyx; teeth of the pubescent calyx lanceolate; filaments bearded.

Bombay.

(13) P. AMARANTOIDES. *(Benth.)*

Ident. Dec. prod. XII. p. 153.

SPEC. CHAR. Very slightly pubescent or glabrate: leaves large, ovate or oblong, cut and serrated, narrowed at the base: verticils nearly equal, interruptedly spiked, panicled: bracts minute, scarcely any: teeth of the puberulous calyx short, lanceolate.

Khasia hills. Assam.

(14) P. ATROPURPUREUM. *(Benth.)*

Ident. Dec. prod. XII. p. 154.

SPEC. CHAR. Stem erect, clothed with reversed hairs: leaves petioled, ovate, duplicato-crenate, round-cuneate at the base, villous on both sides: verticils equal, approximated: bracts very short, bristly: teeth of the tubular villous calyx lanceolate, shortly subulate-acuminate: filaments long-exserted, somewhat naked; flowers dark-purple.

In moist places on the Neilgherries.

(15) P. MOLLE. *(Benth.)*

Ident. Benth. Lab. p. 155.—Dec. prod. XII. p. 154.

SPEC. CHAR. Softly tomentosely villous: stem ascending: leaves broad-ovate, crenated, round-cuneate at the base; floral ones somewhat longer than the calyx: racemes simple: verticils equal; lower ones remote; uppermost ones approximated; bracts linear-subulate: teeth of the villous calyx lanceolate-subulate: filaments bearded.

Pulney Hills.

(16) P. ROTUNDATUM. *(Benth.)*

Ident. Benth. in Wall. Pl. As. Rar. I. p. 31.—Dec. prod. XII. p. 155.

SPEC. CHAR. Villous: stem ascendent: leaves round, duplicato-crenate, truncated or cordate at the base; upper floral ones shorter than the calyx: racemes simple: verticils equal, somewhat remote; bracts linear-subulate; teeth of the villous calyx lanceolate: filaments bearded; flowers small, white.

Peninsula. Neilgherries. Flowering in March and April.

(17) P. VESTITUM. *(Benth.)*

Ident. Benth. in Wall. Pl. As. Rar. I. p. 31.—Dec. prod. XII. p. 155.

SPEC. CHAR. Tomentosely woolly : stem ascending or erect : leaves ovate, crenated, subcordate at the base, wrinkled, softly tomentose, upper floral ones shorter than the calyx : racemes simple : verticils secund, approximated : bracts linear-subulate : teeth of the villous calyx lanceolate : filaments bearded : flowers lilac.

Peninsula.

(18) P. SPECIOSUM *(Benth.)*

Ident. Benth. in Wall. Pl. As. Rar. I. p. 31.—Dec. prod. XII. p. 155.

Engrav. Wight's Icon. t. 1443.

SPEC. CHAR. Pilosely bispid : stem erect : leaves broad-ovate, cordate at the base, duplicato-crenate : racemes simple : verticils terete, loose, approximate : bracts minute : teeth of the glabrous tubular suberect calyx subulate : filaments naked : anthers yellowish.

Mountains of the Peninsula, flowering nearly all the year.

(19) P. BRACHYSTACHYUM. *(Benth.)*

Ident. Dec. prod. XII. p. 156.

SPEC. CHAR. Stem procumbent, hispid with reversed hairs : leaves short-petioled, ovate, crenate, rounded at the base, villous on both sides : verticils equal, densely approximated in short spikes or lowest ones somewhat remote : teeth of the hirsute calyx subulate, erect : filaments naked or slightly bearded.

Khasia hills. Assam.

(20) P. STRIGOSUM. *(Benth.)*

Ident. Dec. prod. XII. p. 155.

Syn. Dysophylla strigosa, Benth. in Wall. Pl. As. Rar. I. p. 30.

SPEC. CHAR. Stem erect, clothed with adpressed hairs : leaves very shortly petioled, ovate-lanceolate, coarsely serrated, cuneate at the base, villous on both sides : verticils equal, approximated in spikes, lowest ones somewhat remote : teeth of the hirsute calyx subulate, erect, somewhat bilabiately unequal : filaments bearded.

Upper Assam. Khasia.

(21) P. HIRSUTUM. *(Benth.)*

Ident. Benth. Lab. p. 155.—Dec. prod. XII. p. 154.
Engrav. Wight's Icon. t. 1442.

SPEC. CHAR. Clothed with adpressed hairs: stem ascending: leaves petioled, ovate, acuminate, serrated, rounded at the base: floral ones shorter than the calyx: racemes simple: verticils equal, distinct: bracts linear-subulate: teeth of the calyx lanceolate, acute, hispid: filaments bearded.

Neilgherries.

(22) P. PATCHOULY. *(Pellet.)*

Ident. Pellet. Descr. in mem. Soc. Sc. Orleans v. 5. cum. ic. —Dec. prod. XII. p. 153.—Hook. Journ. Bot. I. p. 329.
Engrav. Hook. l. c. t. VI.—Rheede Mal. X. t. 77.

SPEC. CHAR. Suffruticose, pubescent, stem procumbent at the base: leaves petioled, rhombeo-ovate, rather obtuse, cuneate at the base, coarsely toothed: spikes terminal and axillary, long-peduncled, interrupted at the base, cymes dense, bracts (floral leaves) longer: bracteoles (bracts) about a half longer than the calyx: tube of the corolla exserted, upper lip spotted, teeth of the fructiferous calyx connivent, lanceolate, filaments bearded: flowers whitish with red stamens.

Silhet. This yields the famous patchouly perfume which is yielded by the dried tops of the plant. The *P. Heyneanum* (Benth) is probably a mere variety of this species with longer and looser spikes.

GENUS XII. DYSOPHYLLA.

Didynamia Gymnospermia. *Sex: Syst*

Deriv. From *Dysodes*, fetid, and *Phyllon*, a leaf.

GEN. CHAR. Calyx ovate, shortly 5-toothed, throat naked: corolla tube enclosed, limb 4-cleft, uppermost lobe entire or emarginate, lowest one somewhat spreading: stamens 4, exserted: filaments bearded in the middle: style 2-cleft, lobes subulate. —Herbs or undershrubs?: leaves opposite or verticilled: verticils many-flowered, dense approximated or imbricated in terminal spikes.

(1) D. RUPESTRIS. (Dalz.)

Ident. Dalz. in Hook. Journ. Bot.—Bomb. flor. p. 208.
Syn. Mentha quadrifolia, *Roxb.*
SPEC. CHAR. Perennial, erect: stems round, woody, 3 to 4 feet high: leaves 4-fold, spreading, short petioled, linear-lanceolate, serrated, rugose downy: spikes terminal, solitary, cylindric, covered with innumerable small rose-coloured flowers; corolla-tube twice the length of the calyx: segments reflexed.

Near Vingorla.

(2) D. TOMENTOSA. (*Dalz.*)

Ident. Dalz. in Hook. Journ. Bot. II. 337.—Bomb. flor. p. 208.
SPEC. CHAR. Softly tomentose all over, with spreading hairs: stem creeping: branches several, simple, erect: leaves verticilled, 6 to 9 together, linear acute, quite entire, much longer than the internodes, covered beneath with scattered glands, margins revolute: floral leaves of the same shape: calyx tuberculate, glandular, densely tomentose: segments triangular, ovate-obtuse, shorter than the hairs.

In rice fields near Malwan.

(3) D. ERECTA. *(Dalz.)*

Ident. Dalz. l. c. Bomb. flor. p. 208.
SPEC. CHAR. Stem erect, branched, rather hispid: leaves verticilled, 9 to 12 together, narrow, linear-obtuse at the apex, papillose and rough on both side, glandular-dotted beneath, equalling the internodes; floral leaves filiform, with a thick oblique head, as long as the calyx: calyx villous; segments erect, obtuse, 7 to 8 inches high.

Near Malwan.

(4) D. GRACILIS. *(Dalz.*

Ident. Dalz. l. c. Bomb. flor. p. 208.
SPEC. CHAR. Stem erect, straight, 9 inches high, sparingly branched above, rough with soft spreading hairs: leaves verticilled in sevens, narrow, linear-acute, longer than the internodes, distantly and minutely toothed towards the apex; floral leaves linear-acute, densely ciliated, longer than calyx and corolla; upper tooth of the corolla the smaller and quite entire.

On the Ghauts.

(5) D. Myosuroides. *(Benth.)*

Ident. Benth. in Wall. Pl. As. Rar. I. p. 30.—Dec. prod. XII. p. 156.

Syn. Mentha Myosuroides, *Roth.*

Spec. Char. Tomentose and silky: stem erect: leaves opposite, shortly-petioled, oblong or lanceolate; floral ones minute: spikes dense: calyx tomentose; the teeth very short, straight: corolla minute, red.

Beds of water courses at Mahableshwur.

(6) D. stellata. *(Benth.)*

Ident. Benth. in Wall. Pl. As. Rar. I. p. 30.—Dec. prod. XII. p. 158.

Syn. Mentha quaternifolia, *Roth.*

Spec. Char. Stem creeping: branches erect: leaves verticilled, 6 to 8 together, narrow linear, almost equal to the internodes, quite entire; floral ones subulate: calyx villous, the segments erect, rather acute.

About Belgaum. Banda, in rice fields.

(7) D. auricularia. *(Blume.)*

Ident. Blume bijdr. p. 826.—Dec. prod. XII. p. 158.

Syn. Mentha fœtida, *Burm.*—M. auricularia, *Linn. Mant.* p. 81. *Roxb. flor. Ind.* III. p. 4.

Engrav. Rumph. Amb. VI. t. 16. fig. 2.

Spec. Char. Hirsute with soft spreading hairs: stems procumbent: leaves opposite, subsessile, ovate-oblong, coarsely serrated; floral ones ovate-lanceolate, nearly equalling the flowers: spikes very thick : teeth of the villous calyx ovate, connivent after flowering.

Ditches and damp places in the Peninsula. Eastern Bengal. Assam. Silhet.

(8) D. cruciata. *(Benth.)*

Ident. Benth. in Wall. Pl. As. Rar. I. p. 30.—Dec. prod. XII. p. 157.

Syn. Mentha quadrifolia, *D. Don. prod. flor. Nep.* p. 113. (not Roxb.)

Spec. Char. Hirsute with soft spreading hairs: stems long, procumbent, somewhat simple: leaves in fours, rarely in fives or

ture, lanceolate or sublinear, obtuse, quite entire, revolute at the margin; floral ones lanceolate, nearly equalling the flowers; teeth of the villous calyx rather obtuse, suberect.

Peninsula.

(9) D. QUADRIFOLIA. (*Benth.*)

Ident. Benth. in Wall. Pl. As. Rar. I. p. 30.—Dec. prod. XII. p. 157.

Syn. Mentha quadrifolia, *Roxb.* (not Don.)

SPEC. CHAR. Tomentosely villous: stem erect: leaves in fours, elliptic-linear, quite entire or remotely serrated, narrowed at both ends; floral ones linear, shorter than the verticils: spikes long, somewhat interrupted at the base: teeth of the ovate calyx narrow-lanceolate, acute, erect: flowers small, rose-coloured.

Circar Mountains. Stagnant waters near Calcutta. Assam. Khasia hills. Flowering in the rainy season.

(10) D. LINEARIS. (*Benth.*)

Ident. Dec. prod. XII. p. 157.

SPEC. CHAR. Stem erect, here and there slightly hairy: leaves verticilled in fours, linear or sublanceolate, quite entire, not dilated at the base, here and there ciliated: spikes short, rather loose: teeth of the pubescent calyx suberect: corolla twice as long as the calyx: flowers purplish.

Assam.

(11) D. VERTICILLATA. (*Benth.*)

Ident. Benth. in Wall. Pl. As. Rar. II. p. 30.—Dec. prod XII. p. 157.

Syn. Mentha stellata, *Lour.*—M. verticillata, *Roxb.*

SPEC. CHAR. Glabrous: stem erect or ascending: branches and leaves 6–10-verticilled: leaves long-linear, narrowed at both ends: spikes very dense: teeth of the fructiferous calyx ovate, stellately spreading: flowers small, purple.

Damp places near Calcutta. Silhet. Flowering in the rainy season.

(12) D. CRASSICAULIS. (*Benth.*)

Ident. Benth. in Wall. Pl. As. Rar. I. p. 30.—Dec. prod. XII. p. 158.

SPEC. CHAR. Glabrous: stem ascending, branched: leaves in fours, sometimes in fives or sixes, lanceolate-linear, dilated at the base, almost quite entire; floral ones lanceolate, exceeding the flowers: spikes long: teeth of the pubescent calyx rather obtuse, suberect.

Bengal. Silhet. Assam.

(13) D. TETRAPHYLLA. *(R. W.)*

Ident. Wight's Icon. Vol. IV. & I. 1444.

SPEC. CHAR. Densely pilose: stem ascending, simple or sparingly branched: leaves in fours, sessile, linear-subulate, entire, revolute at the margin; floral ones lanceolate-spathulate, pubescent, about the length of the flowers; spikes elongated: calyx pubescent, teeth short, pointed: filaments long, the exserted part bearded.

Malabar ?

GENUS XIII. COLEBROOKIA.

Didynamia Angiospermia. *Sex: Syst.*

Deriv. In honour of *H. T. Colebrooke,* an accomplished Botanist.

GEN. CHAR. Calyx campanulate, 5-partite, plumose, pappose when ripe, adhering to the nuts: corolla tube equalling the calyx, limb very short, 4-cleft, upper lobe emarginate: stamens 4, equal, distant: anthers orbiculate: cells confluent into one, valves reflexed: style deeply 2 cleft, lobes subulate.

(1) C. TERNIFOLIA. *(Roxb.)*

Ident. Dec. prod. XII. p. 159.—Dalz. Bomb. flor. p. 209.

Engrav. Roxb. Cor. III. t. 245.

SPEC. CHAR. Small shrub: leaves oblong-elliptic, narrow at both ends, serrulate, softly pubescent above, tomentose beneath, with the branches and spikes verticilled in threes; very dense: flowers very minute, white.

Common on the Ghauts. Mysore. Flowering in February and March.

GENUS XIV. ELSHOLZIA.

Didynamia Gymnospermia. *Sex: Syst.*

Deriv. In honor of *J. P. Elsholz,* a Prussian Botanist.

GEN. CHAR. Calyx ovate or campanulate, 5-toothed, fruit-bearing often elongated: corolla tube equalling the calyx or rarely exserted, limb shortly 4-cleft, oblique or sub-bilabiate, uppermost lobe suberect, somewhat concave, emarginate, lower ones spreading: stamens four, usually exserted, ascending, diverging or distant, lower ones somewhat longer: filaments naked: cells of the anthers divergent or divaricate, at length confluent: style 2-cleft: nuts ovoid, smoothish.—Herbs or undershrubs: verticils many-flowered, arranged in loose or in densely imbricated spikes.

(1) E. BLANDA. (*Benth.*)

Ident. Benth. Lab. p. 162.—Dec. prod. XII. p. 160.

Syn. Perilla elata, *Don. prod. flor. Nep.*—Aphanochylus blandus, *Benth. in Wall. Pl. As. Rar.* I. p. 19.

Engrav. Bot. Mag. t. 3091.

SPEC. CHAR. Herbaceous, hoary: branches 4-cornered, erect: leaves oblong or lanceolate, narrowed at both ends; floral ones subulate: verticils loose, secund: spikes panicled: fructiferous calyx ovate-inflated, membranaceous, pubescent: teeth very sharp: flowers white.

Silhet. Assam.

(2) E. FLAVA. (*Benth.*)

Ident. Benth. Lab. p. 161.—Dec. prod. XII. p. 160.

Syn. Aphanochylus flavus, *Benth. in Wall. Pl. As. Rar.* I. p. 28. & t. 34.

SPEC. CHAR. Erect: leaves petioled, broad-ovate, acuminate; floral ones bract-shaped, ovate or lanceolate: spikes loose, axillary and terminal: corolla nearly twice longer than the calyx: fructiferous calyx inflated, tubular: flowers yellow.

Khasia hills.

(3) E. POLYSTACHYA. (*Benth.*)

Ident. Benth. Lab. p. 161.—Dec. prod. XII. p. 160.

Syn. Perilla fruticosa, *Don. prod. flor. Nep.* p. 115.—*Benth. in Wall. Pl. As. Rar.* I. p. 28. t. 33.

SPEC. CHAR. Erect: branches pubescent or tomentose: leaves short-petioled, oblong, narrowed at both ends; floral ones bract-shaped, minute: spikes loose, panicled: corolla twice longer than the calyx: fructiferous calyx narrow tubular.

Khasia. hills.

GENUS XV. PERILLA.

Didynamia Gymnospermia. *Ser: Syst.*

GEN. CHAR. Calyx campanulate, 5-cleft, fructiferous one nodding, gibbous at the base, bilabiate, upper lip dilated, 3-cleft, middle tooth less, lower lip 2-cleft, throat naked : corolla obliquely campanulate, limb shortly 5-cleft, lower lobe a little longer : stamens 4, nearly equal, distant, erect, equalling the corolla: anthers 2-celled, cells parallel, at length diverging or somewhat divaricate : style deeply 2-cleft : nuts dry, smooth.—Erect or decumbent herbs : flowers shortly pedicelled at the axils of the floral bract-shaped leaves, solitary, opposite, arranged in axillary or panicled secund racemes.

(1) P. OCIMOIDES. (*Linn.*)

Ident. Linn. Gen. p. 578.—Dec. prod. XII. p. 684.

Syn. P. macrostachya, *Benth. in Wall. cat.*—Ocimum frutescens, *Linn.*—Mentha perilloides, *Lam.* (not *Linn.*) Roxb. flor. Ind.

Engrav. Bot. Mag. t. 2395.

SPEC. CHAR. Hairy : leaves broad-ovate, coarsely serrato-crenate: fructiferous calyx very hairy at the base, teeth of the upper lip ovate, acute : flowers whitish.

Silhet. Assam. Flowering in the cold season.

GENUS XVI. MICROMERIA.

Didynamia Gymnospermia. *Ser: Syst.*

Deriv. From *Micros*, small, and *Meris*, a part.

GEN. CHAR. Calyx tubular, 13 or 15 striated, 5-dentate, teeth about equal, straight or scarcely 2-lipped, throat usually villous within : tube of the corolla equal, straight, naked within, usually shorter than the calyx ; limb 2-lipped, upper lip erect, entire or emarginate ; lower one spreading, 3-lobed, lobes about equal, or the middle one broader, entire or emarginate : stamens 4, didynamous, the inferior ones longer, ascending, arcuately-connivent at the apex, shorter than the corolla or rarely exserted : anthers 2-celled, the connectivum often thickened, cells diverging or at length divaricate, connective adnate : lobes of the style sometimes equal, subulate, sometimes the upper one shorter the lower, elongated, recurved, flattened : nuts dry, smooth.—Undershrubs or herbs : verticillasters axillary or spicate, rarely cyme-like or subpanicled : flowers usually small, purplish or white.

(1) M. MALCOLMIANA. *(Dalz.)*

Ident. Dalz. Bomb. flor. p. 209.—Hook. Journ. Bot. IV. p. 109.

Spec. Char. Herbaceous: branches elongated, simple, slender, villous: leaves small, shortly-petioled, ovate-obtuse, crenated, pubescent on both sides: verticils of flowers distant, dichotomously cymose, peduncled, few-flowered, contracted into a kind of umbel: flowers minute.

On the banks of the Yeena, Mahableshwur.

(2) M. CAPITELLATA. *(Benth.)*

Ident. Dec. prod. XII. p. 218.

Spec. Char. Stem elongated, softly pubescent: leaves short-petioled, ovate, somewhat flat, pale puberulous on both sides, upper floral ones small: cymes densely many-flowered, subglobose, lower ones longish peduncled: bract small: calyx small, ovate, subsessile, pubescent, throat villous within: teeth short, erect, spreading.

Neilgherries.

(3) M. BIFLORA. *(Benth.)*

Ident. Benth. Lab. p. 378.—Dec. prod. XII. p. 220.

Syn. Thymus biflorus, *Ham. in Don. prod. flor. Nep. p.* 112.

Engrav. Wight's Icon. t. 1446.—Decane in Jacq. voy. t. 134.

Spec. Char. Suffruticose, much branched, cæspitose, branches ascending, pubescent or pilose: leaves sessile, ovate, acute, flat or revolute at the margins, rigid, glabrous, subcordate at the base, upper ones shorter than the flowers: verticils loose, few-flowered: bracts equalling the pedicels: calyx pedicelled, subsecund, delicately pubescent, throat villous within: flowers pale-reddish, or pink.

Neilgherries. Khasia. Flowering all the year.

GENUS XVII. CALAMINTHA.

Didynamia Gymnospermia. *Sec: Syst:*

Deris. From *Kalas*, beautiful, and *Minths*, mint.

Gen. Char. Calyx tubular, with thirteen longitudinal parallel ribs (two between the midribs of the lower teeth, and one only between the midribs of the upper teeth), and five pointed teeth, the 3 upper teeth more or less connected at the base into an upper lip, the mouth more or less closed with hairs: corolla tube usually longer than the calyx, upper lip erect and slightly concave, lower one spreading with 3 broad lobes: stamens in pairs under the upper lip, outer ones longest, but not spreading beyond the corolla.—Erect or ascending, branched herbs: leaves ovate, toothed: cymes axillary, sometimes forming dense whorls, occasionally loose and paniculate: flowers purplish.

(1) C. UMBROSA. (*Benth.*)

Ident. Dec. prod. XII. p. 232.

Syn. Clinopodium repens, a. Benth. in *Wall. Pl. As. Rar.* I. p. 66.

SPEC. CHAR. Herbaceous, diffuse, pubescent or villous: leaves petioled, ovate, serrato-crenate, rounded at the base: verticils equal, globose, many-flowered: bracts minute, or outer ones subulate, a half shorter than the calyx: flowers purplish.

Mountains of the Peninsula.

(2) C. ANEURA. (*Benth.*)

Ident. Dec. prod. XII. p. 233.

Syn. Clinopodium repens, Benth. in *Wall,. Pl. As. Rar.* I. p. 66. Roxb. fl. Ind. III. p. 13.—Thymus repens, D. Don. prod. fl. Nep.—Melissa repens, Benth. Lab. p. 392.

SPEC. CHAR. Herbaceous, diffuse, rooting at the base, pubescent or villous: leaves petioled, ovate, serrato-crenate, rounded at the base: verticils globose, many-flowered: bracts subulate, numerous, equalling the calyx: flowers purplish.

Assam, flowering in the rainy season.

GENUS XVIII. HEDEOMA.

Didynamia Gymnospermia. *Ser: Syst:*

Deriv. The Greek term for Mint.

GEN. CHAR. Calyx ovate-tubular, 13-striated, 5-toothed, usually 2-lipped; upper lip 3-toothed, lower 2-cleft: throat villous within: tube of the corolla equalling the calyx, or shortly exserted; limb 2-lipped, upper lip erect, entire, emarginate or somewhat 2-cleft, flat: lower one spreading, 3-cleft: stamens 2, (the lower ones) fertile, ascending: anthers 2-celled: cells diverging or divaricate: rudiments of the two upper ones more or sterile, short, subulate, capitate: lobes of the style nearly equal, or the lower one elongated and involving at the base the very short upper one: nuts dry, smooth.- Herbs or undershrubs: leaves small, quite entire or subdentate: verticils loose, few-flowered, axillary, subapproximate in terminal racemes.

(1) H. NEPAULENSE. *(Benth.)*

Ident. Benth. Lab. p. 366.—Dec. prod. XII. p. 244.

Syn. Cunila Nepaulensis, *Don. prod. flor. Nep.*—Lycopus Dianthera, *Roxb.*—Melissa Nepaulensis, *Benth. in Wall. Pl. As. Rar.* I. p. 66.—C. Buchani, *Spreng.*—Moschosma ocimoides, *Benth. Lab.* p. 25.

Engrav. Decane. in Jacq. voy. t. 138.

SPEC. CHAR. Stem herbaceous, erect, branched; leaves petioled, ovate, narrowed at both ends, serrated, flat; floral ones small; verticils 2-flowered, secund, loosely racemose; flowers pale-rose.

Assam. Khasia. Flowering in February and March.

GENUS XIX. SALVIA.

Diandria Monogynia. *See: Syst.*

Deriv. From *Salvo*, to save; in allusion to the healing qualities of the sage.

GEN. CHAR. Calyx 2-lipped, upper lip entire or with three small teeth, lower one 2-cleft: upper lip of the corolla erect, concave or arched, lower one spreading, 3-lobed, middle lobe often notched or divided: stamens two, anthers with a long slender connectivum appearing like a filament fastened to the centre by short real filaments having at one end a perfect anther-cell under the upper lip of the corolla, and an abortive empty cell at the other end,—Herbs or shrubs: flowers usually in 6 or more whorls, forming terminal racemes or spikes: floral leaves bracteiform.

(1) S. PLEBEIA. *(R. Br.)*

Ident. R. Br. prod. p. 501.—Dalz. Bomb. flor. p. 209.

Syn. S. brachiata, *Roxb. flor. Ind.* I. p. 146.—Ocimum fastigiatum, *Roth.*

SPEC. CHAR. Stem herbaceous, erect, branched, pubescent: leaves petioled, oblong, wrinkled: verticils lax, about 6-flowered, racemose: racemes paniculate: calyx campanulate; upper lip quite entire; teeth of the lower lip obtuse: corolla scarcely longer than the calyx: flowers purple.

Kandalla and Island of Caranjah. Hindostan. Oude. Silhet. Bengal. Flowering in the cold season.

GENUS XX. NEPETA.

Didynamia Gymnospermia. *Sm: Syst:*

Derio. From *Nepet*, a town in Tuscany where the plants were first discovered.

GEN. CHAR. Calyx tubular, 15-ribbed, mouth oblique, 5-toothed, upper teeth usually longest: tube of the corolla elongated, throat enlarged, upper lip erect, slightly concave, notched or 2-lobed: lower lip spreading, 3-lobed: stamens in pairs under the upper lip, upper or inner pair longest.—Erect or creeping herbs: flowers in axillary whorls or terminal spikes, usually blue.

(1) N. BOMBAIENSIS. (*Dalz.*)

Ideut. Dalz. Bomb. flor. p. 209.

SPEC. CHAR. Branched, 1 foot high: stem quadrangular, pubescent: leaves long-petioled, softly villous on both sides, cordate, ovate-obtuse, crenated: flowers axillary, peduncled: peduncle as long as the petiole, with about 5-pedicelled flowers, subtended by a pair of lanceolate-acute bracts: calyx pilose, deeply ribbed, upper lip much longer than the lower, of 3-acute ciliate teeth; lower 2-subulate teeth, increasing with the fruit: corolla small, pale-blue with purple spots.

Old walls and rocks on Sewnere Fort, flowering in July and August.

(2) N. RUDERALIS. (*Ham.*)

Ideut. Benth. in Wall. Pl. As. Rar. I. p. 64.—Dec. prod. XII. p. 381.

Syn. Glechoma erecta, *Roxb.*—G. Hindostana, *Roth.*—G. Indicum, *Spreng.*—Thymus Nepetoides, *Don. prod. flor. Nep.* p. 113.

SPEC. CHAR. Annual, suberect, slightly pubescent: leaves petioled, ovate, obtuse, crenated, cordate at the base, pubescent on both sides, green or scarcely hoary: racemes somewhat simple, secund: cymes dense, lower ones peduncled: bracts subulate, outer ones equalling the calyx or a half shorter: mouth of the pubescent calyx oblique; teeth subulate, the upper ones longer: corolla longer than the calyx: nuts smooth or minutely granular: flowers purplish.

Rajmahal. Hindostan. Foot of the Himalayahs, and plains of Northern India. Deccan. Flowering nearly all the year.

(3) N. CLINOPODIOIDES. (*Royle.*)

Ident. Hook. Bot. Misc. III. p. 79.—Dec. prod. XII. p. 382.

SPEC. CHAR. Ascending, slightly pubescent: leaves petioled, ovate, obtuse, crenated, broadly subcordate at the base, green on both sides, slightly pubescent: racemes somewhat simple: verticils rather loose, many-flowered, subsecund: cymes peduncled, loose: bracts subulate, shorter than the calyx: mouth of the pubescent, tubular calyx oblique, the upper teeth lanceolate, lower subulate: corolla shortly exceeding the calyx.

Banks of the Jumna.

GENUS XXI. PRUNELLA.

Didynamia Gymnospermia. *Sex: Syst:*

Deriv. Altered from *Brunella*, derived from the German *die Braune*, a disorder in the jaws and throat which this plant was supposed to cure.

GEN. CHAR. Calyx tubuloso-campanulate, about 10-nerved and reticulately veined, flat above, bilabiate, the upper lip broad, truncated, shortly 3-toothed, the lower one half bifid with the lobes lanceolate, throat naked within: tube of the corolla large, sub-exserted, ascending within, near the base, annulate with scales or hairs: upper lip erect, galeate, somewhat keeled above, entire, the lower one 3-lobed dependent, the lateral lobes oblong, deflexed, the middle one rounded, concave, crenulate: stamens exserted, filaments edentulate at the base, glabrous, shortly bidentate at the apex, the lower tooth bearing the anthers: anthers approximated by pairs under the upper lip, free, two-celled: cells distinct, divaricated: gynobase, equal, straight: style glabrous, bifid at the apex, lobes subulate: nuts oblong, dry, smooth.—Herbaceous plants: verticillasters 6-flowered, densely spicate: floral leaves bract-like, orbiculate, persistent, equalling the calyxes and imbricated with them.

(1) P. VULGARIS. (*Linn.*)

Ident. Linn. Spec. p. 837.—Dec. prod. XII. p. 410.

Engrav. Wight's Spicil. II. t. 200. Icon. t. 1448.

SPEC. CHAR. Leaves petioled, ovate or oblong, entire, dentate or inciso-pinnatifid: teeth of the upper lip of the calyx truncated, aristate, or submuticous, or rarely sub-lanceolate: corolla from a half to twice as longer the calyx: flower purplish.

Common by road sides and in pastures on the Neilgherries.

GENUS XXII. SCUTELLARIA.

Didynamia Angiospermia. *Ses: Syst:*

Deriv. From *Scutella*, a little saucer; alluding to the form of the calyx.

GEN. CHAR. Calyx divided into two lips, both entire, upper one bearing on its back a hollow scale-like protuberance: corolla tube elongated, lips nearly closed, upper one concave, lower lobed: stamens in pairs, anthers of the lower pair 1-celled: nuts raised on a short oblique or curved stalk.—Herbs, rarely shrubs: flowers solitary in the axil of each leaf, either all in distant axillary pairs, or forming terminal spikes or racemes.

(1) S. DISCOLOR. *(Colebr.)*

Ident. Colebr. in Wall. Pl. As. Rar. I. p. 60.—Dec. prod. XII. p. 417.—Dalz. Bomb. flor. p. 210.

Syn. S. Indica, *Don. flor. Nep.*

SPEC. CHAR. Stem rooting at the base, leafy, ascending, rather naked above: leaves petioled, ovate-obtuse, crenated, rounded or cuneate at the base, strongly nerved and purple beneath; floral leaves minute: racemes elongated, somewhat branched at the base: flowers scattered, secund, pale-blue, violet.

Parwar Ghaut. Mahableshwur. Canara. Khasia hills. Flowering nearly all the year.

(2) S. COLEBROOKIANA. *(Wall.)*

Ident. Wall. Pl. As. Rar. I. p. 67.—Dec. prod. XII. p. 418.

SPEC. CHAR. Stem erect, branched, slightly pubescent: leaves petioled, ovate-rotund, obtuse, crenated, lowest ones cordate at the base, upper ones deltoid, round-truncate at the base, all slightly pubescent on both sides, floral ones minute: racemes simple, loose; flowers opposite, secund: flower-bearing calyx shorter than the pedicel, pubescent, fructiferous one increased, glabrate; lower lip of the corolla very broad.

Peninsula.

(3) S. VIOLACEA. *(Heyne.)*

Ident. Wall. Pl. As. Rar. I. p. 66.—Dec. prod. XII. p. 418.

Syn. S. Wightiana, *Benth. in Wall. l. c. p. 67.*—S. Indica, *Roxb.* (not Linn.)

Engrav. Wight's Icon. t. 1449.

SPEC. CHAR. Stem erect or ascending, pubescent: leaves petioled, cordate-ovate, crenated, hispidulous above, pubescent or glabrous beneath, floral ones sessile, ovate, shorter than the pedicel: raceme loose, somewhat simple: flowers opposite, secund,
Peninsula.

(4) S. RIVULARIS. *(Wall.)*

Ident. Wall. Pl. As. Rar. II. p. 66.—Dec. prod. XII. p. 426.
Syn. S. peregrina, *Roxb.* (not Linn.)—S. barbata, *Don.* prod. *flor. Nep.* p. 109.
Engrav. Wight's Icon. t. 1450.
SPEC. CHAR. Herbaceous (?) quite glabrous: stem procumbent at the base; branches ascending: lowest leaves petioled, ovato-rounded, and with the ovate or narrow-lanceolate middle ones obtuse, crenated, dilated at the base, cordate, upper and floral ones less, narrowed at the base: flowers opposite, secund, subracemose: calyx and corolla glabrous, bluish.
Khasia. Neilgherries. Flowering in the cold season.

GENUS XXIII. CRANIOTOME.

Didynamia Gymnospermia. *Ser. Syst.*

Deriv. From *Kranion*, a helmet, and *Temno*, to cut; alluding to the shape of the corolla.
GEN. CHAR. Calyx ovate, equal, 5-toothed; throat villous within: corolla tube exserted: limb 2-lipped, upper lip very short, concave, entire: lower one longer, spreading, 3-cleft: lateral segments short: stamens ascendent under the helmet, scarcely exserted from the tube; anthers 2-celled: cells at length divaricate: style 2-cleft: nuts dry, smooth.

(1) C. VERSICOLOR. *(Reich.)*

Ident. Reichb. Icon. Bot. exot. I. p. 39. & t. 54.—Dec. prod. XII. p. 455.
Syn. Anisomeles Nepaulensis, *Spreng.*—Ajuga furcata, *Link.*
SPEC. CHAR. Erect, herbaceous, branched, hairy: leaves petioled, ovate, acuminate, cordate at the base, hispid on both sides: lowest floral leaves agreeing with the stem ones, uppermost ones minute, bract-shaped: cymes loose, many-flowered, racemose: racemes slender, panicled: flowers numerous, elegantly varied with white rose and purple.
Khasia. Silhet.

GENUS XXIV. ANISOMELES.

Didynamia Gymnospermia. *Ser; Syst:*

Deriv. From *Anisos*, unequal, and *Melos*, a member.

GEN. CHAR. Calyx ovate-tubular, 5-toothed : corolla tube equalling the calyx with a hairy ring inside, limb 2-lipped, upper lip erect, oblong, entire, lower longer, spreading, lateral lobes ovate, obtuse, middle one emarginate, somewhat 2-cleft: stamens exserted: anthers approximated by pairs, of the longer stamens dimidiate, of the lower ones 2-celled, cells parallel: style 2-cleft: nuts dry, smooth.—Herbs: verticils sometimes densely many-flowered racemose, or very lax consisting of long-peduncled cymes, or axillary, few-flowered: flowers purplish.

(1) A. HEYNEANA. *(Benth.)*

Ident. Benth. in Wall. Pl. As. Rar. I. p. 59.—Dec. prod. XII. p. 455.

SPEC. CHAR. Glabrous or very slightly pubescent: leaves ovate or oblong-lanceolate, narrow at the base: cymes long peduncled, secund, few-flowered: calycine teeth lanceolate acute: branches elongated, slender, acutely quadrangular; leaves pale-green on both sides, serrate-crenate: cymes unilateral at the apex of the peduncles.

Bombay. Salsette.

(2) A. OVATA. *(R. Br.)*

Ident. R. Br. in Ait. Hort. Kew. II. p. 364.—Dec. prod. XII, p. 455.

Syn. A. disticha, *Heyne in Roth.*—Nepeta Amboinica, *Linn.*—Ballota disticha, *Linn.*—Ajuga disticha, *Roxb.*—Marrubium Indicum, *Burm.*—B. Mauritiana, *Pers.*

Engrav. Wight's Icon. t. 865.

SPEC. CHAR. Hirsute, more rarely subglabrous: leaves ovate, acuminated or rounded, truncate, subcordate or rounded at the base, broadly crenate: verticils many-flowered, dense: calycine teeth lanceolate-acute; corolla purple, the lip darker in colour.

Coromandel. Bombay. Bengal. Western Coast. Travancore. Flowering nearly all the year.

(3) A. MALABARICA. *(R. Br.)*

Ident. Dec. prod. XII. p. 450.

Syn. Nepeta Malabarica, *Linn.*—Ajuga fruticosa, *Roxb.*

Engrav. Wight's Icon. t. 864.—Bot. Mag. t. 2071.—Hook. Journ. Bot. I. t. 127.—Rheede X. t. 93. ?

SPEC. CHAR. Tomentose and villous: leaves oblong-lanceolate, narrow at the base, serrato-crenate in the upper part soft, tomentose, or woolly: verticils many-flowered, dense, or cymes large, at length elongated; floral leaves and bracts subulate, very soft: corolla rosy or purple, the throat hairy within.

On the Ghauts. Peninsula. Flowering nearly all the year.

(4) A. INTERMEDIA. *(R. W.)*

Ident. Wight in Benth. Lab. p. 703.—Dec. prod. XII. p. 456.

SPEC. CHAR. Tomentosely pubescent or villous: leaves ovate-lanceolate, cuneate at the base: verticils many-flowered, dense or cymes large afterwards elongated: bracts subulate; teeth of the hirsute calyx lanceolate at the base, subulate at the apex, very soft.

Pulney hills.

GENUS XXV. STACHYS.

Didynamia Gymnospermia. *Sex: Syst:*

Deriv. From *Stachys*, a spike, the mode of inflorescence.

GEN. CHAR. Calyx 5-10-ribbed with 5-nearly equal erect or spreading pointed teeth: upper lip of the corolla erect, concave, entire, lower lip longer, spreading, 3-lobed, lateral lobes often reflexed; stamens 4, in pairs under the upper lip: nuts smooth, rounded at the top.—Herbs or shrubs: leaves often cordate: flowers in verticils forming terminal racemes, spikes or heads.

(1) S. OBLONGIFOLIA. *(Benth.)*

Ident. Benth. in Wall. Pl. As. Rar. L p. 64.—Dec. prod. XII. p. 474.

SPEC. CHAR. Erect or ascendent, herbaceous, pubescent or villous: leaves short-petioled, oblong-lanceolate, serrated, subcordate at the base, floral ones longer than the calyx: verticils somewhat 6-flowered, distant: calyx pubescent, teeth lanceolate, very acute, scarcely spinous: corolla a half longer than the calyx, tube enclosed: flowers violet.

Khasia hills. Assam.

GENUS XXVI. LEUCAS.

Didynamia Gymnospermia. *Sex: Syst:*

Derio. From *Leukos*, white; alluding to the whiteness of the flowers.

GEN. CHAR. Calyx tubular or tubuloso-campanulate, striated, straight or recurved at the apex, mouth equal or obliquely elongated either above or below, 8 or 10-toothed: tube of the corolla within the calyx, annulate or naked within, limb bilabiate, the upper one concavo, erect, entire or rarely emarginate, very hairy above, the lower one longer, spreading, trifid, the middle lobe the largest: stamens under the helmet ascending: filaments naked or sometimes pubescent at the base: anthers under the upper lip approximated by pairs, somewhat 2-celled: cells divaricating, confluent: upper lobe of the style very short, inferior, subulate: nuts 3-angular, obtuse.—Herbs or under shrubs: leaves entire or dentate, the floral ones conformable: verticillasters sometimes few, sometimes densely many-flowered: corolla usually white, rarely purplish.

(1) L. LONGIFOLIA. (*Benth.*)

Ident. Benth. Lab. p. 744.—Dec. prod. XII. p. 527.

SPEC. CHAR. Stem herbaceous, erect, villous: leaves linear, subentire, rather glabrous: verticils 6-10-flowered: bracts minute: calyx turbinate, tubular, the mouth equal; teeth very short, setaceous, straight; leaves with one or two teeth, sessile, narrow at the base.

Poona.

(2) L. BIFLORA. (*R. Br.*)

Ident. R. Br. prod. p. 504.—Dec. prod. XII. p. 527.

Syn. Phlomis biflora, *Vahl.* (not Roxb.)

Engrav. Wight's Icon. t. 866.—Burm. Zeyl. t. 63. fig. 1.

SPEC. CHAR. Herbaceous, diffuse: leaves ovate, coarsely toothed, pubescent on both sides, half an inch long: verticils 2-flowered; bracts minute: calyx tubular, mouth equal, teeth subulate.

Concans.

(3) L. COLLINA. (*Dalz.*)

Ident. Dalz. in Hook. Journ. Bot. II. p. 337.—Bomb. flor. p. 211.

SPEC. CHAR. Suffruticose, erect: branches quadrangular, tomentose, with adpressed hairs: leaves petioled, ovate-lanceolate, acute, cuneate at the base, coarsely crenate-serrate, softly pubescent and green above, hoary and tomentose beneath: verticils 10-flowered: bracts linear or narrow-spathulate, hirsute, ciliated, half the length of the calyx: calyx tomentose, turbinate, tubular, mouth equal, teeth erect, subulate, alternately shorter.

Southern Concan.

(4) L. STELLIGERA. *(Wall.)*

Ident. Wall. Pl. As. Rar. I. p. 61.—Dec. prod. XII. p. 529.

SPEC. CHAR. Herbaceous, erect, a little hoary: stem hirsute: leaves oblong, lanceolate-obtuse, serrated, scabrous and hispid above: calyx tomentose, mouth villous within; teeth (10) and bracts subulate, soft, spreading, their apices revolute; leaves green above, pale beneath: flowers white.

Western Ghauts. Aurungabad. Flowering in the cold season.

(5) L. CILIATA. *(Benth.)*

Ident. Benth. in Wall. Pl. As. Rar. I. p. 61.—Dec. prod. XII. p. 530.

SPEC. CHAR. Herbaceous: stem erect, adpressed, pubescent, or rough with reflexed hairs: leaves ovate-lanceolate or oblong, serrate-crenate, green on both sides, hairy and pubescent: bracts linear, ciliate, hairy: calyx tubular, hirsute, mouth truncate, equal, teeth elongated, subulate, hairy, spreading like a star.

Near Banda, between Roha and Thul. Peninsula. Silhet. Khasia.

(6) L. ASPERA. *(Spreng.)*

Ident. Spr. Syst. II. p. 743.—Dec. prod. XII. p. 532.

Syn. Phlomis aspera, *Willd.*—P. Plukenetii, *Roth.*—P. esculenta, *Roxb.*—P. Wightiana, *Benth. in Wall. Pl. As. Rar.* I. p. 60.

Bagrov. Rheede Mal. X. t. 91.

SPEC. CHAR. Herbaceous: hairy and pubescent: leaves oblong or linear, subcrenate, green: verticils dense, equal: bracts oblong-linear or subulate, hairy: calyx smooth at the base, striated at the apex, subincurved, mouth oblique, teeth short: flowers white.

On the sea-shore at Alibaug. Peninsula. Bengal. Northern India. Flowering nearly all the year.

(7) L. CEPHALOTES. (*Spreng.*)

Ident. Spreng. Syst. II. p. 743.—Dec. prod. XII. p. 532.
Syn. Phlomis cephalotes, *Roth.*—Leucas capitata, *Desf. Mem. Mus. Par.* II. p. 8. t. 4.
Engrav. Wight's Icon. t. 337.
SPEC. CHAR. Herbaceous: hairy and pubescent: leaves ovate, or oblong, subserrate, green: verticils subsolitary, large, globose, densely many-flowered: bracts ovate-lanceolate, acute, imbricated: calyx striated and subvillous at the apex; mouth oblique, teeth subulate, short; uppermost leaves coming out of the top of the verticil.

Coast of Kattywar. Ahmedabad. Peninsula. Hindostan. Flowering nearly all the year.

(8) L. LINIFOLIA. (*Spreng.*)

Ident. Spreng. Syst. II. p. 743.—Dec. prod. XII. p. 533.
Syn. Phlomis linifolia, *Roth.*—P. Zeylanica, *Roxb.*—Leonurus Indicus, *Burm.*—L. lavandulæfolia, *Sm. in Rees. Cycl.*
Engrav. Jacq. ic. Rar. I. t. 111.—Rumph. Amb. 6. t. 16. fig. 1.
SPEC. CHAR. Herbaceous, erect, slightly pubescent or tomentose: leaves oblong, linear-entire, or remotely serrated: verticils dense, subequal, many-flowered: bracts linear, hoary: calyx elongated above, mouth very oblique, lower teeth very short, upper largest: flowers white.

Peninsula. Bengal. Flowering in the cold season.

(9) L. URTICÆFOLIA. (*R. Br.*)

Ident. R. Br. prod. p. 504.—Dec. prod. XII. p. 524.
Engrav. Wight's Icon. t. 1431.
SPEC. CHAR. Herbaceous, finely tomentose and hoary: verticils many-flowered, globose: calyxes hairy, membranaceous: mouth oblique, lengthened below, split above; teeth 8 to 10, very short, setaceous: verticils nearly 1 inch in diameter, distant; leaves petioled, broadly ovate, coarsely serrate-crenate, rounded or cuneate at the base: flowers white.

Cambay. Coimbatore. Flowering in December and January.

(10) L. CHINENSIS. (*R. Br.*)

Ident. R. Br. prod. p. 504.—Dec. prod. XII. p. 524.
Syn. Phlomis chinensis, *Retz.*

SPEC. CHAR. Suffruticose? branches silky tomentose: leaves broad-ovate, coarsely toothed, wrinkled, villous, whitish beneath: verticils few-flowered: bracts minute: calyx tomentose, funnel-shaped, limb equal at length dilated, acutely 10-toothed.

Peninsula.

(11) L. MONTANA. *(Spreng.)*

Ident. Spreng. Syst. II. p. 742.—Dec. prod. XII. p. 525.
Syn. Phlomis montana, *Roth.*
SPEC. CHAR. Suffruticose: branches silky-tomentose: leaves ovate, serrate-crenated, wrinkled, above green pubescently hairy, beneath white woolly: verticils many-flowered: bracts minute: calyx silky woolly; throat equal, teeth very short, briefly erect: flowers white.

Near Madras.

(12) L. ANGULARIS. *(Benth.)*

Ident. Benth. in Wall. Pl. As. Rar. 1. p. 62.—Dec. prod. XII. p. 526.
SPEC. CHAR. Herbaceous, procumbent, pubescent: branches acutely tetragonal, angles ciliated: leaves short-petioled, ovate, crenated, wrinkled, strigosely hispid or villous on both sides: verticils few-flowered: bracts minute: calyx tubular campanulate, villous, throat equal, teeth bristly acuminate, straight or slightly recurved.

Thickets round Ootacamund.

(13) L. PILOSA. *(Benth.)*

Ident. Benth. in Wall. Pl. As. Rar. I. p. 63.—Dec. prod. XII. p. 526.
Syn. Phlomis pilosa, *Roxb.*
SPEC. CHAR. Perennial, erect: branches hairy pubescent: leaves ovate, serrato-crenate, green on both sides, hairy, or hoary pubescent beneath: verticils many-flowered: bracts minute: calyx pubescent, mouth equal, teeth very short, bristly erect: flowers white or pale purplish.

Bengal flowering in the rainy season.

(14) L. NEPETÆFOLIA. *(Benth.)*

Ident. Benth. in Wall. Pl. As. Rar. I. p. 62.—Dec. prod. XII. p. 527.

SPEC. CHAR. Herbaceous? slightly pubescent: leaves petioled, ovate-rounded, crenate, green on both sides or scarcely hoary: verticils few-flowered: bracts minute: calyx tubular campanulate, glabrous, throat equal, teeth short, acute.

Peninsula.

(15) L. PROCUMBENS. (Desf.)

Ident. Desf. Mem. Mus. Par. XI. p. 7. t. 3. fig. 2.—Dec. prod. XII. p. 527.

Syn. Phlomis biflora, *Roxb.* (not Vahl.)—Nepeta Indica, *Burm. flor. Ind.*

SPEC. CHAR. Herbaceous: diffuse: leaves ovate-lanceolate, sub-serrate, glabrous: verticils 2-6-flowered: bracts minute: calyx tubular, ; throat equal, teeth lanceolate-subulate, erect: flowers white.

Peninsula. Bundelkund. Flowering nearly all the year.

(16) L. PUBESCENS. (*Benth.*)

Ident. Benth. Lab. p. 610.—Dec. prod. XII. p. 528.

Syn. L. marrubioides, *Var.* glabrior, *Benth. in Wall.. Pl. As. Rar.* I. p. 61.

SPEC. CHAR. Herbaceous: branches slightly pubescent: leaves ovate, crenated, truncated at the base, thin, green on both sides or pale pubescent beneath: verticils densely many-flowered: bracts subulate, shorter than the calyx: calyx tubular, throat equal; teeth subulate, straight: flowers white.

Peninsula.

(17) L. MARRUBIOIDES. (*Desf.*)

Ident. Desf. Mem. Mus. Par. XI. p. 6. t. 3. fig. 1.—Dec. prod. XII. p. 528.

SPEC. CHAR. Herbaceous: branches pubescently woolly: leaves broad-ovate, white-woolly beneath: verticils densely many-flowered: bracts subulate, nearly equalling the calyx; throat of the calyx equal; teeth subulate, straight: flowers white.

Peninsula. Near Ootacamund.

(18) L. SUFFRUTICOSA. (*Benth.*)

Ident. Benth. Lab. p. 611.—Dec. prod. XII. p. 528.

Engrav. Wight's Icon. t. 1454.

SPEC. CHAR. Suffruticose: branches red-villous, leafy at the base: leaves sessile, oblong-lanceolate, linear, quite entire, hispid above, white-tomentose beneath: bracts subulate; calyx red-villous; throat truncated, villous: teeth very short, somewhat spreading.

Neilgherries, flowering in the Autumn.

(19) L. ROSMARINIFOLIA. (*Benth.*)

Ident. Benth. in Wall. Pl. As. Rar. I. p. 61.—Dec. prod. XII. p. 528.

Engrav. Wight's Icon. t. 1455.

SPEC. CHAR. Suffruticose: branches villous with adpressed hairs: leaves sessile, linear, quite entire, scabrously hairy above, white-tomentose or hoary beneath, hispid at the rib; bracts lanceolate-linear; throat of the villous calyx truncated, villous; teeth very short, somewhat spreading.

Peninsula. Neilgherries. Flowering nearly all the year.

(20) L. HELIANTHEMIFOLIA. (*Desf.*)

Ident. Desf. Mem. Mus. Par. XI. p. 2. t. 1. f. 1.—Dec. prod. XII. p. 528.

SPEC. CHAR. Stem fruticose at the base: branches silky-woolly, somewhat rufescent: leaves opposite, sessile, oblong-elliptic or lanceolate, quite entire, silky on both sides, greener above, very white beneath: bracts linear, shorter than the calyx: calyx silky-villous; throat truncated, villous; teeth very short, somewhat spreading.

Neilgherries.

(21) L. TERNIFOLIA. (*Desf.*)

Ident. Desf. l. c. p. IV. t. 1. f. 2.—Dec. prod. XII. p. 529.

Engrav. Wight's Icon. t. 1453.

SPEC. CHAR. Stem fruticose at the base: branches densely tomentose: leaves verticilled in threes, sessile, oblong-lanceolate, quite entire, silky on both sides, very white beneath: bracts linear: calyx silky-woolly; throat truncated, villous; teeth very short, somewhat spreading: helmet of the white corolla very densely beset with white villi.

Neilgherries.

(22) L. HAMATULA. (*Arn.*)

Ident. Arn. in Nov. act. nat. cur. XVIII. p. 355.—Dec. prod. XII. p. 529.

SPEC. CHAR. Erect, rufous-canescent: leaves linear-lanceolate, obtuse, serrated, hoary-pubescent above, white-tomentose beneath: calyx tubular, hoary-pubescent, a little inflated above the middle; throat equal, villous within; teeth short, and together with the bracts revolute at the apex.

Peninsula.

(23) L. LANCEÆFOLIA. (Desf.)

Ident. Desf. Mem. Mus. Par. XI. p. 5. t. 2. fig. 2.—Dec. prod. XII. p. 529.

Engrav. Wight's Icon. t. 1432.

SPEC. CHAR. Stem erect, rufous-tomentose: leaves oblong-lanceolate, almost quite entire, green and pubescent above, canescent and tomentosely pubescent beneath: bracts equalling the calyx: calyx rufous-villous, mouth truncated, pubescent; teeth very short and together with the bracts stiffly mucronulate.

Neilgherries.

(24) L. LAMIIFOLIA. (Desf.)

Ident. Desf. l. c. t. 2. fig. 1.—Dec. prod. XII. p. 529.

SPEC. CHAR. Herbaceous: rufous-villous: leaves cordate-ovate, rufous-villous above, hoary tomentose or villous beneath: bracts lanceolate-linear equalling the calyx: mouth of the silky calyx truncated, equal, villous within; teeth subulate, afterwards stellately spreading.

Neilgherries.

(25) L. HIRTA. (Spreng.)

Ident. Spreng. Syst. II. p. 743.—Dec. prod. XII. p. 530.

Syn. Phlomis hirta, *Heyne in Roth.*—Leucas belianthemifolia, *Benth. in Wall. Pl. As. Rar.* I. p. 61. (not Desf.)

SPEC. CHAR. Suffruticose? branches rufescent villous: leaves ovate or oblong, suberenate, green and hairy on both sides: bracts equalling the calyx: calyx villous; throat truncated, very villous; teeth subulate, stellately spreading, muticous.

Peninsula.

(26) L. VESTITA. (Benth.)

Ident. Benth. in Wall. Pl. As. Rar. I. p. 61.—Dec. prod. XII. p. 530.

Engrav. Wight's Icon. t. 338.

SPEC. CHAR. Herbaceous, erect: stem rufescent, very hirsute; leaves ovate-lanceolate or oblong, crenated, hairy, green or scarcely canescent beneath; bracts linear, ciliately hairy; calyx hairy; throat truncated, nearly equal, very hirsute within; teeth subulate, scarcely equal, stellately spreading.

Pulney hills near Madura.

(27) L. STRICTA. *(Benth.)*

Ident. Benth. in Wall. Pl. As. Rar. I. p. 61.—Dec. prod. XII. p. 531.

SPEC. CHAR. Herbaceous, erect, straight, pubescently hairy: leaves oblong-lanceolate or linear, quite entire: verticils many-flowered, solitary or few, terminal: bracts subulate, hairy: calyx hairy, substriated, mouth very oblique; teeth subulate, straight, afterwards stellately spreading.

Peninsula.

(28) L. ZEYLANICA. *(R. Br.)*

Ident. R. Br. prod. p. 504.—Dec. prod. XII. p. 531.

Syn. Phlomis Zeylanica, *Linn.* (not Roxb.)—Leonurus marrubiastrum, *Burm. flor. Ind.* (not Linn.)

SPEC. CHAR. Herbaceous, erect, pubescently hairy: leaves oblong-lanceolate or linear, quite entire or few-crenated: verticils dense, many-flowered, dimidiate or nearly equal: bracts linear, hairy: calyx glabrous at the base, somewhat veinless, striated at the apex, slightly incurved, mouth oblique; teeth abbreviated, uppermost ones longer: flowers white.

Assam.

(29) L. DIFFUSA. *(Benth.)*

Ident. Benth. Lab. p. 615.—Dec. prod. XII. p. 531.

Syn. L. dimidiata, *Benth. in Wall. Pl. As. Rar.* (not Roth.)

SPEC. CHAR. Herbaceous, diffuse, pubescently hairy: leaves oblong or linear, quite entire: verticils few-flowered, dimidiate, remote: bracts linear, hairy: calyx glabrous at the base, somewhat veinless, striated at the apex, hairy, slightly incurved, mouth oblique; teeth short, scarcely unequal.

Peninsula.

(30) L. NUTANS. *(Spreng.)*

Ident. Spreng. Syst. II. p. 743.—Dec. prod. XII. p. 532.

Syn. Phlomis nutans, *Roth.*—L. decurva, *Benth. in Wall.*

SPEC. CHAR. Herbaceous, pubescent: leaves ovate-oblong, subcrenate, green: verticils many-flowered, distant: bracts oblong or lanceolate: calyx pubescent, striated: fructiferous ones much elongated, incurved, nodding and reflexed at the apex, mouth oblique; teeth linear, short, scarcely unequal.

Peninsula.

GENUS XXVII. LEONOTIS.

Didynamia Gymnospermia. *Ser : Syst.*

Deriv. From *Leon*, a lion, and *Ous*, a ear ; from a fancied resemblance in the Corolla.

GEN. CHAR. Calyx ovate-tubular, 10-nerved, incurved at the top, mouth oblique, somewhat 10-toothed, uppermost tooth longer: corolla tube usually exserted, naked or incompletely ringed inside, limb 2-lipped, upper one concave, erect, long, entire, lower one short, spreading, 3-cleft, middle segment scarcely longer: stamens ascending under the helmet : filaments exappendiculate at the base: anther approximated by pairs under the upper lip, 2-celled, cells divaricate, acute ; upper lobe of the style very short : nuts obtuse at the top.—Herbs or shrubs ; verticils many-flowered, usually very dense ; bracts very numerous, subulate: corolla showy, scarlet or yellowish.

(1) L. NEPETAEFOLIA. (*R. Br.*)

Ident. R. Br. prod. p. 504.—Dec. prod. XII. p. 535.
Syn. Phlomis Nepetaefolia, *Linn.*—Leonurus globosus, *Maech.*
Engrav. Wight's Icon. t. 867.—Bot. Reg. t. 281.
SPEC. CHAR. Herbaceous, 6 feet high : leaves membranaceous, ovate crenate : verticils large, globular ; teeth of calyx spinous, uppermost largest, ovate ; corolla orange-coloured, about twice the length of the calyx.

Peninsula. Western coast. Silhet. Flowering in the cold season.

GENUS XXVIII. GOMPHOSTEMMA.

Didynamia Gymnospermia. *Ser : Syst.*

Deriv. From *Gomphos*, a club, and *Stemma*, a crown ; in reference to the tube of the corolla being inflated above the middle.

GEN. CHAR. Calyx ovate-campanulate or tubular, 5-toothed, throat naked : corolla tube straight, usually exserted, exannulate within, inflated above the middle, bilabiate, lips nearly equal, upper one erect, entire, forked, lower one spreading, 3-cleft : anthers approximated by pairs, 2-celled : cells parallel, transverse : style 2-cleft. —Perennial herbs : stems usually erect, simple or procumbent at the base, rooting : leaves usually large, thick villous or tomentose : flowers largish : verticils spicate or axillary, remote.

(1) G. HEYNEANUM. (*Wall.*)

Ident. Benth. in Wall. Pl. As. Rar. II. p. 12.—Dec. prod. XII. p. 551.
Engrav. Wight's Icon. t. 1456.
SPEC. CHAR. Stem erect: leaves elliptic-ovate, wrinkled above, softly beneath, densely floccoso-tomentose: verticils congested into a terminal spike, or the lower ones somewhat remote, sub-axillary: floral leaves bract-like, broad, ovate, longer than the calyx: calyx campanulate, softly tomentose, teeth ovate, lanceolate, scarcely shorter than the corolla: flowers bluish with a purple tinge.

Walliar. Coimbatore. Flowering in July and August.

(2) G. OBLONGUM. (*Wall.*)

Ident. Benth. in Wall. Pl. As. Rar. II. p. 12.—Dec. prod. XII. p. 551.
Engrav. Wight's Icon. t. 1457.
SPEC. CHAR. Stem erect: leaves oblong, elliptic, wrinkled, hispidulous above, densely tomentose beneath, floral ones conformable: verticils axillary, remote, few-flowered: calyx campanulate, tomentose, with long linear-lanceolate acute teeth: corolla thrice the length of the calyx.

Courtallum.

(3) G. LUCIDUM. (*Wall.*)

Ident. Benth. in Wall. Pl. As. Rar. II. p. 12.—Dec. prod. XII. p. 551.
SPEC. CHAR. Stem erect: leaves elliptic-ovate, shining above, pilosely hispid, densely rufus-tomentose beneath, floral ones conformable: verticils axillary, remote: calyx tubular, campanulate, with short, lanceolate teeth: corolla 3 times longer than the calyx: flowers yellow, densely tomentose outside.

Pundua. Assam. Flowering in the rainy season.

(4) G. PARVIFLORUM. (*Wall.*)

Ident. Benth. in Wall. Pl. As. Rar. II. p. 12.—Dec. prod. XII. p. 551.
SPEC. CHAR. Stem erect: leaves elliptic-ovate, pubescent above, tomentose beneath, floral ones conformable: verticils axillary, remote, loosely many-flowered: bracts lanceolate or linear, exceeding the calyx: calyx campanulate with lanceolate, linear teeth: corolla slender, twice longer than the calyx: flowers yellow.

Silhet.

(5) G. MULTIFLORUM. (*Benth.*)

Ident. Dec. prod. XII. p. 552.

SPEC. CHAR. Stem erect: leaves elliptic, scarcely pubescent above, tomentose beneath, floral ones conformable: verticils axillary, remote, loosely many-flowered: bracts ovate or obovate, leafy, exceeding the calyx: calycine teeth lanceolate-linear: corolla slender, many times longer than the calyx.

Assam, Silhet.

(6) G. ERIOCARPUM. (*Benth.*)

Ident. Benth. in Wall. Pl. As. Rar. II. p. 12.—Dec. prod. XII. p. 552.

SPEC. CHAR. Stem ascending: leaves elliptic-oblong, coarsely toothed, hispidulous above, tomentose beneath, floral ones conformable: verticils remote, few-flowered: bracts ovate, denticulate: calyx campanulate, at length inflated, tomentose: teeth linear-bristly: corolla villous, twice longer than the calyx: nuts tomentose.

Courtallum.

(7) G. MELISSÆFOLIUM. (*Wall.*)

Ident. Benth. in Wall. Pl. As. Rar. II. p. 13.—Dec. prod. XII. p. 552.

SPEC. CHAR. Stem creeping at the base, rooting: branches ascending: leaves ovate, glabrous or hispid above, sparingly tomentose beneath, floral ones conformable: verticils axillary, remote, loosely few-flowered: bracts ovate, the outer ones exceeding the calyx: calyx hispid, teeth lanceolate-linear: corolla thrice exceeding the calyx: flowers largish, yellow.

Silhet. Assam. Flowering in the rainy season.

(8) G. VELUTINUM. (*Benth.*)

Ident. Benth. in Wall. Pl. As. Rar. II. p. 13.—Dec. prod. XII. p. 552.

SPEC. CHAR. Stem creeping at the base, rooting: branches ascending: leaves ovate, thick, densely velvety, pubescent on both sides, floral ones conformable: verticils axillary, remote, few-flowered: bracts oblong, equalling the calyx: calyx tomentose: teeth ovate-lanceolate: corolla twice longer than the calyx: tube much dilated at the apex: flowers yellow.

Silhet.

GENUS XXIX. TEUCRIUM.

Didynamia Gymnospermia. *Ser: Syst:*

Deriv. Named after *Teucer*, son of Scamander, father-in-law of Dardanus, king of Troy.

GEN. CHAR. Calyx tubular or campanulate, rarely inflated, 5-toothed, teeth equal or the upper one often broader: tube of the corolla short, exannulate within, the 4 upper lobes of the limb about equal, or the upper ones longer and broader, sometimes oblong, declining, sometimes very short, nearly erect, the lower one large, roundish or oblong, often concave: stamens 4, protanding between the upper lobes, didynamous, the inferior pair longer: cells of the anthers confluent: style equally bifid at the apex: nuts in most of the species coarsely reticulato-rugose, in a few however with the reticulations scarcely elevated, in all obliquely attached by the interior side of the base.—Herbs or under shrubs variable in habit and inflorescence.

(1) T. TOMENTOSUM. (*Heyne.*)

Ident. Benth. in Wall. Pl. As. Rar. I. p. 58.—Dec. prod. XII. p. 582.

Engrav. Wight's Icon. t. 1458.—Spicil. II. t. 202.

SPEC. CHAR. Suffruticose, erect, branches tomentosely pubescent: leaves ovate, rounded at the base, villous above, tomentosely pubescent, whitish beneath or rarely sub-glabrous: racemes paniculately branched: calyx declinate, pilose, bilabiate, the upper tooth broadest: flowers pale-rose or nearly white.

Neilgherries, flowering after the rains.

(2) T. MACROSTACHYUM. (*Wall.*)

Ident. Wall. in Benth. Lab. p. 664.—Dec. prod. XII. p. 574.

Syn. Leucosceptrum canum, *Smith.*—Clerodendron Leucosceptrum, *D. Don. flor. Nep.* p. 103.

Engrav. Sm. Exot. Bot. t. 116.

SPEC. CHAR. Fruticose: branches hoary: leaves oblong, densely white, tomentose beneath: verticils many-flowered, densely spiked: calyx somewhat equally dentate, incurved: uppermost segments of the corolla very short, erect: floral leaves bract-shaped, ovate-rotund, sessile, shorter than the calyx.

Khasia hills.

(3) T. WALLICHIANUM. (*Benth.*)

Ident. Benth. in Wall. Pl. As. Rar. II. p. 19.—Dec. prod. XII. p. 580.

SPEC. CHAR. Herbaceous, ascending: leaves elliptic-ovate, acuminate, long-narrowed at the base, green on both sides, hispid, serrato-crenate; spike simple: calyx declinate, teeth scarcely unequal: lowest floral leaves petioled, oblong, longer than the flowers, upper ones and bracts minute, lanceolate.

Silhet.

(4) T. STOLONIFERUM. (*Ham.*)

Ident. Hamilt. Benth. in Wall. Pl. As. Rar. I. p. 59.—Dec. prod. XII. p. 583.

SPEC. CHAR. Herbaceous, ascending or erect, glabrous or slightly pubescent: leaves ovate or rounded at the base or cuneate, green on both sides, scarcely wrinkled: racemes paniculately branched: calyx declinate, slightly hairy, ovoid, fructiferous one inflated, uppermost tooth broader: corolla enclosed: flowers pink.

Silhet.

(5) T. QUADRIFARIUM. (*Ham.*)

Ident. Hamilt. in Don. prod. flor. Nep. p. 108.—Dec. prod. XII. p. 583.

SPEC. CHAR. Herbaceous, erect: branches tomentosely villous or hispid: leaves short-petioled, ovate, serrated, cordate at the base, wrinkled, villous, somewhat hoary beneath: racemes branched: floral leaves broad ovate, acuminate, exceeding the calyx: declinate, hispid, somewhat 2-lipped, uppermost tooth broader: corolla tube enclosed: flowers purple.

Silhet.

GENUS XXX. AJUGA.

Didymamia Angiospermia. *Lin: Syst:*

Deris. Said to be from *A*, not, and *Zugos*, a yoke, as the calyx is unequal, the lips not being a pair.

GEN. CHAR. Calyx 5-cleft: corolla with a distinct tube, upper lip very short, erect, entire or nearly so, lower longer, spreading: stamens in pairs projecting beyond the upper lip or tooth of the corolla: nuts rough or wrinkled.—Herbs: flowers verticillate in the upper axils, often forming terminal leafy spikes: corolla withering, not deciduous, purplish-blue or yellow.

(1) A. REMOTA. *(Benth.)*

Ident. Benth. in Wall. Pl. As. Rar. I. p. 50.—Dec. prod. XII. p. 597.

SPEC. CHAR. Without suckers, branched at the base, procumbent, floriferous branches ascending, villous : leaves oblong-elliptic, or ovate, narrowed at the base, floral ones ovate-concave exceeding the flowers, all thickish, sub-coriaceous, villous, coarsely few-toothed : verticils remote, or the uppermost ones approximate : corolla tube exserted, lowest segment of the lower lip scarcely emarginate : flowers violet, rose or whitish.

Oude. Hurdwur.

(2) A. MACROSPERMA. *(Wall.)*

Ident. Benth. in Wall. Pl. As. Rar. I. p. 58.—Dec. prod. XII. p. 599.

SPEC. CHAR. Without suckers, erect, or ascending, branched : stem leaves large, ovate, coarsely toothed, long narrowed at the base, membranaceous, glabrous or sparingly hairy, floral ones somewhat agreeing, uppermost ones or even all bract-shaped, shorter than the leaf : verticils spicate, or lowest ones somewhat remote, calycine teeth obtuse : corolla tube shortly exserted.

Khasia hills. Assam.

(3) A. REPENS. *(Roxb.)*

Ident. Roxb. flor. Ind. III. p. 3.—Dec. prod. XII. p. 602.

SPEC. CHAR. Biennial, procumbent, villous : leaves oblong-lanceolate, narrowed at the base, coarsely and unequally serrated : spikes terminal, cylindric : bracts oblong, ventricose, 5-flowered (verticil 10 flowered :)

Chittagong, flowering in January and February.

ORDER CXXIII. PLUMBAGINACEÆ.

Calyx tubular, persistent, sometimes coloured, corolla (of very thin texture) monopetalous, with a narrow tube, or composed of 5 petals, which have a long narrow claw : stamens definite, opposite the petals, in the monopetalous species hypogynous [in Plumbago seated on the very bottom of the corolla but not truly hypogynous], in the polypetalous rising from the petals: ovary superior composed of 5 (or 3 or 4) valvate carpels, 1-celled, 1-seeded: ovule anatropal, pendulous from the point of an umbilical cord arising from the bottom of the cavity: styles 5! seldom 3 or 4: stigmas the same number: fruit a nearly indehiscent utricle: seed inverted with a rather small quantity of mealy albumen: testa simple: embryo straight; radicle superior.

GENUS I. ÆGIALITIS.

Pentandria Pentagynia. *Sex: Syst.*

Deriv. From the Greek adjective, *Aigialos*, on the shore; alluding to the habitat of the species.

GEN. CHAR. Calyx tubular, herbaceo-coriaceous, 5-angled, ribbed, very narrowly plicato-membranaceous between the ribs, shortly 5-toothed at the apex: petals five, oblong-linear, subcoriaceous, connate at the base round the ovary into a short urceolus somewhat pointed at the point of connexion: stamens five, perigynous, connate at the base of the petals, connected and afterwards free: anthers adnate to the filament by the base: ovary linear-pentagonal: styles distinct: stigmas capitate: capsule coriaceous, somewhat angularly cylindric, without valves, broken at the apex by the germinating seed.

(1) Æ. ANNULATA. (*R. Br.*)

Var. rotundifolia.

Ident. R. Br. prod. p. 426.—Dec. prnd. XII. p. 621.—Roxb. flor. Ind. II. p. 111.

SPEC. CHAR. Undershrub: stems ringed with the scars of the fallen petioles: leaves approximated at the upper part of the branches, ovate, quite entire, rounded at the base, long-petioled, petioles slightly auricled above, dilated below into a long amplexical sheath: spikes terminal, paniculately branched, scarcely exceeding the leaves, rachis flexuose pointed: flowers alternate, remotish, appressed to the rachis, short pedunclad between the bracts and pointed in the persistent peduncle, largish, pale-yellow or white: bracts oblong, obtuse, concave, outer one longer and twice as broad.

Shores of the Sounderbuns.

GENUS II. PLUMBAGO.

Pentandria Monogynia. *Lin. Syst.*

Deriv. From *Plumbum*, a disorder of the eyes, which some species were formerly said to cure.

GEN. CHAR. Calyx tubular, 5-toothed at the apex: corolla hypocrateriform, gamopetalous, limb rotate, 5-parted: stamens five, hypogynous, filaments connivent at the base into a lobed disk: anthers linear, 2-cleft at the base: ovary ovate or oblong, surmounted by the filiform style: stigmas five, filiform, inner side densely beset with glands in many rows: utricle membranaceous, broken irregularly at the lowest base, afterwards valvately cleft from the base to the middle, valves cohering at the apex: seed ovate or oblong.—Perennial herbs or shrubs: flowers sub-sessile arranged in spikes, each with three bracts.

(1) P. ZEYLANICA. (*Linn.*)

Ident. Dec. prod. XII. p. 692.—Linn. Spec. I. 215.—Roxb. flor. Ind. 1. p. 463.

Syn. P. flaccida, *Moench.*—P. sarmentosa, *Lam.*

Engrav. Rheede Mal. X. t. 8.—Wight's Ill. II. t. 178.

SPEC. CHAR. Stems shrubby, subscandent, striated, much branched: leaves ovate or oblong, rather acute, shortly and abruptly attenuated into a short stem-clasping petiole: rachis glandular: petals cuneate, retuse: flowers white.

Rocky places in the Concan. Travancore. Bengal. Guzerat. Flowering nearly all the year.

(2) P. COCCINEA. (*Boissier.*)

Ident. Dec. prod. XII. p. 693.

Syn. P. rosea, *Linn.*—Thela coccinea, *Lour.*

Engrav. Rheede Mal. X. t. 9.—Bot. Mag. t. 230.

SPEC. CHAR. Stems herbaceous, erect, branched above: leaves oblong, petiole very short by clasping the stem: flowers in terminal spikes which are long, twiggy, loose, elongated after flowering and rise also from the upper axils: bracts ovate, cuspidate: shorter than the calyx: calyx reddish, short cylindric, shortly and acutely 5-toothed beset at its five ribs the whole length with stipitate bifarious glands, some sub-sessile intermixed: tube of the corolla much longer than the calyx: flowers scarlet, or bright red.

Southern India. Flowering nearly all the year.

GENUS III. VOGELIA.

Pentandria Monogynia. *Sex. Syst.*

Deriv. Named after *Herr Vogel*, a German author and botanist.

GEN. CHAR. Sepals ovate or oblong, rib broad, rufescent: corolla gamopetalous, funnel-shaped, limb 5-parted ; stamens five, hypogynous, filaments free: anthers oblong, emarginate at the base : ovary linear : style puberulous, scarcely equalling the stigmas which are long and filiform and glanduliferous on the inner side: utricle linear, pentagonal, separating in five valves, and sustained by a funiculus.

(1) V. ARABICA. (*Boissier.*)

Ident. Dec. prod. XII. p. 690.—Dals. Bomb. flor. p. 220.
Syn. V. Indica, *Gibson in Wight's Icon. & t. 1075.*

SPEC. CHAR. Erect undershrub of a singular whitish glaucous hue: branches twiggy, striated, dichotomous: leaves ovate or obovate, sessile or perfoliate, coriaceous, smooth: sepals lanceolate, undulated: flowers small on long slender spikes, petals emarginate, mucronulate.

Hummunt Ghaut. Mount Aboo. Deccan.

ORDER CXXIV. PLANTAGINACEÆ.

Calyx imbricated in æstivation, 4-parted, persistent: corolla membranous, monopetalous, hypogynous, persistent, with a four parted limb: stamens 4, inserted into the corolla, alternately with its segments: filaments filiform, flaccid, doubled inwards in æstivation: anthers versatile 2-celled: ovary composed of a single (?) carpel, sessile, without a disk, 2-very seldom, 4-celled, the cells caused by the angles of the placentæ: ovules peltate or erect, solitary, twin or indefinite: style simple, carpellary: stigma hispid, simple, rarely half bifid: capsule membranous, dehiscing transversely, with a loose placenta bearing the seeds on its surface: seeds sessile, peltate, or erect solitary, twin or indefinite: testa mucilaginous: embryo lying across the hilum in the axis of fleshy albumen: radicle remote from the hilum, inferior, or in some cases centrifugal.—Herbaceous plants, usually stemless, occasionally with a stem: leaves forming rosettes, or in the caulescent species both alternate and opposite, flat and ribbed, or taper and fleshy: flowers in spikes, rarely solitary: usually bisexual, seldom, by abortion, with the male and female in separate flowers.

GENUS I. PLANTAGO.

Tetrandria Monogynia. *Ser : Syn :*

Derio. From *Planta*, the sole of the foot, from a resemblance in the leaves.

GEN. CHAR. Flowers hermaphrodite, spiked or capitate, each with a bract : calyx 4-leaved, leaflets nearly equal : corolla tubular, 4-lobed, scarious, persistent : stamens four, exserted or enclosed, filaments flaccid : anthers cordate : ovary 2-4-celled : cells 1-8-ovuled : style simple : capsule membranaceous, dehiscing circularly at the base, pyxidate, dissipiment at length free, seed-bearing on the faces : seeds in many-seeded capsules small, angular, in two-seeded ones boat-shaped, testa mucilaginously pale, olive or brown.

(1) P. EROSA. (*Wall.*)

Ident. Wall. Cat. No. 6412.—Dec. prod. XIII. p. 1. p. 696.—Roxb. flor Ind. Ed. car. 1, p. 423.

Syn. P. Asiatica, *Linn.*

Engrav. Wight's Ill. 2. t. 177.

SPEC. CHAR. Herbaceous : leaves oval or ovate-elliptic, entire, or crusely toothed, 5-nerved, scattered with slight pubescence, glabrate, petioled : petioles channelled, dilated at the base, membranaceous, a little bearded : peduncles ascending, exceeding the leaves compressed or channelled below : spikes long loose-flowered below : bracts shorter than the calyx, ovate glabrous or ciliolate : calycine leaflets ovate-rotund, glabrous broadly membranaceous : lobes of the corolla ovate, acutish, reflexed : ovary small, 2-celled, 12-18-ovuled, cells 8-ovuled : flowers small, whitish.

Khasia hills. Neilgherries. Flowering in the cold season.

INDEX OF TAMIL SYNONYMS.

Acha marum	...Diospyros Ebenaster.
Adatodey	...Adhatoda Vasica.
Amei-nerunshil	...Pedalium murex.
Amkoolang	...Physalis flexuosa.
Anaueringie	...Pedalium murex.
Castavalie	...Cerbera odallam.
Castmallica	...Jasminum angustifolium.
Ceatsiragum	...Vernonia anthelmintica.
Cairata	...Andrographis paniculata.
Caroo-noochie	...Gendarussa vulgaris.
Chivan-amelpodie	...Ophloxylon serpentinum.
Chittramoolum	...Plumbago Zeylanica.
Codagam	...Tylophora asthmatica.
Coommy	...Gmelina arborea.
Gooringa	...Tylophora asthmatica.
Oooruvingie	...Ehretia buxifolia.
Cunjah-koray	...Ocimum album.
Eloopie	...Bassia longifolia.
Eroomuthie	...Anisomeles ovata.
Kat-mielli	...Vitex altissima.
Karantageris	...Eclipta erecta.
Karoo-comatay	...Datura fastuosa.
Kodga-saleh	...Bungia repens.
Kodie-palay	...Hoya viridiflora.
Kodaivaylie	...Plumbago Zeylanica.
Kolcuttay tek	...Premna tomentosa.
Kreata	...Andrographis paniculata.
Kromela	...Gmelina Asiatica.
Kursalankunnie	...Eclipta erecta.
Kulimitan	...Ocimum hirsutum.
Maghadam	...Mimusops elengi.
Manay-poongnng-kai	...Sapindus emarginatus.
Mashiputri	...Artemisia Indica.
Mogalunga	...Schrebera Swietenoides.
Moollie	...Solanum Indicum.
Moonnee	...Premna integrifolia.
Mungil-comatay	...Datura Metel.
Naraga	...Ehretia ovalifolia.
Narvillie	...Cordia Rothii.
Neer-moollie	...Asteracantha longifolia.
Neer-nochie	...Vitex trifolia.
Neer-plumie	...Herpestis Monniera.
Nettay peymaruttie	...Anisomeles Malabarica.
Noochie	...Vitex Negundo.
Nundeavettie	...Tabernaemontana coronaria.

Ootamunnie	...Dæmia extensa.
Padris	...Stereospermum Chelonoides or
Pala	...Wrightia tinctoria. (suaveolens)
Paltan	...Mimusops hexandra.
Panawoodachie	...Calosanthes Indica.
Panechikai	...Embryopteris glutinifera.
Paymoosley	...Ipomœa Malabarica.
Pay maruttie	...Anisomeles Malaberica.
Podootalie	...Lippia Nodiflora.
Poochacottay	...Sapindus emarginatus.
Poonakapoondoo	...Justicia Tranqueharensis.
Poovandie	...Sapindus emarginatus.
Portalaykaiantagherie	...Wedelia calendulacea.
Ponpadyrae	...Bignonia chelonoides.
Poum or Pouvum	...Schleichera trijuga.
Ray-pullay	...Spathodea arcuata.
Samutra-cheddie	...Argyreia speciosa.
Sangkhaphulie	...Vinca pusilla.
Shangam-cooppy	...Clerodendron inerme.
Shayracet-coochie	...Agathotes chirayta.
Shemmoolie	...Barleria Prionitis.
Sheecodie-vaylie	...Plumbago rosea.
Sbevadi	...Ipomœa Turpethum.
Sukkaray-mullie	...Batatas edulis.
Taita	...Strychnos potatorum.
Taloo-dalei	...Clerodendron phlomoides.
Tovashoo-morunghie	...Justicia Tranqueharensis.
Talyl-kodugboo	...Tiaridium Indicum.
Tettan-kotlay	...Strychnos potatorum.
Toodavullay	...Solanum trilobatum.
Toombie	...Diospyros glutinosa.
Tumhali	...Melanoxylon, sp.
Tulasee	...Ocimum sanctum.
Turnoot-patchie	... Do. basilicum.
Vaag-marum	...Calosanthes Indica.
Vadencoonie-marum	...Bignonia Xylocarpa.
Vara-moolie	...Barleria Prionitis.
Vaylie-partie	...Dæmia extensa.
Vela-parhrie	...Stereospermum chelonoides.
Veppalie	...Wrightia antidysenterica.
Vidi-marum	...Cordia myxa.
Vistacokrandie	...Evolvulus alsinoides.
Vackana-marum	...Diospyros cordifolia.
Vul-ademhoo	...Ipomœa grandiflora.
Vaillay-oomatay	...Datura alba.
Vaydoo-boorikeu	...Anisomeles ovata.
Vuttie-paymarutie	... Do. do.
Wangkari	...Solanum Melongena.
Yetti	...Strychnos nux vomica.
Yercum	...Calotropis gigantea.

INDEX OF MALAYALIM SYNONYMS.

Acatsja-valli	...	...Cassyta filiformis.
Ada-kodien	...	...Holostemma Rheedii.
Adamboe	...	...Lagerstrœmia reginæ.
Adel-odagam	...	...Adhatoda vasica.
Amelpodi	...	...Ophioxylon serpentinum.
Ana-collappa	...	...Lippia nodiflora.
Ana-schunda	...	...Solanum ferox.
Appel	...	...Premna integrifolia.
Bahel-shulli	...	...Asteracantha longifolia.
Bahel-tajulli	...	...Artanema sesamoides.
Ballel	...	...Ipomœa reptans.
Bel-adamboe	...	... „ rufosa.
Bel-ericu	...	...Calotropis gigantea.
Bella-mudagam	...	...Scævola taccada.
Belluta-amelpodi	...	...Ophioxylon, sp.
Belluta-kakakodi	...	...Chonemorpha macrophylla.
Bemharini	...	...Adhatoda Betonica.
Bem-nosi	...	...Vitex Negundo.
Bea-tirtali	...	...Aniseia uniflora.
Bena-pataja	...	...Tiaridium Indicum.
Brami	...	...Herpestis Monneira.
Caca-mulla	...	...Pedalium murex.
Cajengam	...	...Eclipta erecta.
Caniram	...	...Strychnos nux vomica.
Canajan-cora	...	...Canscora perfoliata.
Car-elu	...	...Sesamum Indicum.
Cara-nosi	...	...Vitex trifolia.
Cara-shulli	...	...Barleria obovata.
Cara-carinam	...	...Andrographis paniculata.
Carim-carini	...	...Justicia Ecbolium.
Carim tumba	...	...Anisomeles Malabarica.
Cattu-scheragam	...	...Vernonia anthelmintica.
Chunda	...	...Solanum Jacquini.
Codaga pala	...	...Wrightia antidysenterica.
Collato-vestla	...	...Barleria Prionitis.
Corosinam	...	...Torenia cordifolia.
Cottam	...	...Pogostemon Heyneanum or P
Cupa vela	...	...Catharanthus pusillus. (cho
Cumbalu	...	...Gmelina arborea.
Curutu pala	...	...Tabernæmontana crispa.
Elengi	...	...Mimusops Elengi.
Ericu	...	...Cr— —is gigantea.
Hummajd	□	...Datura alba.

INDEX OF MALAYALIM SYNONYMS.

Inota Inodien	...Physalis minima.
Kaka-kodi	...Gymnema Nepaulense.
Kaka-pu	...Torenia Asiatica.
Kalupolapan	...Striga euphrasioides.
Kalu-tali	...Rhyncoglossum Bheedii.
Kametti-valli	...Chonemorpha cristata.
Kapa-kelungu	...Batatas paniculatus.
Karil	...Vitex Leucoxylon.
Katu-mall-elou	... Do. altissima.
Kurka	...Anisochilos carnosum.
Kattu-kelangu	...Argyreia Malabarica.
Kattu-tjettipu	...Artemisia Indica.
Kattu-mulla	...Jasminum, sp.
Kattu-pitsjegam-mulla	... ,, angustifolium.
Kattu-pal-valli	...Cryptolepis Buchanani.
Katru-tajavegam mulla	...Jasminum hirsutum.
Kattu-pootajanga-puspam	...Vandellia crustacea.
Kuda-mulla	...Jasminum sambac.
Kudici-kodie	...Plectranthus rotundifolius
Kurka	...Aganosma lævigata.
Kurka	...Plectranthus rotundifolius.
Kulbree vahl	...Porana racemosa.
Mail-elou	...Vitex alata.
Mala-elengi	...Mimusops, sp.
Manga-nari	...Limnophila gratioloides.
Manja kurini	...Cromandra infundibuliformis.
Manil-kara	...Mimusops kanki.
Molago-marum	...Schmidelia cobbe.
Mudala-nelu-hummater	...Datura fastuosa.
Muel-echery	...Emilia sonchifolia.
Munam podam	...Pogostemon paniculatum.
Munjapoo marum	...Nyctanthes arbortristis.
Nala tirtava	...Ocimum sanctum.
Nandi-ervatam	...Tabernæmontana coronaria.
Nasschera-canachapu	...Torenia minuta.
Nanajeva-pataja	...Hoya pendula.
Nedel-ambel	...Villarsia Indica.
Nelam-pala	...Grangea Madraspatensis.
Nelem-pala	...Wrightia tomentosa.
Nelen-tajunda	...Solanum incertum.
Neli-pu	...Utricularia graminifolia.
Nur-notajilil	...Clerodendron inerme.
Niir-pongelion	...Spathodea Bheedii.
Nila-hummato	...Datura fastuosa.
Nila-harudana	...Solanum Melongena.
Nir-schulli	...Hygrophila obovata.
Opela	...Avicennia officinalis.
Pallay	- Wrightia tinctoria, Alstonia scholaris.
Pooven koodoonghei	...Conyza pinnea,

Padri ...	— Stereospermum chelonoides.
Palna schulli	...Dilivaria ilicifolia.
Pajanelli ...	...Calosanthes Indica.
Pala ...	...Alstonia scholaris.
Palega-pajaneli	...Calosanthes Indica.
Pal-moodera	...Batatas paniculata.
Pal-valli ...	...Chonemorpha Malabarica.
Panitajaca-marum	...Embryopteris glutinifera.
Parparam	...Pentatropis macrophylla.
Poolajanga-palpam	...Bonnaya brachiata.
Poo-tumba	...Andrographis echioides.
Poo-cajennain	...Wedelia calendulacea.
Poo-inota inodjen	...Physalis minima.
Poeraga ...	...Clerodendron infortunatum.
Perim-tolassi	...Plectranthus cordifolius.
Pitajigum-mulla	...Jasminum grandiflorum.
Poerinsi ...	...Sapindus laurifolius.
Pola-tajera	...Limnophila Roxburghii.
Putumba ...	...Adenostemma viscosum or Decaneurum molle.
Puam-curundala	...Conyza cinerea.
Pul-colli ...	...Rhinacanthus communis.
Puli-schovadi	...Ipomœa pes tigridis.
Samudra-sjogam	...Argyreia speciosa.
Scheru-katu-valli caniram	...Strychnos colubrina.
Schero-valli-caniram	...Cansjera scandens.
Schetti-codiveli	...Plumbago rosea.
Schit-elu ...	...Sesamum Indicum.
Schovanna-adamboe	...Ipomœa pes capræ.
Sendera-klandi	..., ,, tridentata.
Sjovanna-amelpodi	...Ophioxylon serpentinum.
Soladi tirtava	...Ocimum basilicum.
Schem-chunda	...Solanum Indicum.
Tala-neli ...	...Ipomœa filicaulis.
Theka ...	...Tectona grandis.
Tiru-tali ...	...Ipomœa sepiaria.
Tondi-teregam	...Callicarpa lanata.
Trjadaen ...	...Anisomeles ovata.
Trjanga-paspam	...Ilysanthes rotundifolia.
Tejera pu-palvalli	...Aganosme caryophyllata.
Tejerou caniram	...Cansjera scandens.
Tejeria manga nari	...Limnophila gratioloides.
Tejeroea cil arobel	...Villarsia cristata.
Tejerou-theka	...Clerodendron serratum.
Tejeru-vallei	...Hydroles Zeylanica.
Tejeragum mulla	...Jasminum undulatum.
Tejude-marum	...Graptophyllum hortense,
Tjnare cranti	...Quamoclit vulgaris.
Tumba-codiveli	...Plumbago Zeylanica.

Upa-dall	...	...Asystaria Coromandeliana.
Vada-kodi	...	...Gendarussa vulgaris.
Valli upa-dall	...	...Asystasia Coromandeliana.
Vallia-mange-nari	...	...Wollastonia biflora.
Vidimaram	...	...Cordia myxa.
Visnu-crandi	...	...Evolvulus alsinoides.
Walla kaka codie	...	...Hoya viridiflora.
Watton valla	...	...Cosmostigma racemosum.

INDEX OF
BENGALEE & HINDOOSTANEE SYNONYMS.

Acaspawan	...	H	Cuscuta reflexa.
Ada-beorna (or birni)	...	B	Herpestis Brownei.
Agnee	...	B	Plumbago Zeylanica.
Ak	...	H	Calotropis gigantea.
Akasha-vullee	...	B	Cassyta filiformis.
Akand	...	B	Calotropis gigantea.
Anderjow	...	H	Wrightia antidysenterica.
Anis	...	...	Adhatoda vasica.
Ananto-mool	...	B	Hemidesmus Indicus.
Arko	...	...	Calotropis gigantea.
Arus	...	H	Adhatoda vasica.
Asrund	...	...	Do.
Ashpbota	...	B	Jasminum sambac.
Babai	...	H	Ocimum pilosum.
Baberung or Babreng	...	B & H	Embelia Ribes.
Baboot-toolsee	...	H	Ocimum pilosum.
Bacchi	...	H	Vernonia anthelmintica.
Bacul	...	...	Mimusops elengi.
Bacumba	...	...	Anisomeles ovata.
Badanjan	...	...	Solanum Melongena.
Bagoon or Begoon	...	B & H	Do.
Bahoar	...	H	Cordia myxa or latifolia.
Baingan	...	...	Solanum Melongena.
Bakas	...	B & H	Adhatoda vasica.
Bamunhatee	...	B	Clerodendron siphonanthus.
Bang	...	H	Hyoscyamus niger.
Banstara	...	...	Barleria cœrulea.
Bantulas	...	...	Ocimum, sp.
Bartakoo	...	B	Solanum Melongena.
Bascha	...	...	Adhatoda vasica.
Bastra	...	H	Callicarpa lanata.
Batis-rung or Beta-rung	...	B	Peristrophe tinctoria.
Beegoon	...	...	Solanum Melongena.
Balphool	...	...	Jasminum sambac.
Bhangra	...	H	Eclipta erecta.
Bhant	...	B	Clerodendron infortunatum.
Bhatestaid	...	H	Solanum Jacquini.
Bhool-jam	...	B	Ardisia solanacea.
Bhooi-jamba	...	...	Premna herbacea.
Bhool-koomra	...	B & H	Batatas paniculatus.
Bhooi-okra	...	H	Lippia nodiflora.
Bhoomi-jumboca	...	...	Premna herbacea.
Bhoomi-nim	...	B	Bonnaya brachiata.
Bhool-chiravee	...	...	Premna integrifolia.

Bhreeng or Bhreenga	... ...	Wedelia calendulacea.
Bhringraj...	... II	Eclipta erecta.
Bina	... B & H	Avicennia tomentosa.
Bish-tarak	... II	Argyreia speciosa.
Bokenakoo	... ...	Lippia nodiflora.
Bong	... B	Solanum Melongena.
Boorans ...	... II	Rhododendron arboreum.
Brrehuti ...	... B	Solanum ferox.
Buckchie ...	... II	Vernonia anthelmintica.
Buhooari ...	... B	Cordia myxa.
Bun baboori	... II	Salvia brachiata.
Bun gab ...	... II	Diospyros cordifolia.
Bun jam ...	... B	Ardisia solanacea.
Bun jama...	... ...	Clerodendron inerme.
Bun mullika	... B & II	Jasminum sambac.
Bun nowaree	... ...	,, attenuatum.
Bun nurukhalee	... II	Ardisia Roxburghiana.
Bura chooli	... B & II	Villarsia Indica.
Bura koonda	... ...	Jasminum arborescens.
Bura koosuma	... ...	Blumea lacera.
Bura hasora	... ...	Cordia latifolia.
Bura tugur	... B	Tabernæmontana coronaria.
Byakool ...	... B & H	Solanum Indicum.
Bygun ...	... II	Solanum Melongena.
Caladana ...	... II	Ipomœa Nil.
Calapunth	... ...	Andrographis paniculata.
Cala-toolsee	... ...	Ocimum sanctum.
Calmi ...	... ...	Ipomœa repens.
Canac ...	... ...	Datura metel.
Casni ...	... ...	Cichorium Endivia.
Cawa ...	... ...	Ipomœa Nil.
Chagal banter	... ...	Urmia extensa.
Chagal patee	- ...	Cynanchum pauciflorum.
Chatin ...	... B	Alstonia scholaris.
Chebiera ...	... ...	Justicia bicalyculata.
Cherayti, Chevaita	... B & H	Agathotes cherayta.
Cheretta ...	... B	Do. do. ..
Chil-hinge	... II	Strychnos potatorum.
Chiti ...	... ...	Plumbago Zeylanica.
Chitra ...	... B	Do. do.
Chitturmool	... ...	Do.
Chooli ...	... II	Villarsia cristata.
Choola-deed-lata	... II	Gymnema sylvestre.
Choota-kookshima	... II	Vernonia cinerea.
Choota micheta	... B	Hemidesmis polysperma.
Chota bish-tarak	... B & H	Argyreia speciosa.
Chota jhanjn	... II	Utricularia bifiora.
Chota kulpa	... B	Trichodesma Indicum.
Chudra ...	... B & II	Solanum Jacquini.
Chukraai...	... H	Chickrassia tabularis.
Chumbell...	... II	Jasminum grandiflorum.
Chundra ...	... B & H	Ophioxylon serpentinum.
Chandra mallika	... B	Chrysanthemum Indicum.
Chundruhasa	— ...	Solanum hirsutum.

Chundruka	...	...	—	Ophioxylon serpentinum.
Creat	...	...	H	Andrographis paniculata.
Carayia	...	...	...	Echites antidysenterica.
Cutaja	...	...	...	Do. do.
Dandi	...	...	H	Chrysanthemum Indicum.
Dela	...	...	H	Jasminum multiflorum.
Deshi-mullika	...	...	B	„ sambac.
Dhae	...	...	H	Grislea tomentosa.
Dhangaphul	...	...	B	Do.
Dhatura	...	...	H	Datura alba.
Dhari	...	...	B	Grislea tomentosa.
Dhootura	...	...	...	Datura alba.
Dohutee-luta	...	...	—	Ipomœa pes capræ.
Doodh-kulmee	...	...	...	„ turpethum.
Doodh luta	...	...	...	Oxystelma esculentum.
Dood luta	...	...	...	Echites marginata.
Dorla	...	...	H	Solanum Jacquinii.
Duna	...	...	...	Artemisia Indica.
Dunkoni	...	...	B	Pladera decussata.
Ecki-lagar	...	...	H	Tabernæmontana coronaria.
Gab	...	...	B & H	Embryopteris glutinifera.
Ghunia	...	...	B	Stereospermum suaveolens.
Gi'auada	...	...	H	Dassia latifolia.
Gohhi	...	...	—	Emilia sonchifolia.
Gobura	...	...	...	Anisomeles ovata.
Gochru	...	...	H	Asteracantha longifolia.
Onkshura	...	...	...	Do. do.
Gendi	...	...	...	Cordia, sp.
Gonial	...	...	B	Diospyros racemosa.
Gonli-bagooa	...	...	...	Solanum longum.
Gonli-sham	...	...	H	Eranthemum pulchellum.
Gone-karaai	...	...	D	Solanum Indicum.
Gota-begoon	...	...	B	Do. stramonifolium.
Gumhar	...	...	B & H	Gmelina arborea.
Gandhumar	...	...	H	Artemisia Indica.
Ganjaree	...	...	B	Premna spinosa.
Hachitte	...	...	B	Myriogyne minuta.
Hapur-maleo	...	...	H	Vallaris dichotomus.
Harcuchila	...	...	B	Strychnos colubrina.
Harkat	...	...	H	Dilivaria ilicifolia.
Harkee	...	...	B	Chonemorpha macrophylla.
Harkuch-kanta	...	...	H	Dilivaria ilicifolia.
Harsingahar	...	...	...	Nyctanthes arbortristis.
Hatta-kuna	...	...	H	Clerodendron hastatum.
Helencha	...	...	B	Enhydra flingcha.
Hingoolee	...	...	B	Solanum melungena.
Huldialgoon-luta	...	...	B	Cascuta reflexa.
Hulsi	...	...	B	Ægiceras fragrans.
Haraman	...	...	...	Compositum mellisæfolium
Jhahugauda	...	...	H	Asteracantha longifolia.
Jadurjaw	...	...	H	Wightia antidysenterica.

Iach-picha	...	...	...	Quamoclit vulgaris.
Ispaguol	...	...	B & H	Plantago Asiatica.
Jad	...	...	H	Jasminum grandiflorum.
Jatee	...	...	B & H	Do. do.
Jhanj	...	...	B	Utricularia fasciculata.
Jhintee	...	...	B	Barleria cristata.
Jitee	...	...	B & H	Maradenia tenacissima.
Joci	...	...	B	Jasminum auriculatum.
Joci-pana	...	...	...	Rhinacanthus communis.
Joothipooshpika	...	...	B & H	Jasminum auriculatum.
Jugut-madan	...	...	B	Gendarussa vulgaris.
Jurad-kuleeee	...	...	...	Hewittia bicolor.
Kabuter-ke-jar	...	...	H	Rhinanthus communis.
Kakathontee	...	...	B	Ardisia solanacea.
Kaladhooura	...	...	B & H	Datura fastuosa.
Kalajatee	...	...	B	Eranthemum pulchellum.
Kalameeh	...	...	B	Andrographis paniculata.
Kalacja	...	...	...	Ehretia serrata.
Kaleeshumhali	...	...	H	Gendarussa vulgaris.
Kafteseeria	...	...	...	Vernonia anthelmintica.
Kama-luta	...	...	B & H	Quamoclit vulgaris.
Kanta-jatee	...	...	H	Barleria prionitis.
Kanta-koolika	...	...	...	Asteracantha longifolia.
Kanth-mullika	...	...	B	Jasminum sambac.
Kantha-karus	...	...	B & H	Solanum Jacquini.
Kathbel	...	...	B	Jasminum pubescens.
Keerat	...	...	B	Agathotes cherayta.
Kendoo	...	...	B	Diospyros melanoxylon.
Keshooriga	...	...	B & H	Eclipta erecta.
Keshoorie	...	...	B	Wedelia calendulacea.
Krishanoo	...	...	...	Plumbago zeylanica.
Khrishna-dhathura	...	...	H	Datura fastuosa.
Kiew	...	...	B	Diospyros melanoxylon.
Koamoora	...	...	B & H	Callicarpa Wallichiana.
Kookhura-shoonga	...	...	B	Blumea lacera.
Kolsi	...	...	H	Solanum Indicum.
Koochila	...	...	B & H	Strychnos nux-vomica.
Koochila-luta	...	...	B	„ Colubrina.
Koochuri	...	...	...	Exacum tetragonum.
Kooli-beguon	...	...	B	Solanum longum.
Koouda	...	...	B	Jasminum pubescens.
Koorchi	...	...	H	Wrightia antidysenterica.
Koostoola	...	...	H	Hypoestes verticillaris.
Koosum, Koosumbha	...	...	...	Carthamus tinctorium.
Kootaya	...	...	H	Solanum Jacquini.
Kulmi-luta	...	...	B	Ipomoea bonanox.
Kulmi-shak	...	...	...	Do. reptans.
Kungeee	...	...	B	Congea pentandra.
Kunja-luta	...	...	...	Pergularia odoratissima.
Kural	...	...	...	Ocimum sanctum.
Kurulka	...	...	...	Prunus spinosa.
Lahera	...	...	H	Cordia myxa.

HINDOOSTANEE SYNONYMS. 585

Lal-chirchiri	...	Plumbago rosea.
Lal-chita	... B & H	Do. do.
Lal-kamaluta	... B	Quamoclit vulgaris.
Lal-sbukurkund-aloo	... B & H	Batatas paniculata;
Lal-turoo-lata	... B & H	Quamoclit vulgaris.
Langolilata	... B	Ipomœa pestigridis;
Lemoora, Lisora	... H	Cordia myxa.
Luia-bichittee	...	Erycibe paniculata.
Lutiam	... B	Willughbeia edulis.
Macao	... H	Physalis Peruviana;
Maco	...	Solanum nigrum.
Madar	...	Calotropis gigantea.
Mabus-moura	...	Bassia latifolia.
Malutee	... B	Jasminum grandiflorum.
Maintee	...	Agunosma caryophyllata.
Mane	... H	Ocimum pilosum.
Massandari	... B	Callicarpa incana.
Matura	...	Do. do.
Meeta-tootle	... B	Ipomœa sepiaria.
Mogra, Mogri	... H	Jasminum sambac.
Moho	...	Bassia longifolia.
Mooa	... B	Do. do.
Moog	...	Stereospermum suaveolens;
Moorga	...	Jasminum sambac.
Moola-bela	...	Do. do.
Motea	...	Do. do.
Moula	... B & H	Bassia latifolia.
Mudar	... H	Calotropis gigantea.
Muha-tita	...	Andrographis paniculata.
Mubootee	... B	Solanum melongena;
Mukurundoo	...	Jasminum pubescens
Mullika	...	„ sambac.
Mustaroo	... H	Artemisia Indica.
Nam-bagba	... B	Justicia bicalyculata.
Nasbo	... H	Ocimum pilosum.
Neel-kulmee	...	Pharbitis Nil.
Neel-vasooka	... B	Eranthemum strictum.
Nicchikri	... H	Myriogyne minuta.
Nirgundi	... B	Vitex negundo.
Nirmullee, Nirmullies	... B & H	Strychnos potatorum.
Nisnda	... H	Vitex negundo.
Nisot	...	Ipomœa Turpethum.
Nukcheknie	...	Iloya viridiflora;
Nuwa-mallaka	... B	Jasminum arborescens.
Ognee	... B	Plumbago Zeylanica.
Odoo-jatm	... H	Justicia echolium.
Oojootee	... B	Barleria ciliata.
Oouso-braha	...	Echites cymosa.
Palo	... H	Ehretia buxifolia.
Palak, or Paleh-jobbis	...	Rhinacanthus communis.
Pani-ke-sbumhalie	... H	Vitex trifolia.

Pantah	...	...	B	Jasminum elongatum.
Paral	...	...	U	Bignonia chelonoides.
Parool	...	...	B	Stereospermum suaveolens.
Pathur choor	...	...	U	Coleus aromaticus.
Patuli	...	...	B	Do. do.
Peela-bhungara	...	...		Wedelia calendulacea.
Pendaloo	...	...	...	Batatas edulis.
Phub-wara	...	...	...	Bassia butyracea.
Pia-hausa	...	...	...	Barleria cristata.
Pida	...	...	...	Cordia monoica.
Puch-pat, or Patchouly		...	...	Pogostemon patchouly.
Purula	...	...	...	Stereospermum suaveolens.
Rachan	...	...	U	Ocimum pilosum.
Raba	...	...	...	Do. do.
Rakhal-phul	...	...	B	Schmidelia serrata.
Ram-bagoon	...	...	...	Solanum hirsutum.
Ramtlie	...	...	...	Guizotia oleifera.
Ram-toolsee	...	...	D & U	Ocimum gratissimum.
Ram utti	...	...	D	Cordia acuminata.
Reetha	...	...	...	Sapindus detergens.
Rihan	...	...	U	Ocimum pilosum.
Rishta	...	...	...	„ emarginatum.
Rujunta	...	...	B	Pogostemon villosum.
Sada-jatee	...	...	B	Barleria dichotoma.
Sadul-ruma	...	...	U	Melodinus monogynus.
Sagnon	...	...	B & U	Tectona grandis.
Sagwaunee	...	...	U	Dæmia extensa.
Saksa	...	...	B	Ichnocarpus frutescens.
Samandr-sokh		}	U	Argyreia argentea.
Samoodr-sokh				
Saro-dullum	...	...	...	Elephantopus scaber.
Saom	...	...	B	Fagræa obovata.
Seduari	...	...	U	Vitex trifolia.
Shah-puansd	...	...	...	Amherstia moschata.
Sham-dullun	...	...	B	Elephantopus scaber.
Shephalika	...	...	...	Nyctanthes arbortristis.
Shiooli	...	...	...	Do. do.
Shona	...	...	B & U	Calosanthes Indica.
Shudi-mudi	...	...	B	Emilia sonchifolia.
Shumhalie	...	...	U	Vitex negundo.
Shwet (or sada) akund		...	B	Calotropis gigantea.
Shwet-chamni	...	...	...	Herpestis monniera.
Shwet-kamaluta	...	...	B	Quamoclit vulgaris.
Shwet-turoolata	...	...	H	Do. do.
Shyama-luta	...	...	B & U	Ichnocarpus frutescens.
Sinhain	...	...	H	Vitex trifolia.
Singarhar	...	...	...	Nyctanthes arbortristis.
Sitaka-punjeene	...	...	B	Anisochilus carnosus.
Soma-luta	...	...	U	Sarcostemma acidum.
Somraj	...	...	B	Vernonia anthelmintica.
Soom	...	...	H	Sarcostemma viminale.
Subye	...	...	B & A	Ocimum basilicum.
Suffet-shukurkund-aloo		...		Batatas edulis.

HINDOOSTANEE SYNONYMS.

Saffet-turoo-lata	...	...		Quamoclit vulgaris.
Saffaid-toolsia	...	...		Ocimum album.
Sukkur-kanda-aloo	—	...		Batatas edulis.
Sungkoopie	...	...		Clerodendron inerme.
Tab-machana	...	...	H	Asteracantha longifolia.
Tan	...	...	—	Lippia nodiflora.
Taporeea...	...	...	B	Physalis Peruviana.
Teak-chama	...	...	H	Microrhynchus asplenifolius.
Tevt-conga	...	...	...	Hoya viridiflora.
Tauris	...	...	B	Ipomea Turpethum.
Tapariya	...	...	...	Physalis Peruviana.
Tikura	...	...	...	Ipomea Turpethum.
Toogee	...	...	H	Peristrophe lanceolaria.
Toolati-pari	...	...	...	Physalis minima.
Toolsis	...	...	B & H	Ocimum sanctum.
Tugura	...	...	B & H	Tabernæmontana coronaria.
Talidun	...	...	H	Solanum nigrum.
Tumali	...	...	B	Diospyros tomentosa.
Turbad	...	...	H	Ipomea Turpethum.
Turoo-lata	...	...	B	Quamoclit vulgaris.
Ungootee...	—	...	B	Holmskioldia coccinea.
Unta-mool	...	...	H	Tylophora asthmatica.
Urka	...	...	B	Calotropis gigantea.
Urash, Urusa	...	...	B	Solanum verbascifolium.
Usgund	...	...	H	Physalis flexuosa.
Ushwagundha	...	—	B	Do. do.
Usoola	...	...	...	Vitex alata.
Ustabanda	—	...	H	Premna spinosa.
Vasocka...	...	...	B	Adhatoda vasica.

INDEX OF TELUGU SYNONYMS.

Ada bukkuda	...Ehretia laevis.
Adavijilakarra	...Vernonia anthelmintica.
Adavi malle	...Jasminum latifolium.
Adavi mollé	...Jasminum auriculatum.
Adavj nelli kura	...Premna, sp.
Adavi pala tige	...Cryptolepis reticulata.
Addasaram	...Adhatoda vasica.
Agaru chettu	...Aquilaria agallocha.
Agni mata	...Plumbago zeylanica.
Alachata	...Ipomea dentata.
Ala pala	...Pergularia pallida.
Alaranji	...Convolvulus parviflorus.
Alarantu	...Rostellaria diffusa.
Alumukada	...Ipomœa filiformis.
Amkudu	...Wrightia tinctoria.
Andabeerakoo	...Anisomeles ovata.
Antara baruara	...Villarsia Indica.
Antara valli tige	...Cassyta filiformis.
Atmaghandhi	...Physalis somnifera.
Bach-chali manda	...Ceropegia tuberosa.
Baggapatti	...Limnophilla racemosa.
Balbandi tige	...Ipomœa pescaprae.
Banka nakkara	...Cordia myxa.
Bapanaburi	...Ehretia buxifolia.
Beda tige	...Ipomœa pescaprae.
Begati kanda	...Amberboa Indica.
Bella goda	} Ceropegia juncea.
Bella manda	
Bharangi chettu	...Clerodendron. sp.
Bharamaia mari	...Clerodendron serratum.
Buchakara gadda	...Batatas paniculatus.
Biburundi	...Tiaridium Indicum.
Bodasaram	...Stemodia viscosa.
Boddi kura	...Rivea hypocrateriformis.
Boddu malle	...Jasminum sambac.
Bokkadi	...Ehretia, sp.
Bokkena	...Llippia nodiflora.
Bottu kuru chettu	...Cordia polygama.
Botuka	...Cordia, sp.
Bramhi chettu	...Clerodendron, sp.
Budda busara	...Physalis peruviana.
Budide chatta	...Heliotropium coromandelianum.
Busara kaya	...Physalis peruviana.
Busi	...Vitex arborea.

INDEX OF TELUGU SYNONYMS.

Challa gummudu	...	...Gmelina parvifolia.
Chamti	...	...Chrysanthemum Roxburghii.
Chandrapoda	...	...Argyreia speciosa.
Chata katta tivva	...	...Ipomœa cymosa.
Chavalapuri kada	...	...Andrographis echioides.
Chebira	...	...Peristrophe bicalyculata.
Chekuti tiva	...	...Pentatropis microphylla.
Chemanti	...	...Chrysanthemum Roxburghii.
Chevukurti chettu	...	...Elevogtia verticillata.
Chevullpilli tige	...	...Ipomœa pescaprae.
Chilagada dumpa	...	...Batatas edulis.
Chilla chettu	...	...Strychnos potatorum.
Chinna bolaku	...	...Cordia angustifolia.
Chinna ranta	...	...Rhaphidospora glabra.
Chiragadam	...	...Batatas edulis.
Chiri alli	...	...Villarsia cristata.
Chiri gummudu	...	...Batatas paniculata.
Chiri malle	...	...Jasminum angustifolium.
Chiri tika	...	...Clerodendron, sp.
Chiri vanga	...	...Solanum melongena.
Chira pala	...	...Onysteima esculentara.
Chitramulum	...	...Plumbago rosea.
Chitta tumaki	...	...Diospyros tomentosa.
Chitti sukudu	...	...Wrightea tinctoria.
Davanamu	...	...Artemisia vulgaris.
Donka burra chettu	...	...Ehretia, sp.
Dudipala	...	...
Dundilapu chettu	...	...Calosanthes Indica.
Dusa tapa chettu	...	...Dœmia extensa.
Duturamu	...	...Datura alba.
Edakula ariti	...	} Alstonia scholaris.
Edakula pala	...	
Edakula ponna	...	
Eddumata chettu	...	...Nelsonia tomentosa.
Eddu nalike chettu	...	...Elephantopus scaber.
Emoga palleru	...	...Pedalium murex.
Erra adavi molla	...	...Jasminum auriculatum.
Erra charanti	...	...Chrysanthemum Roxburghii.
Erra chitramulum	...	...Plumbago rosea.
Erra goda	...	...Diospyros montana.
Erra kanraucki	...	...Solanum rubrum.
Erra kuta	...	...Argyreia aggregata.
Erra pula pedla goranta		...Bariera ciliata.
Eru sumiki	...	...Diospyros, sp.
Erra valambram		...Crossandra infundibuliform
Eru malle	...	...Jasminum, sp.
Eru pichcha	...	...Clerodendron inerme.
Eru vanga	...	...Solanum, sp.
Eti chilla	...	...Dilivaria ilicifolia.
Eti pisinika		...Clerodendron inerme.
Etrinta	...	...Sonchus ciliatus.
Gabbu nelli	...	...Premna longifolia.

Gaju chettu	...Solanum rubrum.
Golagara	...Eclipta prostrata.
Gandharasamu	...Gendarussa vulgaris.
Gandu gannera	...Alstonia venenata.
Gantu bharangi	...Clerodendron, sp.
Gariti kamma	...Vernonia cinerea.
Garuda malle	...Jasminum, sp.
Gote	...Diospyros sylvatica.
Genusugadda	...Batatas edulis.
Girimellika	...Wrightia antidysenterica.
Gobli	...Asteracantha longifolia.
Gontema gomaru chetta	...Ipomœa filicaulis.
Gorre chemidi	...Andrographis echioides.
Grandi lagarapu chetta	...Tabernemontana coronaria.
Gubbadara	...Symphorema involucrata.
Guggilam chettu	...Aegiceras fragrans.
Gulla gila gaddi	
Gumudu chetta	...Gmelina Asiatica.
Gumudu teku	... Do. arborea.
Gumudu tige	...Batatas paniculata.
Gnota galijeru	...Eclipta prostrata.
Gnota kalagara	
Gunta kaminam	...Stemodia viscosa.
Gurropu gotteaku	...Clerodendron viscosum.
Gurugu palæ tige	...Cryptolepis reticulata.
Guriti chettu	...Docmia extensa.
Gutti nemaladugu manu	...Vitex, sp.
Guvva gutti	...Trichodesma Indicum,
Hamsa padi	{ Heliotropium Coromandelianum or Coldenia procumbens.
Hanjika	...Clerodendron, sp.
Hasti sundi	...Tiaridium Indicum.
Hemapushpika	...Jasminum chrysanthemum,
Illinda	...Diospyros chloroxylon.
Illu katte	...Ichnocarpus frutescens.
Indra tige	...Thunbergia fragrans.
Indupa chettu	...Strychnos potatorum.
Ippa or Ippe chettu	...Bassia latifolia.
Iriki	...Cordia myxa.
Iaspa gala vettulu	...Plantago ispaghula.
Istaraku pala	...Holarrhena antidysenterica.
Jaji	...Jasminum grandiflorum,
Jaka tige	...Maredenia sp.
Jiddu	...Solanum diffusum.
Jilledu	...Calotropis gigantea.
Jimandra tige	...Thunbergia fragrans,
Juttupaku	...Docmia extensa.
Kachi	...Solanum rubrum.
Kakamachi	...Solanum Indicum.
Kaka pala	...Tylophora vomitoria.
Kaka tundamu	...Aquilaria, sp,

INDEX OF TELUGU SYNONYMS.

Kaka alimora	...Diospyros cordifolia.
Kaki alli	...Diospyros, sp.
Kaki neredu	...Ardisia humilis.
Kakkita	...Argyreia speciosa.
Kaligottu	...Bignonia chelonoides.
Kalingamu	...Wrightia antidysenterica.
Kaliva chettu	...Carissa diffusa.
Kamanachi chettu	...Solanum rubrum.
Kanakambram	...Crossandra infundibuliformis.
Kappa tivva	...Ipomœa cymosa.
Kari vemu	...Andrographis paniculata.
Karnika	...Premna spinosa.
Karallamu	...Carallums adscendens.
Karu chiya	...Nyctanthes arbortristis.
Karu navvulu	...Artanema sesamoides.
Kasiratmalu	...Quamoclit phoeniceum.
Kasmaryamu	...Gmelina arborea.
Kata kasu	...Strychnos potatorum.
Kavvagummudu	...Gmelina parvifolia.
Kicha virigi chettu	...Cordia latifolia.
Kodi mursu	...Petalidium barlerioides.
Kodisa pala	} Wrightia antidysenterica.
Kodisa chettu	
Kookkita	} Argyreia speciosa.
Kokkiru	
Kola mukki chakka	...Wrightia antidysenterica.
Kokissa chettu	...Bignonia, sp.
Kolli	...Pharbitis nil.
Komma manda	...Ceropegia acuminata.
Konda ganneru	...Alstonia venenata.
Konda gobbi	...Barleria prionitis.
Kondgummudu gadda	...Batatas pentaphylla.
Konda mayuru	...Ardisia humilis.
Konda pala	...Sarcostemma acidum.
Konda sita sanaram	...Ipomœa filicaulis.
Konda tekkali	...Symphorema involucrata.
Kousu kandira	...Strychnos bicirrhosa.
Koyila mokiri	...Wrightia tomentosa.
Krishna agaru	...Aquilaria agallocha.
„ tulasee	...Ocimum album.
Kuberakshi	...Stereospermum suaveolens.
Kukka pala	...Tylophora vomitoria.
„ tulasee	...Ocimum album.
Kunda	...Jasminum, sp.
Kuntana chettu	...Ardisia humilis.
Kura nelli	...Premna esculenta.
Kuta jamu	...Wrightia antidysenterica.
Lanjasavaramu	...Ipomœa filicaulis.
Linga malle	...Jasminum, sp.
Loduga	—Symplocos racemosa.
Mabera	...Anisomeles ovata.
Machi patri	...Artemisia Indica.

Mada chettu	...Avicennia tomentosa.
Madanamu	...Datura, sp.
Madana erku	...Cryptolepis reticulata.
Madi tige	...Argyreia cymosa.
Makkam	...Schrebera swietenioides.
Malati	...Aganosma Roxburghii.
Malati yariau	...Cryptolepis pauciflora.
Malle	...Jasminum sambac.
Manchi manda	...Ceropegia bulhosa.
Mandara	...Calotropis gigantea.
Mande	...Ceropegia, sp.
Manduka brahmi	...Clerodendron viscosum.
Manu pala	...Wrightia antidysenterica.
Manu pairi	...Dipteracanthus dejectus.
Mayuramu	...Ardisia humilis.
Meda	...Tetranthera Roxburghii.
Mehamu adugu	...Ipomœa pestigridis.
Metta pala	...Batatas paniculata.
Metta toti	...Ipomœa sinista.
Metta vanke	...Solanum melongena.
Mogalinga maram	...Schrebera swietenioides.
Mohanam	...Batatas edulis.
Mokka vepa	...Bignonia, sp.
Molla chettu	...Jasminum, sp.
Mollalu	...Jasminum auriculatum.
Moogaheerakoo	...Anisomeles Malabarica.
Muchchi tanki	...Diospyros, sp.
Mukodi	...Schrebera swietenoides.
Mukkapu kokkesa	...Bignonia, sp.
Makkedam	...Striga euphrasoides.
Mukk mungera	...Asystasia Coromandeliana.
Mulla muste	...Solanum trilobatum.
Mule goranta	...Barleria prionitis.
Musta gajjamu	...Ichnocarpus frutescens.
Mushkam	...Bignonia, sp.
Muridi	...Strychnos nux vomica.
Nach chu	...Utricularia fasciculata.
Naga danti	...Tiaridum Indicum.
Naga malle	...Rhinacanthus communis.
Nagaru chettu	...Premna tomentosa.
Nakkera	...Cordia myxa.
Nalla gunta kalagara	...Eclipta, sp.
Nalla jilledu	...Calotropis procera.
Nalla kaka mushti	...Diospyros sp.
Nalla kamanchi	...Solanum rubrum.
Nalla kokkita	...Ipomœa obscura.
Nalla mada	...Avicennia tomentosa.
Nalla mulu goranta	...Barleria obovata.
Nalla nela gummadu	...Batatas paniculata.
Nalla nilambari	...Eranthemum nervosum.
Nalla peddu goranta	...Barleria cristata.
Nalla taputa	...Sonchus Orixensis.
Nalla tige	...Ichnocarpus frutescens.

INDEX OF TELUGU SYNONYMS.

Nalla tumiki	...Diospyros, sp.
Nalla vavili	...Vitex negundo.
Nalla vishnu	...Evolvulus alsinoides.
Nalla nimera	...Diospyros cordifolia.
Nalla ummetta	...Datura fatuosa.
Nalla urimida	...Diospyros cordifolia.
Nandi vardhana chettu	...Tabernæmontana coronaria.
Nava malika	...Jasminum sambac.
Navuru	...Premna tomentosa.
Nela gulimidi	...Elevogia verticillata.
Nela guruguda	...Batatas paniculatus.
Nela guruguda	...Elevogia verticillata.
Nela kalikotta	...Bignonia, sp.
Nela mulaka	...Solanum Jacquini.
Nela neredu	...Premna herbacea.
Nela pala	...Oxystelma esculentum.
Nela pippale	...Lippia nodiflora.
Nela yakuda	...Solanum Jacquini.
Nela vavili	...Gendarussa vulgaris.
Nela vamu	...Andrographis paniculata.
Nelli chettu	...Premna esculenta.
Neva ledi	...Vitex leucoxylon.
Novali adugu	...Vitex arborea.
Nilambaram	...Barleria coerulea.
Nimma tayi	...Ceropegia bulbosa.
Niru boddi	...Rivea hypocrateriformis.
Niru gobbi	...Asteracantha longifolia.
Niru goranta	...Barleria cristata.
Niru tumiki	...Diospyros, sp.
„ toolsee	...Ocimum pilorum.
Niru vanga	...Solanum melongena.
Niti gannera	...Limnophila racemosa.
Niti tumiki	...Diospyros, sp.
Nulu tega	...Cassyta filiformis.
Nuvva	...Sesamum Indicum.
Oka chettu	...Carissa carandas.
Pachcha adavi molla	...Jasminum chrysanthemum.
Pachcha botuku	...Cordia polygama.
Pachcha chamanti	...Chrysanthemum, sp.
Pachcha mulu goranta	...Barleria prionitis.
Pachcha vadambaram	...Justicia dentata.
Pachi tige	...Cassyta filiformis.
Padari chettu	...Stereospermum suaveolens.
Padma kashtam	...Sarcostemma acidum.
Pala chettu	...Mimusops hexandra.
Pala chuckhandara	...Hemidesmus indicus.
Pala dantam	...Ehretia lævis.
Pala garuda	...Alstonia scholaris.
Pala gurugu	...Holostemma Rheedianum.
Pala kura	...Oxystelma esculentum.
Pala malle tivva	...Vallaris dichotoma.
Pala nela gummudu	...Batatas pentaphylla.

INDEX OF TELUGU SYNONYMS.

Pala samudra	...Argyreia speciosa.
Pala sugandhi	...Hemidesmus Indicus.
Pala tige	...Leptadenia reticulata.
Pam budda or budama	...Physalis Peruviana.
Pampena	...Calosanthes Indica.
Pandiți vankaya	...Calonyction Roxburghii.
Panu giri	...Cordia monoica.
Pari jatamu	...Nyctanthes arbortristis.
Pasara gunna	...Diospyros, sp.
Palala gandhi	...Ophioxylon serpentinum.
Patali	...Stereospermum suaveolens.
Pedda sukudn chettu	...Wrightia antidysenterica.
Padd botuku	...Cordia myxa.
Pedda gummudu teku	...Gmelina arborea.
Pedda illinda	...Diospyros chloroxylon.
Pedda mulu goranta	...Barleria buxifolia.
Pedda nelikura	...Premna latifolia.
Pedda pala	...Wrightia tomentosa.
Pedda pallera	...Pedalium murex.
Pedda pulimera	...Ehretia loevis.
Pedda rantu	...Stenosiphonium confertum.
Pedda tiku	...Tectona grandis.
Pedda vara goki	...Salvadora Indica.
Pedda vemu	...Andrographis, sp.
Pedda whimera	...Diospyros chloroxylon.
Pruneru	...Physalis somnifera.
Pilti vendrom	... Do. do.
Pinna goranta	...Barleria, sp.
Pinna ippa	...Bassia, sp.
Pinna mulaka	...Solanum Jacquini.
Pinna nilli	...Premna biırsa.
Pinna pala	...Oxystelma esculentum.
Pinna varagogu	...Salvadora, sp.
Pisangi	...Clerodendron inerme.
Pisinika	...Maba buxifolia.
Pita pisinika	...Ehretia buxifolia.
Poda pairu	...Gymnema sylvestre.
Pogada chettu	...Mimusops elengi.
Poka banti	...Ageratum conyzoides.
Polla nuvvulu	...Sesamum, sp.
Potu bokada	...Ehretia sp.
Potu malle	...Jasminum, sp.
Potu nela vemu	...Justicia, sp.
Pula pala	...Pentatropis michrophylla.
Pullamanda	...Ceropegia tuberosa.
Pulla tige	...Sarcostemma acidum.
Puriti tige	...Ipomœa hispida.
Puta jilledu	...Wrightia tomentosa.
Putta julledu	... Do. do.
Putta podara yarala	...Vallaris dichotoma.
Rachaba jilledu	...Calotropis gigantea.
Ramaswara asta	...Solanum pubescens.
Rama gadi maan	... Do. do.

INDEX OF TELUGU SYNONYMS.

Sahadevi chettu	...Echites fruteseens.
Salaras	...Ophelia elegans.
Samhrani chettu	...Herpestis monniera.
Samudra pala	...Argyreia speciosa.
Sanna jajulu	...Jasminum auriculatum.
Sarasmati aku	...Clerodendron viscosum.
Sepa chettu	...Oxystelma esculentum.
Silajita	...Ophelia elegans.
Sitamanóháram	...Pergularia odoratissima.
Sitamana pogu metu	...Cuscuta reflexa.
Sitamma vari savaram	...Ipomœa filicaulis.
Sitaeavaram	... Do. do.
Soma lata	...Sarcostemma acidum.
Sugandhi pala	...Hemidesmus indicus.
Sukka bommi	...Catharanthus pusillus.
Suradu	...Symphorema involucrata.
Surya ratnald	...Quamoclit pennatum.
Tagada	...Bignonia chelonoides.
Takkedu chettu	...Premna, sp.
Talantu tige	...Ipomœa dentata.
Tappeta	...Anystama Coromandeliana.
Tavit chettu	...Caralluma adascendens.
Todla pala	...Wrightia tinctoria.
Tegada	...Ipomœa turpethum.
Teggummudu	...Gmelina arborea.
Takkali chetta	...Clerodendron phlomoides.
Teku chettu	...Tectona grandis.
Tella adavi molla	...Jasminum auriculatum.
Tella chitramalam	...Plumbago zeylanica.
Tella jiladu	...Calotropis gigantea.
Tella juvvi	...Ehretia buxifolia.
Tella loddugu	...Symplocos, sp.
Tella mulu goranta	...Barleria obovata.
Tella mulaka	...Solanum Indicum.
Tella nela gummudu	...Batatas, sp.
Tella nela mulaka	...Solanum Jacquini.
Tella nelambari	...Barleria dichotoma.
Tella pippali	...Symphorema involucratum.
Tella rasta	...Adhatoda betonica.
Tella sugandhipala	...Hemidesmus Indicus.
Tella tegada	...Ipomœa turpethum.
Tella vakudu	...Solanum Jacquini.
Tella vavili	...Vitex trifolia.
Tella vishnu kranta	...Evolvulus pilosus.
Tella usta	...Solanum trilobarum.
Telu maal	...Tiaridium Indicum.
Tige jemudu	...Sarcostemma acidum.
Tilaka	...Clerodendron phlomoides.
Tiyya mandi	...Ceropegia bulbosa.
Tota nela vemu	...Raphidospora glabra.
Tota vemu	...Dicliptera parvibracteata.
Tubiki	...Embryopteris glutinifera.
Tumida	...Diospyros melanoxylon.

INDEX OF TELUGU SYNONYMS

Tumiki	...Embryopteris glutinifera.
Tumeda chettu	...Diospyros melanoxylon.
Tummika	... Do. do.
Tuti kura	...Ipomœa reptans.
Uchchinta	...Solanum trilobatum.
Ulsi	...Guizotia oleifera.
Ullind	...Diospyros chloroxylon
Ummetta	...Datura alba.
Uste	...Solanum trilobatum.
Utti chettu	...Maba buxifolia.
Vadambram	...Eranthemum nervosum.
Vada teddu aku	...Microrhynchus sarmentosus.
Vaka chettu	...Carissa carandas.
Vakudu	...Solanum Jacquini.
Valise chettu	...Guizotia oleifera.
Vallari	...Sievogtia verticillata.
Vanga chettu	...Solanum melongena.
Varshakala malle	...Jasminum, sp.
Vavili chettu	...Vitex trifolia.
Vayn velangram chettu	...Embelia ribes.
Vedala chettu	...Gœrtnera racemosa.
Venna katte tigo	...Asystasia Coromandeliana.
Vepoodipatsa	...Ocimum basilicum.
Verri tala noppi	...Xanthium orientale.
Verripala	...Tylophora vomitoria.
Veru malle	...Ipomœa cymosa.
Virajaji	...Jasminum, sp.
Vurige chettu	...Cordia sebestena.
Vishnu kranta	...Evolvulus alsinoides.
Vis'a rakula pala	...Holarrhena antidysenterica.
Vodi	...Spathodea Rheedii.
Vodite	...Diospyros, sp.
Vulusi	...Guizotia oleifera.
Yaddu melte	...Nelsonia tomentosa.

INDEX OF NAMES.

	Page		Page
ACANTHACEÆ	392	ASCLEPIADACEÆ	211
Acanthus	435	Asteracantha	447
Acrocephalus	519	Asystasia	422
Adenoon	2	Avicennia	513
Adenosma	400	AZIMACEÆ	174
Adenostemma	13	Azima	174
Adhatoda	465		
Ægialitis	503	Babactes	285
ÆGICERACEÆ	143	Barleria	440
Ægiceras	144	Bassia	140
Æginetia	388	Batatas	302
Æschynanthus	286	Berniera	87
Ætheilema	453	Berthelotia	25
Aganosma	204	Beaumontia	199
Agapetes	111	Bidaria	231
Ageratum	12	Bidens	51
Alseuosmia	86	BIGNONIACEÆ	275
Ajuga	571	Bignonia	270
Alstonia	200	Blainvillea	46
Amberboa	83	Blepharis	453
Amblyanthus	137	Blepharispermum	19
Amphirapis	19	Blumea	28
Anagallis	128	Bonnaya	374
Anaphalis	64	BORAGINACEÆ	324
Andrographis	464	Bothriospermum	333
Andromeda	116	Boucerosia	252
Anisia	317	Bouchea	489
Anisochilus	533	Brachylepis	214
Anisomeles	557	Brachyrampus	90
Anodendron	208	Breweria	319
Antennaria	63	Buchnera	360
Antistrophe	136	Buddleia	378
Aplotaxis	81		
APOCYNACEÆ	186	Cæsulia	45
AQUIFOLIACEÆ	161	Calamintha	550
Ardisia	139	Caligula	113
Argyreia	296	Callicarpa	498
Arisæma	366	Callistephus	16
Artemisia	86	Calonyction	304

	Page		Page
Calophanes	410	Cynoglossum	339
Calosanthes	270	CYRTANDRACEÆ	285
Calotropis	221		
CAMPANULACEÆ	100	Dœmia	225
Campanula	103	Datura	352
Campanumœa	102	Decalepis	215
Camphellia	390	Decaneurum	9
Cancora	266	Dichrocephala	20
Caraliuma	251	Dicliptera	477
Carissa	186	Dicoma	85
Carpesium	67	Didymocarpus	287
Celsia	355	Dilivaria	455
Centranthera	381	Dipteracanthus	418
Cephalostigma	101	Dischidia	236
Cerbera	104	Diospyros	153
Ceropegia	243	Dupatisium	365
Chondrospermum	175	Doronicum	71
Chonemorpha	203	Dysehoriste	415
Christisonia	389	Dysophylla	458
Chrysanthemum	53		
Chrysophyllum	145	EBENACEÆ	152
Cicdum	85	Ebermaiera	402
Clerodendron	501	Ecdysanthera	206
Corionacanthus	410	Echaltium	205
Codonopsis	103	Echinacanthus	424
Coldenia	337	Echinops	61
Colebrookia	547	Echites	209
Coleus	529	Eclipta	46
COMPOSITÆ	1	Ebretia	326
Congea	494	Elephantopus	12
CONVOLVULACEÆ	294	Ellertonia	210
Convolvulus	315	Elsholzia	547
Conyza	25	Elytraria	396
Cordia	325	Embelia	137
Cosmostigma	228	Emilia	70
Craniotome	550	Endopogon	411
Crawfurdia	270	Enhydra	55
Crocea	320	Epaltes	41
Cremandra	456	Epigynium	114
Cryptolepis	253	Epithema	293
Cryptophragmium	408	Eranthemum	473
Cryptostegia	213	Erigeron	17
Curanga	367	Eriopetalum	231
Cuscuta	321	Eryelbe	323
Cyathocline	23	Erythracanthus	401
Cyclocodon	103	Erythrœa	265
Cyrtonium	219	Eubulis	3

601

	Page		Page
Eupatorium	15	Heterophragma	279
Evolvulus	321	Heterostemma	238
Exacum	262	Hewittia	318
		Hexacentris	396
Fagræa	260	Hieracium	95
Filago	82	Holarrhena	201
Finlaysonia	213	Holmskioldia	311
Franciscuria	44	Holochilus	139
Fraxinus	169	Holostemma	221
Frerea	254	Hoya	240
		Huoteria	193
Gardneria	259	HYDROLEACEÆ	342
Geudarussa	469	Hydrolea	342
Geniosporum	517	Hygrophila	406
GENTIANACEÆ	201	Hymenandra	137
Gentiana	269	Hyoscyamus	353
Glossocardia	54	Hypoestes	482
Glossogyne	54		
Glossostigma	377	Ichnocarpus	207
Gmelina	505	Ilex	161
Gnaphalium	60	Ilysanthes	372
Goldfussia	427	Ionia	42
Gomphostemma	567	Ipomæa	305
Gongronema	234	Isanthera	291
Goniostemma	216	Isonandra	147
GOODENIACEÆ	98	Ixeris	90
Grangea	22		
Graptophyllum	460	JASMINACEÆ	174
Gualthieria	116	Jasminum	176
Gulzotia	50	Jerdonia	292
Gymnema	232	Justicia	469
Gymnostachyum	405		
Gynura	66	Klugia	290
Halenia	272	LABIATÆ	514
Haplanthus	483	Lactuca	89
Hedeoma	551	Lantana	490
Helichrysum	60	Lavandula	536
Heligme	210	LENTIBULACEÆ	118
Hollophytum	336	Leonotis	567
Heliotropium	334	Lepidagathis	448
Hemiadelphis	403	Lepistemon	304
Hemichoriste	460	Leptacanthus	426
Hemidesmus	213	Leptadenia	238
Hemigraphis	419	Leucas	559
Hemigymnia	511	Ligustrum	172
Herpestis	364	Limnanthemum	273

	Page		Page
Limnophila	360	Olea	169
Linaria	356	Oligolepis	22
Lindenbergia	358	Oligopholis	391
Linociera	172	Ophelia	270
Lippia	490	Ophioxylon	190
Lobelia	107	Orthanthera	236
LOGANIACEÆ	253	Orthosiphon	520
Loxanthus	439	Oxystelma	224
Lysimachia	129		
		Pajanelia	281
Mala	160	Parsonsia	198
Machlis	59	PEDALIACEÆ	282
Madascarpus	73	Pedalium	284
Mæsa	130	Pedicularis	385
Mandenia	229	Pentasacme	236
Marus	357	Pentatropis	223
Melodinus	187	Peplidium	376
Mesona	518	Peracarpa	105
Meyenia	395	Pergularia	231
Micrargeria	382	Perilla	549
Microcarpæa	377	Peristrophe	479
Micrommeria	549	Pharbitis	303
Microrhynchus	91	Phlebophyllum	410
Micropyxis	121	Phlogacanthus	457
Millingtonia	277	Physalis	350
Mimusops	150	Physichilus	404
Mitrasacme	256	Picris	88
Mitreola	256	Piddingtonia	107
Monosis	3	PLANTAGINACEÆ	575
Moonia	48	Plantago	576
Moschosma	520	Plectranthus	521
Mulgedium	96	Pluchea	40
Myriactis	18	PLUMBAGINACEÆ	572
Myriogyne	58	Plumbago	574
MYRSINACEÆ	130	Pogostemon	537
Myrsine	138	Pouzia	207
		Premna	495
Nelsonia	399	Prenanthes	95
Nepeta	553	PRIMULACEÆ	126
Neuracanthus	447	Primula	127
Nomaphila	405	Priva	488
Notonia	80	Pterostelma	239
Nyctanthes	184	Pterostigma	358
		Prunella	551
Ocimum	515	Pulicaria	44
Oiospermum	1	Pyrethrum	56
OLEACEÆ	168		

	Page		Page
Quamoclit	301	Sphenodesma	492
		Spilanthes	32
Raphistemma	219	Stachys	358
Rhabdia	331	Stemodia	359
Rhamphicarpa	382	Stenosiphonium	414
Rhaphidospora	481	Stereospermum	280
Rhinacanthus	471	Streptocaulon	215
RHODORACEÆ	117	Striga	380
Rhododendron	117	Strobilanthes	430
Rhyncoglossum	290	Strophanthus	202
Rhynchospermum	204	Strychnos	257
Rivea	294	STYRACACEÆ	162
Ruellia	420	Styrax	168
Rungia	474	STYLIDIACEÆ	97
		Stylidium	97
SALVADORACEÆ	185	Sutera	357
Salvadora	185	Symphorema	492
Salvia	552	Symplocos	163
Samara	136		
SAPOTACEÆ	144	Tabernæmontana	195
Sapota	146	Taraxacum	89
Sarcolobus	235	Tecoma	281
Sarcostemma	233	Tectona	494
Scævola	90	Teucrium	570
Schmidia	398	Thespis	24
Schrebera	185	Thunbergia	393
Sclerocarpus	52	Thyrsacanthus	459
SCROPHULARIACEÆ	354	Torenia	367
Scutellaria	555	Tournefortia	332
Sesamone	216	Toxocarpus	217
Senecio	75	Triænanthus	485
Serratula	86	Trichodesma	340
Sesamum	283	Tricholepis	84
Sideroxylon	147	Tylophora	225
Siegesbeckia	47		
SIPHONANDRACEÆ	111	Utricularia	119
Skinneria	318		
Stevogtia	288	Vallaris	197
SOLANACEÆ	343	Vandellia	369
Solanum	343	Verbascum	355
Sonchus	92	VERBENACEÆ	487
Sopubia	383	Verbena	512
Spathodea	278	Vernonia	3
Sphæranthus	21	Vicoa	43
Sphæromorphæa	59	Vinca	197
SPHENOCLEACEÆ	100	Vitex	507
Sphenoclea	100	Vogelia	573

	Page		Page
Wahlenbergia	104	Xanthium	47
Wedelia	48	Ximenesia	83
Willughbeia	187		
Withania	351	Youngia	93
Wollastonia	48		
Wrightia	191		

www.ingramcontent.com/pod-product-compliance
Lightning Source LLC
Chambersburg PA
CBHW021228300426
44111CB00007B/462